Polymer Membranes for Fuel Cells

S. M. Javaid Zaidi • Takeshi Matsuura
Editors

Polymer Membranes for Fuel Cells

 Springer

Editors
S. M. Javaid Zaidi
King Fahd University of Petroleum
   and Minerals
Dhahran
Saudi Arabia
zaidismj@kfupm.edu.sa

Takeshi Matsuura
University of Ottawa
Ottawa
Canada
matsuura@eng.uottawa.ca

ISBN: 978-1-4419-4462-7   e-ISBN: 978-0-387-73532-0
DOI: 10.1007/978-0-387-73532-0

springer.com

Foreword

From the late-1960's, perfluorosulfonic acid (PFSAs) ionomers have dominated the PEM fuel cell industry as the membrane material of choice. The "gold standard' amongst the many variations that exist today has been, and to a great extent still is, DuPont's Nafion® family of materials. However, there is significant concern in the industry that these materials will not meet the cost, performance, and durability requirementsnecessary to drive commercialization in key market segments – especially automotive. Indeed, Honda has already put fuel cell vehicles in the hands of real end users that have home-grown fuel cell stack technology incorporating hydrocarbon-based ionomers.

"Polymer Membranes in Fuel Cells" takes an in-depth look at the new chemistries and membrane technologies that have been developed over the years to address the concerns associated with the materials currently in use. Unlike the PFSAs, which were originally developed for the chlor-alkali industry, the more recent hydrocarbon and composite materials have been developed to meet the specific requirements of PEM Fuel Cells. Having said this, most of the work has been based on derivatives of known polymers, such as poly(ether-ether ketones), to ensure that the critical requirement of low cost is met. More aggressive operational requirements have also spurred the development on new materials; for example, the need for operation at higher temperature under low relative humidity has spawned the creation of a plethora of new polymers with potential application in PEM Fuel Cells.

Working with its development partner, Ebara Research, Ballard Power Systems has developed and demonstrated pilot scale semi-continuous manufacturing for a new class of ionomers based on radiation induced graft polymerization. This technology leaves the two key requirements of a fuel cell membrane; namely, gas separation and proton conductivity, to be met by separate chemistries. In this approach, a commercial preformed membrane with inherent gas separation capabilities is irradiated with high-energy gamma rays. This, in turn, induces reactive sites which are graft polymerized with a monomer, of a different chemistry than the preformed membrane; when post-sulfonated the required proton conductivity is attained. Many others have taken a similar approach as described in detail in Chapter 5.

Given the ever-increasing demands on the membrane to facilitate the use of fuels other than direct gaseous hydrogen (e.g., DMFC), in addition to meeting new operational specifications such as performance at temperatures greater than 90°C, many researchers have taken the approach of blending different materials and using heteropolyacids to produce composite membranes with enhanced performance. Chapters 8–9 and 13–15 discusses many of the key activities going on in these areas.

While the proton exchange membrane is a key functional component of a PEM Fuel Cell, the capability of the membrane to meet PEM fuel cell product requirements is very dependent on how the material is processed into a Membrane Electrode Assembly. This critical area is described in some detail in Chapter 11.

The balance of the review deals with some of the latest state-of-the-art development activities in polymers and membrane forming technologies that will continue to push the boundaries of innovation and creativity essential to sustain the growth and commercial potential of PEM Fuel Cells.

"Polymer Membranes in Fuel Cells" provided the reader with a comprehensive overview of PEM Fuel Cells focusing on the membrane as a key component, while describing in detail, with excellent reference materials, many of the key technologies of the recent past and the future.

Dr. Charles Stone
Vice President, Research & Development
Ballard Power Systems

Preface

Fuel cell is considered to be one of the most promising clean energy sources since it does not generate toxic gases and other hazardous compounds. It is currently an important research topic in all leading automobile and energy industries. Fuel cells will provide an urgently needed solution to the increasing impact of vehicles pollution. Among various kinds of fuel cells, polymer electrolyte membrane fuel cells (PEMFC) are easy to be miniaturized and suited as energy sources for automobiles as well as domestic applications and potable devices. Fuel cell is also currently considered as reaching the threshold of commercialization. The center of PEMFC is the polymer electrolyte membrane, as it defines the properties needed for other components of fuel cell and is the key component of a fuel cell system. The membrane component of a fuel cell stack may account for as much as 30% of the total material cost. Their properties are paramount to the successful operation and commercialization of fuel cells. The most widely used solid polymer electrolyte membranes in fuel cells are perfluorosulfonic acid membranes such as Nafion, Acipex and Flemion. However, they have some drawbacks that should be overcome before their practical applications will be achieved. The most significant drawbacks of these membranes are their relatively high cost in the range of US $ 800/m^2, and their limited stability at temperatures substantially in excess of 100°C. Moreover, there is a problem of methanol crossover when used in direct methanol fuel cell. Hence, there is a need to develop less expensive polymeric membranes with improved performance. More specifically, it is currently targeted to develop membranes with high proton conductivity (above 10^{-2} S cm^{-1} at 120°C), preventing excessive methanol crossover, and with durability of 5000 hours for transportation and 4000 hours for stationary devices.

During the past two decades many attempts have been made all over the world to improve the performance of presently available membranes and develop new ones for PEMFC. The fuel cell membrane field is growing with such a fast pace that it will emerge as one of the most important membrane technologies. A literature search on fuel cell membranes over the years 1990–2006 revealed more than 2500 patents only in addition to thousands of journal publications, which clearly indicate the importance of the subject. It seems most timely to summarize the results of such research efforts. Hence, this book is focused on the development of polymeric and polymeric/inorganic hybrid membranes for PEMFC.

   Although each chapter is independent from the other chapters, attempts were made
to give some cohesiveness between chapters.

   The first five chapters are general description of the strategies adopted to
improve the membrane properties and status of the PEMFC technology.

   Chapter 1 and 2 outline the principle of PEMFC and its desired properties. The
general trend in recent R & D efforts and the future outlook is also summarized. It
would be a good start to read these two chapters before entering any other chosen
chapter.

   Chapter 3 presents an overview of fuel cell technology, potential applications of
fuel cell technology, current research and development in fuel cells, key technology
players in fuel cells, and provides directions for fuel cell research.

   Chapter 4 presents an overview of the synthesis, chemical properties, and polymer electrolyte fuel cell applications of new proton-conducting polymer electrolyte
membranes based on sulfonated poly(arylene ether ether ketone) polymers and
copolymers.

   In chapter 5, a comprehensive review was made of the attempts to prepare
alternative proton conductive membranes (PCMs) by radiation-induced graft
polymerization.

   Chapters 6 and 7 deal with the hydrocarbon polymers and composites targeted
for high temperature PEM fuel cell applications. Specifically, chapter 6 deals with
a series of high molecular weight, highly sulfonated poly(arylenethioethersulfone)
(SPTES) polymers synthesized by polycondensation. They were characterized by
different methods and tested for proton conductivity. Finally, membrane electrode
assemblies (MEAs) were fabricated.

   Chapter 7 emphasizes polymer/inorganic composite membranes to increase
thermal stability. More specifically, polymers include perfluoronated polymers,
sulfonated poly(arylene ether)s, polybenzimidazoles (PBIs), and others. The inorganic proton conductors are silica, heteropolyacids (HPAs), layered zirconium
phosphates, and liquid phosphoric acid.

   In the next five chapters more specific classes of macromolecules and other
materials are shown for the design of improved PCMs.

   Chapter 8 describes a strategy to improve the membrane properties by blending
polymer electrolytes with other polymers. Different types of interactions are
involved when cross-linking between polymers is formed.

   In chapter 9 the general strategy involved in the preparation of organic-inorganic
membranes for fuel cell applications is described and its advantages and disadvantages are discussed.

   In Chapter 10 efforts have been made to highlight the response of thermal and
mechanical properties with variation of different parameters characteristic of a typical fuel cell environment.

   Chapter 11 also includes the polymer inorganic membranes consisting of Nafion
and zirconium phosphates, heteropolyacids, metal hydrogen sulfates and metal
oxides. Moreover, this chapter includes the design of thin film electrodes for MEA.

Chapter 12 deals with carbon nanotube (CNT) filled membranes for PCMs. CNT-filled polyethylene terephthalate was blended with various polymers, injection molded and characterized by different methods.

The next four chapters are for direct methanol fuel cells (DMFC).

In Chapter 13 critical issues for the commercialization of DMFCs are discussed thoroughly. Functions, current status and technical approaches have been discussed in terms of proton conductivity, methanol permeability, water permeability, life cycle and processing cost as well as the interaction with other compartments.

Chapter 14 presents a brief literature survey of such modifications, along with recent experimental results (membrane properties and fuel cell performance curves) for: (i) thick Nafion films, (ii) Nafion blended with Teflon®-FEP or Teflon®-PFA, and (iii) Nafion doped with polybenzimidazole.

Chapter 15 is a general overview of DMFC research to develop membranes with low methanol permeability without sacrificing other important qualities.

Chapter 16 presents a unique membrane design and development using the concept of pore-filling. The membranes are used for both polymer electrolyte membrane fuel cells (PEMFCs) and direct methanol fuel cells (DMFCs).

And finally, chapter 17 was written to summarize the whole chapters. Attempts were also made to show the future direction of the fuel cell R & D.

The editors believe that this book is the first book exclusively dedicated on fuel cell membranes in which the experts of the field are brought together to review the development of polymeric membranes for PEFC in all their aspects. The book was written for engineers, scientists, professors, graduate students as well as general readers in universities, research institutions and industry who are engaged in R & D of synthetic polymeric membranes for PEMFC. It is therefore the editors' wish to contribute to the further development of PEMFC by showing the future directions in its R & D.

S.M. Javaid Zaidi
Takeshi Matsuura

Contents

About the Editors

S.M. Javaid Zaidi, Ph.D.
Associate Professor, Chemical Engineering
Ph.D. Chemical Engineering, Laval University, Canada (2000)
M.Sc. (Eng.): King Fahd University, Saudi Arabia (1991)
B.Sc. (Eng.): Aligarh University, India (1986)

Dr. Zaidi received his Ph.D. degree in Chemical Engineering from Laval University, Canada. He obtained M.S. in Chemical Engineering from King Fahad University of Petroleun & Minerals (KFUPM), Saudi Arabia and B.S. in Chemical Engineering from Aligarh Muslim University, India.

He joined the Chemical Engineering department of KFUPM in fall 2000 as Assistant Professor. He has worked on various fuel cell membrane development projects both in Canada and Saudi Arabia funded by Natural Science and Engineering Research Council of Canada (NSERC), Canadian Mineral and Energy Technology (CANMET) in Canada, KFUPM, King Abdulaziz City of Science and Technology and University Research Council, Saudi Arabia. Prior to joining Chemical Engineering department of KFUPM, Dr. Zaidi worked as Guest Researcher at National Research Council of Canada, Research Fellow at the University Laval and as Research Engineer in Research Institute of KFUPM. In the summer 2001 Dr. Zaidi was visiting Professor in the Department of Materials Engineering, Ecole Polytechnique of University of Montreal, and in 2003 as visiting professor in the chemical Engineering department, Ottawa University, Canada. He has published over 85 papers in refereed international journals and international/national conferences and symposia. He has completed many research projects and directed 10 M.S. theses in the field of polymer membranes for fuel cell. He is member of university Research Advisory Board, chaired by H.E. The Rector of the University. He received FCAR award from Government of Quebec, Canada for outstanding performance and excellence in research (1998). In 2006 he was awarded the prestigious "Distinguished Research Award' from the King Fahad University of Petroleum & Minerals and in 2005 he received "Best Researcher Award" from the College of Engineering Sciences, KFUPM. In May 2007 he is invited by UNESCO to participate in the ministerial-level conference on Energy in the Changing World, in Paris.

Matsuura Takeshi, D.-Ing.
Professor, Chemical Engineering
B.Sc. (Eng.): University of Tokyo (1961)
M.Sc. (Eng.): University of Tokyo (1963)
D.-Ing.: Technical University of Berlin (1965)

Professor Matsuura was born in Shizuoka, Japan, in 1936. He received his B.Sc. (1961) and M.Sc. (1963) degrees from the Department of Applied Chemistry at the Faculty of Engineering, University of Tokyo. He went to Germany to pursue his doctoral studies at the Institute of Chemical Technology of the Technical University of Berlin and received Doktor-Ingenieur in 1965.

After working at the Department of Synthetic Chemistry of the University of Tokyo as a staff assistant and at the Department of Chemical Engineering of the University of California as a postdoctoral research associate, he joined the National Research Council of Canada in 1969. He came to the University of Ottawa in 1992 as a professor and the chairholder of the British Gas/NSERC Industrial Research Chair. He served as a professor of the Department of Chemical Engineering and the director of the Industrial Membrane Research Institute (IMRI) until he retired in 2002. He was appointed to professor emeritus in 2003. He served also at the Department of Chemical and Environmental Engineering of the National University of Singapore as a visiting professor during the period of January to December 2003.

Dr. Matsuura received the Research Award of International Desalination and Environmental Association in 1983. He is a fellow of the Chemical Institute of Canada and a member of the North American Membrane Society. He has delivered many lectures at international conferences on invitation. He was also invited in October 1986 by the Institute of Oceanography, Hangzhou, China, to give a series of lectures on membrane separation processes at various institutes of China. In September 1991, he was invited by Japan Industrial Technology Association to deliver a series of lectures at industry, government and universities in Japan. He was also invited to Korea in July 1996 to deliver a lecture at the 4th Workshop of the Membrane Society of Korea. In 2002 he was invited by Nihon University as an Overseas Visiting Professor to give a series of lectures at various University campuses. He has published more than 250 papers in refereed journals, authored and co-authored 3 books and edited 4 books. Recently, he contributed a chapter on "Membrane Separation Technologies" to Encyclopedia of Life Support Systems at the invitation of UNESCO. A symposium of membrane gas separation was held at the Eighth Annual Meeting of the North American Membrane Society, May 18-22, 1996, Ottawa, to honour Dr. Matsuura together with Dr. S. Sourirajan for their life-long contribution to the membrane research. He received George S. Glinski Award for Excellence in Research from the Faculty of Engineering of the University of Ottawa in 1998.

Contributors

Zongwu Bai
University of Dayton Research Institute, University of Dayton, Dayton,
OH 45469, USA

M. Bello
Department of Chemical Engineering, King Fahd University
of Petroleum & Minerals, Dhahran, Saudi Arabia

Hyuk Chang
Samsung Advanced Institute of Technology, P.O. Box 111, Suwon, Korea

Yeong Suk Choi
Samsung Advanced Institute of Technology, P.O. Box 111, Suwon, Korea

Thuy D. Dang
Air Force Research Laboratory/MLBP, Wright-Patterson Air Force Base,
OH 45433, USA

Michael D. Guiver
Institute for Chemical Process and Environmental Technology, National Research
Council, 1200 Montreal Road, Ottawa, ON, Canada K1A 0R6

Ibnelwaleed A. Hussein
Department of Chemical Engineering, KFUPM, Dhahran 31262, Saudi Arabia

A.F. Ismail
Membrane Research Unit, Faculty of Chemical and Natural Resources
Engineering, Universiti Teknologi Malaysia, 81310 Skudai, Johor, Malaysia

Jochen Kerres
Universität Stuttgart, Institut für Chemische Verfahrenstechnik, Böblinger Str. 72,
70199 Stuttgart, Germany

Dae Sik Kim
Institute for Chemical Process and Environmental Technology, National Research
Council, 1200 Montreal Road, Ottawa, ON, Canada K1A 0R6

Haekyoung Kim
Samsung Advanced Institute of Technology, P.O. Box 111, Suwon, Korea

Wonmok Lee
Samsung Advanced Institute of Technology, P.O. Box 111, Suwon, Korea

Jun Lin
Department of Chemical Engineering, Case Western Reserve University,
Cleveland, OH 44106-7217, USA

Yuxiu Liu
Polymer Program, Institute of Materials Science and Department of Chemical,
Materials, and Biomolecular Engineering, University of Connecticut, Storrs,
CT 06269-3136, USA

Takeshi Matsuura
Industrial Membrane Research Laboratory, Department of Chemical Engineering,
University of Ottawa, Ottawa, ON, Canada K1N6N5

R. Naim
Membrane Research Unit, Faculty of Chemical and Natural Resources
Engineering, Universiti Teknologi Malaysia, 81310 Skudai, Johor, Malaysia

Mohamed Mahmoud Nasef
Department of Chemical Engineering, Faculty of Chemical and Natural Resources
Engineering, Universiti Teknologi Malaysia, 81310 UTM Skudai, Johor, Malaysia

Suzana Pereira Nunes
Institute of Polymer Research, GKSS Research Centre, Max-Planck-Str. 1,
21502 Geesthacht, Germany

Peter N. Pintauro
Department of Chemical Engineering, Case Western Reserve University,
Cleveland, OH 44106-7217, USA

S.U. Rahman
Department of Chemical Engineering, King Fahd University
of Petroleum & Minerals, Dhahran, Saudi Arabia

Vijay Ramani
Department of Chemical and Environmental Engineering,
Illinois Institute of Technology, 10 W 33rd Street; 127 PH,
Chicago IL 60616, USA

Dipak Rana
Industrial Membrane Research Laboratory, Department of Chemical Engineering,
University of Ottawa, Ottawa, ON, Canada K1N6N5

M. Abdur Rauf
Department of Chemical Engineering, King Fahd University
of Petroleum & Minerals, Dhahran, Saudi Arabia

Leon L. Shaw
Department of Materials Science and Engineering, University of Connecticut,
Storrs, CT, USA

Panagiotis Trogadas
Department of Chemical and Environmental Engineering,
Illinois Institute of Technology, 10 W 33rd Street; 127 PH, Chicago,
IL 60616, USA

Ryszard Wycisk
Department of Chemical Engineering, Case Western Reserve University,
Cleveland, OH 44106-7217, USA

Takeo Yamaguchi
Department of Chemical Systems Engineering, The University of Tokyo,
7-3-1, Bunkyo-ku, Tokyo 113-8656, Japan

Mitra Yoonessi
Air Force Research Laboratory/MLBP, Wright-Patterson Air Force Base,
OH 45433, USA

S.M. Javaid Zaidi
Department of Chemical Engineering, King Fahd University
of Petroleum & Minerals, Dhahran, Saudi Arabia

Lei Zhu
Polymer Program, Institute of Materials Science and Department of Chemical,
Materials, and Biomolecular Engineering, University of Connecticut, Storrs,
CT 06269-3136, USA

Xiuling Zhu
Polymer Program, Institute of Materials Science and Department of Chemical,
Materials, and Biomolecular Engineering, University of Connecticut, Storrs,
CT 06269-3136, USA
and
Department of Polymer Science & Materials, Dalian University of Technology,
Dalian 116012, P.R. China

N.A. Zubir
Membrane Research Unit, Faculty of Chemical and Natural Resources
Engineering, Universiti Teknologi Malaysia, 81310 Skudai, Johor, Malaysia

Chapter 1
Fuel Cell Fundamentals

S.M. Javaid Zaidi and M. Abdur Rauf

Abstract In this chapter fuel cell introduction and general concepts about fuel cell are presented. Types of fuel cell and their classification are given and desired properties of the polymeric membranes for use in PEMFC are described. At the end challenges facing the fuel cell industry and their future outlook are briefly discussed.

1.1 Introduction

Fuel cells are considered environment-benign technology providing solutions to a range of environmental challenges, such as harmful levels of local pollutants, in addition to providing economic benefits due to their high efficiency. Because of their potential to reduce the environmental impact and geopolitical consequences of the use of fossil fuels, fuel cells have emerged as potential alternatives to combustion engines.

The main advantages of the fuel cells are their pollution-free operation and high energy density [1]. They also have high energy conversion efficiency, low noise, and low maintenance costs [2]. Fossil fuel reserves are limited, and it has been predicted that the production of fossil fuels will peak around 2020 and then decline [3]. To meet the demand of food and energy of the ever-increasing population and to protect the environment from the ill effects of fossil fuel use [4], research and development in the area of fuel cells needs to be significant. The reported estimate of worldwide environmental damage is around $5 trillion annually [5]. Fuel cells are electrochemical devices that convert chemical energy of the fuel and oxidant directly into electricity and heat with high efficiency. Electrochemical processes in fuel cells are not governed by Carnot's cycle; therefore, their operation is simple and more efficient compared with internal combustion (IC) engines. High efficiency makes fuel cells an attractive option for a wide range of applications, including transportation, stationary, and portable electronic devices such as consumer electronics, laptop computers, video cameras, etc.

All fuel cells consist of an anode, to which the fuel is supplied, a cathode, to which oxidant (e.g., oxygen) is supplied, and an electrolyte, which allows the flow of ions between the anode and cathode. The electrolyte should be highly resistive

S.M.J. Zaidi, T. Matsuura (eds.) *Polymer Membranes for Fuel Cells*,
doi: 10.1007/978-0-387-73532-0, © Springer Science+Business Media, LLC 2009

to the electron current. The net chemical reaction is exactly the same as if the fuel were burnt, but by spatially separating the reactants, the fuel cell intercepts the stream of electrons that flow spontaneously from the fuel to the oxidant and diverts it for use in an external circuit. The fundamental driving force behind this process of ion migration is the concentration gradient between the two interfaces (electrode–electrolyte). The difference between a fuel cell and a conventional battery is that the fuel and oxidant are not integral parts of a fuel cell, but instead are supplied as needed to provide power to the external load. A fuel cell is "charged" as long as there is a supply of fuel to the cell, so self-discharge is absent. However, these are complicated systems and the voltage output of the typical polymer electrolyte membrane fuel cell (PEMFC) is around 0.7 V only [6].

For many years it was attempted to develop fuel cells as a power source. In the beginning they were developed mainly for space and defense applications. Attempts to develop earth-based systems were made during 1980s and 1990s. The recent drive for more efficient and environmentally friendly electrical generation technologies has resulted in substantial resources being diverted to fuel cell development. Presently fuel cell technology is maturing toward commercialization, but work still needs to be done in many fields. For commercial applications, component materials need to be developed and optimized to improve performance and lower costs. Long-term testing has to be done to obtain information on fuel cell performance. Furthermore, as fuel cell research moves on from laboratory-scale single cell studies to the development of application-ready stacks, new in situ measurement methods are needed for characterization and diagnostics.

1.2   Fuel Cell Classification

Fuel cells are classified based on the type of electrolyte they use. The six most common fuel cell types are:

1. Polymer electrolyte membrane fuel cells (PEMFC)
2. Direct methanol fuel cells (DMFC)
3. Alkaline fuel cells (AFC)
4. Phosphoric acid fuel cells (PAFC)
5. Molten carbonate fuel cells (MCFC)
6. Solid oxide fuel cells (SOFC)

DMFC is the fuel cell that is similar to PEMFC, except that it uses methanol as the fuel directly on the anode instead of hydrogen or hydrogen-rich gas. The characteristics of these types of fuel cells are summarized in Table 1.1.

Among the various types of fuel cells listed in the preceding, the PEMFC, which uses a sheet of polymer membrane as the solid electrolyte, is the subject of this book. The solid polymer electrolyte membrane is the most important constituent of these fuel cells that allows protons, but not electrons, to pass through. These fuel cells currently use perfluorinated Nafion® membranes (Du Pont), which exhibit a number of shortcomings to optimum efficiency. Also, the polymer membranes represent approximately 30% of the material cost of the fuel cell. The membrane in

Table 1.1 Types of fuel cells and their characteristics

Fuel cell type	Operating temperature (°C)	Efficiency	Suitable applications			
			Domestic power	Small scale power	Large scale cogeneration	Transport
AFC	50–90	50–70	√	√	X	√
PEMFC	50–120	40–50	√	√	X	√
PAFC	175–220	40–45	X	√	X	X
MCFC	600–650	50–60	X	√	√	X
SOFC	800–1,000	50–60	√	√	√	X

PEMFC functions twofold, it acts as the electrolyte that provides ionic communication between the anode and cathode, and also serves as a separator for the two reactant gases. Both optimized proton and water membrane transport properties and proper water management are crucial for efficient fuel cell operation. Dehydration of the membrane reduces proton conductivity, and excess water can flood the electrodes. Both conditions may result in poor cell performance.

A typical PEM fuel cell uses hydrogen as the fuel and oxygen/air as the oxidant. For hydrogen, a separate reformer reactor is required. Some fuel cells use methanol as fuel. In this case there is no need of a reformer, and the fuel cell is called a DMFC. In effect, DMFC is a special iteration of PEMFC. Its low temperature and pressure operation coupled with the low cost of methanol are attributes that makes DMFC a promising energy source [7].

1.3 Membrane Materials

Ever since the invention of ion exchange membranes as electrolyte for fuel cells in the 1950s, researchers were challenged to find the ideal membrane material to withstand the harsh fuel cell operating environment. Since then numerous attempt have been made to optimize membrane properties for application in fuel cells. The desired properties for use as a proton conductor in DMFC essentially include the following:

1. Chemical and electrochemical stability in the cell operating environment (high resistance to oxidation, reduction, and hydrolysis)
2. Mechanical strength and stability in the cell operating environment
3. Chemical properties compatible with the bonding requirements of the membrane electrode assembly (MEA)
4. Extremely low permeability to reactant gases to minimize columbic inefficiency
5. High water transport to maintain uniform water content and prevent localized drying
6. High proton conductivity to support high currents with minimal resistive losses and zero electronic conductivity
7. Production costs compatible with the application

1.4   Fuel Cell Challenges

Although great improvements in fuel cell design and components have been made over the past years, several issues remain to be addressed before PEM fuel cells can become competitive enough to be used commercially. A major hurdle is the fuel, which is mostly hydrogen. PEMFCs run best on very pure hydrogen. Hydrogen obtained from hydrocarbons tends to contain small amount of carbon monoxide (CO), which have disastrous effects on the efficiency of anode reaction. Moreover, the onboard storage of hydrogen is problematic, whereas alternatives such as the on-board reformation of methanol complicate the system and reduce efficiency. The simpler DMFC, in which methanol is used as a fuel instead of hydrogen, suffers from the poor kinetics of methanol oxidation reaction at low temperatures and a high methanol permeation rate through the membrane. Finally, the overall cost of the fuel cell with its platinum-based catalyst, and problems with the currently used membranes (e.g., methanol permeation, high temperature stability, and high membrane cost) thwart the commercialization of fuel cells. Cost is a significant factor hindering the full commercialization of PEM fuel cells. In 2002, the cost of the catalyst itself was $1,000 per kW of electrical power output. This was cut down to $30 by 2007.

To address PEMFC design issues, research in PEM fuel cells is being conducted by various scientific groups and automobile companies worldwide. It is today's hot topic. The research in polymer fuel cells has been very impressive, as can be seen by the number of patents issued every year. There has been exponential growth in the number of PEM fuel cell patents over the last few years. Many new materials have been developed to improve performance of the fuel cell stack. Enhanced modeling activities have facilitated the design of better-performance components for fuel cell. With the growing number of patents in PEM fuel cell area, one can guess how much research had been done and technological advances have been achieved over the last few years (Fig. 1.1).

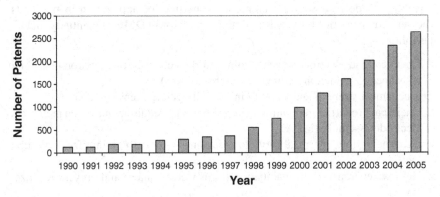

Fig. 1.1 Growth in research activity of PEM fuel cell represented as a plot of number of patents per year

1.5 Future Outlook

The future of the PEM fuel cell strongly depends on the technological solution of the challenges facing the fuel cell industry. Academic organizations are looking into the technological aspects, while industries are looking into the involved cost factors. Right now the automotive industry is the largest investor in PEM fuel cell development. Other main areas of PEM applications are distributed power generation and portable electronics [8]. PEM fuel cells were tested in the buses of Vancouver and Chicago at the end of the last century. The fuel cell market is growing, and between 2003 and 2005 an annual growth rate of about 18.7% was reported in the literature. The recent Worldwide Fuel Cell Industry Survey shows that sales remain at $331 million with an increase in R&D expenditure by 9% to $716 million from 2003 to 2004 [9]. Proton exchange membrane fuel cells dominated the industry as a key focus area. There has been a tremendous increase in R&D funding from virtually zero 30 years ago to about $400 million now spent annually by the U.S. government alone on hydrogen and fuel cell—related programs. In the United States, the year 2006 marked the launch of the President's Advanced Energy Initiative, which provided for a 22% increase in funding for clean energy technology research, including hydrogen and fuel cell technologies, at the U.S. Department of Energy (DOE). Industry has already sought roughly 600 fuel cell vehicles to date. Stationary fuel cell systems also continue to make progress with record-setting installations worldwide.

It is projected that worldwide around 550 GW of generating capacity will be installed during 2000–2010 [2]. Buses running on PEM fuel cells are being demonstrated in North America, Japan, Australia, and Europe. These were shown to have smooth acceleration with heavy passenger loads. China, one of the largest potential fuel cell markets in the world, is considering deploying fuel cell buses for the Beijing Olympic Games in 2008. Buses are being built by New Flyer for which Hydrogenics is the fuel cell system using 180 kW PEM fuel cells. With crude oil reserves being limited, and various environmental protection agencies imposing stricter environmental regulations, the highly efficient PEM fuel cell can hope to have a good market in the near future.

References

1. Yan, Q., H. Toghiani, H. Causey, "Steady state and dynamic performance of proton exchange membrane fuel cells (PEMFCs) under various operating conditions and load changes", *J. Power Sources* 161 (2006) 492–502.
2. Smitha, B., S. Sridhar, A. A. Khan, "Solid polymer electrolyte membranes for fuel cell applications — A review", *J. Membr. Sci.* 259 (2005) 10–26.
3. Sopian, K., A. H Shamsuddin, T. N. Veziroglu, "Solar hydrogen energy option for Malaysia Proceeding of the international conference on advances in stratic technology, June 1995", *UKM, Bangi* (1995) 209–220.

4. Kordesch, K. V., G. R. Simader, "Environmental impact of fuel cell technology", *Chem. Rev.* 95 (1995) 191–207.
5. Barbir, F., "PEM fuel cells: Theory and practice", *California: Elsevier Academic Press*, (2005).
6. Iojoiu, C., F. Chabert, M. Marechal, N. El. Kissi, J. Guindet, J. Y. Sanchez, "From polymer chemistry to membrane elaboration — A global approach of fuel cell polymeric electrolytes", *J. Power Sources* 153 (2006) 198–209.
7. Ge, J., H. Liu, "Experimental studies of a direct methanol fuel cell", *J. Power Sources* 142 (2005) 56–69.
8. Wang, L., A. Husar, T. Zhou, H. Liu, "A parametric study of PEM fuel cell performances", *Int. J. Hydrogen Energy* 28 (2003) 1263–1272.
9. Satyapal, S., Proc: Fuel Cell Seminar 2006 Abstracts, November 13–17, Hawai, USA (2006).

Chapter 2
Research Trends in Polymer Electrolyte Membranes for PEMFC

S.M. Javaid Zaidi

Abstract In this chapter research trends followed by various scientific groups for the development of polymeric membranes have been described and reviewed. Most notably, the developments made at Ballard Advanced Materials (BAM) and some of their results are discussed. In general three different approaches have been followed worldwide by various research groups for the development and conception of alternative membranes. These approaches include: (1) modifying perfluorinated ionomer membranes; (2) functionalization of aromatic hydrocarbon polymers/membranes; and (3) composite membranes based on solid inorganic proton conducting materials and the organic polymer matrix or prepare acid-base blends and their composite to improve their water retention properties. The current trend is for the composite and hybrid membranes, which combines the properties of both the polymeric component and inorganic part. The most widely studied polymer after Nafion is the sulfonated polyether-ether ketone (SPEEK), as it has a high potential for commercialization. A number of research projects are currently undergoing dealing with the SPEEK polymer in various research labs.

2.1 Introduction

Fuel cells have been in development for over 150 years, ever since their invention in 1839 by William Robert Grove. The fuel cell became a real option for a wider application base in the late 1980s and early 1990s, although structural improvements are still needed to accommodate the increasing demands of fuel cell systems for specific applications. The research in this area gained momentum in the 1980s due to increased awareness of energy and environmental concerns. Also, some pivotal innovations (e.g., low platinum catalyst loading, thin film electrodes) drove down the cost of the fuel cell, which made the development of PEMFC more realistic.

The center of the fuel cell is the polymer electrolyte membrane, as it defines the properties needed for other components of the fuel cell. However, fuel cells' efficiency and power density also strongly depend on the conductance of electrolytes, and only acidic electrolyte can be used to aid carbon dioxide rejection for DMFC. The performance of polymer electrolyte fuel cells is closely related to the

S.M.J. Zaidi, T. Matsuura (eds.) *Polymer Membranes for Fuel Cells*,
doi: 10.1007/978-0-387-73532-0, © Springer Science+Business Media, LLC 2009

performance of MEA. The price of MEA ranks first in market position, roughly 75% of the overall price of PEMFC. Historically, the progress in PEMFC performance in terms of efficiency and lifetime was related to the development of proton-conducting membranes. Currently, efforts concentrate on the development of new proton-conducting polymer membranes, although a large number of scientific contributions still deal with Nafion membrates.

The desired properties for a membrane to be used as a proton conductor in a fuel cell are listed in the following:

1. Chemical and electrochemical stability in fuel cell operating conditions
2. Elevated proton conductivity to support high currents with minimal resistive losses and zero electronic conductivity
3. Good water uptakes at high temperatures of approximately 100°C
4. Thermal and hydrolytic stability
5. Chemical properties compatible with the bonding requirements of membrane electrode assembly
6. Extremely low permeability to reactant species to maximize efficiency
7. Mechanical strength and stability in the operating conditions. (The membrane must be resistant to the reducing environment at the anode as well as the harsh oxidative environment at the cathode.)
8. Resistance of fuel transport through it. (This is a concern in a DMFC, in which methanol crossover takes place, and gets oxidized at the cathode. This reduces the cell voltage by formation of mixed potential at the cathode.)
9. High durability
10. Facilitation of rapid electrode kinetics
11. Flexibility to operate with a wide range of fuels
12. Production cost compatible with the commercial requirements of the fuel cell

In addition to the preceding properties, hydration of the membrane (water management) and thickness also play important roles in affecting the overall performance of fuel cells.

The advances made in fuel cell performance have been closely associated with advances in polymer electrolyte technology. Until now, perfluorinated ionomer (PFI) membranes, Nafion (DuPont de Nemours), and Dow membranes (Dow Chemical Co.) have been useful in practical fuel cell systems and are currently the only ones being used commercially. The performance of Dow membranes is superior to that of Nafion 117, but they are more expensive than Nafion. In spite of the outstanding properties of these membranes, such as high proton conductivity and high chemical inertness, these membranes cost US\$800–2,000 m^{-2} and suffer from serious drawbacks, such as high methanol permeation and water balance problems. Moreover, lack of safety during manufacturing and use, requirements of supporting equipment, and temperature-related limitations are some other drawbacks. Degradation of Nafion membranes at high temperatures is a serious drawback. Conductivity at 80°C is reduced by more than 10-fold relative to that at 60°C. Safety concerns rise from evolution of toxic intermediates and corrosive gases liberated at elevated temperatures above 150°C. Degradation products could be a concern during the manufacturing process or vehicle accidents and could limit fuel cell recycling options [1]. Thus PFI membranes, which have a number of limitations, are not the ideal choice today, and their cost is a major drawback for the commercialization of PEMFC and DMFC technology. For fuel cells to be commercially feasible for transportation devices, the projected membrane cost has to be reduced significantly, to a range of US$5–15 ft^{-2}.

Thus began the search for an alternative low-cost polymer material. The current challenge is to improve the membrane properties in terms of thermal stability and proton conductivity while reducing methanol crossover. This may be achieved by the creation of new low-cost membranes. It has been mentioned by Smitha et al. [1] in their study that Gilpa and Hogarth identified 60 alternatives to the PFI membranes. Among these, 15 membranes showed potential for replacing Nafion membranes. To develop new polymer membrane with similar and improved properties by a less expensive route, the properties of Nafion polymers need to be understood; consequently, research on Nafion membranes was carried out.

2.2 Early Developments: Nafion Membranes

The concept of using ion exchange membranes as electrolytes was first reported by General Electric (GE) in 1955. The idea of using organic cation exchange membranes as solid electrolytes was first described by William Thomas Grubb and Lee Niedrach in 1959. NASA's interest in fuel cells as power sources for space applications gave great impetus to polymer fuel cell development with the testing of phenolic membranes [2]. These membranes showed power densities of 0.05–0.1 kW m^{-2} and lifetimes of 300–1,000 h, as well as low mechanical strength. Later on, GE improved the power density by developing partially sulfonated poly(styrene sulfonic acid) membranes, which showed improved power densities of 0.4–0.6 kW m^{-2}. The first PEMFC used in an operational system was built by GE as a primary power source for the GEMINI series of spacecraft during the mid-1960s [3]. It was a 1-kW power plant. At that time extremely expensive materials were used and the fuel cells required very pure hydrogen and oxygen. The polymer membranes used as electrolytes were based on poly(styrene sulfonic acid). However, these membranes exhibited brittleness in the dry state and were later replaced with crosslinked polystyrene-divinyl benzene sulfonic acid membranes. This material also lacked stability and underwent degradation and suffered other problems. Also, the main problem encountered with these membranes was that proton conductivity was not sufficiently high to reach a power density even as low as 100 mW cm^{-2}. Therefore, in 1966 they were replaced by PFI Nafion membranes; this was a real breakthrough in membrane developments for PEM fuel cells. At this early stage of development the most improved membranes showed lifetimes of up to 3,000 h at low current densities and temperatures of 50°C [4, 5]. The demonstrated cell life as a function of operating temperature of various membranes is illustrated in Fig. 2.1 [4, 5]. Results are also given for a perfluorosulfonic acid membrane Nafion. The

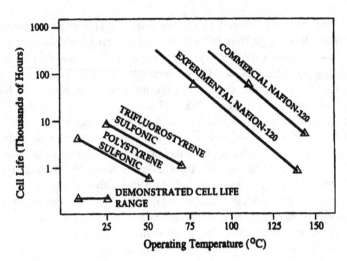

Fig. 2.1 PEM fuel cell life capacity [4]. (With the permission of Prof. Savadogo, editor and publisher of the Proceedings of the First International Symposium of New Materials for Electrochemical Systems, Montreal, Canada, 2005.)

most commonly used membrane is Nafion, which relies on liquid water for humidification of the membrane to transport proton. Nafion possessed inherent chemical, thermal, and oxidative stability and it displaced the unstable polystyrene sulfonic acid membranes. The advent of Nafion in the late 1960s gave an impetus to the PEM industry. The second GE PEFC unit, a 350 W module, powered the Biosatellite spacecraft in 1969. An improved DuPont manufactured Nafion membrane was used as an electrolyte. Nafion, a sulfonated tetrafluorethylene copolymer discovered in the late 1960s by Walther Grot (DuPont) [6], is the first of a class of synthetic polymers with ionic properties called ionomers. From the late 1960s onward, membrane requirements were best met by the Nafion family of membranes. Nafion's unique ionic properties result from incorporating perfluorovinyl ether groups terminated with sulfonate groups onto a tetrafluoroethylene (Teflon) backbone. Nafion received considerable attention as a proton conductor for PEM fuel cells because of its excellent thermal and mechanical stability. Nafion is a perfluorosulfonic acid (PFSA) membrane, chemically synthesized in four steps according to the DuPont process [7].

It is similar to Teflon, to which sulfonic acid groups are attached. The acid molecules are fixed to the polymer and cannot leak out, but the protons of these acid groups are free to migrate through the membrane. The chemical structure of Nafion is in Fig. 2.2.

With commercial Nafion 120, a lifetime of over 50,000 h has been achieved [7]. Nafion 120 has an equivalent repeat unit molecular weight of 1200 ($x = 6–10$ and $y = z = 1$) and a dry state thickness of 260 μm, whereas Nafion 117 and 115 have equivalent repeat unit molecular weights of 1,100 and thicknesses in the dry state of 175 and 125 μm, respectively. The Nafion family of membranes extended the lifetime by four orders of magnitude, and soon became standard for PEMFC, which it remains to this today. The Dow and Asahi Chemical companies also synthesized

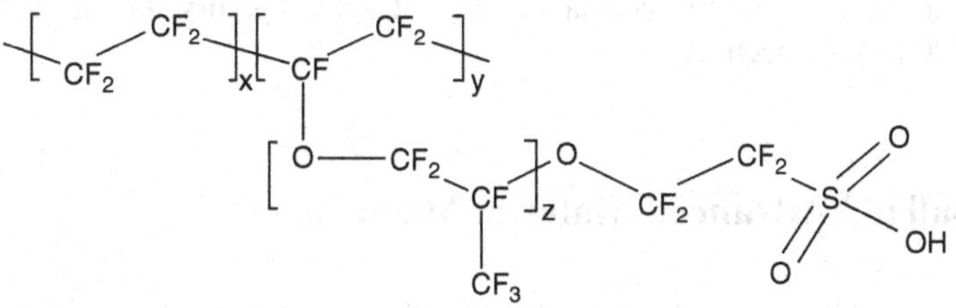

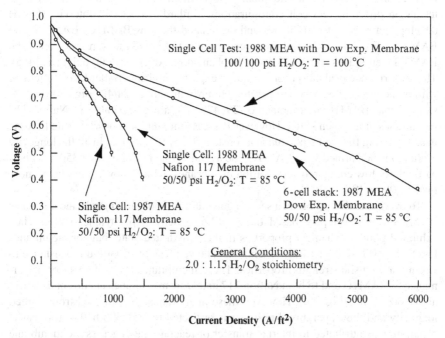

Fig. 2.2 Chemical structure of PFSA Nafion membrane

Fig. 2.3 Polarization data for Nafion and Dow membranes in Ballard Fuel cells. (With the permission of Prof. Savadogo, editor and publisher of the Proceedings of the First International Symposium on New Materials for Electrochemical Systems, Montreal, Canada, 2005.)

advanced perfluorosulfonic acid membranes with shorter side chains. The Dow polymer membrane has higher conductivity than Nafion. The Dow membranes are structurally and morphologically similar to Nafion membranes, but differ with respect to their lower equivalent weights, which are typically in the range of 800–850, and have shorter side chains (Dow $z = 0$, Nafion $z = 1$). The conductivity of 800 and 850 EW Dow membranes have been reported to be 0.2 and 0.12 S cm^{-1} [7, 8]. The best performance of Nafion in a six-cell MK 4 stack was 0.5 V at 1,400 A ft^{-2}, whereas the best performance of the DOW membrane was 0.5 V at 5,000 A/ft^2 as shown in Fig. 2.3 Although the large-side chain-perfluorinated polymer electrolytes prolonged service life, there is no large-scale industrial electrochemical system using Dow membranes. Asahi Glass Company developed Flemion R, S, T, which has equivalent repeat unit molecular weights of 1,000 and dry state thicknesses of 50, 80, 120 μm, respectively.

2.3   Ballard Advanced Material Membranes

The research efforts made at Ballard Power Systems for the development of low cost PEM membranes for fuel cell are encouraging. These membranes are the only ones among membrane development works which have gone through extensive testing in different fuel cell configurations. Ballard Advanced Materials (BAM) developed a series of membranes and designated them as BAM1G, BAM2G, and BAM3G—first-, second-, and third-generation BAM membranes, respectively. The BAM1G membranes were based on poly(phenylquinoxalene) (PPQ) polymers [4, 5]. The performance evaluation carried out in experimental size Ballard MK4 fuel cell with an active area of 50 cm^2 operated on air/hydrogen at 24/42 psig and at 70°C showed that BAM1G membranes of 39–420 EW are comparable to Nafion 117 membranes or are even better. However BAM1G membranes showed finite lifetime in an operating fuel cell when run at a constant 500 A ft^{-2} at 70°C. In all the longevity evaluations performed with BAM1G, the average time to failure was 350 h. This is particularly low compared with Nafion membranes, which show a lifetime of more than 10,000 h [8].

To overcome this problem, a second generation of polymer membrane known as BAM2G was developed, based on poly(2,6-diphenyl-4-phenylene oxide). They exhibited good mechanical properties in the dehydrated state, but for membranes less than 450 EW the hydrated membrane showed less than optimum resistance to tearing and tensile strength properties. These membranes showed superior performance to BAM1G and to both Nafion 117 and Dow membranes at current densities above 600 A ft^{-2} [Fig. 2.4]. However, these membranes also suffered from limited longevity and their operation lifetime was restricted to 500–600 h. The root cause of failure was attributed to internal transfer of reactant gases across the membrane electrode assembly.

Using α, β, β-trifluorostyrene (TFS), a novel family of sulfonated copolymer incorporating TFS series and a series of substituted TFS comonomers provided the group of materials referred to as BAM3G. Their performance was evaluated in a standard membrane electrode assembly (MEA) in an experimental size Ballard MK4 single cell hydrogen/air fuel cell at 85°C with an active area of 50 cm^2. These membranes maintained high efficiency as that of BAM2G and increased lifetime. The longevity achieved from these new membranes exceeded 15,000 h of operation. These BAM3G membranes are still in the testing stage and have not yet been commercialized as low-cost PEM membranes, so their cost information is not available, only some projections are there. Also, their performance in a DMFC, and the permeation behavior of methanol through these membranes are not known.

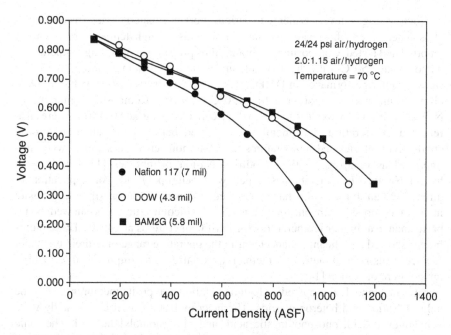

Fig. 2.4 Polarization data for Nufion 117, Dow, and BAM2G. (With the permission of Prof. Savadogo, editor and publisher of the Proceedings of the First International Symposium on New Materials for Electrochemical Systems, Montreal, Canada, 2005.)

Moreover, BAM3G being a fluorinated product will not be environmentally friendly, as when mass produced these membranes will pose a problem for their safe disposal. Apart from that, a great deal of information relating to their ionic conductivity, thickness, exact chemical composition, and mechanical strength, hydraulic permeation is proprietary and is not available in the literature.

2.4   Modification of Nafion Membranes

Some of the pioneering works will be discussed here, with important results reported in the literature. Primarily attempts have been made to modify the morphology of Nafion membranes themselves by using different processes such as plasma etching and palladium sputtering to modify the Nafion membranes. Plasma etching of Nafion membranes increases the roughness of the membrane surface and decreases the methanol permeation. The sputtering of palladium on plasma-etched Nafion further decreased methanol permeation. Apart from the decrease in methanol permeation, the open circuit voltages and current-voltage performance of fuel cells fabricated with membranes which had undergone plasma etching and palladium sputtering were also improved significantly.

Compositing Nafion polymer with various inorganic materials, such as zirconium phosphate, boron phosphate, and heteropolyacids at high temperature has been reported and tested. Composite membranes thus prepared at operating temperatures up to about 150°C with dry oxidant, under mild preheating conditions (85°C) showed better performance in DMFC. Typical cell resistances of 0.08 Ω-cm^2 were observed under cell operation at 140–150°C [6, 9, 10]. Composite membranes of Nafion with silicon oxide for use in fuel cell operating at 80–140°C were also reported in literature. The membranes showed better water retention, proton conductivity at elevated temperatures, and thermomechanical stability were also improved as compared with unmodified Nafion membranes [11–13]. Further, hybrid Nafion-silica membranes doped with heteropolyacids for application in direct methanol fuel cells have also been reported. Nafion-silica composite membranes doped with phosphotungstic and silicotungstic acids showed better performance at higher temperatures for DMFC operations at 145°C. These membranes showed significant enhancement in the operating range of a direct methanol fuel cell; also, the kinetics of methanol oxidation was improved due to high temperature operation [14].

Savadogo and Tazi prepared and studied Nafion composite membranes with the help of Nafion and heteropolyacids. The heteropolyacids used in this study were silicotungstic acid, phosphotungstic acid, and phosphomolybdic acid. The ionic conductivities of composite membranes were found to be higher than those of pure Nafion membranes. The composite membrane prepared from Nafion and silico-tungstic acid was found to be most conductive of all other membranes prepared [15]. In another study, Ramani et al. has prepared and investigated Nafion/HPA composite membranes for high temperature and low relative humidity fuel cell operation. The decomposition temperature of the composite membrane was extended to 150°C, permitting more stringent operating conditions. The protonic conductivities of the composite membranes at 120°C and 35% RH were on the order of the 0.015 S cm^{-1} [13, 16]. The HPAs were water soluble and studies of their long-term stability in the membrane matrix were not investigated.

Tricoli and Nanetti [17] prepared a novel zeolite-Nafion composite membrane by embedding zeolite fillers in Nafion. The zeolites used in this study were chabazite and clinoptilolite. The presence of zeolites in the membranes caused notable changes in conductivity, methanol permeability, and selectivity with respect to pure Nafion. In another interesting study, Holmberg et al. synthesized and characterized zeolite-Y nanocrystals for Nafion-zeolite-Y composite proton exchange membranes. The composite membranes were found to be more hydrophilic and proton conductive than the base-unmodified membranes at high temperatures [18].

Blending Nafion polymer with other polymeric materials has also been tried by some researchers. In one such study, poly(1-methylpyrrole) has been impregnated with commercial Nafion membrane by in situ polymerization. A decrease of more than 90% in the permeability of the membranes to methanol is reported, although the ionic resistance of such heavily loaded membranes became too high for high-power fuel cells. At lower poly(1-methylpyrrole) loadings, a decrease in methanol permeability by as much as 50% could be realized without a significant increase in ionic resistance [19]. In another study Nafion/polytetrafluoroethylene (PTFE) membranes have also been prepared using porous PTFE membranes as support material. The membranes were synthesized by impregnating porous PTFE membranes with Nafion solutions. Resulting membranes were found to be mechanically and thermally stable. The composite membranes thus prepared were also cost effective [20, 21].

Membrane prepared by blending sulfonated polybenzimidazole (PBI) with Nafion polymer showed a conductivity of 0.032 S cm^{-1}. The methanol permeability of the composite membrane was found to be 0.82×10^{-6} cm^2 s^{-1} as compared to Nafion, which is around 2.21×10^{-6} cm^2 s^{-1} [22]. Addressing the problem of methanol permeation, a composite membrane of Nafion with polyvinyl alcohol (PVA) for direct methanol fuel cell has been reported. It is concluded that at the weight ratio of 1:1 in PVA and Nafion, the thin film-coated Nafion membrane exhibited low methanol crossover, and the membrane protonic conductivity could be improved by the sulfonation treatment [23]. Recently, Zaidi et al. [24] prepared composite membranes of PFSA ionomer with boron phosphate and showed the conductivity of 6.2×10^{-2} S cm^{-1} at 120°C.

2.5 Hydrocarbon Composite Membranes

The approach of making composite membranes from the inorganic modification of the polymer matrix has gained momentum recently due to the exemplary success achieved by some people working on this approach. The composite membrane approach represents one of the ways to improve the properties of the polymer electrolyte membranes as the desired properties of the two components can be combined in one composite. Various polymers have been used in this approach with different inorganic materials. One of the important polymers that has drawn much attention is the sulfonated polyether ether ketone (SPEEK) polymer whose structure is shown in Fig. 2.5. SPEEK falls into the category of aromatic sulfonic acid polymers. Its advantages include low methanol permeation, high conductivity, and very good mechanical properties. Good mechanical stability provides the membrane enough flexibility, thus making the membrane thin enough to decrease the resistance offered by the thickness of the membranes [25–27].

SPEEK polymer is supposed to be a noble substitute to Nafion membranes as far as its use in DMFC is concerned. SPEEK has potential to substitute Nafion membranes but only after modification of certain properties. Mainly these properties include protonic conductivity of SPEEK membranes as compared with Nafion membranes; otherwise, SPEEK has comparable methanol permeation, high temperature stability, and low cost as compared with commercial Nafion membranes. The

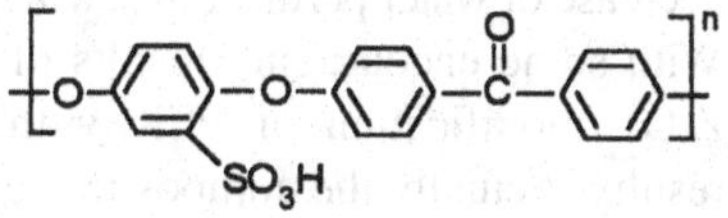

Fig. 2.5 Chemical Structure of SPEEK

electrochemical properties of a series of composite membranes prepared by incorporation of boron phosphate into the polymeric matrix of SPEEK were studied. The proton conductivity of the composites was found to be higher than the pure SPEEK polymer. The mechanical stability was also in the satisfactory range for use in DMFC at moderately high temperatures [26, 28].

SPEEK polymer was blended with the polyetherimide (PEI) polymer and then doped with HCl and H_3PO_4 to get a better solution to the DMFC problem. Results with these membranes had mixed success. Doping of HCl was found to be more significant than that of H_3PO_4 [29]. Protonic conductivity increased moderately with boron phosphate and PEI, but the incorporation of various heteropolyacids into the SPEEK structure increased proton conductivity significantly. Room temperature conductivities on the order of 10^{-2} S/cm were reported, while the conductivity values were raised up to 10^{-1} s cm^{-1} with the same composite membrane at high temperatures of around 100°C. The composite membranes were found to be thermally stable up to 250°C [25]. Although these composite membranes showed high conductivities at elevated temperature, they had the problem of leaching of the solid heteropolyacids (HPAs). The problem of the high leaching of HPAs was found to be a detriment to the long-term use of these membranes. In order to avoid the problem of leaching, HPAs are loaded onto molecular sieve zeolite Y and mesoporous MCM-41, respectively, so that the HPAs incorporated onto these molecular sieves does not leach out of the structure. These HPA-loaded molecular sieve Y-zeolite and HPA-loaded molecular sieve MCM-41 powdered solids are embedded into SPEEK polymers to make composite membranes. These membranes exhibited conductivities on the order of 10^{-2} s cm^{-1} at 140°C and were found to be thermally and mechanically stable [30, 31].

Zirconium oxide-modified SPEEK membranes have been proposed recently for direct methanol fuel cell applications. The zirconium oxide modification affected its water swelling, chemical and mechanical stability, methanol and water permeations and, finally, proton conductivity. Depending on the amount of the inorganic component in the membrane, a good balance between high proton conductivity, good chemical stability, and low methanol permeability could be reached [32, 33]. Composite polymer membranes have prepared by embedding layered silicates (laponite and montmorillonite) into SPEEK. While the SPEEK polymer contributed partially to conductivity, layered silicates incorporated into SPEEK significantly helped to reduce swelling in hot water. Also, methanol crossover was reduced without an alarming reduction in the proton conductivity [34].

New organic-inorganic composite membranes based on sulfonated polyetherketone (SPEK) and SPEEK were synthesized with SiO_2, TiO_2, and ZrO_2. The modification of SPEK and SPEEK with ZrO_2 reduced the methanol flux by 60-fold. On the other hand, there was a big compromise on conductivity, which was reduced by 13-fold, while modification of PEK and SPEEK with silane (SiO_2) led to a 40-fold decrease of water permeability without a large decrease of protonic conductivity [35]. With some encouraging results of the modification of PEK with SiO_2, TiO_2, and ZrO_2, modification of PEK with heteropolyacid further yielded some notable results. Actually the composite membranes were prepared using an organic matrix of SPEK, different heteropolyacids, and an inorganic network of ZrO_2 and $RSiO_{3/2}$. The bleeding out of the heteropolyacid from the membranes was also measured, in addition to water and methanol permeation and protonic conductivity tests. The presence of ZrO_2 decreased water and methanol permeability and reduced the bleeding out of heteropolyacid [36].

Apart from SPEEK polymers, some other polymer materials have also been used in order to get the appropriate candidate material for DMFC applications. Sulfonated poly(arylene ether sulfone) polymer has been incorporated with HPA. The membranes thus prepared showed excellent thermal stability (about 300°C) and good proton conductivity, especially at elevated temperatures (130°C). Infrared and dynamic thermo-gravimetric data showed that the composite membrane had much higher water retention (100–280°C) than pure sulfonated copolymer. These results also suggested that the incorporation of HPA into these proton-conducting copolymers should be good candidates for elevated temperature operation of DMFC [37]. However, these membranes showed bleeding out of heteropolyacids into the poly(arylene ether sulfone) polymer matrix. Novel composite membranes based on PVA with embedded phosphotungstic acid were prepared and measured for their protonic conductivity and methanol permeation. A marginal conductivity of the order of 6.27×10^{-3} S cm^{-1} was obtained, while the values of methanol permeation were found to be in the range of 1.28×10^{-7} and 4.54×10^{-7} cm^2 s^{-1}. From the values for methanol permeation obtained, these composite membranes had the potential to use them in direct methanol fuel cell [38].

Polybenzimidazole doped with phosphoric acid has been used to prepare the composite membranes with inorganic proton conductors such as zirconium phosphate, phosphotungstic acid, and silicotungstic acid. The conductivity of the phosphoric acid–doped PBI and PBI composite membranes was found to be dependent on the acid doping level, relative humidity (RH), and temperature [29, 39]. Apart from the direct incorporation of one component into another, a new synthetic route to synthesize organic/inorganic nanocomposites hybrid polymer membrane using SiO_2 and polymer such as modified PBI, polyethylene oxides, polypropylene oxide, polyvinylidene fluoride, etc., the composite membranes was prepared through sol-gel processes. The methanol permeation through the membranes decreased significantly and membranes showed excellent proton conductivity [40].

Different hybrid membranes with the help of inorganic/organic or organic/ organic components have been reported in literature with some exciting results. Hybrid polyaryls ether ketone membranes with the help of zirconium phosphate and modified silica have been prepared and characterized for fuel cell applications. The examples were chosen to illustrate the in situ formation of inorganic particles, either on a prepared membrane, or in a polymer solution. SPEEK modified silica and SPEEK zirconium phosphate membranes provided power densities of 0.62 W cm^{-2} at 100°C. In all cases, the presence of the inorganic particles led to an increase in proton conductivity of the polymer membrane, without any harm to its flexibility [41, 42]. Zirconium carboxybutylphosphonate was synthesized to prepare inorganic/ organic composite membranes based on PBI. The membrane thus prepared showed promising performance and relatively high protonic conductivities under the given conditions. Membranes were also highly thermally stable [43]. A series of organic/inorganic composite materials based on polyethylene glycol (PEG)/SiO$_2$ for use as electrolytic membrane in DMFC have been synthesized through sol-gel processes. Acidic moieties of 4-dodecylbenzene sulfonic acid (DBSA) were doped into the network structure at different levels to provide the hybrid membrane with proton conducting behavior. An increasing trend of proton conductivity with increasing DBSA doping was obtained, while the presence of an SiO$_2$ framework in the nanocomposite hybrid membrane provided enhanced thermal stability. Some of the hybrid membranes exhibited low methanol permeability without sacrificing their conductivities significantly, and were thus proposed to be potentially useful in DMFC [44].

The blending of two organic components to get a hybrid membrane has always been a point of attraction for almost all researchers. Basically the flexibility of playing with the microstructure of the organic polymer candidates and easy handling have paved the way to explore a suitable hybrid membrane. As in one study, novel acid-base polymer blends have been characterized for application in membrane fuel cells. The membranes synthesized are composed of SPEEK Victrex or polyether sulfone (PES) as well as sulfonated polysulfone (sPSU) Udel as the acidic compounds, and of PSU CELAZOLE, or poly(ethyleneimine) PEI (Aldrich) as the basic compounds. The membrane showed good proton conductivities and excellent thermal stabilities and showed good performance [45]. In another contribution, different types of acid-base composite membranes have been prepared and characterized for their use in DMFC at high temperatures. In this study, sulfonated polyetherketone and sulfonated polysulfone are used as acidic blend components, while PSU(NH$_2$)$_2$, poly (4-vinylpyridine), and polybenzimidazole are used as basic components [46]. Multilayered polyphosphazene membranes have been suggested as a new series of hybrid membranes, with improved protonic conductivity and low methanol permeation. A phosphagenic polymer was used for membrane synthesis, in which the polyphosphazene was sulfonated, blended with an uncharged polymer, and then cross-linked. Poly[bis(3-methylphenoxy)phosphazene] has also been reported as a promising material for fuel cell applications. Polymer cross-linking was carried out by the use of UV light and photoinitiator. The results showed that there was a significant decrease in methanol crossover (the methanol flux was about 10 times lower than Nafion 117) [47]. Membranes from polybenzimidazole/sulfonated polysulfone have been studied and compared with homopolymer membranes made from sulfonated polysulfone, blends of polyether sulfone with sulfonated polysulfone, and Nafion 117. Also an improved behavior of these membranes toward methanol permeation was observed [48]. In another study, blended membranes were prepared by the blending of SPEEK and polyether sulfone. The transport properties of membranes with SPEEK content in the range of 50–80 wt% were found to be comparable to those known for commercial ion exchange membranes [49].

Novel methanol barrier polymer electrolyte membranes for direct methanol fuel cells were proposed and characterized from PVA blend–polystyrene sulfonic acid (PSSA). The effects of curing temperature, methanol concentration, and membrane composition on the ionic conductivity and the methanol permeability of the membranes were also investigated [50]. In an interesting study, a new proton exchange membrane from the sulfonation of poly(phthalazinone ether ketone) has been synthesized. Membrane performances were directly related to the degree of sulfonation (DS). Proton conductivity increased with degree of sulfonation and temperature up to 95°C, reaching up to 10^{-2} S cm^{-1} [51]. Composite membranes were synthesized with the help of polyvinyl alcohol membranes loaded with mordenite or tin mordenite. It was observed that a single layer of these materials had poor mechanical strength with noticeable cracks and also with poor conductivity, but some encouraging results were obtained when they sandwiched these layers, and membranes were prepared with many layers of the polyvinyl alcohol and tin mordenite [52].

2.6 Other Relevant Developments in the Last Decade

Several studies reported the preparation of membranes for PEM fuel cells by the radiation grafting technique. Most of these studies involve grafting of styrene or α, β, β-trifluorostyrene on to a fluorine-containing polymer, followed by sulfonation of the grafted film. The results obtained are encouraging, but these membranes are partially fluorinated and are not environmentally friendly. Furthermore, these membranes showed good performance in hydrogen fuel cells, but no information is available for methanol fuel cell. The sulfonic acid-based polymer membranes prepared by Lee et al. [53] showed a specific resistivity of 24 Ωcm, which is comparable to Nafion 117 (16 Ωcm). The data for stability of these membranes are not available and these are not tested in actual fuel cells, so nothing can be said about their long-term stability and methanol permeation [54].

Membranes having grafted poly(styrene sulfonic acid) and three different backbone polymers, and low-density polystyrene, poly(tetrafluoroethylene), and a copolymer of tetrafluoroethylene and hexafluoroethylene showed similar conductivities to those found for Nafion and Dow membranes [7]. However, the oxidative stability of these membranes was poor. Only poly(tetrafluoroethylene) showed some promise as a candidate material. The stability of these membranes in fuel cell systems for applications above 60–70°C was not investigated.

There is considerable methanol permeation through Nafion, which affects the fuel cell performance in a DMFC. Using doped PBI the same proton conductivity as Nafion can be maintained while virtually eliminating the crossover of methanol. PBI is doped with a conducting solid, usually phosphoric acid, to make it suitable for DMFC applications [55, 56]. In another attempt PBI is modified by sulfonation to make it an intrinsic proton conductor and is deposited onto a layer of Nafion membrane. This gives a composite polymer electrolyte that is a reasonable proton conductor and reduces the crossover of methanol [57].

Several materials have been studied on the goal of producing cost-effective PEMs [55–62]. Some of these are PBI-based membranes, polysterene membranes, sulfonated polyimide, cross-linked poly(vinyl alcohol), and phosphobenzene, sulfonated poly(aryl ether ketone) —based membranes. Sulfonation of aromatic thermoplastics such as polyether sulfone, polybenzimidazole, polyimides, and poly(ether ether ketone) makes them proton conductive suitable for fuel cell applications. In a recent study, Zaidi [63] used blends of SPEEK and PBI with solid boron phosphate to prepare composite membranes. These membranes showed conductivity on the order of 0.5×10^{-2} S cm^{-1} and are thermally stable up to 150°C.

Although PBI or phosphoric acid membranes can reach up to 220°C without using any water management, higher temperatures allow for better efficiencies, power densities, ease of cooling, and controllability. These types of membranes are not common; most researchers and labs still use Nafion. It has been pointed out by Smitha et al. [1] that hydrogen permeation data and diffusion parameters at elevated temperatures are important for the selection of new materials for fuel cells operating at low temperatures. To investigate hydrogen permeation rates with varying temperatures across polymer membranes, hydrogen radiotracers such as tritium were used. Permeation and diffusion coefficients were determined for Viton, Teflon, etc., and these gave very low values. Today in most cases the membrane is made by a perfluorosulfonic ionic polymer, whereas the electrodes with a mixture of Pt supported on carbon and a dispersion of generally the same ionomer of the membrane. Making durable membranes and cost reduction of MEA is one of the main targets of fuel cell research today. Solvey Solexis is developing Hyflon ion ionomers for producing membranes and dispersing for MEA manufacture [64]. Hyflon ion is similar in structure to Nafion ionomer, except that it contains a shorter side chain compared with that of Nafion [Fig. 2.6].

The precursor can be used in the production of extruded films in the thickness range of 20–200 microns. These films are hydrolyzed by acid exchange to reach the final functional form [Fig. 2.7].

This is dispersed in polar solvents, usually water-alcohol mixtures for film processing. The concentration, viscosity, and particle size of the dispersion is controlled by changing the operating parameters (temperature, mixing, and solvent composition) during dissolution. The ionomer dispersion is then used for the production of membranes by the casting or impregnation process. This is also used for the preparation of catalyst inks that are used for the preparation of fuel cell electrodes or catalyst-coated membranes. The thickness of the cast membranes is usually in the range of 10–50 microns. These membranes are partly

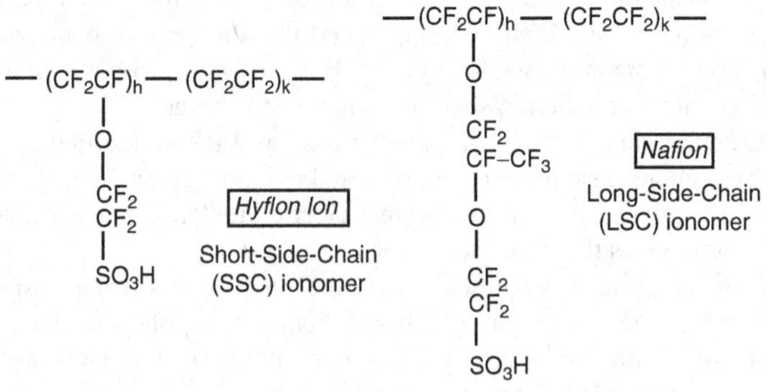

Fig. 2.6 Chemical structure of Hyflon Ion and Nafion

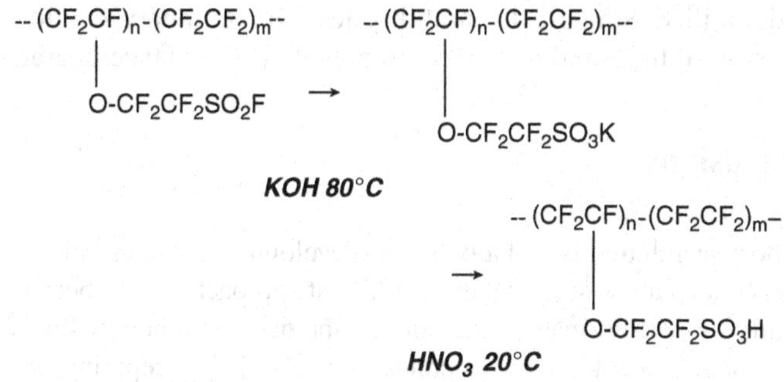

-- (CF2CF)n-(CF2CF2)m--
O-CF2CF2SO2F
-- (CF2CF)n-(CF2CF2)m--
O-CF2CF2SO3K
KOH 80°C
-- (CF2CF)n-(CF2CF2)m–
O-CF2CF2SO3H
HNO3 20°C

Fig. 2.7 Hydrolysis of the precursor

1000
900
800
700
600
500
400
Cell voltage [mV]
90°C
Hyflon
Nafion
0
2000
4000
6000
8000
10000
12000
14000
16000
18000
20000
Current density [A/m2]

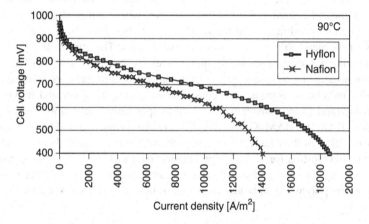

Fig. 2.8 Polarization curve comparison between Hyflon and Nafion at 90°C. (From: V. Arcella, A. Ghielmi, L. Merlo and M. Gebert, Membrane electrode assemblies based on perfluorosulfonic ionomers for an evolving fuel cell technology, *Desalination* **199**, 6–8 (2006).)[1]

crystalline (responsible for mechanical stability), and partly amorphous (responsible for proton conduction). Although Hyflon ion has lower equivalent weight (gram of polymer per mole of functional group) than Nafion, both have the same crystallinity and therefore mechanical properties. Sometimes crystallinity is expressed in terms of heat of fusion. That is why Hyflon ion shows more proton conductivity and therefore higher fuel cell performance. A 60°C increase in glass transition temperature for Hyflon ion gives the option of using this membrane at higher temperatures without mechanical failure. The resistance of Hyflon ion is lower than that of Nafion at all operating conditions [Fig. 2.8]. Requirements of 5,000 and 50,000 h are specified for automotive and stationary applications, and

[1] This figure was published in V. Arcella, A. Ghielmi, L. Merlo and M. Gebert, Membrane electrode assemblies based on perfluorosulfonic ionomers for an evolving fuel cell technology, Desalination 199, 6–8 (2006), copyright Elsevier.

long-term durability. Single-cell durability tests in stationary conditions (fixed current) for several thousand h showed no notable performance degradation.

2.7 Conclusions

Following the research trends globally for the development of alternative membranes for PEM fuel cell, it can be seen that three different approaches have been used. These are: (1) modifying perfluorinated ionomer membranes; (2) functionality of aromatic hydrocarbon membranes to improve conductivity; and (3) preparing new polymer electrolyte composite membranes based on solid inorganic proton–conducting materials or prepare acid-base blends and their composite to improve their water retention properties at temperature above 100°C. The most studied polymeric material after Nafion is the sulfonated polyether-ether ketone, which holds strong potential to replace Nafion membranes. And a number of studies have been reported using SPEEK or its blend to prepare composite membranes. The current focus is on the development of temperature stable membranes in the range of 100–150°C and methanol-resistant membranes. To meet these goals, most of the work is based on the composite membrane approach to reduce the methanol crossover and improve the conductivity and water management at high temperature. The membranes prepared by BAM—BAM1G, BAM2G, and BAM3—are also promising and have undergone extensive testing in their hardware. Most probably these are the only ones that have undergone field testing for different applications, but most of the literature is patented and proprietary. In spite of the extensive research efforts worldwide, the need still exists for new polymer membranes that could satisfactorily replace Nafion.

References

1. B. Smitha, S. Sridhar and A. A. Khan, Solid polymer electrolyte membranes for fuel cell applications—a review, *J. Membr. Sci.* **259**, 10–26 (2005).
2. K. Kordesch and G. Simader, "Fuel Cells and Their Applications", *New York: VCH Publishers*, (1996).
3. M. Rikukawa and K. Sanui, Proton-conducting polymer electrolyte membranes based on hydrocarbon polymers, *Prog. Polym. Sci.* **25**, 1463–1502 (2000).
4. A. E. Steck, "Membrane materials in fuel cells", *First International Symposium on New Materials for Electrochemical systems, Montreal,* July (1995), pp. 74–94.
5. A. E. Steck and C. Stone, Development of the BAM membrane for fuel cell applications, in "New Materials for Fuel Cell and Modern Battery Systems II", ed. O. Savogado and P. R. Roberge, *Montreal, Quebec: Ecole Polytechnique de Montreal,* (1997), p. 792.
6. A. S. Arico, S. Srinivasan and V. Antonucci, DMFCs: From fundamental aspects to technology development, *Fuel Cells* **1**(2), 133–161 (2001).
7. O. Savadogo, Emerging membranes for electrochemical systems: (I) Solid polymer electrolyte membranes for fuel cell systems, *J. New Mater. Electrochem. Syst.* **1**(1), 47–66 (1998).
8. A. J. Appleby and E. B. Yeager, Solid polymer electrolyte fuel cells (SPEFCs), *Energy* **11**(1–2), 137–152 (1986).

9. C. Yang, S. Srinivasan, A. S. Arico, P. Creti, V. Baglio and V. Antonucci, Composite Nafion/ Zirconium phosphate membranes for direct methanol fuel cell operation at high temperature, *Electrochem. Solid State Lett.* **4** (4), A31–A34 (2001).

10. P. Costamagna, C. Yang, A. B. Bocarsly and S. Srinivasan, Nafion® 115/zirconium phosphate composite membranes for operation of PEMFcs above 100°C, *Electrochim. Acta* **47**, 1023–1033 (2002).

11. K. T. Adjemian, S. J. Lee, S. Srinivasan, J. Benziger and A. B. Bocarsly, Silicon oxide Nafion composite membranes for proton-exchange membrane fuel cell operation at 80–140°C, *J. Electrochem. Soc.* **149**(3), A256–A261 (2002).

12. D. H. Jung, S. Y. Cho, D. H. Peck, D. R. Shin and J. S. Kim, Performance evaluation of a Nafion/silicon oxide hybrid membrane for direct methanol fuel cell, *J. Power Sources* **106**, 173–177 (2002).

13. P. Dimitrova, K. A. Friedrich, U. Stimming and B. Vogt, Modified Nafion-based membranes for use in direct methanol fuel cells, *Solid State Ionics* **150**, 115–122 (2002).

14. P. Staiti, Proton conductive membranes constituted of silicotungstic acid anchored to silica-polybenzimidazole matrices, *J. New Mater. Electrochem. Syst.* **4**, 181–186 (2001).

15. O. Savadogo and B. Tazi, New cation exchange membranes based on Nafion and heteropoly-acids with and without thiophene, *J. New Mater. Electrochem. Syst.* **1**, 15–20 (2003).

16. V. Ramani, H. R. Kunz and J. M. Fenton, Investigation of Nafion®/HPA composite membranes for high temperature/low relative humidity PEMFC operation, *J. Membr. Sci.* **232**, 31–44 (2004).

17. V. Tricoli and F. Nannetti, Zeolite-Nafion composites as ion conducting membrane materials, *Electrochim. Acta* **48**, 2625–2633 (2003).

18. B. Holmberg, H. Wang, J. Norbeck and Y. Yan, Synthesis and characterization of Zeolite Y Nanocrystals for Nafion Zeolite Y composite proton exchange membranes, *AIChE, Spring National Meeting*, New Orleans, Los Angeles, USA, March 10–14 (2002).

19. P. G. Pickup, M. C. Lefebvre, J. Halfyard, Z. Qi and J. Nengyou, Modification of Nafion proton exchange membranes to reduce methanol crossover in PEM fuel cells, *Electrochem. Solid State Lett.* **3**(12), 529–531 (2000).

20. F. Liu, B. Yi, D. Xing, J. Yu and H. Zhang, Nafion/PTFE composite membranes for fuel cell applications, *J. Membr. Sci.* **212**, 213–223 (2003).

21. J. Shim, H. Y. Ha, S. A. Hong and I. H. Oh, Characteristics of the Nafion ionomer-impregnated composite membrane for polymer electrolyte fuel cells, *J. Power Sources* **109**, 412–417 (2002).

22. G. Deluga, B. S. Pivovar and D. S. Shores, Composite membranes in liquid feed direct methanol fuel cells, *DOE/NRL Workshop*, University of Minnesota, October 6–8 (1999).

23. Z. G. Shao, X. Wang and I. Hsing, Composite Nafion/polyvinyl alcohol membranes for the direct methanol fuel cell, *J. Membr. Sci.* **210**, 147–153 (2002).

24. S. M. J. Zaidi and S. U. Rahman, Perfluorinated ionomer-boron phosphate composite membranes for polymer electrolyte membranes for fuel cell applications, *J. Electrochem. Soc.* **152**(8), A1590–A1594 (2005).

25. S. M. J. Zaidi, S. D. Mikhailenko, G. P. Robertson, M. D. Guiver and S. Kaliaguine, Proton conducting composite membranes from polyether ether ketone and heteropolyacids for fuel cell applications, *J. Membr. Sci.* **173**, 17–34 (2000).

26. S. M. J. Zaidi, S. D. Mikhailenko and S. Kaliaguine, Sulfonated polyether ether ketone based composite polymer electrolyte membranes, *Catal. Today* **67**, 225–236 (2001).

27. L. Li, J. Zhang and Y. Wang, Sulfonated poly(ether ether ketone) membranes for direct methanol fuel cell, *J. Membr. Sci.* **226**, 159–167 (2003).

28. S. M. J. Zaidi, S. D. Mikhailenko and S. Kaliaguine, Electrical conductivity of boron ortho-phosphate in presence of water, *J. Chem. Soc. Faraday Trans.* **94**(11), 1613–1618 (1998).

29. S. M. J. Zaidi, S. D. Mikhailenko and S. Kaliaguine, Electrical properties of sulfonated poly-ether ether ketone/polyetherimide blend membranes doped with inorganic acids, *J. Polymer Sci. B: Polym. Phys.* **38**, 1386–1395 (2000).

24

30. S. M. J Zaidi and M. I. Ahmed, Novel SPEEK/heteropolyacids loaded MCM-41 composite membranes for fuel cell applications, *J. Membr. Sci.* **270**, 548–557 (2006).

31. M. I. Ahmed, S. M. J. Zaidi and S. U. Rahman, Proton conductivity and characterization of novel composite membranes for medium temperature fuel cells, *Desalination* **193**, 387–397 (2006).

32 V. Silva, B. Ruffmann, H. Silva, H. Mendes, A. Madeira and S. P. Nunes, Zirconium oxide modified sulfonated polyether ether ketone membranes for direct methanol fuel cell applications, *Materials 2003-II International Materials Symposium*, Caprica, Portugal, April 14–16 (2003).

33. B. Ruffmann, H. Silva, B. Schulte and S. P. Nunes, Organic/inorganic composite membranes for application in DMFC, *Solid State Ionics* **162–163**, 269–275 (2003).

34. J. H. Chang, J. H. Park, G. G. Park, C. S. Kim and O. O. Park, Proton conducting composite membranes derived from sulfonated hydrocarbon and inorganic materials, *J. Power Sources* **124**, 18–25 (2003).

35. S. P. Nunes, B. Ruffmann, E. Rikowski, S. Vetter and K. Richau, Inorganic modifications of proton conductive polymer membranes for direct methanol fuel cells, *J. Membr. Sci.* **203**, 215–225 (2002).

36. M. L. Ponce, L. Prado, B. Ruffmann, K. Richau, R. Mohrand and S. P. Nunes, Reduction of methanol permeability in polyetherketone-heteropolyacid membranes", *J. Membr. Sci.* **217**, 5–15 (2003).

37. Y. S. Kim, F. Wang, M. Hickner, T. A. Zawodzinski and J. E. McGrath, Fabrication and characterization of heteropolyacid ($H_3PW_{12}O_{40}$)/directly polymerized sulfonated poly(arylene ether sulfone) copolymer composite membranes for higher temperature fuel cell applications, *J. Membr. Sci.* **212**, 263–282 (2003).

38. L. Li, L. Xu and Y. Wang, Novel proton conducting composite membranes for direct methanol fuel cell, *Mater. Lett.* **57**, 1406–1410 (2003).

39. R. He, Q. Li, G. Xiao and N. J. Bjerrum, Proton conductivity of phosphoric acid doped polybenzimidazole and its composites with inorganic proton conductors, *J. Membr. Sci.* **226**, 169–184 (2003).

40. X. F. Xie, H. Guo, Z. Q. Mao and J. M. Xu, Organic/inorganic nanocomposites for direct methanol fuel cell, *AICHE, Spring National Meeting*, New Orleans, Los Angeles, USA, March 10–14 (2002).

41. D. J. Jones, L. T. Bouckary and J. Roziere, Hybrid polyetherketone membranes for fuel cell applications, *Fuel Cells*, **2**(1), 1–6 (2002).

42. D. J. Jones, B. Baur, J. Roziere, L. Tchicaya, G. Alberti, M. Casciola, L. Massinelli, A. Peraio, S. Besse and E. Ramunni, Electrochemical characterization of sulfonated polyetherketone membranes, *J. New Mater. Electrochem. Syst.* **3**, 93–98 (2000).

43. M. Y. Jang and Y. Yamazaki, Preparation, characterization and proton conductivity of membrane based on zirconium tricarboxybutylphosphonate and polybenzimidazole for fuel cells, *Solid State Ionics* **167**, 107–112 (2004).

44. H. Y. Chang and C. W. Lin, Proton conducting membranes based on PEG/SiO_2 nanocomposites for direct methanol fuel cells, *J. Membr. Sci.* **218**, 295–306 (2003).

45. J. Kerres, A. Ullrich, F. Meier and T. Haring, Synthesis and characterization of novel acid-base polymer blends for application in membrane fuel cells, *Solid State Ionics* **125**, 243–249 (1999).

46. J. Kerres, A. Ullrich, T. Haring, M. Baldauf, U. Gebhardt and W. Preidal, Preparation, characterization and fuel cell application of new acid base blend membranes, *J. New Mater. Electrochem. Syst.* **3**, 129–139 (2000).

47. P. Pintauro, H. Yoo, R. Wycisk, J. Lee and R. Carter, Direct methanol fuel cell performance with multilayered polyphosphazene membranes, *AIChE, Spring National Meeting*, New Orleans, Los Angeles, USA, March 10–14 (2002).

48. C. Manea and M. Mulder, New polymeric electrolyte membranes based on proton donor-proton acceptor properties for direct methanol fuel cells, *Desalination* **147**, 179–182 (2002).

49. F. G. Wilhelm, I. G. M. Punt, N. F. A. Van der vagt, H. Strathmann and M. Wessling, Cation permeable membranes from blends of sulfonated Polyether ether ketone and polyether sulfone, *J. Membr. Sci.* **199**, 167–176 (2002).

50. Y. Wang, H. Wu and S. Wang, A methanol barrier polymer electrolyte membrane in direct methanol fuel cells, *J. New Mater. Electrochem. Syst.* **5**, 251–254 (2002).

51. Y. Gao, G. P. Robertson, M. D. Guiver and X. Jian, Synthesis and characterization of sulfonated poly(pthalazinone ether ketone) for proton exchange membrane materials, *J. Polym. Sci. A: Polym. Chem.* **41**, 497–507 (2003).
52. B. Libby, Composite multi-layered membranes for direct methanol fuel cells, *PhD Thesis*, University of Minnesota, Department of Chemical Engineering and Materials Science, Minneapolis (2001).
53. W. Lee, A. Shibasaki, K. Saito, K. Okuyama and T. Sugo, Proton transport through polyethylene-tetrafluoroethylene-copolymer-based membrane containing sulfonic acid group prepared by RIGP, *J. Electrochem. Soc.* **143**(9), 2795–2799 (1996).
54. P. L. Antonucci, A. S. Arico, P. Creti, E. Ramunni and V. Antonucci, Investigation of a direct methanol fuel cell based on a composite Nafion®-silica electrolyte for high temperature operation, *Solid State Ionics* **125**, 431–437 (1999).
55. Q. Guo, P. N. Pintauro, H. Tang and S. O'Connor, Sulfonated and crosslinked polyphosphazene-based proton-exchange membranes, *J. Membr. Sci.* **154**(2), 175–181 (1999).
56. J. J. Fontanella, M. C. Wintersgill, J. S. Wainright, R. F. Savinell and M. Litt, High pressure electrical conductivity studies of acid doped polybenzimidazole, *Electrochem. Acta* **43**, 1289–1294 (1998).
57. G. A. Deluga, S. C. Kelley, B. Pivovar, D. A. Shores and W. H. Smyrl, Composite membranes to reduce crossover in PEM fuel cells, *Battery Conference on Applications and Advances, 2000, The fifteenth annual*, 51–53 (2000).
58. L. Xiong and A. Manthiram, High performance membrane-electrode assemblies with ultra-low Pt loading for proton exchange membrane fuel cells, *Electrochem. Acta* **50**, 3200–3204 (2005).
59. H. Pu, Q. Liu and G. Liu, Methanol permeation and proton conductivity of acid-doped poly (N-ethylbenzimidazole) and poly(N-methylbenzimidazole), *J. Membr. Sci.* **241**, 169–175 (2004).
60. B. Bae and D. Kim, Sulfonated polysterene grafted polypropylene composite electrolyte membranes for direct methanol fuel cells, *J. Membr. Sci.* **220**, 75–87 (2003).
61. Y. Woo, S. Y. Oh, Y. S. Kang and B. Jung, Synthesis and characterization of sulfonated poly-imide membranes for direct methanol fuel cell, *J. Membr. Sci.* **220**, 31–45 (2003).
62. J. W. Rhim, H. B. Park, C. S. Lee, J. H. Jun, D. S. Kim and Y. M. Lee, Crosslinke poly(vinyl alcohol) membranes containing sulfonic acid group: proton and methanol transport through membranes, *J. Membr. Sci.* **238**, 143–151 (2004).
63. S. M. J. Zaidi, Preparation and characterization of Composite membranes are using blends of SPEEK/PEI with Boron Phosphate, *Electrochim. Acta* **50**, 4771–4777 (2005).
64. V. Arcella, A. Ghielmi, L. Merlo and M. Gebert, Membrane electrode assemblies based on perfluorosulfonic ionomers for an evolving fuel cell technology, *Desalination* **199**, 6–8 (2006).

Chapter 3
Fuel Cell Technology Review

A.F. Ismail, R. Naim, and N.A. Zubir

Abstract The scope of this chapter is to give a brief introduction about fuel cells, types of applications in fuel cell technology, characteristics of fuel cells, potential applications in fuel cell technology, and current research and development and key technology players in fuel cells.

3.1 Introduction

Fuel cells play a significant role in the strategy to effect positive global change. Smaller fuel cell plants ideally suited for distributed power are cost competitive with other competitive technologies, and larger conventional centrally located power plants. Battery and hybrid systems (non—fuel cell) do not meet overall requirements for transportation with respect to energy density and range. However, fuel cell technology alone has reached a level that meets or exceeds targets identified by automakers and hybrids may offer even further advantages. Power density, energy density, dynamic and operational response, cost potential, and operation on multiple fuels has been clearly demonstrated even for the most technically challenging transportation applications. However, significant challenges and opportunities still remain for improvement in fuel cell technology. The fuel cell will find applications that lie beyond the reach of the internal combustion engine. Once low-cost manufacturing becomes feasible, this power source will transform the world and bring great wealth potential to those who invest in this technology. It is said that the fuel cell is as revolutionary in transforming our technology as the microprocessor has been. Once fuel cell technology has matured and is in common use, our quality of life will improve and the environmental degradation caused by burning fossil fuel will be reversed. It is generally known that the maturing process of the fuel cell will not be as rapid as that of microelectronics. Fuel cells will play a key role as an element of a future, sustainable energy supply infrastructure because they are energy efficient and environmentally friendly. There are significant global environmental issues with existing energy paths today. Global emission and fuel regulations, global fuel and power structures, and costs are driving new technologies and unconventional approaches. Certain fuel cell types promise significant performance advances based on tests and demonstrations held worldwide by active participating technology players. Under the rapid advance of fuel cell technology research and development by players around the world, the mechanism exists for fuel cells to make the leap into future commercial markets.

The objective of this chapter is to provide a general overview of fuel cell technology and its current status worldwide in order to assess the future viability and feasibility of this technology.

3.2 Fuel Cells: Introduction

3.2.1 Potential of the Fuel Cell

For decades, the internal-combustion engine has been a hallmark in the history of the automotive industry and stand-alone energy supply. To most users, it has been the only appropriate solution so far to drive cars or generate power at remote sites. For the first time fuel cells offer the chance to replace the combustion engine in a number of applications and thereby avoid harmful emissions. For the energy industry, they open up the option of sustainable, resource-saving supply, and—thanks to their ecological satisfactoriness—many diverse applications. This includes applications in the mobile sector and all areas of the energy industry [1].

The fuel cell looks back on a long track record. An Englishman, Sir William Robert Grove (1811–1896), constructed the first fuel cell in 1839. Its further development proved such an arduous task that for nearly 100 years Grove's concept was only used in isolated applications. His fuel cells featured electrodes made of platinum sitting in a glass tube with their lower end immersed in dilute sulfuric acid as an electrolyte and their upper part exposed to hydrogen and oxygen inside the tube. This was sufficient to produce a voltage of 1 V. To turn the fuel cell into a really efficient source of power, substantial technical efforts had to be made.

Over 160 years have elapsed since the fuel cell was invented. Its true potential as the energy converter of the future has only recently manifested. Today, it is on the point of commercial use.

3.2.2 Operating Principle of the Fuel Cell

Fuel cells generate electricity from hydrogen and oxygen—without any harmful emissions and therefore in an extremely environmentally friendly way. Heat is produced in varying amounts, as well as the by-product water.

Figure 3.1 shows how the fuel cell works. A proton exchange membrane is coated with a thin platinum catalyzer layer and a gas-permeable electrode made of graphite paper. Hydrogen fed to the anode side ionizes into protons and electrons at the catalyzer. The protons pass the catalyzer layer, while the electrons remaining behind give a negative charge to the hydrogen-side electrode. During the proton migration, a volt-

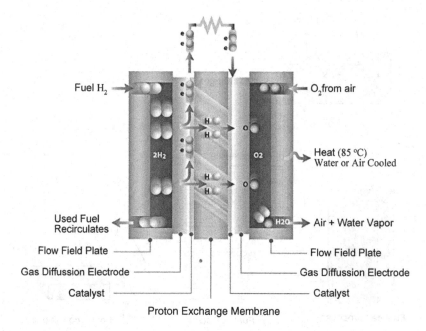

Fuel H$_2$

O$_2$from air

2H$_2$

O2

Heat (85 °C)
Water or Air Cooled

Used Fuel
Recirculates

H2O

Air + Water Vapor

Flow Field Plate

Flow Field Plate

Gas Diffussion Electrode

Gas Diffussion Electrode

Catalyst

Catalyst

Proton Exchange Membrane

Fig. 3.1 How fuel cell works [2]

age difference builds up between the electrodes. When these are connected, this difference produces a direct current that can drive an engine, for example.

Finally, the protons recombine with the electrons and oxygen into water at the cathode. Besides the recovered electric energy, the only reaction product is water. Additionally, heat is produced by the electrochemical reactions and the contact resistances in the fuel cell, which can be used for space or service water heating. The voltage of a single non-operated cell is about 1.23 V. In operation, this level falls to about 0.6–0.7 V under load. As this level is too low for practical applications; a sufficient number of cells are connected in series to obtain a usable voltage. They may add up to 800 cells in larger-sized plants. The line-up of cells is equivalent to a stack; this word has become a technical term generally used for this arrangement. It is characteristic of fuel cells that they generate DC voltage. To allow practical use, it has to be transformed into an AC signal. This is done by downstream DC/AC converters [3].

3.2.3   The Fuel Cell System: Design and Function

Figure 3.2 shows a basic fuel cell system. A fuel cell system consists of:

Fuel Processor. A fuel processor converts readily available hydrocarbon fuels (e.g., natural gas, propane, gasoline, diesel fuel, methanol, etc.) to a hydrogen-rich gas that is fed to a fuel cell. This process adds complexity to the system, but has the

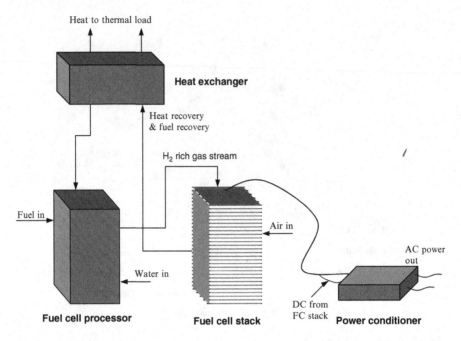

Fig. 3.2 Basic fuel cell system

advantage of using fuels, which have existing infrastructures and distribution networks. When hydrogen fuel is used, the fuel processor becomes unnecessary [4].

Power Section. The power section includes a fuel cell stack. The stack is a series of electrode plates interconnected to produce a desired amount of DC power.

Power Conditioner. A fuel cell produces DC power that must be converted to AC. The power conditioning section also reduces voltage spikes and harmonic distortions.

The design and function of a fuel cell can most easily be explained by using the concept of the low temperature polymer membrane fuel cell, also called proton exchange membrane fuel cell (PEMFC).

3.2.4 Component Technology

3.2.4.1 Electrodes

The electrodes consist of a conducting catalyst support material (often a porous form of carbon), which is impregnated with a platinum or platinum alloy catalyst. These gas diffusion electrodes with large catalytically active surface areas and correspondingly large interfaces with the electrolyte allow to attain high levels of power for a given volume (power density), while retaining good efficiencies, because they permit rapid transport of gases to fuel cell electrodes and fast electrochemical reaction at the electrodes [5].

3.2.4.2 The Membrane

The membrane has two functions. First, it acts as the electrolyte that provides ionic conduction between the anode and the cathode but is an electronic insulator. Second, it serves as a separator for the two-reactant gases. Some sources claim that solid polymer membranes (e.g., sulfonated fluorocarbon acid polymer) used in PEMFC are simpler, more reliable, and easier to maintain than other membrane types. Since the only liquid is water, corrosion is minimal. Pressure balances are not critical. However, proper water management is crucial for efficient fuel cell performance [6]. The fuel cell must operate under conditions in which the by-product water does not evaporate faster than it is produced, because the membrane must be hydrated. Dehydration of the membrane reduces proton conductivity. On the other hand, excess of water can lead to flooding of the electrodes.

3.2.4.3 Membrane-Electrode Assembly

Membrane-electrode assembly (MEA) is an anode-membrane-cathode composite structure in which each layer is made as thin as possible without losing the mechanical integrity of the composite structure or the electrochemical activity of the two electrodes. The functional advantages of such a structure include short diffusion paths of air and fuel gas to the electrochemical reaction sites, intimate contact of electrodes with the polymer electrolyte and low ionic resistance of the membrane. All these properties are needed to sustain high current and power densities [7].

3.2.4.4 Separator Plate

Separator plates contain channels on both sides of a gas-impervious layer, which distribute fuel gas and air to the anode and cathode of two adjacent cells, respectively. They must be conducting and in electric contact with the electrodes to collect the current and transmit it between adjacent cells. Furthermore, they must be impervious to contain the reactant gases within each half-cell. The gas inlets and outlets of the separator plates have to be connected separately to air and fuel gas inlet and outlet manifolds. The separator plate is another key component of the fuel cell stack for achieving high power densities [8].

3.2.5 Fuel Cell Types and Applications

Five fuel cell technologies are at present being developed. They differ in their electrolyte structure, working temperature, and fuel requirements. Their designations refer to the electrolyte used. Table 3.1 illustrates the characteristics of fuel cells.

Table 3.2 shows the advantages and disadvantages of the various fuel cell systems. The most common are the previously mentioned and most widely developed, proton exchange membrane fuel cell using a polymer electrolyte. This system is

Table 3.1 Characteristics of fuel cells

Fuel cell type	Electrolyte type	Operating temperature (°C)	Efficiency	Special features	Applications
Solid oxide fuel cell (SOFC)	Yttria & zirconium oxides	800–1,000	45–55% (FC only) 65–75% (hybrid)	Direct power production from natural gas, ceramics	Central and stand-alone CHP generation
Molten carbonate fuel cell (MCFC)	Lithium, potassium carbonate salt	600–650	45–55% (FC only) 65–75% (hybrid)	Complex process control, corrosion problems	Central and stand-alone CHP generation
Phosphoric acid fuel cell (PAFC)	Phosphoric acid	200–220	35–45%	Limited efficiency, corrosion problems	Stand-alone CHP generation
Alkaline fuel cell (AFC)	Aqueous alkaline	80–100	Not reported	High efficiency, pure H_2 and O_2 only	Space operations, defense
Proton exchange membrane fuel cell (PEMFC)	Fluorinated-sulfonic acid polymer membrane	70–80	35–45%	High flexibility in operation, high power density	Vehicles, stand-alone CHP generation (small scale)

aimed at vehicles and portable electronics. Several developers are also targeting stationary applications. The alkaline fuel cell, which uses a liquid electrolyte, is the preferred fuel cell for aerospace applications, including the space shuttle. Molten carbonate, phosphoric acid, and solid oxide fuel cells are reserved for stationary applications, such as power generating plants for electrical utilities. Among these stationary systems, the solid oxide fuel cell is the least developed but has received renewed attentions due to breakthroughs in cell material and stack designs [9].

3.2.6   Potential Applications of the Fuel Cell

Fuel cells were developed for and have long been used in the space program to provide electricity and drinking water for the astronauts. Terrestrial applications can be classified into categories of transportation, stationary or portable power uses. Polymer electrolyte membrane fuel cells are well suited to transportation applications because they provide a continuous electrical energy supply from fuel at high levels of efficiency and power density. They also offer the advantage of minimal maintenance because there are no moving parts in the power generating stacks of the fuel cell system. The utility sector is expected to be an early arena where fuel cells will be widely commercialized. Today, only about one third of the energy

Table 3.2 Advantages and disadvantages of the various fuel cell systems [5]

Type of fuel cell	Applications	Advantages	Limitations	Status
Proton exchange membrane (PEMFC)	Mobile (buses, cars), portable power, medium to large-scale stationary power generation (homes, industry)	Compact design; relatively long operating life; adapted by major automakers; offers quick start-up, low temperature operation, operates at 50% efficiency	High manufacturing coasts, needs heavy auxiliary equipment and pure hydrogen, no tolerance for contaminates; complex heat and water management	Most widely developed; limited production; offers promising technology
Alkaline (AFC)	Space (NASA), terrestrial transport (German submarines)	Low manufacturing and operation costs; does not need heavy compressor, fast cathode kinetics	Large size; needs pure hydrogen and oxygen; use of corrosive liquid electrolyte	First generation technology; had renewed interest due to low operating cost
Molten carbonate (MCFC)	Large-scale power generation	Highly efficient; utilizes heat to run turbines for con-generation	Electrolyte instability; limited service life	Well developed; semi-commercial
Phosphoric acid (PAFC)	Medium- to large-scale power generation	Commercially available; lenient to fuels; utilizes heat for co-generation	Low-efficiency, limited service life, expensive catalyst	Mature but faces competition from PEMFC
Solid oxide (SOFC)	Medium- to large-scale power generation	High efficiency, lenient to fuels, takes natural gas directly, no reformer needed. Operates at 60% efficiency; utilizes heat for co-generation	High operating temperature; requires exotic metals, high manufacturing costs, oxidation issues; low specific power	Least developed. Breakthroughs in cell material and stack design sets off new research
Direct methanol (DMFC)	Suitable for portable, mobile and stationary applications	Compact design, no compressor or humidification needed; feeds directly off methanol in liquid form	Complex stack structure, slow load response times; operates at 20% efficiency	Laboratory prototypes

consumed reaches the actual user because of the low energy conversion efficiencies of power plants. Using fuel cells for utility applications can improve energy efficiency by as much as 60% while reducing environmental emissions [10].

Phosphoric acid fuel cells have been generally used in the initial commercialization of stationary fuel cell systems. These environmentally friendly systems are simple, reliable, and quiet. They require minimal servicing and attention. Natural gas is the primary fuel; however, other fuels can be used, including gas from local landfills, propane, or fuels with high methane content. All such fuels are reformed to hydrogen-rich gas mixtures before feeding to the fuel cell stack. Over 200 (phosphoric acid fuel cells) units, 200 kW each are currently in operation around the world [11].

Fuel cell manufacturers are now developing small-scale polymer electrolyte fuel cell technology for individual home utility and heating applications at the power level of 2–5 kW because the potential for lower materials and manufacturing costs could make these systems commercially viable [12]. Like the larger fuel cell utility plants, smaller systems will also be connected directly to natural gas pipelines—not the utility grid. In addition to these small-scale uses, polymer electrolyte fuel cell technology is also being developed for large-scale building applications.

Distributed power is a new approach utility companies are beginning to implement—locating small, energy-saving power generators closer to the need. Because fuel cells are modular in design and highly efficient, these small units can be placed on site. Installation is less of a financial risk for utility planners and modules can be added as demand increases. Utility systems are currently being designed to use regenerative fuel cell technology and renewable sources of electricity. Figure 3.3 shows the potential applications in fuel cell technology.

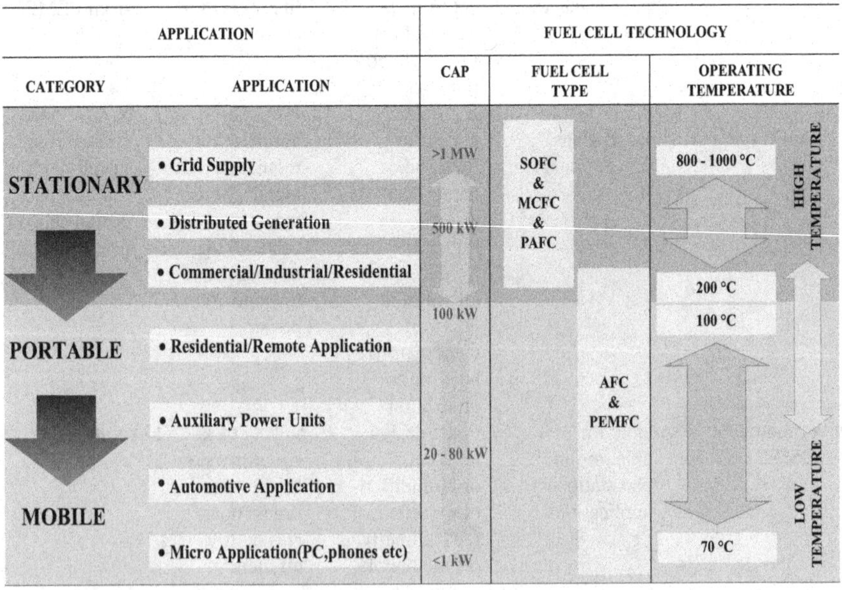

APPLICATION		FUEL CELL TECHNOLOGY		
CATEGORY	APPLICATION	CAP	FUEL CELL TYPE	OPERATING TEMPERATURE
STATIONARY	• Grid Supply	>1 MW	SOFC & MCFC & PAFC	800 - 1000 °C
	• Distributed Generation	500 kW		
	• Commercial/Industrial/Residential			200 °C
PORTABLE	• Residential/Remote Application	100 kW		100 °C
	• Auxiliary Power Units		AFC & PEMFC	
MOBILE	• Automotive Application	20 - 80 kW		
	• Micro Application(PC,phones etc)	<1 kW		70 °C

Fig. 3.3 Potential applications for fuel cells

3.3 Current Status in the Development and Application of Fuel Cells

3.3.1 Low Temperature Fuel Cells

3.3.1.1 Alkaline Fuel Cells

While there remains some interest in alkaline fuel cells (AFCs) for commercial applications, notably by Zetek as a range extender for battery-powered vehicles, this is at a low level compared with that for PEMFC systems. Modified anode and cathode catalysts have now been developed that supersede the high concentrations of Pt and Pt/Au used in the space Orbiter vehicle. However, there is no consensus on whether precious or base metals are the preferred system. The choice depends on the balance between the required performance and acceptable cost. With increasing interest in the use of hydrogen for fuel cell powered vehicles, interest in the AFC system may increase. On the other hand, it is reported that the new fuel cells for the space shuttle Orbitor vehicles will be PEMFC in preference to the established AFC [12].

3.3.1.2 Proton Exchange Membrane Fuel Cells

By 1989 and the first Grove Fuel Cell Symposium, the early work by General Electric that established the proton exchange membrane fuel cell (PEMFC) system for use on the Gemini space missions had been re-evaluated and developed by Ballard Power Sources to become one of the most significant fuel cell developments of the 1980s. It is suggested that General Electric chose not to pursue commercial applications for the PEMFC as, in its then-existing form, it required electrodes with a high platinum content and was more sensitive to CO poisoning than the established PAFC system.

In addition, it was potentially more expensive as a result of the cost of the polymer electrolyte and the high Pt metal loaded electrodes. While the sensitivity to CO poisoning remains, although less so with anode catalysts such as Pt/Ru, rapid developments were achieved in improving power density while reducing metal loadings [13]. For example, Nafion had been identified as a suitable proton-conducting membrane electrolyte. Substituting other sulfonated fluorocarbon polymer materials gave a fourfold increase in current density at the same operating cell voltage. Further, by optimizing catalyst and electrode structure, Pt loadings were reduced from 28 mg cm^2 to 0.2 mg Pt cm^2 while maintaining current densities.

3.3.1.3 PEMFC Fuel Cells for Transport Applications

These developments attracted the interests of the major car companies who were seeking ways to eliminate CO, HC, and NOx emissions from vehicles, while at the same time reducing fuel consumption and the associated CO_2 emissions.

In 1994, California introduced the "zero emissions mandate." The time scales were relaxed, largely as a result of the failure of battery powered vehicles to meet the required performance and range targets. However, the zero emissions vehicle (ZEV) standards, together with the US Government's Partnership for a New Generation of Vehicles (PNGV) initiative to develop light duty vehicles with improved fuel economy remain. Together, they have provided the stimulus for major investments by the car companies in fuel cell technology. Developments in PEMFC and its application to vehicles have been rapid in the last decade [14]. In 1993, Ballard demonstrated a PEMFC-powered bus. Following the announcement of the first fuel cell stack with a power density of 1 kW L^{-1} a Ballard phase 2 buses were demonstrated powered entirely by a 200 kW unit. This bus had no reduction in performance or passenger seating compared with a standard bus. At present, Xcellsis, the DaimlerChrysler/Ballard/Ford consortium developed and demonstrated buses in North America and Europe with the aim of making them commercially available in 2005. Table 3.3 shows several prototypes of fuel cell cars.

Progress in developing fuel cell—powered light duty vehicles has been equally impressive but significant technical challenges remain to be addressed. These include the fuel to be used and the cost of the fuel cell system and drive train. However, with today's reformer technology, methanol is the preferred choice and has been demonstrated by DaimlerChrysler in their prototype NECAR3 vehicle. Although gasoline or other petroleum-derived fuel is significantly more difficult to reform, the challenge to solve the problems is being addressed with the aim of bringing fuel cell vehicles into widespread use prior to the dawn of the "hydrogen economy."

3.3.1.4 Polymer Electrolyte Membrane Fuel Cells for Stationary Applications

In the early 1990s, as interest in PEMFC systems for transport applications gathered pace, it became apparent that the performance and cost objectives set by the car manufacturers would provide an attractive and competitive fuel cell system for stationary applications. Since then, a 250 kW stationary unit has been developed by Ballard Generation Systems operating on natural gas. Demonstration units in America, Japan, and Europe are now in place with some having achieved over 1 year of operation. Cell voltage decay rates are reported to be less than 0.3% per 1,000 h. Some significant projects for PEFC commercialization are illustrated in the Table 3.4.

The applications for 250 kW units in distributed power systems are well established. An exciting and potentially new development with major implications for reducing energy consumption in the housing sector is the adoptions of the PEMFC units for mobile applications to smaller systems in the 2–10 kW range. Up to 100 units have now been built and demonstrated by companies such as Plug Power, H Power, and Sanyo.

Table 3.3 Several prototypes of fuel cell cars

Automaker	Vehicle type	Year shown	Fuel type
BMW	Series 7 Sedan	In development	Hydrogen
Daimler Chrysler	Necar (van)	1993	Gaseous hydrogen
	Necar 2 (minivan)	1995	Gaseous hydrogen
	Necar 3	1997	Liquid methanol
	Necar 4	1999	Liquid hydrogen
	Jeep commander 2 (SUV)	2000	Methanol
	Necar 5	2000	Methanol
	DMFC (one person vehicle)	2000	Methanol
Energy Partners	Green car (Sports car)	1993	Hydrogen
Ford Motor Company	P2000 HFC (sedan)	1999	Hydrogen
	P2000 SUV	1999 (concept only)	Methanol
	TH!NK FC5	2000	Methanol
General Motors/Opel	Zafira (minivan)	1998	Methanol
	Precept	2000	Hydrogen
	Hydrogen1	2000	Hydrogen
Honda	FCX-V1	1999	Hydrogen
	FCX-V2	1999	Methanol
	FCX-V#	2000	Hydrogen
H Power	New Jersey Venturer	1999	Hydrogen
	New Jersey Genesis	2000	Hydrogen (from sodium borate or "Borax")
Hyundai	Santa Fe (SUV)	2000	Hydrogen
Mazda	Demio	1997	Hydrogen (stored in metal hydride)
Nissan	R'nessa (SUV)	1999	Methanol
	Xterra (SUV)	2000	Methanol
Renault	FEVER (station wagon)	1997	Liquid hydrogen
	Laguna Estate	1998	Liquid hydrogen
Toyota	RAV 4 FCEV (SUV)	1996	Hydrogen (stored in a metal hydride)
	RAV 4 FCEV (SUV)	1997	Methanol
Volkswagon/Volvo	Bora HyMotion	1999	Hydrogen

3.3.1.5 Direct Methanol Fuel Cells

Fuel cell technology, particularly for transport applications, would take a leap forward if a viable system were developed that could use a liquid fuel without reforming. The development of anode catalysts with the activity to operate on simple hydrocarbon fuels is unlikely. However, Shell and other oil companies established in the 1960s that methanol, with anode catalysts such as Pt/Ru, had potential. Early work utilized sulfuric acid as the electrolyte.

Table 3.4 Some significant projects for PEFC commercialization [10]

Producer	Power (kW)	Fuel	Prototype availability	Application
Ballard	1	Processed NG	Yes	Micropower
Ballard	40	—	—	—
Ballard	250	Processed NG	Yes	Commercial
Nuvera	1	Hydrogen	Not yet	Premium
Nuvera	1	Processed propane	2002	Premium
Nuvera	5	Processed NG	Not yet	Residential
H-Power	4–5	Processed NG	Not yet	Domestic
Vaillant—Plug Power	4–6	Processed NG	Yes	Domestic
GE-FCS	4–5	Processed NG	Yes	Domestic
Toshiba	30	NG/Propane/Biogas	Yes	Residential
Energy	20–50	Processed NG	—	Residential
Fuji Electric	1	Processed NG	Not yet	Domestic
Sanyo	2–3	Processed NG	Yes	Residential
Shatz Energy	4	Hydrogen	Yes	Remote power

With the introduction of proton conducting membranes, interest in direct methanol fuel cells (DMFC) systems in the 1990s has been renewed with projects in North America, Japan, and Europe. Of particular significance has been the work of Los Alamos National Laboratory. If the power density required for vehicle applications is to be achieved, further improvements to anode performance are necessary. Existing membranes also allow "methanol crossover," which in turn contributes to poor cell performance. In this context, it is interesting to speculate on how high-temperature membranes such as that developed by Celanese would perform in a direct methanol fuel cell.

3.3.1.6 Phosphoric Acid Fuel Cells

Following the successful development of alkaline fuel cells for space exploration, attention turned to commercial applications, initially in the United States in 1967 with the target program for small-scale units, followed by the FCG1 program for larger multi-MW units. The phosphoric acid system using platinum-containing electrodes was chosen as the most viable technology at the time for use with hydrocarbon fuels such as natural gas.

The decision by Pratt & Whitney to use phosphoric acid as the electrolyte and adapt steam-reforming technology to produce hydrogen was ground breaking at the time. For these developments, William Podolni, head of fuel cell development at Pratt & Whitney (later to become United Technologies) was awarded the Grove Medal in 1995.

Much of the technology relating to PEMFC systems originates from the development of PAFC and its demonstration [15]. Notable in this are the use of highly dispersed Pt metal catalysts, Teflon-bonded electrode structures, graphite bipolar plates, and also reformer technology and the associated balance of plant. PAFC systems, such as the ONSI Corp. PC25 200 kW combined heat and power (CHP) unit, are still the only units that are commercially available, albeit at a price that is still not competitive with established CHP systems. They have established the viability and reliability for on site electricity and heat generation. Table 3.5 illustrated some of significant projects for MCFC and SOFC commercialization.

There are now some 65 MW of PAFC systems worldwide. Most of the plants are in the 50–200 kW capacity range, but large plants of 1 and 5 MW have been built, including an 11 MW plant for Tokyo Electric Power. Details on performance, reliability, and cost reduction initiatives are contained in several reviews.

3.3.2 High-Temperature Fuel Cell

These high-temperature fuel cells operating at 650 and 1,000°C, respectively, were developed largely with the intention of overcoming the limitations of the low temperature PEMFC and PAFC systems. Their two main advantages are that their performance is not affected by carbon monoxide and the residual heat, which is available at temperatures in excess of 600°C, makes them applicable to industrial as well as commercial uses. Although neither system with existing anode catalyst technology is capable of truly operating in a direct fuel mode; hence, hydrocarbon

Table 3.5 Some significant projects for MCFC and SOFC commercialization [10]

Producer	Power (kW)	Fuel	Prototype availability	Application
MCFC	250	NG	Yes	Commercial
FCE/MTU	3000	NG	Not yet	Subpower
Ansaldo	100	NG	Yes	Commercial
	500	NG	Not yet	Commercial
Hitachi	250	NG	Yes	Commercial
L.H.I	250	NG	Not yet	Commercial
SOFC Siemens-Westinghouse	1,000 hybrid	Processed NG	Yes	Subpower
Siemens-Westinghouse	300 hybrid	Processed NG	Yes	Commercial
Siemens-Westinghouse	250	Processed NG	Yes	Commercial
Siemens-Westinghouse	25	Processed NG	Yes	Residential
ZTEK	1–25	—	Yes	Domestic
Sulzer/Hexis	1–5	NG	Yes	Domestic
Sulzer/Hexis	200	NG	Yes	Commercial
Mitsubishi Heavy	5	—	Yes	Domestic
Mitsubishi Heavy	25	—	Yes	Residential
Fuji	1	Hydrogen	Yes	Domestic
SOFC	1–4	NG	Yes	Domestic
SOFC	10–50	Diesel	Yes	Residential

reforming is necessary, the stack operating temperatures are such that internal reforming is possible. In fact, the systems that today have been developed and demonstrated all employ internal reforming either directly within the cell or in a separate reformer contained within the stack module. This simplifies the system with cost and power efficiency benefits.

3.3.2.1 Molten Carbonate Fuel Cells

The largest demonstration of an MCFC system was a proof of concept natural gas-fueled 2-MW unit operated from 1996 to 1997 in California. The plant was built by Energy Research Corp. (now Fuel Cell Energy, Inc.), and incorporated internal reforming in what is known as the direct carbonate fuel (DFC) cell TM. The plant-operated grid connected for over 4,000 h. Based upon the technology and experience gained from this 2 MW demonstration, 250 kW units have been designed and demonstrated by FCE and its partner, MTU, in Germany.

Several MCFC developers are also active in Japan and Europe, including BCN in the Netherlands, Ansaldo in Italy, and Hitachi, IHI, Mitsubishi Electric, and Toshiba in Japan. The recent progress in demonstrating MCFC systems is being supported by an extension to the existing FCE production plant to increase its annual production capacity from 50 MW in 2001 to 400 MW in 2004 [16].

3.3.2.2 Solid Oxide Fuel Cells

The sixth fuel cell system that makes up today's range is the solid oxide fuel cell (SOFC). It is less developed than the rest of the fuel cell range, although with a wider potential. Until recently, SOFC was seen as only having application in multi-MW stationary plants. Now, SOFC systems have been developed and demonstrated by Sulzer Hexis, Ltd. at the 1-kW level for residential CHP applications. These are primarily intended to be fueled with natural gas but have been demonstrated using low-sulfur home heating oil. Although an SOFC system operating at 1,000°C is not a first choice for transport applications, small-scale systems are now being developed for use as auxiliary power units (APU) in cars. A duel fuel hydrogen/gasoline vehicle with as SOFC auxiliary power unit is being demonstrated by BMW. The unit provides electric power for on road use as well as for accessories, heating and cooling when the car is stationary.

The first demonstration of SOFC for cogeneration was carried out in the Netherlands. The 100 kW unit built by Siemens Westinghouse and operated by Elsam and EDB (a consortium of Dutch utility companies) began operation in 1998. At the end of the demonstration project in December 2000, the plant had operated for over 16,000 h. Although SOFC has the potential for a wider range of applications than any of the other fuel cell systems, there remain significant materials, cell and stack assembly challenges to be met before confidence in the technology is justified. Experience from more advanced systems such as PAFC and MCFC, has shown that extensive demonstrations in real situations are required before operating deficiencies become apparent [17].

3.4   Fuel Cell R&D Direction

Multi-stakeholder consultations have identified the following fuel cell and related systems as priorities where R&D actions have the best prospects of meeting the targets set out. Low temperature fuel cell systems (PAFC, PEMFC, and DMFC), which have a potential for a very low cost per kW and which, in the medium term, may be commercialized in stationary (buildings, industrial, commercial), mobile or portable applications. Meanwhile, the high temperature fuel cell (SOFC and MCFC) has the potential for high-capacity power generation. Table 3.6 below summarizes the current R&D of fuel cell technology and direction for major countries involves in fuel cell development.

Some of the research and development in fuel cell have focused on the improving and further developing several areas related to fuel cell technology which are fuel processing and fuel cell stack component as illustrated in various tables in this chapter.

3.5   Technology Players and Users

Several technology players are illustrated in Table 3.7. For more information on these companies, please refer to Tables 3.8. and 3.9.

3.6   Conclusions

Since the first fuel cell discovery, progress in developing fuel cell technology has been steady if not spectacular. The most significant has been the rapid development of the PEMFC system for transport applications, together with the realization that success in this field would provide the basis of a low temperature stationary unit for CHP applications.

Following the successful demonstration of fuel cell powered buses and light duty vehicles by all of the major vehicle manufacturers, attention has now turned to choice of fuel and fuel infrastructure. Technical solutions to two challenging areas will go a long way to resolving the question. The first challenge is the development of a hydrogen storage system. It should provide a vehicle range of at least 300 miles with no significant increase in volume or weight compared with that of the equivalent gasoline tank. Second, a reformer and gas clean up system for use with gasoline with a start-up and response time giving a fuel cell—powered vehicle similar performance to that of the conventional internal combustion engine (ICE) vehicle.

Table 3.6 Current R&D and direction of fuel cell technology

	Current R&D	Direction	Active participants
United States	– Overcoming the major technical, economic infrastructure-related hurdles with most R&D resources devoted to light duty vehicle application – Research on direct methanol fuel cells and SOFCs at federal laboratories, university and private industry – Actively conducted demonstration program in stationary power plants	Improve fuelling infrastructure, i.e., production, storage, transportation and dispensing for commercialization of automotive fuel cell applications Increase the installation and production of PAFC commercial stationary power plant	Government agencies, i.e., DOE, PNGV, Universities, etc. GM, Ford, IFC, DaimlerChrysler, etc.
Europe (including the UK and Germany)	– Current research focus on – Low-temperature fuel cells (PAFC, PEMFC, and DM FC) in stationary, mobile or portable applications – High-temperature fuel cell system (MCFC and SOFC) for utility/industrial/commercial/ building cogeneration, large-scale electricity production and possibly transportation – Development of advanced fuel processor for the production of hydrogen from variety of fuels (esp. for PEMFC application)	Further increase the technology in transportation applications to be competitive in the automotive market	Siemens and DaimlerChrysler in Germany, EON in the Netherlands and Imperial College in the UK
Canada	– Research actively involved in PEMFC technology for transit buses demonstration and field test car. – Increase the development in stationary power system	Further improve the PEMFC system in fleet applications and stationary power system	Ballard with major companies such as Dow Chemical, Johnson Matthey, DaimlerChrysler, etc.
Japan	– Major development focus for both stationary and transportation applications in PEMFC, PAFC, and SOFC	Spreading market for PAFC, marketing development of stationary PEMFC	Mitsubishi, Fuji Electric, Toshiba, Sanyo Electric, Toyota, and Honda

Fuel processing

Participants	R&D activities	Target
Pacific Northwest National Laboratory (PNNL)*	• Technical feasibility has been demonstrate	• Ro develop bench-scale system (5 kW)
H2 fuel and Argonne National Laboratory*	• To develop small, low coast hydrogen generator	• To further enhance the generator's performance
Catalytica Energy Systems*	• To develop flexible-fuel processor for PEM	• To produce a multi-fuel processor
Johnson Matthey*	• To develop HotSpot reformer	• Plans for mass manufacturing
* Under US DOE funding		

Bipolar Plate

Participants	R&D activities	Target
Institute of Gas Technology	• Developing low cost composite bipolar plates	• Production of much thinner plates
Provair Fuel CEll Technology*	• Development carbon/carbon composite bipolar plates for PEMFCs.	• To develop high-volume production of composites bipolar plates
SGL Carbon	• Developing cost-effective bipolar plates	• To develop efficient manufacturing technology
* Under US DOE funding		

Membrane

Participants	R&D activities	Target
DeNora North America*	• To improve cathodes performance and to develop high temperature membranes	• To develop high-volume production of composites bipolar plates.
Celanese	• To develop high temperature membrane	• Mass production of MEAs
DuPont	• To develop new products of MEA	• Production of advanced materials
* Under US DOE funding		

Membrane electrode assembly (MEA)

Participants	R&D activities	Target
Southwest Research Institute (SwRI) teamed with W.L. Gore*	• To develop low cost MEA	• Mass production of MEA
3M Company*	• To developed MEA's with improved cathodes,	• Design and fabrication of MEAs
E-TEK	• To develop gas diffusion electrodes that are less expensive with higher performance	• 20,000 sq ft production automation

* Under US DOE funding

Table 3.7 Some of technology players in fuel cell

Company	Location	Mfg facilities	Technology (type of fuel cell)	Applications (stationary, transportation, portable)	Power (kW)	Known customers	Partners (equity or others)
Anuvu, Inc	Sacramento, CA	Sacramento, CA (2001)	Proton exchange membrane (PEM)	Residential, transportation, stand-by	N/A	N/A	Small start-up using some state incentives
Avista Corp.	Spokane, WA	Logan Industries	PEM	Commercial, residential	50–300	N/A	Logan Industries
Ballard	Burnby, BC, Canada	Opened in 4Q 2000–2001	PEM	Stationary, transportation	250	Ford, DCX, GM, Nissan, Honda, VW	Ford, DCX
Ceramic fuel cells limited (CFCL)	Australia, Asia	Australia	Solid oxide	Residential	40	N/A	N/A
Delphi	Troy, MI	Troy, MI	Solid oxide	Auxiliary	5	BMW, Renault	Global Thermoelectric (fuel cell developer)
Fuel cell energy	Danbury, CT	Torrington in spring 2001	Carbonate fuel cell	Stationary	225–250	DCX, US Navy, US Coast Guard	MTU/DCX
H-Power	Clifton, NJ		PEM	Stationary, mobile, portable	Up to 1,000	N/A	ECO fuel cells
International fuel cell	South Windsor, CT		PEM	Residential, transportation stationary (commercial)	500–11,000	N/A	A division of United Technology
Nuvera fuel cells	Cambridge, MA		PEM	Residential, transportation	N/A	Several automotive customers	DeNora Fuel Cells (Italy), Epyx Corp. (A.D. Little), Amerada Hess
Plug Power	Latham, NY	Netherlands	PEM	Residential	7–15	N/A	DTE
Texaco Ovonic fuel cell LLC	Rochester Hills, MI	Troy and Rochester Hills, MI	Ovonic proprietary	Residential, stationary, UPS, transport	1–1,000	N/A	Energy conversion devices, Texaco

Table 3.8 Key players, suppliers and users profiles

Participant		Fuel cell technology			
		PAFC	PEMFC	MCFC	SOFC
United States of America	**Stationary power**				
	International Fuel Cells (IFCs)	•	•	•	
	Fuel Cell Corporation of America	•			
	Westinghouse				•
	Energy Research Corporation			•	
	M-C Power			•	
	Ballard Power Systems 1 (Canada)				
	Energy Partners		•		
	Allied Signal (AI Research)		•		•
	SOFCo (Ceramatec/ Babcock & Wilcox)				•
	Ztek (Waltham, MA)				•
	Analytic Power (Boston)		•		
	Transportation				
	DaimlerChrysler		•		
	General Motors		•		
	Ford Motor Co.		•		
	Energy Partners		•		
	Toyota		•		
	Honda		•		
	H-Power		•		
	Hyundai		•		
	Mazda		•		
	Nissan		•		
	Renault		•		
	Volkswagen/Volvo		•		
	Portable Power				
	H- Power		•		
	Nuvera Fuel Cell		•		
	ElectroChem (Woburn, MA)		•		
	Giner (Waltham, MA)2		•		
Material/System Producers/Suppliers					
	Johnson Matthey Fuel Cells				
	Nextech Materials				
	Arthur D. Little				
	McDermott Technology				
	Foster-Miller, Inc.				
	Spectracorp, Ltd.				
	Idatech Corp.				
	Stuarts Energy Systems				
	3M				
	DuPont				
	Vairex Corp.				
	Celanese AG				
	Thermo Technologies				
	Mechanology, Inc.				

(continued)

Table 3.8 (continued)

	Fuel cell technology			
Participant	PAFC	PEMFC	MCFC	SOFC
Nuvera Fuel Cells				
Electrochem				
Plug Power				
Praxair				
National Laboratories				
Argonne National Laboratory (ANL)				
Oak Ridge National Laboratory				
Los Alamos National Laboratory				
National Renewable Energy Laboratory				
Lawrence Livermore				
Laboratory				
Pacific Northwest National Laboratory				
Lawrence Berkeley National laboratory				
Ames National Laboratory				
Brookhaven National Laboratory				
Sandia National Laboratories				
Idaho National Engineering and				
Environmental Laboratory				
Universities				
Sponsoring and Support Organizations				
Department of Energy (DOE)				
Office of Fossil Energy, Department of Energy				
The Office of Transportation Technologies				
The Office of Advanced Automotive Technologies				
The Online Fuel Cell Information Center				
The American Hydrogen Association (AHA)				
AQMD's Technology Advancement Office				
The Electric Power Research Institute				
Federal Energy Technology Center				
Gas Research Institute				
The Hydrogen and Fuel Cell Letter				
Partnership of New Generation Vehicles				

1. Partners for stationary fuel cells include GPU International, GEC Alsthom, and Ebara.
2. Direct methanol fuel cell technology

(continued)

In the 1980s projections for stationary fuel cells anticipated there being at least 2,000 MW of capacity in use by 2000. Not only have these projections been over-optimistic, but opinions on the size of the units have changed. For example, much of the market was seen to require large stationary multi-MW units. The smallest size that was thought to be commercially viable was 200 kW. While this may still be the case in some circumstances, both low- and high-temperature micro-CHP fuel cell units are seen to have wide-scale application in domestic applications. Why are there these changes of direction? What are the driving forces? In part, interest in

Table 3.8 (continued)

Participant		Fuel cell technology			
		PAFC	PEMFC	MCFC	SOFC
Europe	Ansaldo	•		•	
	Siemens		•	•	•
	DeNora Permalec		•		
	Dornier				•
	Vickers Shipbuilding & Engineering		•		
	Sulzer				•
	Brandstofel				
	Nederland				
	Deutsche Aerospace AG				•
	Daimler Benz (with Ballard and Ford)		•	•	
	British Nuclear Fuels			•	•
	MTU			•	
	Bewag		•		
	Zetek Power				
	SGL Carbon				
Japan	Fuji Electric	•	•	•	•
	Toshiba	•	•	•	
	Mitsubishi Electric	•	•	•	•
	Hitachi	•	•	•	
	Sanyo Electric	•	•	•	•
	Mitsubishi Heavy Industries		•		•
	Matsushita Electric Industrial			•	
	Ishikawajima- Harima Heavy Industries			•	
	Kawasaki Heavy Industries			•	
	Tonen (with Sanyo Electric)			•	•
	NKK Corp.				•
	Fujikura				•
	Murata Mfg. Co.				•
	Sumitomo Electric		•		
	Tanaka		•		
	Toyota		•		
	Aisin Seiki/Equos Research		•		
	Honda		•		
	Yamaha		•		

micro-CHP fuel cell units results from technical developments in stack and reformer systems.

However, deregulation of the energy utilities and environmental issues, including Kyoto climate change aims, have stimulated a re-evaluation of the relative benefits of large base load power stations versus distributed power, including heat, power, and cooling for individual buildings. In all, the drive to develop the zero emission vehicle with improved fuel economy, together with the contribution of fuel cells have long been predicted to boost stationary applications in the next 5–7 years.

Table 3.9 Electrolyte membrane and membrane electrode assembly (MEA) manufacturers

Company	Electrolyte membrane	MEAs	Comments
Ballard [2]	Developmental stages	Yes	Currently purchases membrane, but is considering internal production capability
DuPont [2] Reference from http://www2.dupont.com	Yes	Yes	Manufacturer of Nafion. Recently announces plans to move into MEA and fuel cell stack manufacturing. Initial goal is stationary, but interested in long-term automotive applications
Gore & Associates Reference from http://www.wlgore.com	Yes	Yes	DOE funding for high-volume electrode manufacturing. Chemical and material manufacturer
Johnson Matthey Reference from www.matthey.com	No	Yes	Manufacturer of MEA. Supplies halt the "world demand," including Ballard's for catalyst. R&D facilities in UK, production facilities in the UK and Pennsylvania
3M 3M Fuel Cell Technology by Andreas Graichen, 3M slides presentation.www.brennst-offzelle-nrw.de/fileadmin/daten/jahrestreffen/Vortraege/15-Graichen-3M.pdf	Yes	Yes	Began R&D in 1995, but has established itself as a major player. 3M does not sell the membrane as a separate component, but delivers it only as part of the MEA. Has manufacturing facilities in Menomonie, Wisconsin, and St. Paul, MN

Acknowledgement Authors wish to thank Ballard Corporation for the courtesy of using original figure and data to be republished in this review.

References

1. M. Warshay and P. R. Prokopius, The fuel cell in space: yesterday, today and tomorrow, *J. Power Sources* **29** (1–2), 193–200 (1990).
2. http://www.ballard.com
3. L. Blomen and M. Mugerwa, *Fuel Cell Systems*, Plenum Publishing, New York (1993).
4. P. H. Eichenberger, The 2 MW Santa Clara Project, *J. Power Sources* **71** (1–2), 95–99 (1998).
5. F. Panik, Fuel cells for vehicle applications in cars — bringing the future closer, *J. Power Sources* **71** (1–2), 36–38 (1998).
6. R. A. Lemons, Fuel cells for transportation, *J. Power Sources* **29** (1–2), 251–264 (1990).
7. T. R. Ralph, G. A. Hards, J. E. Keating, S. A. Campbell, D. P. Wilkinson, M. Davis, J. St-Pierre and M. C. Johnson, Low cost electrode for proton exchange membrane fuel cells, *J. Electrochem. Soc.* **144** (11), 3845–3857 (1997).
8. P. Jung, Technical and economic assessment of hydrogen and methanol powered fuel cell electric vehicles, Master Thesis, Chalmers University of Technology, January 1999
9. X. Ren, P. Zelenay, S. Thomas, J. Davey and S. Gottesfeld, Recent advances in direct methanol fuel cells at Los Alamos National Laboratory. *J. Power Sources* **86**, (1–2) 111–115 (2000).
10. G. Cacciola, V. Antonucci and S. Freni, Technology up date and new strategies on fuel cells. *J. Power Sources* **100** (1–2), 67–79 (2001).
11. M. Ishizawa, S. Okada and T. Yamashita, Highly efficient heat recovery system for phosphoric acid fuel cells used for cooling telecommunication equipment, *J. Power Sources* **86** (1–2), (2000).
12. M. Walsh, The importance of fuel cells to address the global warming problem, *J. Power Sources* **29** (1–2), 13–28 (1990).
13. B. D. McNicol, D. A. J. Rand and K. R. Williams, Fuel cells for road transportation purposes — yes or no?, *J. Power Sources* **100** (1–2), 47–59 (2001).
14. K. B. Prater, Solid polymer fuel cells for transport and stationary applications, *J. Power Sources* **61** (1–2), 105–109 (1996).
15. B. Baker, Grove Medal acceptance address, *J. Power Sources* **86** (1–2), 9–15 (2000).
16. M. Nomura, S. Namic, K. Okano, T. Kobayashi, K. Koseki and H. Komaki, Current status of development and technical subjects of fuel cells, *The Bulletin of MESJ* **23** (1), 12–26 (1995).
17. http://www.pg.siemens.com/en/fuelcells/demo/index.cfm

Chapter 4
Development of Sulfonated Poly(ether-ether ketone)s for PEMFC and DMFC

Dae Sik Kim and Michael D. Guiver

Abstract During the last two decades, extensive efforts have been made to develop alternative hydrocarbon-based polymer electrolyte membranes to overcome the drawbacks of the current widely used perfluorosulfonic acid Nafion. This chapter presents an overview of the synthesis, chemical properties, and polymer electrolyte fuel cell applications of new proton-conducting polymer electrolyte membranes based on sulfonated poly(arylene ether ether ketone) polymers and copolymers.

Primary attention has been paid to the basic properties of the sulfonated polymer prepared by post-sulfonation and direct copolymerization. This chapter attempts to summarize the preparation of sulfonated poly(arylene ether ether ketone) polymers with high proton conductivity, including synthesis from monomers containing sulfonic acid groups and hybrid membranes containing inorganic materials, and fuel cells derived from new proton-conducting polymer electrolytes that have been made during the past decade.

4.1 Introduction

The challenge of continually meeting the world's increasing energy needs will be one of the most important tasks that we will face in the twenty-first century. Current energy sources are being depleted at high rates due to both the world population's growth and its desire to live at higher levels of comfort. Petroleum is the world's most prevalent fuel for transportation. However, fossil fuels are becoming scarcer and their burning produces emissions that pollute the air. Furthermore, fossil fuels are not a renewable energy source.

Renewable and environmentally friendly energy sources will be essential for an ever-changing and populous planet. Solar power, hydropower, and wind power systems have been employed to complement current electric power sources. In recent years, clean, efficient energy conversion techniques have increasingly become the focal point of public interest, and are increasing in importance. The main causes of this are the increase in global energy requirements, the increase in environmental

S.M.J. Zaidi, T. Matsuura (eds.) *Polymer Membranes for Fuel Cells*,
doi: 10.1007/978-0-387-73532-0, © Springer Science+Business Media, LLC 2009

awareness, and the perceptible shortage of raw materials. One of the most attractive and convenient alternative energy supply devices are fuel cells [1].

Fuel cells have the potential to become an important energy conversion technology. Research efforts directed toward the widespread commercialization of fuel cells have accelerated in light of ongoing efforts to develop a hydrogen-based energy economy to reduce dependence on foreign oil and decrease pollution. The heart of the fuel cell is the polymer electrolyte membrane (PEM), also known as the proton exchange membrane, whose essential function is to act as a barrier to avoid direct contact between the fuel and oxidant and as a proton-conducting medium [2].

The PEM was first deployed in the Gemini space program in the early 1960s using sulfonated polystyrene-divinylbenzene copolymer membranes. These PEM were considered as having too short a lifetime for real-world applications. The commercialization of Nafion by DuPont in the late 1960s helped to demonstrate the potential interest in terrestrial applications for fuel cells, although its primary application and focus were the chloroalkali processes [3]. However, this ionomer is also expensive, and the upper operation temperature limit is considered to be about 80–100°C because of the deterioration of transport, mechanical and electrochemical properties [4]. Additionally a critical drawback of the Nafion membrane associated with its application in direct methanol fuel cells (DMFCs) is its high methanol permeability ($\sim 10^{-6}$ cm^2 s^{-1}), which drastically reduces DMFCs' performance [5]. That is, the methanol crossover is a severe problem in the DMFC application because the methanol fuel at the anode diffuses through the membrane to the cathode and then reacts at the cathode, resulting in a mixed potential without generation of electricity. Desirable properties of the PEM in fuel cells include [6,7]: (1) chemical and electrochemical stability in the fuel cell–operating conditions; (2) good mechanical strength and stability at operating conditions; (3) chemical properties of components compatible with the (interfacial) bonding requirements of the PEMFC or DMFC; (4) low permeability to reactant species; (5) high electrolyte transport to maintain uniform electrolyte content and prevent localized drying; (6) high proton conductivity with minimal resistance and zero electronic conductivity and (7) low methanol or fuel permeability; and (8) low production cost relative to the application. As a consequence, the major research goal would be to satisfy and achieve these prescribed needs, and numerous groups have worked to develop alternative PEMs to replace expensive perfluorinated sulfonic acid copolymers. Nearly all existing membrane materials for PEM fuel cells rely on absorbed water and its interaction with acid groups to produce protonic conductivity. Due to the large fraction of absorbed water in the membrane, both mechanical properties and water transport are key issues. Devising systems that can conduct protons with little or no water is perhaps the greatest challenge for new membrane materials. Specifically, for automotive applications the US Department of Energy has currently established a guideline of 120°C and 50% relative humidity as target operating conditions, and a goal of 0.1 S cm^{-1} for the protonic conductivity of the membrane [3].

Two main types of polymer membranes have dominated research efforts: sulfonated aromatic polymers (e.g., sulfonated poly(ether ether ketone) SPEEK, and poly(ether ketone) SPEK) and perfluorosulfonic acid membranes such as Nafion, which have been the industry benchmark. These membranes both exhibit phase-separated domains consisting of an extremely hydrophobic back-bone that gives morphologic stability and extremely hydrophilic functional groups. These functional groups aggregate to form hydrophilic nanodomains that act as water reservoirs [8]. As schematically illustrated in Fig. 4.1, the water-filled channels in sulfonated PEEKK are narrower compared with those in Nafion. They are less separated and more branched with more dead-end "pockets." These features correspond to the larger hydrophilic/hydrophobic

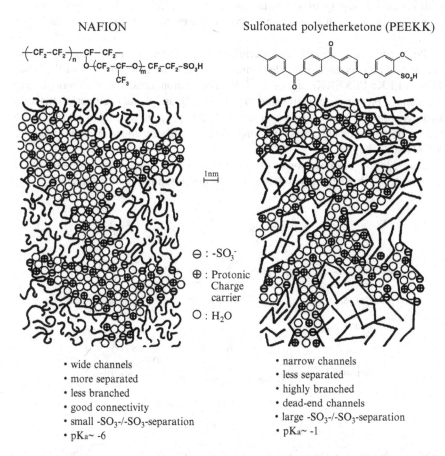

Fig. 4.1 Comparison of structures: Nafion and sulfonated polyetherketone (SPEEK) [9]. Reprinted from the K.D. Kreuer, On the development of proton conducting polymer membranes for hydrogen and methanol fuel cells. *J. Membrane Sci.* 185, 32 (2001) with permission from Elsevier

interface; and therefore, also to a larger average separation of neighboring sulfonic acid functional groups [9].

At present, many sulfonated derivatives of polymers such as poly(ether ether ketone), polysulfone, poly(arylene ether sulfone), poly(styrene), and poly(phenylene sulfide) have been developed for fuel cells [10–14]. More recently, the synthesis of sulfonated poly(arylene ether sulfone) and/or sulfonated poly(arylene ether ketone)s copolymers by direct copolymerization of biphenol, disulfonated-activated aromatic halide monomers, and the precursor—activated aromatic halide monomer for fuel cell membrane applications—were carried out [12,15].

Recently, sulfonated poly(arylene ether sulfone) copolymers were synthesized by McGrath group [4]. Four bisphenols (4,4'-bisphenol A, 4,4'-bisphenol AF, 4,4'-bisphenol, and hydroquinone) were investigated for the synthesis of novel copolymers with controlled degrees of sulfonation as shown in Fig. 4.2. These copolymers are promising candidates for high-temperature proton exchange membranes in fuel cells [12,15,34].

Poly(aryl ether ketone)s are thermostable polymers in which ether (E) and ketone (K) units connect phenylene rings, giving a range of polymers of the types PEK, PEEK, PEKEKK, etc. (Fig. 4.3). The proton conductivity, thermal, and mechanical properties of sulfonated PEEK [16,17], and its fuel cell performance in hydrogen-air and hydrogen-oxygen up to 110°C [17], as well as in DMFC, have been reported in recent years, and long-term tests have claimed lifetimes of up to 4,300 h at 50°C [18].

This section reviews some of the past, present, and proposed sulfonated poly(arylene ether ketone)s copolymer PEM described in the literature.

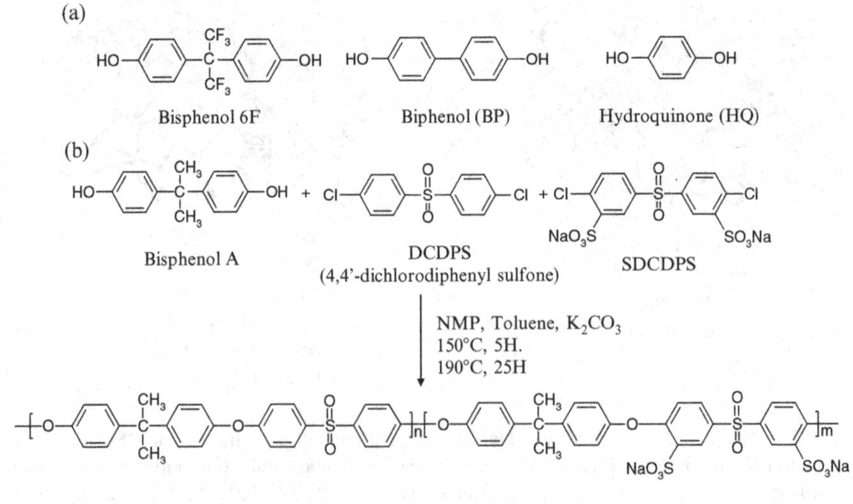

Fig. 4.2 (a) Investigated bisphenol structure by McGrath group. (b) Synthesis of random BPA-based disulfonated poly(arylene ether sulfone) copolymers

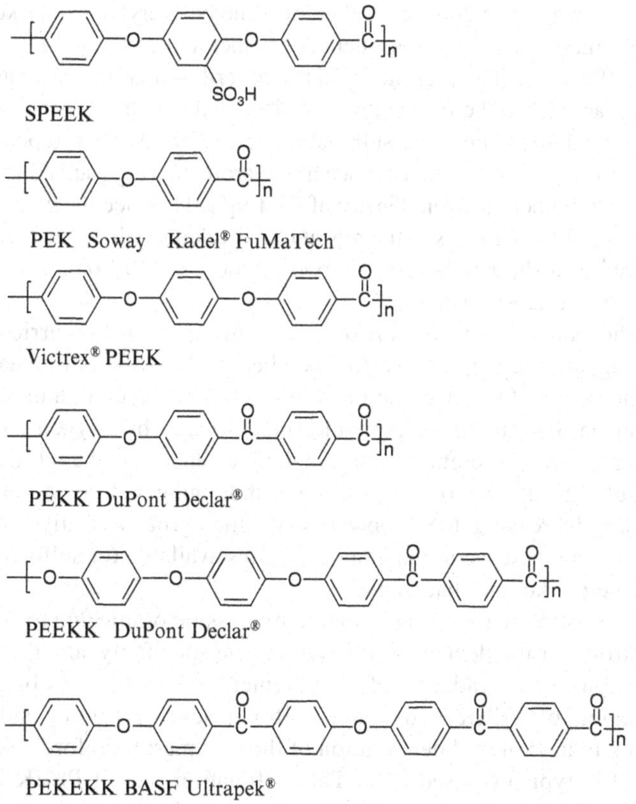

SPEEK

PEK Soway Kadel® FuMaTech

Victrex® PEEK

PEKK DuPont Declar®

PEEKK DuPont Declar®

PEKEKK BASF Ultrapek®

Fig. 4.3 Sulfonated poly(ether ether ketone) and structures of representative membranes of the poly(ether ketone) family

4.2 Synthesis of Poly(ether ether ketone) Membrane for PEMFC

4.2.1 Post-sulfonated PEEK

Poly(ether ether ketone) (PEEK) is an aromatic, high performance, semicrystalline polymer with extremely good thermal stability, chemical resistance, and electrical and mechanical properties. This polymer shows little solubility in organic solvents due to the semicrystalline nature of certain poly(ether ketone)s. The most common way to modify aromatic polymers for application as a PEM is to employ electrophilic aromatic sulfonation. By introducing sulfonic acid groups to the backbone, the crystallinity decreased and solubility increased [19,20].

Poly(arylene ether ketone)s have been modified using various sulfonation agents, such as concentrated sulfuric acid, fuming sulfuric acid, chlorosulfonic acid, or sulfur trioxide (or complexes thereof). The poly(arylene ether ketone)s are a class of polymers consisting of sequences of ether and carbonyl linkage between phenyl rings (Fig. 4.3). The presence of adjacent ortho-directing ether groups confers highest reactivity to the four equivalent sites on the hydroquinone unit situated between the ether segments. The sulfonation of PEEK has been reported to be a second-order reaction, which takes place at the aromatic ring flanked by two ether links, due to the higher electron density of the ring [21]. Since the electron density of the other two aromatic rings in the repeat unit is relatively low due to the electron attracting nature of the neighboring carbonyl group, one sulfonic acid group per repeat unit may be substituted.

Ortho-ether substitution by sulfonic acid groups can be carried out with sulfonation agents, the extent of sulfonation being a function of the reaction time and temperature and SO_3 concentration. The level of sulfonation in these materials is dependant on the number of aromatic rings bridged by oxygen atoms as O–phenyl–O units are preferentially sulfonated, whereas O–phenyl–CO– groups remain unsulfonated due to the electron-withdrawing nature of the carbonyl group. Hence, increasing the proportion of ether groups relative to carbonyl groups leads to an increase in the number of sites available for sulfonation on the poly(arylene ether ketone) backbone.

Post-modification reactions are moderately less favorable due to their lack of precise control over the degree of sulfonation, site specificity, and the possibility of side reactions, or degradation of the polymer backbone. It has been reported that sulfonation of PEEK with chlorosulfonic acid or fuming sulfuric acid causes a mostly unexplored degradation of the polymer; therefore, concentrated sulfuric acid is typically used [22]. The sulfonation rate of PEEK in concentrated sulfuric acid can be controlled by changing the reaction time, temperature, and acid concentration to provide polymers with a sulfonation range of 30–100% without degradation and cross-linking reactions [23]. However, it has been shown that the sulfonation of PEEK in sulfuric acid cannot be used to produce truly random copolymers at sulfonation levels less than 30% because dissolution and sulfonation occur in a heterogeneous environment [24]. Nevertheless, this area of PEM synthesis has received much attention and may be the source of emerging products such as sulfonated Victrex poly(ether ether ketone) [25–27].

Poly(phthalazinone)s (PPs) including poly(phthalazinone ether sulfone) (PPES), poly(phthalazinone ether ketone) (PPEK), and poly(phthalazinone ether sulfone ketone) (PPESK) are new high performance polymers in the early stages of commercialization. Among other advantages, this class of polymers is distinguished by excellent chemical and oxidative resistance, mechanical strength, high thermal stability, and very high glass transition temperatures (295, 263, and 278°C, respectively) [28]. Recently Guiver's group reported the post-synthesis sulfonation of PPs as well as the proton conductivity properties of polymer membranes [28–30]. The structures of PPs are shown in Fig. 4.4. Membrane obtained from highly sulfonated PPs showed proton conductivity above 10^{-2} S cm^{-1} at both room temperature and elevated temperature [28,29].

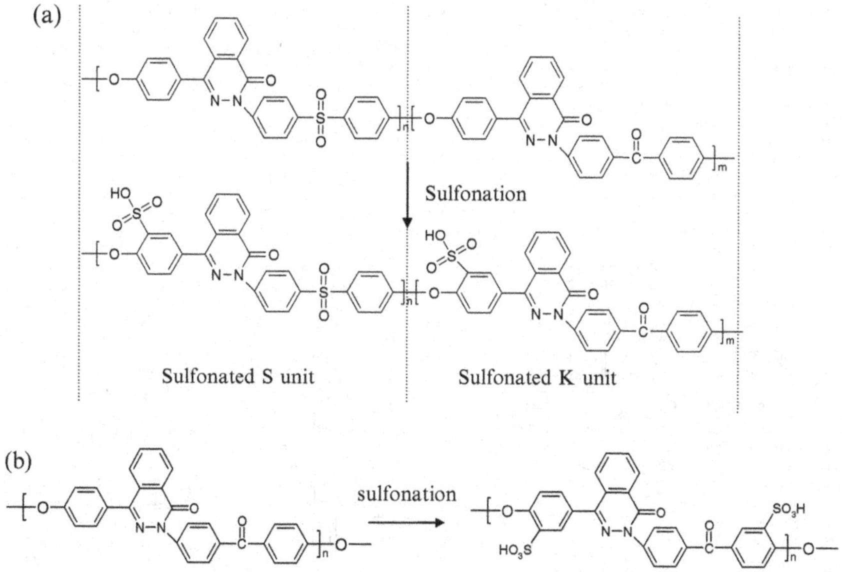

(a)

(b)

Fig. 4.4 Sulfonation reaction of (**a**) PPESK and (**b**) PPEK [28–30]

It is well known that poly(arylene ether)s are highly stable structural materials, which can serve as a skeleton for electrolytes. However, the degradation of these polymers was documented to take place at the ether bonds in the backbone of aromatic polymers when there is a sulfonic acid group attached to the main chain. Meng's group reported the synthesis of the polyaromatics with hindered and bulky groups that allowed position-directed sulfonation as shown in Fig. 4.5. These polymers are claimed to be more stable to hydrolysis than those with acidic groups directly attached onto the main chain [31–33].

4.2.2 Direct Copolymerization

4.2.2.1 Direct Sulfonated Poly(arylene ether ketone)s

A novel approach has been developed by several research groups to obtain sulfonated aromatic copolymers by copolymerization of sulfonated monomers. The direct copolymerization of a sulfonated monomer is an alternative approach with some distinct advantages compared with the modification of a preformed polymer. To some degree, the incorporation of sulfonated or modified monomers allows a closer control of molecular-design of the resulting copolymer [34].

There are several commonly cited drawbacks of post-modification, including the lack of control over the degree and location of functionalization, which is usually a problem when dealing with macromolecules. It has been interesting to investigate the

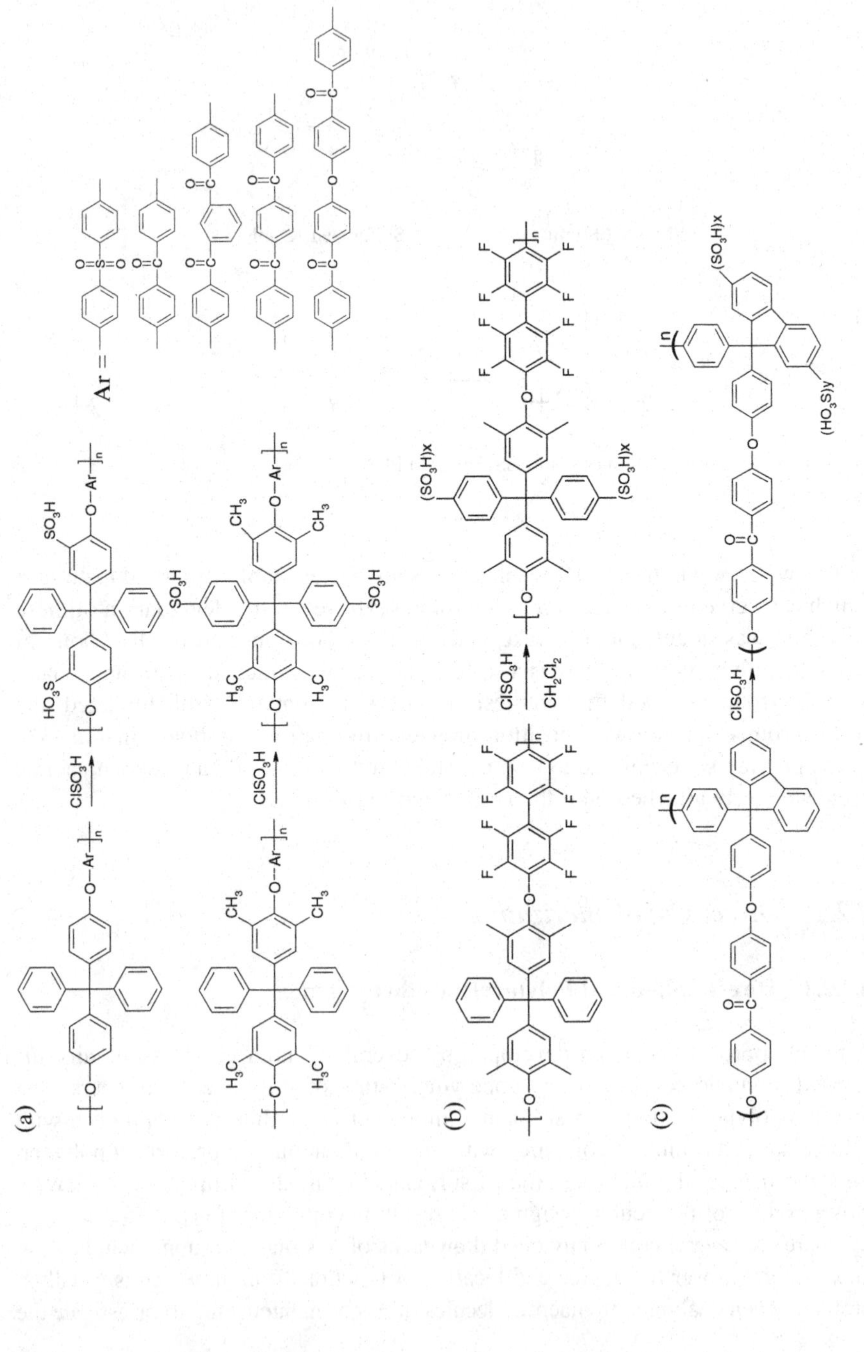

Fig. 4.5 Sulfonation of poly(arylene ether)s [31–33]

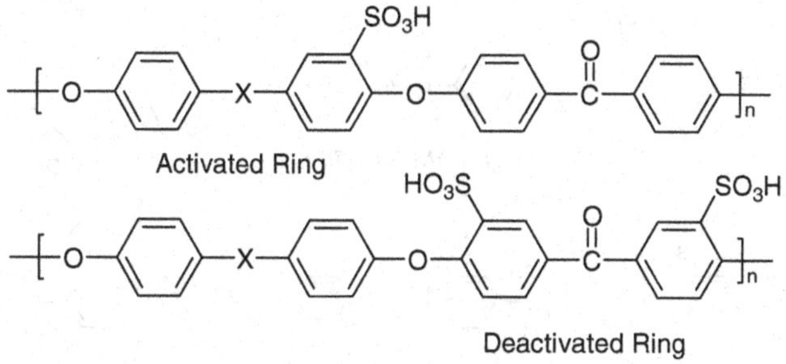

Fig. 4.6 Placement of the sulfonic acid group in post-sulfonation (activated ring) versus direct copolymerization (deactivated ring)

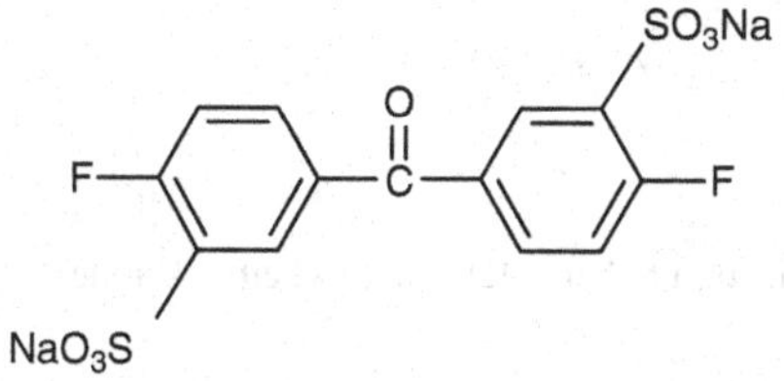

Fig. 4.7 Structure of 3,3'-disulfonated 4,4'-difluorodiphenyl ketone (DFBP)

effect of sulfonation, for example, on the deactivated sites of the repeat units, since one might expect enhanced stability and higher acidity from sulfonic acid groups, which are attached to electron-deficient aromatic rings, rather than from sulfonic acid groups, which are bonded to electron-rich aromatic rings, as in the case of post-modification [35]. The possibilities of controlling and/or increasing molecular weight to enhance durability are not feasible in the case of post-reaction on an existing commercial product. The difference between sulfonic acid placement in typical example of post-sulfonation and direct copolymerization is shown in Fig. 4.6.

The preparation of directly copolymerized sulfonated poly(arylene ether ketone)s PEMs are possible by employing a sulfonated dihalide ketone monomer, as first reported by Wang [36]. Using 3,3'-disulfonated 4,4'-difluorodiphenyl ketone (DFBP) (Fig. 4.7), Wang et al. produced high molecular weight copolymers with a bisphenol and the unsulfonated 4,4'-difluorodiphenyl ketone co-monomers [36,37]. Powerful sulfonation conditions of fuming sulfuric acid and a relatively high temperature (100°C) were necessary to sulfonate the monomer. Figure 4.8 shows a typical polymerization scheme; however, no proton conductivity values were reported for this polymeric sulfonic salt. Hexafluoroisopropylidene bisphenol A (Bisphenol 6F) was used to polymerize more thermally stable amorphous sulfonated poly(ether ketone)s via direct co-polymerization as competitive candidates for PEMFCs (Fig. 4.9) [38]. High proton conductivities (up to 0.08 S cm^{-1}) were reported at 30°C.

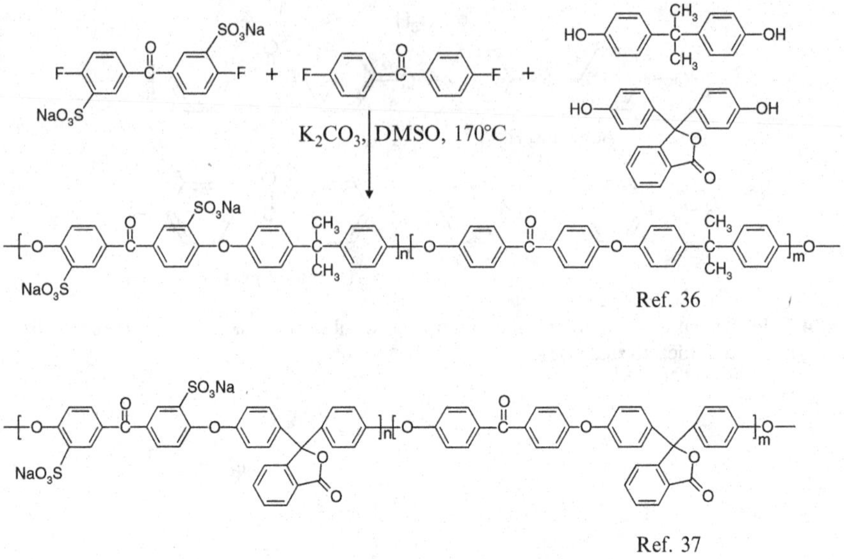

Fig. 4.8 Typical polymerization Scheme for poly(aryl ether ketone)s

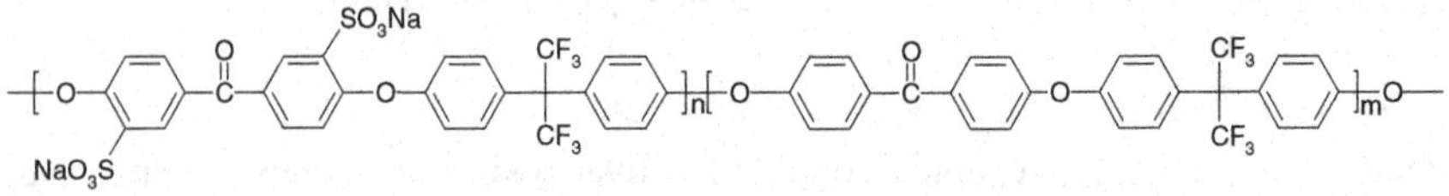

Fig. 4.9 Structure of 6F-containing sulfonated poly(ether ketone) via direct copolymerization [38]

The choice of bisphenol-type monomers for the polymerization of poly(arylene ether ketone)s is large. Guiver's group reported that sulfonated poly(phthalazinone ether ketone) (SPPEK) copolymers and sulfonated poly(phthalazinone ether sulfone) (SPPES) copolymers containing pendant sodium sulfonate groups were prepared by direct copolymerization [39,40]. The bisphenols for the polymerization of these copolymers were 4-(4-hydroxyphenyl)-1(2H)-phthalazinone (DHPZ) and 2,6-dihydroxynaphthalene (NA). Figure 4.10 shows the structure of sulfonated copolymer derived from DHPZ and NA. The phthalazinone derived from phenolphthalein has a N–H group that behaves as a phenolic OH group. The compounds with two phthalazinone groups should react as bisphenols in nucleophilic aromatic substitution reactions. The resulting polymers, such as poly(arylene ether)s, would be expected to be thermally stable at high temperature [41,42]. The rigid planar aromatic NA group was incorporated into the polymers' backbone in order to improve the hot water stability of sulfonated poly(aryl ether ketone)s with a high degree of sulfonation (DS). Compared with SPPEK [39], the sulfonated poly(aryl ether ketone)

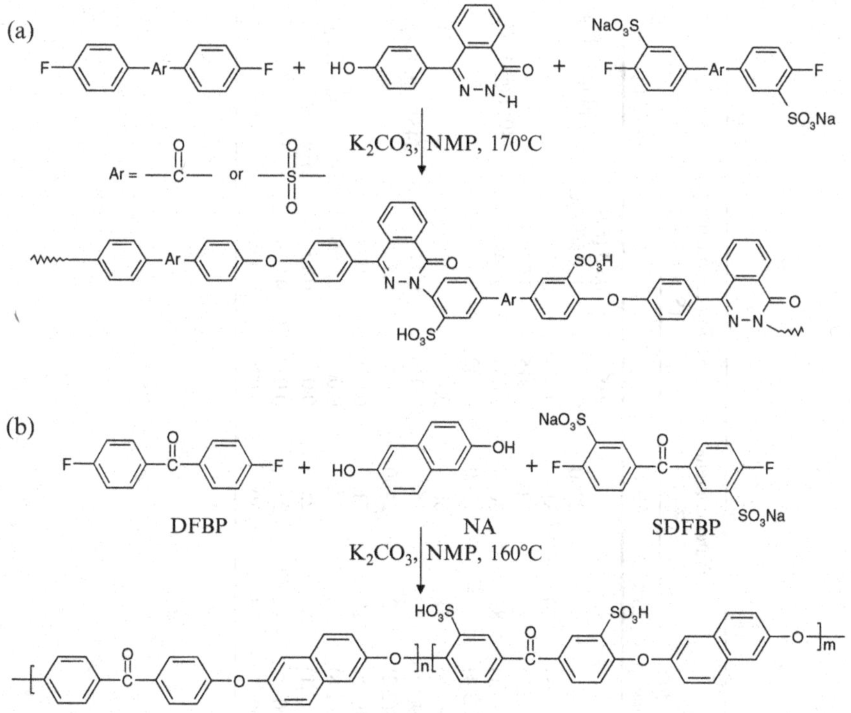

Fig. 4.10 (a) Structure of sulfonated poly(phthalazinone arylene ether)s (SPPEK) (b) Structure of sulfonated poly(aryl ether ketone) containing a naphthalene moiety (SPAEK-NA) [39,40]

–containing naphthalene group (SPAEK-NA) [40] shows lower water uptake at high temperature for similar DS, as listed in Table 4.1.

Meng's group reported that a new approach to the preparation of ionomers from poly(phthalazinone ether ketone)s that have been recently synthesized by a N–C coupling reaction [43,44]. These new polymers are claimed to exhibit improved oxidative resistance by the Fenton's test when compared with other sulfonated polymers. The structures of these polymers are shown in Fig. 4.11.

The Guiver's group recently reported the preparation of sulfonated poly(aryl ether ketone), containing Bisphenol 6F [45,46]. The sulfonated poly(aryl ether ketone)s (SPAEK) were synthesized by replacement of the non-sulfonated monomer 4,4′-difluorobenzophenone (DFBP) with 1,3-bis(4-fluorobenzoyl) benzene (1,3-BFBB) to increase the statistical length of non-sulfonated segments in order to improve the mechanical strength of the membranes in [46]. Figure 4.12 shows the structure of sulfonated poly(aryl ether ketone) copolymers. These membranes exhibit high proton conductivities, very close to that of Nafion 117, and considered to be possible candidates for PEM operation in a fuel cell. The SPAEK membrane [46], containing the 1,3-BFBB, showed initial Young's modulus in the range of 459–767 MPa, which is much higher than that of Nafion 117 (234 MPa) and are

Table 4.1 Selected properties of sulfonated poly(arylene ether ketone)s

Membrane	Monomer (mol ratio)			DS	EW (IEC) mequiv. g⁻¹	Water content (%)		Proton conductivity (S cm⁻¹)		Ref./structure
						25°C	80°C	25°C	80°C	
SPPEK	DFBP(0.5)[a]	SDFBP(0.5)[b]	DHPZ(1)[c]	0.94	–	42	210	0.02	0.03 (89°C)	[39]/Fig. 4.10a
	DFBP(0.4)	SDFBP(0.6)	DHPZ(1)	1.12	–	60	2300	0.031	0.055(88°C)	
SPAEK-NA	DFBP(0.5)	SDFBP(0.5)	NA (1)[d]	0.96	432(2.3)	50	81	0.0071	0.025	
	DFBP(0.4)	SDFBP(0.6)	NA (1)	1.21	359(2.8)	74	137	0.015	0.036	[40]/Fig. 4.10b
	DFBP(0.3)	SDFBP(0.7)	NA (1)	1.37	327(3.1)	87	361	0.024	0.049	
	DFBP(0.2)	SDFBP(0.8)	NA (1)	1.60	291(3.4)	115	1200	0.042	0.11	
SPAEK-6F	DFBP(0.5)	SDFBP(0.5)	6F-BPA(1)	0.98	610(1.6)	32	68	0.039	0.08	
	DFBP(0.4)	SDFBP(0.6)	6F-BPA(1)	1.14	519(1.9)	50	157	0.046	0.11	[45]/Fig. 4.12b
	DFBP(0.3)	SDFBP(0.7)	6F-BPA(1)	1.36	458(2.2)	92	2400	0.080	0.16	
SPAEK	1,3-BFBB(0.5)[e]	SDFBP(0.5)	[f]6F-BPA(1)	0.98	653(1.5)	29	51	0.028	0.069	
	1,3-BFBB(0.4)	SDFBP(0.6)	6F-BPA(1)	1.16	555(1.8)	37	135	0.063	0.085	[46]/Fig. 4.12a
	1,3-BFBB(0.3)	SDFBP(0.7)	6F-BPA(1)	1.36	476(2.1)	67	2500	0.094	0.15	
Nafion 112	–	–		–	1100	12	23	0.033	0.056	[40]
Nafion 117	–	–		–	1100	13	30	0.075	0.096	[46]

[a]4,4'-difluorobenzophenone
[b]Sulfonated DFBP
[c]4-(4-Hydroxyphenyl)-1(2H)-phthalazinone
[d]2,6-dihydroxynaphthalene
[e]1,3-bis(4-fluorobenzoyl) benzene
[f] Bisphenol 6F

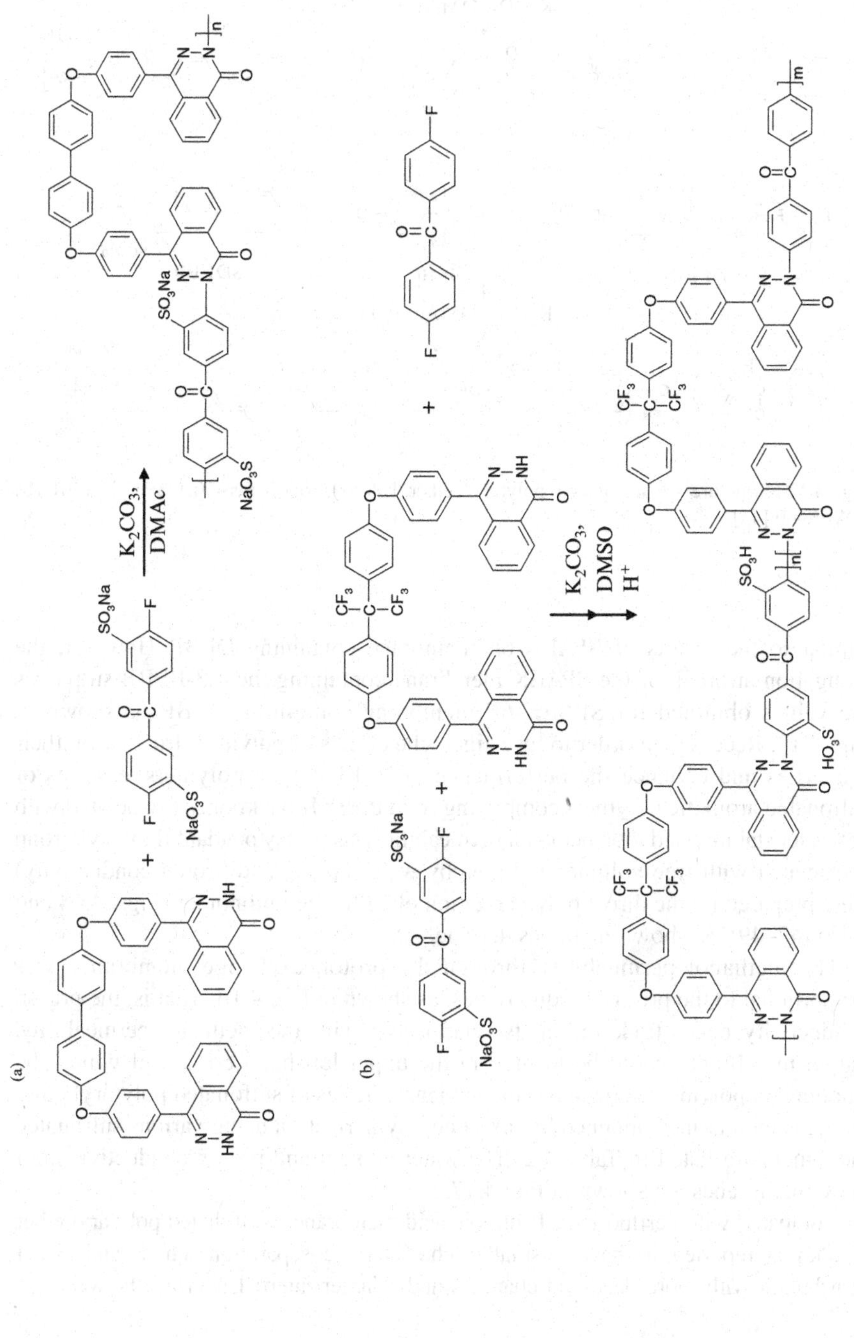

Fig. 4.11 Synthesis of sulfonated poly(phthalazinone ether)s [43,44]

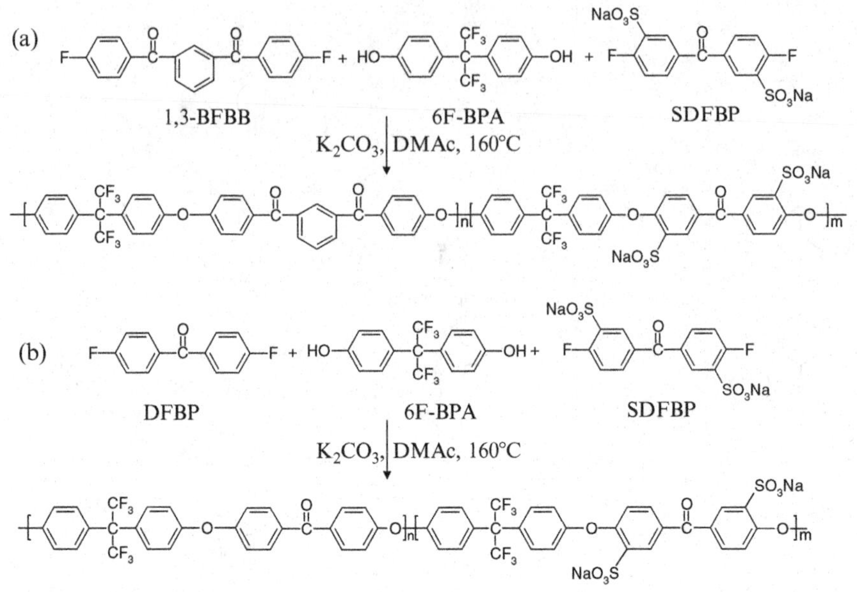

Fig. 4.12 Structure of sulfonated poly(aryl ether ketone) copolymers (**a**) SPAEK [46] (**b**) SPAEK-6F [45]

similar to the results of SPAEK-6F membrane containing DFBP. However, the elongation at break of the SPAEK membrane containing the 1,3-BFBB surpasses the values obtained for SPAEK-6F membrane containing DFBP, as shown in Fig. 4.13. Recently, in order to investigate the effects of polymer structure on their properties and enhance the performance of PEEKK-type polymers, a series of sulfonated aromatic polymers comprising rigid PEEKK backbones (associated with hot water stability and low methanol permeability) and bulky pendant fluorenyl group (associated with free volume, and thereby water uptake and proton conductivity) were prepared by the direct polymerization of sulfonated monomer (Figs. 4.14 and 4.15) [47–49], as shown in Tables 4.1 and 4.2.

The methanol permeability through the proton exchange membranes was proportional to the proton conductivities, as shown in Fig. 4.16. That is, the proton conductivity has a trade-off in its relationship with the methanol permeability. Target membrane would be located in the upper left-hand corner, of which the fluorenyl copolymers show the same tendency. Series of sulfonated poly(aryl ether ketone)s membranes obtained by direct copolymerization using various sulfonated monomers are listed in Table 4.2. The water content and proton conductivities of these membranes are shown in Fig. 4.17.

Compared with perfluorinated sulfonic acid membranes, sulfonated poly(aryl ether ketone) is reported to have a smaller characteristic separation length and wider distribution with more dead-end channels and a larger internal interface between the

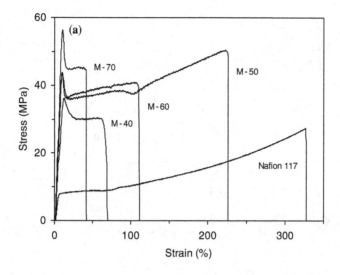

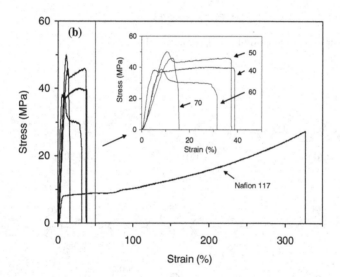

Fig. 4.13 Stress vs. strain curves for sulfonated poly(aryl ether ketone)s (**a**) using 1,3-BFBB (M-x: Sulfonated monomer (SDFBP) mol%) [46] (**b**) using DFBP (x: Sulfonated monomer (SDFBP) mol%) [45]. Reference [46] reprinted from M.D. Guiver et al., Synthesis and characterization of poly(aryl ether ketone) copolymers containing (hexafluoroisopropylidene)-diphenol moiety as proton exchange membrane materials. *Polymer* 46, 3263 (2005), with permission from Elsevier. Reference [45] reprinted with permission from P. Xing et al., Sulfonated poly(aryl ether ketone)s containing the hexafluoroisopropylidene diphenyl moiety prepared by direct copolymerization, as proton exchange membranes for fuel cell application. *Macromolecules* 37, 7966 (2004), with permission from the American Chemical Society

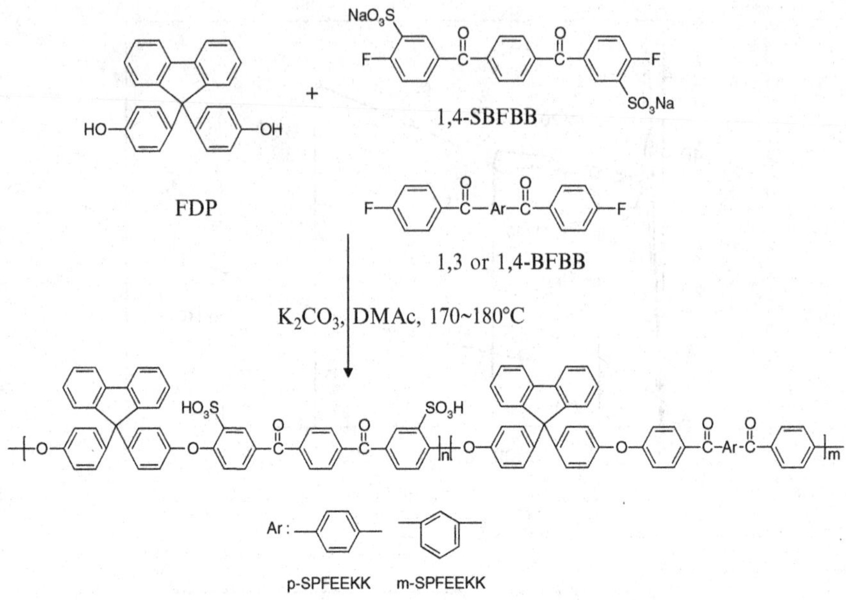

Fig. 4.14 Structure of m-,p-SPFEEKK polymers [47]

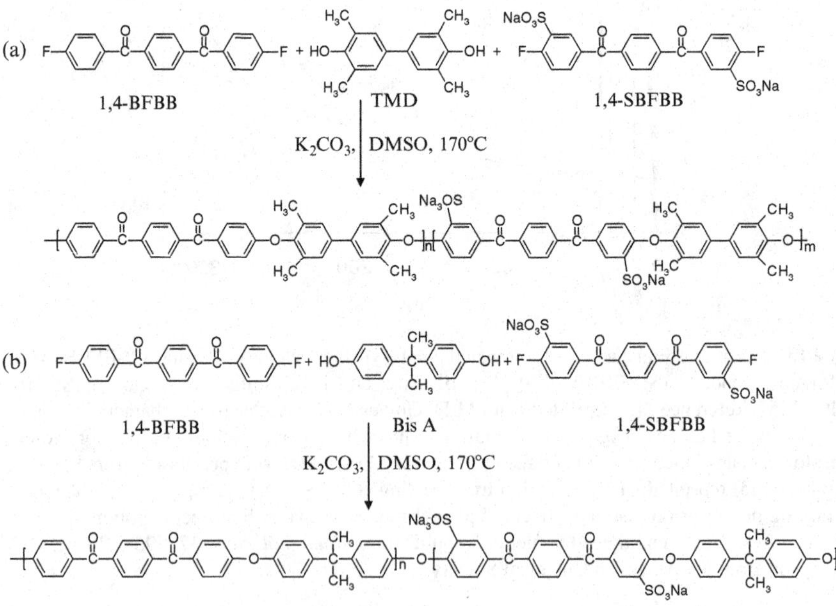

Fig. 4.15 Structure of SPEEKK (**a**) [48] and (**b**) [49]

Table 4.2 Water content and proton conductivities of various sulfonated poly(arylene ether ketone)s

Membrane	Monomer (mol ratio)			DS	EW (IEC mequiv. g⁻¹)	Water content (%)		Conductivity (S cm⁻¹)		Ref./structure
						25°C	80°C	25°C	80°C	
m-SPFEEKK	1,3-BFBB (0.5)	1,4-SBFBB (0.5)	FDP(1)[a]	1.0	714 (1.39)	7.3	7.9	0.01	0.038	[47]/Fig. 4.14
	1,3-BFBB (0.35)	1,4-SBFBB (0.65)	FDP(1)	1.3	568 (1.72)	12.1	15.9	0.033	1.0	
	1,3-BFBB (0.2)	1,4-SBFBB (0.8)	FDP(1)	1.6	476 (2.1)	14.1	16.9	0.043	0.127	
p-SPFEEKK	1,4-BFBB (0.5)	1,4-SBFBB (0.5)	FDP (1)	1.0	714 (1.38)	6.9	8.6	0.011	0.036	[47]/Fig. 4.14
	1,4-BFBB (0.35)	1,4-SBFBB (0.65)	FDP (1)	1.3	568 (1.74)	12.1	14.5	0.03	0.094	
	1,4-BFBB (0.2)	1,4-SBFBB (0.8)	FDP (1)	1.6	476 (2.06)	13.4	22.2	0.041	0.119	
SPEEKK	1,4-BFBB (0.8)	1,4-SBFBB (0.2)	TMD (1)[b]	0.37	1388 (0.72)	9.14	12.07	0.0004	0.0009	[48]/Fig. 4.15(a)
	1,4-BFBB (0.6)	1,4-SBFBB (0.4)	TMD (1)	0.78	735 (1.36)	16.08	20.60	0.013	0.027	
	1,4-BFBB (0.5)	1,4-SBFBB (0.5)	TMD (1)	0.97	606 (1.65)	23.95	41.12	0.03	0.062	
	1,4-BFBB (0.4)	1,4-SBFBB (0.6)	TMD (1)	1.23	518 (1.93)	26.71	48.33	0.048	0.063	
SPEEKK	1,4-BFBB (0.8)	1,4-SBFBB (0.2)	Bis A (1)[c]	0.43	1,333 (0.75)	6.02	11.9	0.007	0.016	[49]/Fig. 4.15(b)
	1,4-BFBB (0.6)	1,4-SBFBB (0.4)	Bis A (1)	0.83	719 (1.39)	9.29	19.26	0.019	0.064	
	1,4-BFBB (0.5)	1,4-SBFBB (0.5)	Bis A (1)	0.97	591 (1.69)	11.48	24.17	0.032	0.08	
	1,4-BFBB (0.4)	1,4-SBFBB (0.6)	Bis A (1)	1.34	505 (1.98)	15.62	30.11	0.04	0.10	
SPAEEKK-H	1,3-BFBB (1)	DHNS (0.5)[d]	HQ(0.5)[e]	0.48	997 (1.00)	13.6	21.9	–	–	[54]/Fig. 4.18
	1,3-BFBB (1)	DHNS (0.6)	HQ(0.4)	–	856 (1.17)	14.2	30.0			
	1,3-BFBB (1)	DHNS (0.7)	HQ(0.3)	0.66	756 (1.32)	20.5	33.4			
	1,3-BFBB (1)	DHNS (0.8)	HQ(0.2)	–	680 (1.47)	25.5	57.2			
SPAEEKK-B	1,3-BFBB (1)	DHNS (0.5)	[f]BP (0.5)	–	1,058 (0.94)	12.4	14.3	–	–	[54]/Fig. 4.18
	1,3-BFBB (1)	DHNS (0.6)	BP (0.4)	0.56	907 (1.10)	18.3	21.2			
	1,3-BFBB (1)	DHNS (0.7)	BP (0.3)	–	788 (1.26)	24.2	28.3			
	1,3-BFBB (1)	DHNS (0.8)	BP (0.2)	0.77	699 (1.43)	26.4	43.4			

[a]4,4-(9-Fluorenylidene)diphenol
[b]3,3',5,5'-tetmethyl diphenyl-4,4'-diol
[c]Bisphenol A
[d]Sodium 6,7-dihydroxy-2-naphthalenesulfonate
[e]Hydroquinone
[f]4,4'-Biphenol

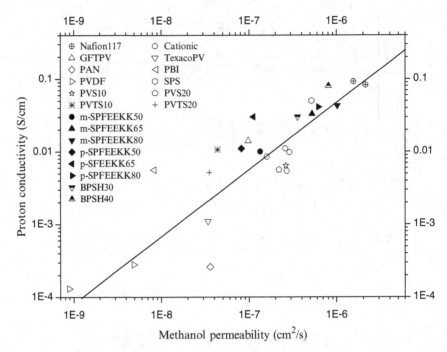

Fig. 4.16 Proton conductivities versus methanol permeabilities at 25°C. Data obtained from [50] (Nafion 117: DuPont, Cationic: CR 61 CMP, CR 61 CXMP, CR 61 CZR, CR 67 HMR: Ionics, Texaco-PV: Texaco's ethanol-dehydration membrane, GFT-PV: GFT PV 1000, *PAN* polyacrylonitrile; *PBI* polybenzimidazole; *PVDF* polyvinylidene fluoride; *SPS* sulfonated polystyrene; *PVS* sulfonated PVA; *PVTS* sulfonated PVA/Silica; *BPSH* sulfonated poly(arylene ether sulfone) data from [35,51])

hydrophobic and hydrophilic domains as measured by small-angle X-ray scattering (SAXS) [9]. However, if short pendant side chains between the polymer main chain and the sulfonic acid groups exist in the polymer structure, the nanophase separation of hydrophilic and hydrophobic domains may be improved and the amount of dead-end pockets may be decreased [52,53]. Rikukawa and co-workers [53] prepared sulfonated PEEK (SPEEK) and sulfonated poly(4-phenoxybenzoyl-1,4-phenylene, Poly-X 2000) (SPPBP) by post-sulfonation reactions of corresponding parent polymers. They found that SPPBP, which has pendant side chains between polymer main chain and sulfonic acid groups, showed higher and more stable proton conductivity than SPEEK. Guiver's group synthesized a new polymer to increase the distance of sulfonic acid groups from the polymer backbone using sodium 6,7-dihydroxy-2-naphthalenesulfonate (DHNS) (Fig. 4.18) [54]. DHNS is a commercially available and inexpensive naphthalenic diol containing a sulfonic acid side group, which is widely used in dye chemistry. The resulting SPAEEKK copolymers are expected to be more thermohydrolytically stable compared with sulfonated poly(aryl ether ketone) obtained by regular post-sulfonation reactions and direct polymerization of sulfonated difluorobenzophenone with biphenols. In many other sulfonated polymers, whether the sulfonic acids groups were

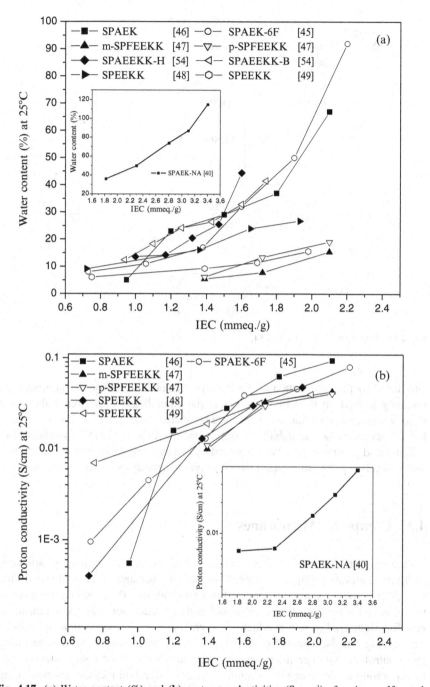

Fig. 4.17 (**a**) Water content (%) and (**b**) proton conductivities (S cm^{-1}) of various sulfonated poly(ether ether ketone)s at 25°C

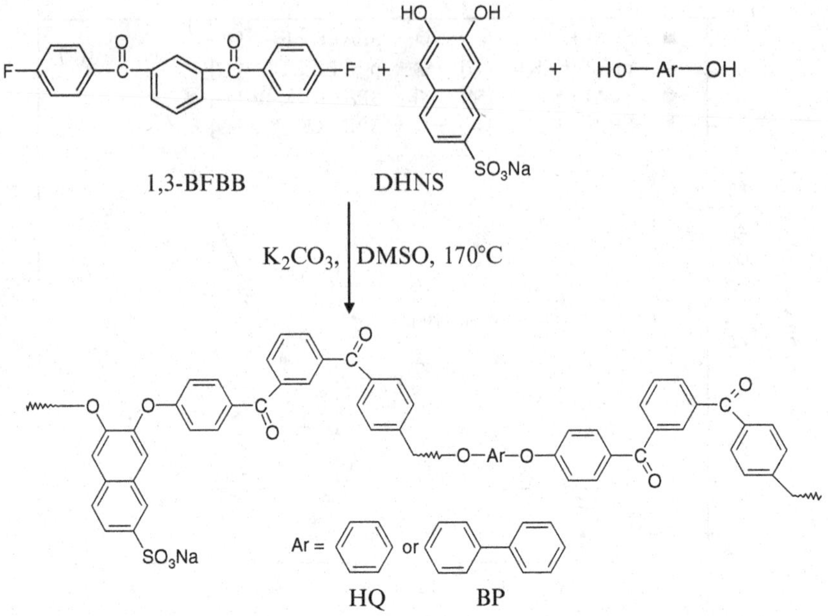

Fig. 4.18 Structure of SPAEEKKs [54]

introduced by post-sulfonation or direct polymerization, the sulfonic acid groups are normally located on the ortho position to the ether linkage. Electron-withdrawing sulfonic acid groups on this site are expected to increase the ease of hydrolysis of ether linkage and decrease the stability [55]. In SPAEEKKs using DHNS monomer [54], sulfonic acid groups are attached on a pendant benzene ring away from the ether linkage, which is expected to decrease the effect on the hydrolysis of ether linkages.

4.3 Composite Membranes

Although improvements have been achieved with aromatic polymers, proton conductivity at elevated temperature is still problematic because of the tendency for the reduction in water retention at temperatures much above 100°C. Recent progress in membranes for medium temperature fuel cells includes not only the synthesis of new functionalized proton conducting polymers or their modification by acid- or base-doping, but also associations of polymers (polymer blends), better understanding and control of polymer microstructure, the development of composite systems incorporating a micro- or macro-reinforcement, and hybrid membranes containing an inorganic component in addition to the polymer matrix [56,57].

In addition, DMFC development still poses considerable technical challenges, which include methanol crossover through the polymer electrolyte and poor kinetics of methanol electro-oxidation. Accordingly, modification of such membranes by the inclusion of inorganic materials has been attempted. There are several factors justifying the increased effort seen in recent years in developing composite inorganic/organic proton electrolyte membranes for fuel cell applications. A homogeneously dispersed hydrophilic inorganic solid assists in improving membrane water management, both by improving self-humidification of the membrane at the anode side by enhancing the back-diffusion of water produced at the cathode and/or by reducing electro-osmotic drag [58]. Beyond the reinforcement or enhancement of proton conduction properties compared with the organic polymer–only system, the presence of inorganic particles may also impede the diffusion of radical species that could contribute to oxidative degradation in nonfluorinated membranes [59]. Recently, composite membranes have been described based on diverse nonsulfonated and sulfonated polymer systems, including PBI and styrene/ethane/butane/styrene copolymer incorporating tungstophosphoric acid, PVDF with silica and alumina, SPEEK with zirconium sulfophenylphosphonate, tungstophosphoric, tungstomolybdic acids, and boron phosphate, sulfonated polystyrene incorporating antimonic acid, and Nafion incorporating silica, tetania, or tungstosilicic acid [60–62]. The following section reviews current advances to formation of blend, hybrid, and composite inorganic-organic systems based on poly(arylene ether ketone)s described in the literature.

The composite membrane approach represents one of the ways to improve the properties of the polymer electrolyte membranes since the desired properties of the two components can be combined in one composite. This approach was employed in previous studies [61–64] in which several heteropolyacids and solid boron phosphate (BPO_4) were used as the second phase in order to improve the proton conductivity of PEEK. Table 4.3 lists a selection of compositions of composite SPEEK-BPO[4] membranes and their reported water content and proton conductivities. It has been shown [57,65–67] that swelling can be reduced by blending with polymers, which are capable of formation of hydrogen bonds. The development of blend membranes of SPEEK with polybenzimidazole Celazole (PBI), which showed a reduction in swelling and methanol permeation, in addition to their high thermal stability and moderate conductivity at higher temperature [67,68] has been reported. It is well known that montmorillonite (Na-MMT) is a type of layered silicate composed of silica tetrahedral and alumina octahedral sheets, and its intercalation into Nafion membrane can successfully decrease methanol permeability and improve mechanical property. Organic/inorganic composite membranes were prepared with SPEEK and layered silicate such as organic-montmorillonite (OMMT) by a solution intercalation technique [69], as shown in Table 4.3. Silica is a widely used inorganic reinforcement for the nanocomposites through sol-gel processes. Both cross-linking and sol-gel techniques involve additional chemical reactions and complicated processes in the preparation of proton exchange membranes. In general, the states of water within a polymer can be classified as free water, freezing bound water, and non-freezing bound water [70]. A low fraction of free water in membranes generally leads to a low electro-osmotic drag under fuel cell operation, resulting in low methanol permeability [71]. It is reported that an important reason for the higher methanol permeability for Nafion is its higher fraction of freezing bound and free water [70]. Kim et al. [72] reported that the silica embedded in sulfonated poly(phthalazinone ether sulfone ketone)

Table 4.3 Water content and proton conductivities of sulfonated poly(arylene ether ketone)s-based composite membranes

Membrane	DS (%)	Thickness (μm)	Inorganic material content (%)	Water content (%)	Proton conductivity (S cm^{-1}) 25°C	100°C	Ref.
SPEEK	72	200	BPO$_4$ (0)	52	0.0028	0.014	[62]
		200	BPO$_4$ (20)	66	0.0032	0.028	
		200	BPO$_4$ (40)	79	0.005	0.033	
		200	BPO$_4$ (60)	105	0.001	0.049 (70°C)	
SPEEK	65	–	BPO$_4$ (0)	20	0.005	0.015	[63]
			BPO$_4$ (10)	25	0.013	0.021	
			BPO$_4$ (20)	35	0.012	0.03	
			BPO$_4$ (30)	38	0.018	0.055	
			BPO$_4$ (40)	51	–	0.065	
SPEEK	70	–	BPO$_4$ (0)	29.5	0.005		[64]
			BPO$_4$ (25)	35.1	0.0069	–	
			BPO$_4$ (40)	42.1	0.0072		
			BPO$_4$ (60)	59.5	0.0098		
SPEEK	65	96	OMMT (5)			(at 90°C) 0.011	[69]
		94	OMMT (10) –		–	0.0088	
		115	–			0.015	

(SPPESK) membranes acted as a material for reducing the fraction of free water and as a barrier for methanol transport through the membrane. Sulfonated poly(phthalazinone ether ketone)/silica hybrid membranes also exhibited improved swelling behavior, thermal stability, and mechanical properties. The methanol crossover behavior of these hybrid membranes was also depressed such that these membranes are suitable for a high methanol concentration in feed (3M) in cell test [73]. A remarkable reduction in methanol and water permeability was achieved by inorganic modification of SPEK and SPEEK with different alkoxides of Si, Ti, and Zr [74].

SPEEK can be conveniently cross-linked through bridging links to the reactive sulfonic acid functions. The first reported cross-linking of SPEEK was carried out using suitable aromatic or aliphatic amines [75]. Later a modification of this preparation method was patented [76] in which it was proposed to use a similar cross-linker also having terminal amide functions, which forms imide functionality through a condensation reaction with the sulfonic acid groups of SPEEK. The imide group is supposed to be acidic and therefore able to participate in proton transfer, contributing to the proton conductivity of the polymer. A new method for the preparation of proton exchange membranes, based on cross-linked sulfonated poly(ether ether ketone)s was reported (Fig. 4.19) [77]. It is based on the thermally activated bridging of the polymer chains with polyatomic alcohols through condensation reaction with sulfonic acid functions. The mechanism of this reaction is still under study. Cross-linking greatly increases polymer mechanical strength and reduces its swelling in water. Although cross-linking decreases the number of sulfonic acid groups available

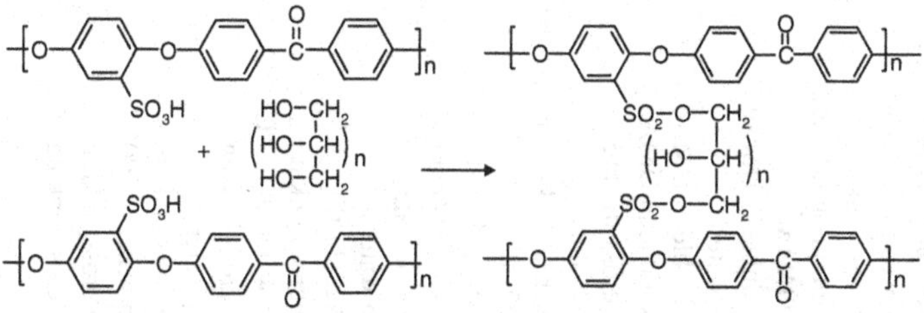

Fig. 4.19 Possible simplified mechanism of SPEEK cross-linking

for proton transfer, SPEEK membrane conductivities are only slightly reduced since the starting polymer is a highly sulfonated water-soluble polymer. It was observed that efficient cross-linking also occurred in films prepared with aqueous solvents in the absence of polar aprotic solvents. Some of the samples exhibited a room temperature conductivity of greater than 2×10^{-2} S cm^{-1} Table 4.4 shows the properties of the SPEEKK membranes after the cross-linking procedure.

4.4 Cell Performances

New membrane materials for PEM fuel cells must be fabricated into a well-bonded, membrane electrode assembly (MEA), as depicted in Fig. 4.20 [78]. MEAs consist of two electrodes, anode and cathode, and the polymer electrolyte. The electrochemical reactions take place at the anode and the cathode catalyst layer, respectively. The gas diffusion layer or electrode substrate (or electrode backing material) at the anode allows hydrogen to reach the reactive zone within the electrode. Upon reacting, protons migrate through the ion-conducting membrane, and electrons are conducted through the substrate layer and, ultimately, to the electric terminals of the fuel cell stack. The anode substrate therefore has to be gas porous as well as electronically conductive. Because not all of the chemical energy supplied to the MEA by the reactions is converted into electric power, heat will also be generated within the MEA. Hence, the gas porous substrate also acts as a heat conductor in order to remove heat from the reactive zones of the MEA.

Figure 4.21 shows a typical fuel cell polarization curve [79]. The curve includes a sharp drop in potential at low current densities due to the sluggish kinetics of the oxygen reduction reaction (ORR). This part of the polarization curve is commonly called the kinetic regime. At moderate current densities, the cell enters an ohmic regime, where the potential varies nearly with current density. At high current densities, mass transport resistance dominates, and the potential of the cell declines rapidly as the concentration of one of the reactants approaches zero at the corresponding catalyst layer. This defines the limiting reactant. In a typical PEM cell operating at temperature below 80°C, much of the water produced by the ORR is liquid, and this liquid water may flood parts of the fuel cell, dramatically increasing the resistance to mass transfer.

Table 4.4 Properties of SPEEK membranes after cross-linking procedure [77]

Sample (DS)[a]	Solvent	Cross-linker	Cross-linker/SPEEK Ratio, mol/repeat unit	T, °C	Time, h	Conductivity S/cm^{-1}	Comment
1a (1.0)	DMAc	G	<1	125	36	–	Soluble in water
1b (1.0)	Water + Acetone	G	>1	130	72	$\leq 1.5 \times 10^{-2}$	Partially soluble in DMAc, H_2O
			<1	125	48	–	Soluble in water
2a (0.63)	DMAc	EG	>1	120	48	$\leq 2.5 \times 10^{-2}$	Complete cross-linking
2b (0.63)	DMAc	G	>1	125	48	–	Partially soluble in DMAc
			<1	125	48	$\leq 2 \times 10^{-3}$	Soluble in DMAc
3a (1.0)	DMAc	G	=1	130	72	$\leq 5.7 \times 10^{-3}$	Partially soluble in DMAc
			>1	130	66	$\leq 7.8 \times 10^{-3}$	Soluble in water
3b (1.0)	DMAc	EG	>1	130	66	$\leq 1 \times 10^{-2}$	Partially soluble in DMAc
			>1	130	66	$\leq 1.5 \times 10^{-2}$	Partially soluble in DMAc
3c (1.0)	Water + Acetone	G	>1	125	60	–	Complete cross-linking
4a (0.94)	Water	Meso-Erythritol	<1	135	60	$\leq 2.2 \times 10^{-2}$	Soluble in water
			>1	135	60	$\leq 2.2 \times 10^{-2}$	Complete cross-linking, brittle
4b (0.94)	Water	EG	=1.4	135	60	$\leq 2.2 \times 10^{-2}$	Soluble in water
			1.4	135	60	$\leq 2.2 \times 10^{-2}$	Complete cross-linking
4c (0.94)	Water	G	2.6	135	60	$\leq 4 \ 10^{-2}$	Complete cross-linking
4d (0.94)	NMP	EG	<2.7	150	60	–	Soluble in hot water
5a (0.96)	Water	Meso-Erythritol	0.5–0.9	140	48	–	Soluble in water
			1–2	140	48	–	Cross-linked, brittle
5b (0.96)	Water	G	>1	140	48	$\leq 1.2 \times 10^{-2}$	Complete cross-linking
5c (0.96)	Water	EG	≤1.5	140	48	$\leq 2.2 \times 10^{-2}$	Soluble in water at 100°C
			>1.5	140	48	$\leq 2.7 \times 10^{-2}$	Complete cross-linking
6a (0.78)	Water + Alcohol	EG	0.6	140	60	–	Soluble in water
			>1.6	140	60	$\leq 2.5 \times 10^{-2}$	Complete cross-linking
6b (0.78)	Water + Acetone	EG	<1.5	140	60	$\leq 1.6 \times 10^{-2}$	Soluble in water at 100°C
			>1.5	140	60	$\leq 2.5 \times 10^{-2}$	Complete cross-linking
7 (0.90)	Water + Alcohol	EG	>1.6	140	60	$\leq 4.2 \times 10^{-2}$	Complete cross-linking

8a (0.78)	DMAc	EG	<1	150	68	—	Soluble in DMAc, hot water
			>1	150	68	—	Partially soluble in DMAc
8c (0.78)	Water + Alcohol	EG	>1.5	150	68	$\leq 2.6 \times 10^{-2}$	Complete cross-linking
9a (0.83)	NMP	EG	1.65–3.3	150	68	$\leq 4.6\ 10^{-2}$	Soluble in water at 100°C
			<1.9	150	68	$\leq 6.5 \times 10^{-2}$	Soluble in water at 100°C
9b (0.83)	Water + Alcohol	EG	>1.9	150	68	$\leq 4 \times 10^{-2}$	Complete cross-linking

G: glycerol, EG: ethylene glycol

[a] measured by NMR before membrane casting

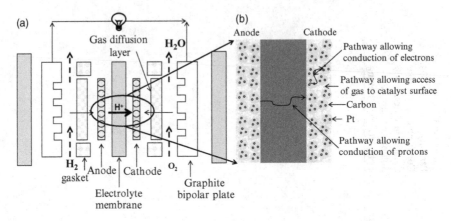

Fig. 4.20 (a) The structure of unit cell (b) polymer electrolyte memrbane with porous electrodes that are composed of platinum paticles uniformly supported on carbon particles [78]

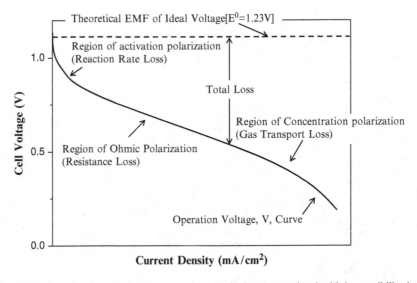

Fig. 4.21 Example of a polarization curve showing the losses associated with irreversibility in a fuel cell [79]

The Teflon-like molecular backbone gives these materials excellent long-term stability in both oxidative and reductive environments. A lifetime of over 60,000 h under fuel cell conditions has been achieved with commercial Nafion membranes. These membranes exhibit proton conductivity as high as 0.1 S cm^{-1} under fully

Table 4.5 Summary of Modifications of Nafion and SPEEK membranes

Polymer	Modifiers	Remarks	Ref.
Nafion	H3PO4	0.05 S cm^{-1} at 150°C	[81]
Nafion	PTA-acetic acid	H$_2$/O$_2$ Cell, 110°C, 660 mA cm^{-2} at 0.6 V, 1/1 atm, humidifier 50/ 50°C	[82]
Nafion	PTA-TBAC	H$_2$/O$_2$ Cell, 120°C, 700 mA cm^{-2} at 0.6 V, 1/1 atm, humidifier 50/ 50°C	[82]
Nafion	SiO2	> 0.2 Scm^{-1}, 100°C, 100% RH	[83]
Nafion	SiO2	DMFC, 145°C, 4.5/5.5 atm (air), 350 mA cm^{-2} at 0.5 V	[84]
Nafion	Teflon + PTA	H2/O2 cell, 120°C, 400 mA cm^{-2} at 0.6 V, 1/1 atm, humidifier 90/ 84°C	[85]
Nafion	ZrP	DMFC, 150°C, 4/4 atm, 380 mW cm^{-2} (O2), 260 mW cm^{-2} (air)	[86]
Nafion	ZrP	H$_2$/O$_2$ cell, 1.5A cm^{-2} at 0.45 V, 130°C, 3 atm	[87]
Nafion	SiO2, PWA-SiO2, SiWA-SiO2	140°C, 3/4 atm, DMFC, 400 mW cm^{-2} (O2), 250 mW cm^{-2} (air)	[88]
SPEEK	SiO2, ZrP, Zr-SPP	0.09 S cm^{-1} at 100°C, 100% RH, H$_2$/O$_2$ fuel cell test at 95°C	[89]
SPEEK	HPA	10^{-1} S cm^{-1} above 100°C	[61]
SPEEK	BPO4	5 × 10^{-1} S cm^{-1}, 160°C, fully hydrated	[62]
SPEEK	SiO2	3–4 × 10^{-2} S cm^{-1} at 100°C, 100% RH	[90]

PTA (PWA) phosphotungstic acid; *TBAC* tetra-*n*-butylammonium chloride; *ZrP* zirconium hydrogen phosphate

hydrated conditions. For Nafion 117 (thickness 175 μm), this conductivity corresponds to a real resistance of 0.2 Ω cm², i.e., a voltage loss of about 150 mV at a practical current density of 750 mA cm^{-2} [80]. Table 4.5 summarizes modified Nafion and SPEEK membranes.

It is reported [91] that the single cell test results indicated that sulfonated poly(phthalazinone ether ketone) performed better than Nafion in terms of higher power density, higher ultimate current density, and higher optimal operating concentration of methanol in feed (Fig. 4.22). It is important to obtain that long-term durability data. This group reported that the membrane with 5 phr (parts per hundred resin) silica nanoparticles showed an open cell potential of 0.6 V and an optimum power density of 52.9 mW cm^{-2} at a current density of 264.6 mA cm^{-2}, which is better than the performance of the pristine of SPPEK membrane and Nafion 117 as shown in Fig. 4.23 [73]. It is reported that the formation of polymer-silica nanocomposite membranes is a convenient and effective approach to improve the properties of highly sulfonated polymers used as PEMs in direct methanol fuel cell. The physical properties as well as cell performance of the membranes are enhanced. Several studies have been reported on SPEEK used as a PEM material in DMFC, including blend polymer membrane and organic-inorganic membrane. The properties of the SPPEK membrane are listed in Table 4.6.

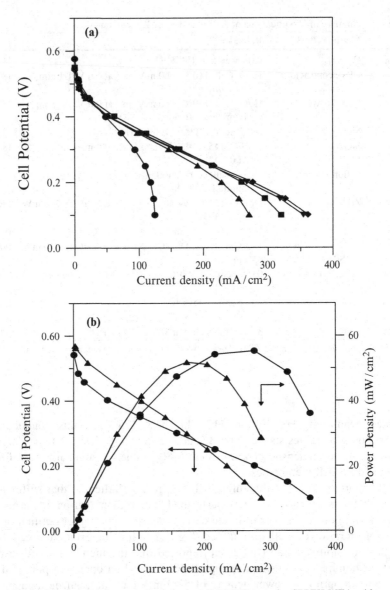

Fig. 4.22 (a) The polarization curves of single cell DMFC tests for SPPEK MEA with various methanol concentrations in feed: (*filled circle*) 1 M, (*triangle*) 2 M, (*square*) 2.5 M, (*diamond*) 3 M and (Δ) 4 M. (**b**)A comparison of the optimal single cell performance with SPPEK and Nafion as the proton exchange membrane: (*filled circle*) MeOH at 3Mfor SPPEK (membrane thickness = 30 μm) cell and (*triangle*) MeOH at 2M for Nafion (membrane thickness = 187 μm) cell; 70°C; MeOH(aq) feeding rate, 2 mL min⁻¹; rate of humidified O₂, 150 mL min⁻¹ [91]. Reprinted from Y.M. Sun et al., Sulfonated poly(phthalazinone ether ketone) for proton exchange membranes in direct methanol fuel cells. *J. Membrane Sci.* 265, 112–113 (2005), with permission from Elsevier

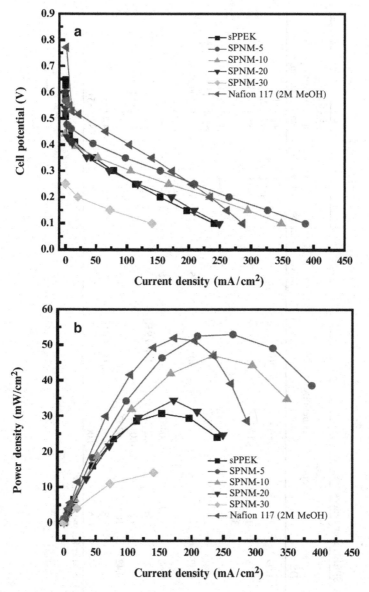

Fig. 4.23 DMFC performance of using various membranes with 3 M methanol in feed: (**a**) polarization curves; and (**b**) the relationship of power density and current density. cell; 70°C; MeOH(aq) feeding rate, 2 mL min^{-1}; rate of humidified O$_2$, 150 mL min^{-1} [73]. Reprinted from Y. L. Liu et al, Using silica nanoparticles for modifying sulfonated poly(phthalazinone ether ketone) membrane for direct methanol fuel cell: A significant improvement on cell performance. *J. Power Sources* (2005), with permission from Elsevier

Table 4.6 Properties of the SPEEK membranes

Membrane	DS	Water content (%) at room temp.	Conductivity (S cm^{-1})	Cell temp (°C)	MeOH conc. (M)	OCV (V)	Maximum power density, mW cm^{-2} (current density, mA cm^{-2})
SPPEK[a]	1.09	25	>10^{-2}	70	2	0.55	47.5 (198)
SPPEK[b]	1.23	45	>10^{-2}	70	3	0.54	55 (276)
SPNM-5	—	38			2	0.65	30.6 (153.6)
SPNM-10	—	25				0.6	52.9 (264.6)
Nafion 117	0.91	25				0.44	47.0 (234.1)
SPPEK/ZrO2 (5 wt%)[c]	0.87	—	—	90	1.5	0.75	52.0 (175.0)
SPPEK/ZrO2 (7.5 wt%)		—	—			0.43	16.4 (109.2)
SPEEK[d]	0.68	17.4	0.0463	110 (138%RH)	1.5	0.60	8.0 (32.1)
/ZrPh/PBI (84.4:10:5.6)		9.4	0.0294			—	4.8 (25)
/ZrPh/PBI (68.8:20:11.2)		12	0.0182			—	14.7 (58.8)
SPEEK	0.42	10.1	0.0202	110 (100%RH)	1.5		10.4 (51.8)
/ZrPh/PBI (84.4:10:5.6)		0.3	0.0115				4.4 (25)
/ZrPh/PBI (68.8:20:11.2)		0.6	0.0028				8.2 (35)

[a]Sulfonated poly(phthalazinone ether ketone) [91],
[b]Data from [73],
[c]Data from [92],
[d]Data from [93]

4.5 Conclusion

Many sulfonated aromatic high-performance polymers combine good durability and high proton conductivity in the wet state, making them interesting low-cost alternatives for perfluorosulfonic polymers (e.g., Nafion) as membranes in hydrogen fuel cells. However, the two current hurdles for polymeric membranes are the high proton conductivity at low water content (e.g., under conditions of 120°C and 50% RH) and long-term durability under fuel cell conditions.

As described, the choice of bisphenols for the polymerization of poly(arylene ether ketone)s is large. Copolymers based on hexafluoroisopropylidene bisphenol (Bisphenol 6F) have been particularly interesting in fuel cell tests. The incorporation of the larger rigid Bisphenol 6F moiety into the backbone is an attempt to increase the hydrophobicity and length of nonsulfonated properties as well as the hot water stability of the membranes. It is thought that the fluorine content in Bisphenol 6F promotes adhesion and electrochemical compatibility with Nafion-based electrodes and reduces swelling. Surface fluorine enrichment of 6F-containing materials may also provide enhanced membrane stability [3].

SPEEK membranes showed proton conductivities higher than 10^{-2} S cm^{-1}, which is close to that of Nafion under humidified conditions, but the methanol permeabilities and cost of the present SPEEKs are much lower than that of Nafion. Therefore, SPEEKs may potentially find application as PEM materials for fuel cells.

References

1. D. S. Kim, Ph.D. Thesis, Novel polymer electrolyte membrane materials: Candidates for proton exchange membrane fuel cell. Hanyang University, Seoul, Korea, 2005.
2. A. J. Appleby, R. L. Foulkes, Fuel Cell Handbook, Van Nostrand Reinhold, New York. (1989).
3. M. A. Hickner, H. Ghassemi, Y. S. Kim, B. R. Einsla, J. E. McGrath, Alternative polymer systems for proton exchange membranes (PEMs). Chem. Rev. 104(10), 4587–4612 (2004).
4. Y. S. Kim, F. Wang, M. Hickner, S. McCartney, Y. T. Hong, W. T. Harrison, A. Zawodzinski, J. E. McGrath, Influence of the bisphenol structure on the direct synthesis of sulfonated poly(arylene ether) copolymers. J. Polym. Sci. A 41(14), 2264–2276 (2003).
5. T. Schults, S. Zhou, K. Sundmacher, Current status of and recent developments in the direct methanol fuel cell. Chem. Eng. Technol. 24(12), 1223–1233 (2001).
6. O. Savadogo, Emerging membranes for electrochemical systems: (I) Solid polymer electrolyte membranes for fuel cell systems. J. New Mater. Electrochem. Syst. 1, 47–66 (1998).
7. L. H. William, Ph.D. Thesis, Synthesis and characterization of sulfonated poly(arylene ether sulfone) copolymers via direct copolymerization: candidates for proton exchange membrane fuel cells. Hampton University, Virginia, USA, 2002.
8. W. H. J. Hogarth, J. C. Diniz da Costa, G. Q. Lu, Solid acid membranes for high temperature (140°C) proton exchange membrane fuel cells. J. Power Sources 142, 223–237 (2005).
9. K. D. Kreuer, On the development of proton conducting polymer membranes for hydrogen and methanol fuel cells. J. Membr. Sci. 185, 29–39 (2001).

10. G. Alberti, M. Casciola, L. Massinelli, B. Bauer, Polymeric proton conducting membranes for medium temperature fuel cells (110–160°C). *J. Membr. Sci.* 185, 73–81 (2001).

11. P. Genova-Dimitrova, B. Baradie, D. Foscallo, C. Poinsignon, J. Y. Sanchez, Ionomeric membranes for proton exchange membrane fuel cell (PEMFC): Sulfonated polysulfone associated with phosphatoantimonic acid. *J. Membr. Sci.* 185, 59–71 (2001).

12. F. Wang, M. Hickner, Y. S. Kim, T. A. Zawodzinski, J. E. McGrath, Direct polymerization of sulfonated poly(arylene ether sulfone) random (statistical) copolymers: Candidates for new proton exchange membranes. *J. Membr. Sci.* 197, 231–242 (2002).

13. N. Carretta, V. Tricoli, F. Picchioni, Ionomeric membranes based on partially sulfonated poly(styrene): Synthesis, proton conduction and methanol permeation. *J. Membr. Sci.* 166, 189–197 (2000).

14. K. Miyatake, E. Shouji, K. Yamamoto, E. Tsuchida, Synthesis and proton conductivity of highly sulfonated poly(thiophenylene). *Macromolecules* 30, 2941–2946 (1997).

15. Y. S. Kim, F. Wang, M. Hickner, T. A. Zawodzinski, J. E. McGrath, Fabrication and characterization of heteropolyacid ($H_3PW_{12}O_{40}$)/directly polymerized sulfonated poly(arylene ether sulfone) copolymer composite membranes for higher temperature fuel cell applications. *J. Membr. Sci.* 212, 263–282 (2003).

16. M. Rikukawa, K. Sanui, Proton-conducting polymer electrolyte membranes based on hydrocarbon polymers. *Prog. Polym. Sci.* 25, 1463–1502 (2000).

17. B. Bauer, D. J. Jones, J. Rozière, L. Tchicaya, G. Alberti, M. Casciola, L. Massinelli, A. Peraio, S. Besse, E. Ramunni, Electrochemical characterisation of sulfonated polyetherketone membranes. *J. New Mater. Electrochem. Syst.* 2, 93 (2000).

18. T. Soczka-Guth, J. Baurmeister, G. Frank, R. Knauf, Method for producing a membrane used to operate fuel cells and electrolyzers, International patent WO 99/29763

19. S. Kaliaguine, S. D. Mikhailenko, K. P. Wang, P. Xing, G. Robertson, M. D. Guiver, Properties of SPEEK based PEMs for fuel cell application. *Catal. Today.* 82, 213–222 (2003).

20. G. P. Robertson, S. D. Mikhailenko, K. Wang, P. Xing, M. D. Guiver, S. Kaliaguine, Casting solvent interactions with sulfonated poly(ether ether ketone) during proton exchange membrane fabrication. *J. Membr. Sci.* 219, 113–121 (2003).

21. P. Xing, G. P. Robertson, M. D. Guiver, S. D. Mikhailenko, K. Wang, S. Kaliaguine, Synthesis and characterization of sulfonated poly(ether ether ketone) for proton exchange membranes. *J. Membr. Sci.* 229, 95–106 (2004).

22. M. T. Bishop, F. E. Karasz, P. S. Russo, K. H. Langley, Solubility and properties of a poly(aryl ether ketone) in strong acids. *Macromolecules* 18, 86–93 (1985).

23. R. Y. M. Huang, P. Shao, C. M. Burns, X. Feng, Sulfonation of poly(ether ether ketone)(PEEK): Kinetic study and characterization. *J. Appl. Polym. Sci.* 82, 2651–2660 (2001).

24. C. Bailly, D. J. Williams, F. E. Karasz, W. J. MacKnight, The sodium salts of sulphonated poly(aryl-ether-ether-ketone) (PEEK): Preparation and characterization. *Polymer* 28, 1009–1016 (1987).

25. J. Rozière, D. J. Jones, Non-fluorinated polymer materials for proton exchange membrane fuel cells. *Ann. Rev. Mater. Res.* 33, 503–555 (2003).

26. H. C. Lee, H. S. Hong, Y. M. Kim, S. H. Choi, M. Z. Hong, H. S. Lee, K. Kim, Preparation and evaluation of sulfonated-fluorinated poly(arylene ether)s membranes for a proton exchange membrane fuel cell (PEMFC). *Electrochim. Acta* 49, 2315–2323 (2004).

27. D. S. Kim, B. Liu, M. D. Guiver, Influence of silica content in sulfonated poly(arylene ether ether ketone ketone)(SPAEEKK) hybrid membranes on properties for fuel cell application. *Polymer* 47, 7871–7880 (2006).

28. Y. Gao, G. P. Robertson, M. D. Guiver, X. Jian, Synthesis and characterization of sulfonated poly(phthalazinone ether ketone) for proton exchange membrane materials. *J. Polym. Sci. A* 41, 497–507 (2003).

29. Y. Gao, G. P. Robertson, M. D. Guiver, X. Jian, S. D. Mikhailenko, K. Wang, S. Kaliaguine, Sulfonation of poly(phthalazinones) with fuming sulfuric acid mixtures for proton exchange membrane materials. *J. Membr. Sci.* 227, 39–50 (2003).

30. Y. Dai, X. Jian, X. Liu, M. D. Guiver, Synthesis and characterization of sulfonated poly(phthalazinone ether sulfone ketone) for ultrafiltration and nanofiltration membranes. *J. Appl. Polym. Sci.* 79, 1685–1692 (2001).

31. L. Wang, Y. Z. Meng, S. J. Wang, A. S. Hay, Synthesis and sulfonation of poly(arylene ether)s containing tetraphenyl methane moieties. *J. Polym. Sci. A* 42, 1779–1788 (2004).

32. L. Wang, Y. Z. Meng, S. J. Wang, M. Xiao, Synthesis and properties of sulfonated poly(arylene ether) containing tetraphenylmethane moieties for proton-exchange membrane. *J. Polym. Sci. A* 43, 6411–6418 (2005).

33. X. Shang, S. Tain, L. Kong, Y. Z. Meng, Synthesis and characterization of sulfonated fluorene-containing poly(arylene ether ketone) for proton exchange membrane. *J. Membr. Sci.* 266, 94–101 (2005).

34. W. L. Harrison, Ph.D. Thesis, Synthesis and characterization of sulfonated poly (arylene ether sulfone) copolymers via direct copolymerization: candidates for proton exchange membrane fuel cells. Virginia Polytechnic Institute and State University, USA, 2002.

35. Y. S. Kim, F. Wang, M. Hickner, S. McCartney, Y. T. Hong, T. A. Zawodzinski, J. E. McGrath, Effect of acidification treatment and morphological stability of sulfonated poly(arylene ether sulfone) copolymer proton-exchange membranes for fuel-cell use above 100°C. *J. Polym. Sci. B* 41, 2816–2828 (2003).

36. F. Wang, T. Chen, J. Xu, Sodium sulfonate-functionalized poly(ether ether ketone)s. *Macromol. Chem. Phys.* 199, 1421–1426 (1998).

37. F. Wang, J. Li, T. Chen, J. Xu, Synthesis of poly(ether ether ketone) with high content of sodium sulfonate groups and its membrane characteristics. *Polymer* 40, 795–799 (1999).

38. R. Hopp, F. Wang, J. E. McGrath, 2001 Summer undergraduate research program reports. Virginia Tech, August 2001.

39. Y. Gao, G. P. Robertson, M. D. Guiver, X. Jian, S. D. Mikhailenko, K. Wang, S. Kaliaguine, Direct copolymerization of sulfonated poly(phthalazinone arylene ether)s for proton-exchange-membrane materials. *J. Polym. Sci. A* 41, 2731–2742 (2003).

40. P. Xing, G. P. Robertson, M. D. Guiver, S. D. Mikhailenko, S. Kaliaguine, Sulfonated poly(aryl ether ketones) containing naphthalene moieties obtained by direct copolymerization as novel polymers for proton exchange membranes. *J. Polym. Sci. A* 42, 2866–2876 (2004).

41. D. S. Kim, H. B. Park, J. Y. Jang, Y. M. Lee, Synthesis of sulfonated poly(imidoaryl ether sulfone) membranes for polymer electrolyte membrane fuel cells. *J. Polym. Sci. A* 43, 5620–5631 (2005).

42. S. J. Wang, Y. Z. Meng, A. R. Hlil, A. S. Hay, Synthesis and characterization of Phthalazinone containing Poly(arylene ether)s, Poly(arylene thioether)s, and Poly(arylene sulfone)s via a novel N-C coupling reaction. *Macromolecules* 37, 60–65 (2004).

43. Y. L. Chen, Y. Z. Meng, A. S. Hay, Direct synthesis of sulfonated poly(phthalazinone ether) for proton exchange membrane via N–C coupling reaction. *Polymer* 46, 11125–11132 (2005).

44. Y. L. Chen, Y. Z. Meng, A. S. Hay, Novel synthesis of sulfonated Poly(phthalazinone ether ketone) used as a proton exchange membrane via N-C coupling reaction. *Macromolecules* 38, 3564–3566 (2005).

45. P. Xing, G. P. Robertson, M. D. Guiver, S. D. Mikhailenko, S. Kaliaguine, Sulfonated Poly(aryl ether ketones) containing the Hexafluoroisopropylidene Diphenyl moiety prepared by direct copolymerization, as proton exchange membranes for fuel cell application. *Macromolecules* 37, 7960–7967 (2004).

46. P. Xing, G. P. Robertson, M. D. Guiver, S. D. Mikhailenko, S. Kaliaguine, Synthesis and characterization of poly(aryl ether ketone) copolymers containing (hexafluoroisopropylidene)-diphenol moiety as proton exchange membrane materials. *Polymer* 46, 3257–3263 (2005).

47. B. Liu, D. S. Kim, J. Murphy, G. P. Robertson, M. D. Guiver, S. D. Mikhailenko, S. Kaliaguine, Y. M. Sun, Y. L. Liu, J. Y. Lai, Fluorenyl-containing sulfonated poly(aryl ether ether ketone ketones) (SPFEEKK) for fuel cell applications. *J. Membr. Sci.* 280, 54–64 (2006).

48. X. Li, C. Liu, H. Lu, C. Zhao, Z. Wang, W. Xing, H. Na, Preparation and characterization of sulfonated poly(ether ether ketone ketone) proton exchange membranes for fuel cell application. *J. Membr. Sci.* 255, 149–155 (2005).

49. X. Li, C. Zhao, H. Lu, Z. Wang, H. Na, Direct synthesis of sulfonated poly(ether ether ketone ketones) (SPEEKKs) proton exchange membranes for fuel cell application. *Polymer* 46, 5820–5827 (2005).

50. D. S. Kim, H. B. Park, J. W. Rhim, Y. M. Lee, Preparation and characterization of crosslinked PVA/SiO^2 hybrid membranes containing sulfonic acid groups for direct methanol fuel cell applications. *J. Membr. Sci.* 240, 37–48 (2004).

51. Y. S. Kim, M. A. Hickner, L. Dong, B. S. Pivovar, J. E. McGrath, Sulfonated poly(arylene ether sulfone) copolymer proton exchange membranes: Composition and morphology effects on the methanol permeability. *J. Membr. Sci.* 243, 317–326 (2004).

52. T. Kobayashi, M. Rikukawa, K. Sanui, N. Ogata, Proton-conducting polymers derived from poly(ether-ether-ketone) and poly(4-phenoxybenzoyl-1,4-phenylene). *Solid State Ionics* 106, 219–225 (1998).

53. B. Lafitte, L. E. Karlsson, P. Jannasch, Sulfophenylation of polysulfones for proton-conducting fuel cell membranes. *Macromol. Rapid Commun.* 23, 896 (2002).

54. Y. Gao, G. P. Robertson, M. D. Guiver, S. D. Mikhailenko, X. Li, S. Kaliaguine, Synthesis of poly(arylene ether ether ketone ketone) copolymers containing pendant sulfonic acid groups bonded to naphthalene as proton exchange membrane materials. *Macromolecules* 37, 6748–6754 (2004).

55. K. Miyatake, K. Oyaizu, E. Tsuchida, A. S. Hay, Synthesis and properties of novel sulfonated arylene ether/fluorinated alkane copolymers. *Macromolecules* 34, 2065–2071 (2001).

56. D. S. Kim, H. B. Park, J. W. Rhim, Y. M. Lee, Proton conductivity and methanol transport behavior of cross-linked PVA/PAA/Silica hybrid membranes. *Solid State Ionics* 176, 117–126 (2005).

57. J. A. Kerres, Development of ionomer membranes for fuel cells. *J. Membr. Sci.* 185, 3–27 (2001).

58. W. Vielstich, A. Lamm, H. Gasteiger (Eds.), Handbook of Fuel Cells, Vol 3: Fuel cell technology and applications: Part 1, Wiley, England, (2003).

59. G. Hubner, E. Roduner, EPR investigation of HO/radical initiated degradation reactions of sulfonated aromatics as model compounds for fuel cell proton conducting membranes. *J. Mater. Chem.* 9, 409 (1999).

60. P. Staiti, M. Minutoli, S. Hocevar, Membranes based on phosphotungstic acid and polybenzimidazole for fuel cell application. *J. Power Sources* 90, 231–235 (2000).

61. S. M. J. Zaidi, S. D. Mikhailenko, G. P. Robertson, M. D. Guiver, S. Kaliaguine, Proton conducting composite membranes from polyether ether ketone and heteropolyacids for fuel cell applications. *J. Membr. Sci.* 173, 17–34 (2000).

62. S. D. Mikhailenko, S. M. J. Zaidi, S. Kaliaguine, Sulfonated polyether ether ketone based composite polymer electrolyte membranes. *Catal. Today* 67, 225–236 (2001).

63. P. Krishnan, J. S. Park, C. S. Kim, Preparation of proton-conducting sulfonated poly(ether ether ketone)/boron phosphate composite membranes by an in situ sol–gel process. *J. Membr. Sci.* 279, 220–229 (2006).

64. A. R. Valencia, S. Kaliaguine, M. Bousmina, Tensile mechanical properties of sulfonated poly(ether ether ketone) (SPEEK) and BPO4/SPEEK membranes. *J. Appl. Polym. Sci.* 98, 2380–2393 (2005).

65. W. C. Cui, J. Kerres, G. Eigenberger, Development and characterization of ion-exchange polymer blend membranes. *Separ. Purif. Technol.* 14, 145–154 (1998).

66. J. A. Kerres, A. Ullrich, F. Meier, Th. Haring, Synthesis and characterization of novel acid–base polymer blends for application in membrane fuel cells. *Solid State Ionics* 125, 243–249 (1999).

67. J. A. Kerres, A. Ullrich, TH. Haring, M. Baldauf, U. Gebhard, W. Preidel, Preparation, characterization, and fuel cell application of new acid-base blend membranes. *J. New Mater. Electrochem. Syst.* 3, 229 (2000).

68. S. M. J. Zaidi, Preparation and characterization of composite membranes using blends of SPEEK/PBI with boron phosphate. *Electrochim. Acta* 50, 4771–4777 (2005).

69. G. Zhang, Z. Zhou, Organic/inorganic composite membranes for application in DMFC. *J. Membr. Sci.* 261, 107–113 (2005).
70. Y. S. Kim, L. Dong, M. A. Hickner, T. G. Glass, V. Webb, J. E. McGrath, State of Water in Disulfonated Poly(arylene ether sulfone) Copolymers and a Perfluorosulfonic acid copolymer (Nafion) and its effect on physical and electrochemical properties. *Macromolecules* 36, 6281–6285 (2003).
71. M. Ise, K. D. Kreuer, J. Maier, Electroosmotic drag in polymer electrolyte membranes: An electrophoretic NMR study. *Solid State Ionics* 125, 213–223 (1999).
72. D. S. Kim, H. S. Kwang, H. B. Park, Y. M. Lee, Preparation and characterization of sulfonated poly(phthalazinone ether sulfone ketone)/silica hybrid membranes for DMFC applications. *Macromol. Res.* 12, 413–421 (2004).
73. Y. H. Su, Y. L. Liu, Y. M. Sun, J. Y. Lai, M. D. Guiver, Y. Gao, Using silica nanoparticles for modifying sulfonated poly(phthalazinone ether ketone) membrane for direct methanol fuel cell: A significant improvement on cell performance. *J. Power Sources* 155, 111–117 (2006).
74. S. P. Nunes, B. Ruffmann, E. Rikowski, S. Vetter, K. Richau, Inorganic modification of proton conductive polymer membranes for direct methanol fuel cells. *J. Membr. Sci.* 203, 215–225 (2002).
75. F. Helmer-Metzmann, F. Osan, A. Schneller, H. Ritter, K. Ledjeff, R. Nolte, R. Thorwirth, Polymer electrolyte membrane, and process for the production thereof, US Patent 5,438,082 (1995).
76. S. S. Mao, S. J. Hamrock, D. A. Ylitalo, Crosslinked ion conductive membranes, US Patent 6,090,895 (2000).
77. S. D. Mikhailenko, K. Wang, S. Kaliaguine, P. Xing, G. P. Robertson, M. D. Guiver, Proton conducting membranes based on cross-linked sulfonated poly(ether ether ketone) (SPEEK). *J. Memb. Sci.* 233, 93–99 (2004).
78. S. Thomas, M. Zalbowits, Green Power, Los Alamos National Laboratory Report, Los Alamos, New Mexico (1999).
79. G. Hoogers (Ed.), Fuel Cell Technology Handbook, CRC Press, FL, (2003).
80. Q. Li, R. He, J. O. Jensen, N. J. Bjerrum, Approaches and recent development of polymer electrolyte membranes for fuel cells operating above 100°C. *Chem. Mater.* 15, 4896–4915 (2003).
81. R. Savinell, E. Yeager, D. Tryk, U. Landau, J. Wainright, D. Weng, K. Lux, M. Litt, C. Rogers, A polymer electrolyte for operation at temperatures up to 200°C. *J. Electrochem. Soc.* 141, L46–L48 (1994).
82. S. Malhotra, R. Datta, Membrane-supported nonvolatile acidic electrolytes allow higher temperature operation of proton-exchange membrane fuel cells. *J. Electrochem. Soc.* 144, L23–L26 (1997).
83. H. Wang, B. A. Holmberg, L. Huang, Z. Wang, A. Mitra, J. M. Norbeck, Y. Yan, Nafion-bifunctional silica composite proton conductive membranes. *J. Mater. Chem.* 12, 834–837 (2002).
84. P. L. Antonucci, A. S. Arico, P. Creti, E. Ramunni, V. Antonucci, Investigation of a direct methanol fuel cell based on a composite Nafion®-silica electrolyte for high temperature operation. *Solid State Ionics* 125, 431–437 (1999).
85. J. C. Lin, H. R. Jnuz, J. M. Fenton, In Handbook of Fuel Cells; W. Vielstich, A. Lamm, H. A. Gasteiger (Eds.), Wiley, New York, (2003), Chap. 3, pp. 457.
86. C. Yang, S. Srinivasan, A. S. Arico, P. Creti, V. Baglio, V. Antonucci, Composite Nafion/ Zirconium phosphate membranes for direct methanol fuel cell operation at high temperature. *Electrochem. Solid-State Lett.* 4, A31–A34 (2001).
87. P. Costamagna, C. Yang, A. B. Bocarsly, S. Srinivasan, Nafion® 115/zirconium phosphate composite membranes for operation of PEMFCs above 100°C. *Electrochim. Acta* 47, 1023–1033 (2002).
88. P. Staiti, A. S. Arico, V. Baglio, F. Lufrano, E. Passalacqua, V. Antonucci, Hybrid Nafion–silica membranes doped with heteropolyacids for application in direct methanol fuel cells. *Solid State Ionics* 145, 101–107 (2001).

89. B. Bonnet, D. J. Jones, J. Rozière, L. Tchicaya, G. Alberti, M. Casciola, L. Massinelli, B. Baner, A. Peraio, E. Ramunni, Hybrid organic-inorganic membranes for a medium temperature fuel cell. *J. New. Mater. Electrochem. Syst.* 2, 87–92 (2000).
90. J. Rozière, D. J. Jones, L. Tchicaya-Bouckary, B. Bauer, WO 02/05370, 2000.
91. Y. M. Sun, T. C. Wu, H. C. Lee, G. B. Jung, M. D. Guiver, Y. Gao, Y. L. Liu, J. Y. Lai, Sulfonated poly(phthalazinone ether ketone) for proton exchange membranes in direct methanol fuel cells. *J. Membr. Sci.* 265, 108–114 (2005).
92. V. S. Silva, J. Schirmer, R. Reissner, B. Ruffmann, H. Silva, A. Mendes, L. M. Madeira, S. P. Nunes, Proton electrolyte membrane properties and direct methanol fuel cell performance: II. Fuel cell performance and membrane properties effects. *J. Power Sources* 140, 41–49 (2005).
93. V. S. Silva, B. Ruffmann, S. Vetter, A. Mendes, L. M. Maderia, S. P. Nunes, Characterization and application of composite membranes in DMFC. *Catal. Today* 104, 205–212 (2005).

Chapter 5
Fuel Cell Membranes by Radiation-Induced Graft Copolymerization: Current Status, Challenges, and Future Directions

Mohamed Mahmoud Nasef

Abstract Radiation-induced graft copolymerization is an attractive technique to prepare alternative proton conducting membranes (PCMs) for fuel cell applications. The purpose of this chapter is to review the latest progress made in the development of various radiation-grafted PCMs for fuel cells. The challenges facing the development of these membranes and their expected future research directions are also discussed.

5.1   Introduction

Fuel cell technology is one of the key technologies that are receiving tremendous efforts to bring about new environmentally friendly and efficient power sources in the twenty-first century. Among all fuel cells, polymer electrolyte fuel cell (PEMFC) and direct methanol fuel cell (DMFC) are promising candidates for low temperature stationary and mobile applications operations. During normal PEMFC or DMFC operations, anodic dissociation of hydrogen in the former and methanol in the latter produces protons that are transported through the hydrated proton-conducting membrane (PCM) to the cathode, in which reduction of O_2 produces water.

Currently, the cost of these fuel cell systems is deemed to be very high mainly due to the excessive cost incurred by some key fuel cell components including PCM [1]. Several PCMs are commercially available including Nafion (DuPont), Aciplex (Asahi Chemicals Co.), Flemion (Asahi Glass Co.), Gore-Tex (Gore and Associates), Ballard Advanced Materials (BAM) (Ballard), and Dais Membranes (Dais Co.) [2]. Of all, Nafion membrane is the most established product that has been widely tested and used in the majority of the available fuel cell systems. However, Nafion is deemed to be expensive and has high methanol permeability (in DMFC). In addition, it is prone to viscoelastic relaxation at high temperatures (low hydrated T_g), which decreases both its mechanical properties and proton conductivity [3]. This situation has triggered rather extensive worldwide efforts to develop alternative cost-effective and highly conductive membranes. Since then, varieties of alternative PCMs have been developed for application in fuel

S.M.J. Zaidi, T. Matsuura (eds.) *Polymer Membranes for Fuel Cells*,
doi: 10.1007/978-0-387-73532-0, © Springer Science+Business Media, LLC 2009

cells. The latest progress on these membranes and their different classes have been reviewed in several articles [2,4–6].

Various approaches have been considered to develop new alternative membranes. The first approach includes formation of Nafion composites [7,8] or modification of Nafion membranes involving the use of surface coatings [9,10]. The second approach involves direct sulfonation of non-fluorinated polymer backbones such as polystyrene [11], polyphosphazene [12], polyphenylene oxide [13], polysulfone [14], polyether sulfone [15], polyether ether ketone [16], polybenzimidazole [17], and polyamides [18]. The challenge in this approach is to achieve sufficient sulfonation for high proton conductivity in the membranes without the polymer becoming soluble. The third approach involves sulfonation of pendent aromatic rings attached to a variety of copolymer (grafted) films obtained by chemical [19,20], plasma [21], thermal [22], or radiochemical graft copolymerization of styrene monomer [23]. Of all, radiochemical (radiation-induced) graft copolymerization of styrene or its substituents onto various fluoropolymer films has been found to be an effective method for preparation of alternative proton conducting membranes for fuel cells [24]. The objective this chapter is to review the current state-of-the-art in using radiation-induced graft copolymerization methods for preparation of PCMs for fuel cells. Challenges and future directions of the research of these membranes are also discussed.

5.2   Radiation-Induced Graft Copolymerization Approach

Radiation-induced graft copolymerization is a simple method for modification of existing polymers and imparting new properties without altering their inherent properties [25]. Therefore, it had been of particular interest for preparation of a variety of functional membranes in past five decades [26]. In this method active cites are formed on the polymer backbone using high energy radiation (γ-radiation, electrons, swift heavy ions) and the irradiated base polymer is allowed to react with monomer units, which then propagate to form side chain grafts when terminated. Two standard methods of radiation-induced graft copolymerization: (1) direct (simultaneous) and (2) preirradiation methods have been developed over the past 50 years. In direct method, the base polymer is irradiated while immersed in the monomer solution, forming free radicals in both polymer and monomer. Alternatively, in preirradiation, the base polymer is irradiated first (in inert atmosphere) to form trapped radicals and subsequently brought into contact with the monomer solution under controlled conditions. If the irradiation step is carried out in the air, the formed radicals react with oxygen-forming peroxides and hydroperoxides, which decompose at elevated temperature to initiate grafting reaction; this method is then called peroxidation. Each one of these grafting methods has its merits, depending on polymer/monomer combination. For example, the direct method is very simple and efficient from the polymer radiation chemistry principle viewpoint and produces higher degrees of grafting due to efficient utilization of radicals. On the other hand, preirradiation methods are very effective, particularly when a highly reactive monomer such as acrylic acid is grafted, and when pilot-scale production is sought. Nonetheless, obtaining desired grafting levels in both grafting methods requires achieving optimum combinations of reaction parameters to vary penetration depth of the monomer into the polymer bulk to eventually allow manipulation of the graft copolymer composition. A detail account for the advantages and disadvantages of both radiation-induced graft copolymerization methods can be found in [27].

Radiation-induced graft polymerization has been found to be advantageous for making membranes in general and PCMs in particular for various reasons, including the simplicity and the flexibility of initiating the reaction using various types of high-energy radiations produced by already available commercial sources such as Co-60, electron beam (EB), or ion beam (IB). The amount of the grafted moiety can also be controlled by appropriate variation of irradiation and reaction parameters. This provides a tool to develop specially designed (tailor made) membranes for particular applications. The graft polymerization can also be initiated in a wide range of temperatures, including low regions in monomers available in bulk, solution, emulsion, or even solid state. Furthermore, this technique shows a superior advantage in which the difficulty of shaping the graft copolymer into a thin foil of uniform thickness could be circumvented by starting the process with thin film already having the physical shape of the membrane. The latest development in using radiation-induced graft copolymerization for preparation of different types of membranes for various applications has been reviewed recently by Nasef and Hegazy [27].

5.3 Radiation-grafted PCMS for Fuel Cells

Strongly acidic (sulfonic acid) membranes have been identified for use as solid polymer electrolytes in fuel cells [28]. Preparation of these membranes by radiation induced graft copolymerization has been reported and reviewed in various occasions [23,24,27]. Historically, the first radiation grafted sulfonic acid membranes were prepared by Chen et al. [29] through grafting of styrene onto polyethylene (PE) films and used for battery separators and dialysis. However, most of the early work on radiation-grafted membranes was carried out and reviewed by Chapiro [30].

The radiation grafted fuel cell membranes (PCMs) are commonly prepared by grafting of styrene or its substituents onto polymer films followed by sulfonation reaction [27]. Figure 5.1 depicts a schematic representation for preparation of PCMs. Both electrons and γ-rays were used to graft styrene onto polymer films using direct irradiation and preirradiation methods [25,27]. The grafting of styrene and subsequent sulfonation instead of grafting of monomer containing sulfonic acid groups (e.g., sodium styrene sulfonate) was motivated by the very low grafting yield obtained with grafting the latter [31]. Due to the stability required in fuel cells, the starting polymers are confined to fluoropolymer films, which are known for their high thermal and mechanical stabilities together chemical inertness. Fluoropolymers commonly used for preparation of radiation-grafted PCMs are listed in Table 5.1. Substituted styrene monomers and cross-linking agents such as those shown in

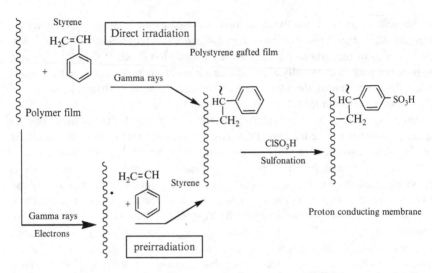

Fig. 5.1 Schematic representation of proton-conducting membrane (PCM) preparation by direct or preirradiation grafting of styrene followed by sulfonation

Table 5.1 List of fluoropolymers commonly used in preparation of radiation-grafted PCMs

Fluoropolymers	Abbreviations	Molecular structure
Poly(vinyl fluoride)	PVF	$-(CF_2-CHF)_n-$
Poly(vinylidene fluoride)	PVDF	$-(CF_2-CH_2)_n-$
Poly(tetrafluoroethylene)	PTFE	$-(CF_2-CF_2)_n-$
Poly(tetrafluoroethylene-co-hexafluoropropylene)	FEP	$-[CF_2-CF_2\text{-}co\text{-}CF_2-CF\,(CF_3)]_n-$
Poly(tetrafluoroethylene-co-perfluorovinyl ether)	PFA	$-[CF_2-CF_2\text{-}co\text{-}CF_2-CF(OC_3F_7)]_n-$
Poly(ethylene-co-tetrafluoroethylene	ETFE	$-[CF_2-CF_2\text{-}co\text{-}CH_2-CH_2]_n-$
Poly(vinylidene fluoride-co-hexafluoropropylene)	PVF-co-HFP	$-[CF_2-CH_2\text{-}co\text{-}CF_2-CF(CF_3)]_n-$

Fig. 5.2 are used to further improve the chemical stability of PCMs. The degree of grafting calculated from the weight increase after grafting is a function of grafting parameters (e.g., monomer concentration, dose, dose rate, solvent, and temperature) and grafting follows the front mechanism, where it starts at the film's surfaces and moves progressively inward with successive monomer diffusion [24,25,27].

The graft copolymer films are subsequently sulfonated to introduce proton-conducting sites. Sulfonation is commonly conducted using strong sulfonating agent such as chlorosulfonic acid diluted with sulfonation resisting solvent (e.g., 1,2-dichloromethane, 1,1,2,2-tetrachloroethane, or carbon tetrachloride) under controlled parameters. Other sulfonating agents such as oleum, sulfonyl chloride, and concentrated sulfuric acid also may be used [28]. Sulfonation of polystyrene grafts

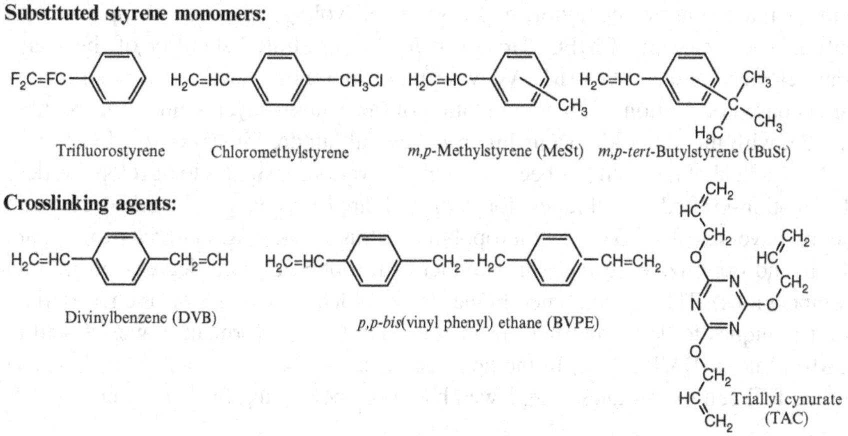

Substituted styrene monomers:

Trifluorostyrene   Chloromethylstyrene   *m,p*-Methylstyrene (MeSt)   *m,p-tert*-Butylstyrene (tBuSt)

Crosslinking agents:

Divinylbenzene (DVB)   *p,p-bis*(vinyl phenyl) ethane (BVPE)   Triallyl cynurate (TAC)

Fig. 5.2 Molecular structures of substituted styrene monomers and cross-linking agents

attached to the film matrix introduces sulfonic acid groups to the aromatic benzene rings mostly in *para*-position; the challenge in this reaction is to obtain 100% degree of sulfonation, where every benzene ring contains one pendant sulfonic acid group, without compromising the physical strength. This issue is addressed by optimizing the reaction conditions. A sulfonation reaction was reported to proceed by the front mechanism, as in the grafting reaction, in which a sulfonation front starts at the surface grafted layers, and then proceeds progressively with the diffusion of more sulfonating agent through the membrane internal layers [32]. To eliminate the influence of side reactions that may form by-products containing sulfonyl chloride or/and a sulfone complex, the membranes are often hydrolyzed with KOH followed by acid treatment, and then washed acid free. The obtained membrane has a structure of polystyrene sulfonic acid grafts bound to the backbone of the polymer film that can be symbolized as film-*g*-PSSA. The properties of these membranes, such as water swelling, ion exchange capacity, and ionic conductivity were found to be functions of the degree of grafting [24,25,27]. The performance of the radiation-grafted membranes was tested in PEMFC and DMFC in various occasions and they proved to be potential alternative materials [6].

Several commercial radiation-grafted PCMs such as Raymion and Permian are available from companies such as Pall Co. (United States) and Solvay (Beligum). The former membranes are prepared by preirradiation grafting of α, β, β-trifluorostyrene onto ETFE films using EB followed by sulfonation reaction [33], whereas the latter are prepared by simultaneous irradiation grafting of styrene onto FEP films or other alternative films such as ETFE using γ-rays followed by sulfonation [34]. The performance of the Raymion membrane is found to be substantially comparable with Nafion 117 membrane in water electrolyzers and is maintained 10,000 h stability [35]. The possibility of using commercially radiated grafted membranes for fuel cell application was explored by Guzman-Garcia et al. [36], and their performance in PEMFC was evaluated by Wang and Capuano [37]. These membranes demonstrated superior behavior, in terms of cell voltage response compared with Nafion when used in PEMFC. However, the demonstrated stability of the membrane electrode assemblies (MEAs) with these membranes was inferior to that of MEAs based on Nafion, with delamination of the catalyst layer from the membrane surface, with the electrode occurring after several hundred hours of use [37].

A great deal of research has been conducted at various institutes to develop a variety of radiation-grafted membranes for fuel cell applications [6,27]. Most research groups have selected specific fluoropolymer films as starting materials to prepare PCMs and optimized the reaction parameters to obtain desired degrees of grafting (compositions). The obtained membranes were subjected to a variety of characterization techniques to determine their properties, and their performances were tested in PEMFC and/or DMFC [25]. In the next sections, the research progress on radiation grafted fuel cell membranes is reviewed based on the starting fluoropolymer films.

5.3.1   Radiation-grafted PCMs Based on PVDF Films

Several research groups have used PVDF films as a base polymer for preparation of PCMs for fuel cell application, with the major contributions made by Sundholm and co-workers [38–40]. This group prepared PVDF-g-PSSA membranes by graft copolymerization of styrene onto preirradiated porous and dense PVDF films followed by sulfonation reaction. Irradiation was performed by a low-energy electron accelerator (175 kV) under N_2 atmosphere, with a total dose up to 200 kGy [38]. The grafted films were sulfonated using chlorosulfonic acid/dichloromethane mixtures and achieved 95–100% degrees of sulfonation. The structure [38,41], the thermal behavior [38,41–43], and the conductivity of the obtained membranes were studied in correlation with the irradiation dose. Conductivity (120 mS cm^{-1}) higher than that of Nafion 117 [38,41,43,44] was reported at room temperature. Structural investigations revealed that styrene grafting was efficiently initiated in the amorphous regions and at the surfaces of the crystallites in the semi-crystalline PVDF backbone [44], leading to high degrees of grafting (in range of 50–86%). Furthermore, polystyrene grafts were found to be formed from both C-H and C-F of PVDF. These findings were supported by intensive structural and morphologic investigations of the membranes by Raman [38], NMR [45], wide-angle X-ray scattering (WAXS), and small-angle X-ray scattering (SAXS) [38,39,45]. Further investigations on structure—property relations showed that the increase in the sulfonic acid content enhances ionic conductivity and water absorbance of these membranes while reducing crystallinity [39,40,46,47]. The overall properties of these membranes were apparently sufficient for PEMFC [48]. However, their performance in PEMFC deteriorated rapidly at a temperature range of 60–70°C because of MEA failure after few hundreds of hours of operation [48,49]. A review of the research of Sundholm and co-workers can be found elsewhere [42].

Similar PVDF-g-PSSA membranes were prepared by Flint and Slade [50] using large-scale EB under atmospheric conditions with a total dose of 150 kGy. The grafting reaction was conducted under reflux in a temperature range of 80–100°C and the subsequent sulfonation of grafted films was conducted using concentrated sulfuric acid (98%).

In their efforts to use various fluoropolymer films to develop PCMs, Scherer and co-workers [51] prepared PVDF-g-PSSA membranes in comparison with their counterparts based on ETFE films. PVDF films were activated from γ-radiation (dose of 20 kGy at dose rate of 5.9 kGy h^{-1}) at room temperature in air and grafting of styrene with peroxidation method interestingly occurred at 60°C. The influence of the base polymer properties on the grafting behavior was addressed [52]. Sulfonation of the grafted films conducted with chlorosulfonic acid/dichloromethane mixture at room temperature. The PEMFC performance of PVDF-based membranes was found to be inferior to their ETFE-based counterparts [52].

Soresi et al. [53] grafted styrene onto PVDF film and its copolymer with hexa-fluropropylene [P(VDF-co-HFP)] and found that membranes based on P(VDF-co-HFP) achieve degree of grafting of 100% with conductivity >60 mS cm^{-1} at 90% relative humidity. Full characterization and fuel cell performance of these membranes was not revealed.

Nasef et al. [54,55] reported the preparation of radiation grafted pore-filled membranes for DMFC by impregnating microporous structures of PVDF films with styrene followed by direct EB irradiation and subsequent sulfonation with chlorosulfonic acid/dichloromethane mixture. Membranes with degrees of grafting in the range of 8–45% were obtained, with those having 40% and 45% PS demonstrating excellent combinations of physicochemical properties compared with Nafion 117. For instance, 45% grafted membrane achieved 61 mS cm^{-1} conductivity (compared with 53 mS cm^{-1} for Nafion 117) and fivefold lower methanol permeability (0.7×10^{-6} cm^2 s^{-1}) than Nafion 117 (3.5×10^{-6}), under the same experimental conditions. The performance of these membranes in DMFC is yet to be addressed.

Scott et al. [56] tested radiation-grafted membranes in DMFC. PVDF-g-PSSA membranes showed superior performance to Nafion under identical conditions at high current density in addition to lower methanol diffusion coefficient by an order of magnitude. However, such membranes suffered performance deterioration after few hundreds of hours due to poor MEA interfacing.

Danks et al. [57,58] prepared radiation-grafted alkaline anion exchange membranes based on PVDF films having the tentative molecular structure shown in Fig. 5.3. The PVDF films were grafted with vinylbenzyl chloride (VBC) followed by amination with trimethylamine and ion-exchanged with aqueous KOH to give benzyltrimethylammonium hydroxide containing (PVDF-g-PVBC) membranes. The obtained membranes were tested in low temperature portable DMFC [57,58]. Previous studies on preparation of radiation-grafted PCMs based on PVDF films and their basic properties are summarized in Table 5.2.

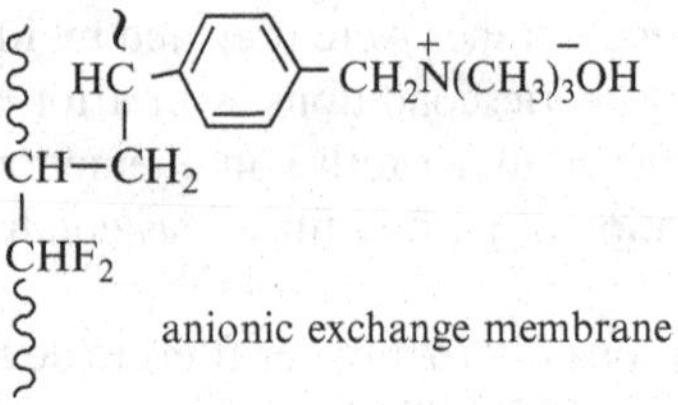

anionic exchange membrane

Fig. 5.3 Tentative molecular structure of PVDF-*g*-PVBC anion exchange membranes

5.3.2 Radiation-grafted PCMs Based on FEP Films

Several research groups have used FEP films as a base polymer for the development of PCMs by radiation-induced grafting, with the Paul Scherer Research Institute (Switzerland) taking the lead since they started their fuel cell program about 15 years ago. The use of FEP has been motivated by its considerably high radiation resistance, efficient radical formation [24]. Scherer and co-workers [59–61] reported the preparation of PCMs by radiation-induced grafting of styrene onto FEP films using direct irradiation method from γ-radiation and subsequent sulfonation. Some of these membranes were cross-linked with DVB [62]. The obtained membranes displayed excellent combinations of physicochemical properties; for instance, specific resistivity values as low as 2 Ω cm (at 20°C) for membranes with a degree of grafting higher than 30%, which is lower than that of Nafion 117 (12 Ω cm). However, their performance in PEMFC was limited to few hundreds of hours, with a power density of 125 mW cm^{-2} at cell voltage of 500 mV under conditions of (H_2 and O_2, 1 atm, 60°C, and 0.8 mg cm^{-2}, Pt loading on electrodes) [59].

Later, this group switched to preirradiation method where they conducted a number of rigorous studies that resulted in identifying important membrane properties for PEMFC application and their correlation with the composition [63,64]. For example, the optimum thickness, cross-linking density, and specific resistance required for fuel cell application were identified [65,66]. Subsequently several modifications aimed at improving membrane properties and interfacing with electrodes in MEA led to significant enhancement in the performance of these membranes in PEMFC [67,68]. Interestingly, membranes based on 25 μm FEP films recorded power densities as high as 500 mW cm^{-2} with an ion exchange capacity of 2.0 mmol g^{-1} [67,68]. These membranes were prepared using an alternative solvent system (2-propanol/water mixture) to toluene, which was conventionally used to dilute styrene during grafting step [69,70]. The new solvent system allowed DVB cross-linked membrane preparations with a grafting level of 20% with an irradiation dose (3 kGy) tenfold lower than that used to prepare them using conventional solvents, i.e., 30 kGy. This led to tremendous improvement in the membrane's mechanical properties (especially percent of elongation), and a MEA stability of 2,500 h was initially reported as a result [71]. These membranes also showed superior properties compared with Nafion 112 membranes and recorded long-term PEMFC

Table 5.2 Summary of previous studies on preparation of radiation-grafted PCMs based on PVDF films and their basic properties

Film nature/thickness (μm)	Monomer/acti-vating agent	Radiation source/ method	Degree of grafting (%)	Ion exchange capacity (mmolg^{-1})	Ionic conductivity (mScm^{-1})	References
Porous PVDF, 125	Styrene Sulfonic acid	EB (175 kV)/ preirradiation	13–275	Variable	Up to130	[38]
Dense PVDF/80	Styrene	EB (175 kV)/ preirradiation	Variable up to 100	Variable	Up to 112.7	[42]
Dense PVDF/80	Styrene Sulfuric acid	EB(1.25 MeV) preirradiation	18–30	0.68–1.70	20–30	[50]
Dense PVDF/80	Styrene/DVB Styrene/ BVPE Sulfonic acid	EB (175 kV)/ preirradiation	18–73 19–103	0.48–2.51 0.19–2.95	<0.1–100 <0.1–68	[39]
Dense PVDF/50	Styrene Sulfonic acid	γ-Radiation/ preirradiation	33	1.51	—	[72,73]
Dense PVDF/80	VBC Triethylamine	γ-Radiation/ preirradiation	Up to 26	0.68–1.70	8–30	[57,58]
Dense PVDF-co-HFP/80	Styrene Sulfonic acid	γ-Radiation/preir- radiation	100	2.51	66	[53]
Porous PVDF/116	Styrene Sulfonic acid	EB (500 kGy)/ direct irradiation	9–45	0.7–2.27	10–61	[54, 55]

(single-cell) performance of more than 7,900 h under operating temperatures of 80–85°C [71]. Membranes based on 75 μm FEP films were found earlier to survive long-term tests in PEMFC for more than 10,000 h at 60°C [67,68].

Other groups, including Horsfall and Lovell [72–75] and Nasef et al. [76,78], developed similar FEP-g-PSSA membranes and established correlations between the various physicochemical properties of the membranes and the degree of grafting, which was controlled by variation of the grafting parameters. The surface properties of FEP-g-PSSA membranes were investigated by XPS analysis [79]. All membrane surfaces were found to be dominated by a hydrocarbon fraction originated from the imparted PSSA grafts [79]. These membranes probably have chemically sensitive surfaces that most likely dictate their interfacial properties and stability.

In their work to develop a polymer electrolyte for DMFC application using alkaline anion exchange membranes, Slade and co-workers [80–82] successfully grafted vinylbenzyl chloride (VBC) onto FEP films using preirradiation method followed by amination with trimethylamine and ion exchanged with aqueous KOH to give benzyltrimethylammonium hydroxide alkaline anion exchange membranes or FEP-g-PVBC with degrees of grafting up to 29%. The ion exchange capacities, the water uptake levels were evaluated and the thermal stability of the membranes was established. These membranes showed stability at 60°C, for at least 120 days. The conductivity values at room temperature achieved satisfactory levels (0.01–0.02 S cm^{-1}). However, these membranes suffered rapid decay in their ion exchange capacity at 100°C and their chemical and thermal stability were also found to be superior compared with PVDF-g-PVBC membranes [80]. Nevertheless, the suitability of these membranes for application in DMFC is found to be below the accepted thermal stability limitation of 60°C for alkaline membrane based on quaternary ammonium functionality [83].

Lappan et al. [84] reported preparation of new FEP-g-PSSA membranes by radiation-induced grafting of styrene onto cross-linked FEP film using the preirradiation method and subsequent sulfonation. The FEP films were cross-linked by irradiation with EB at molten state (290°C). However, the properties of such membranes have not been reported. A summary of various studies on radiation-grafted PCMs membranes based on FEP films and their basic properties is presented in Table 5.3.

5.3.3 Radiation-grafted PCMs Based on ETFE Films

ETFE is one of the interesting polymers that have been frequently used in preparation of radiation-grafted membranes [85]. This is due to its unique combination of properties (toughness, stiffness, high tensile strength, flexural, excellent thermal stability, superior resistance to common solvents, and high resistance to radiation and fatigue) imparted from the presence of alternate fluorocarbon and hydrocarbon units in its molecular structure [86,87].

Table 5.3 Summary of previous studies on preparation of radiation-grafted PCMs based on FEP films and their basic properties

Monomer and activating agent	Base polymer/thickness (μm)	Radiation source/method	Degree of grafting (%)	Ion exchange capacity (mmol g^{-1})	Specific resistivity (Ω cm)	References
Styrene/sulfonic acid	FEP/50	γ-Radiation/simultaneous	13–52	0.72–2.5	22.0–2.0	[60, 61]
Styrene/sulfonic acid	FEP/50	γ-Radiation/simultaneous	6.5–40	0.59–1.26	160–3.2	[5]
Styrene/DVB/sulfonic acid	FEP/125	γ-Radiation/preirradiation	19.4 and 19.6	1.07 and 1.15	36.0 and 10.3	[65]
Styrene/DVB/TAC/sulfonic acid	FEP/75	γ-Radiation/preirradiation	39	1.92	6.2	[66]
Styrene/DVB/sulfonic acid	FEP/50	γ-Radiation/simultaneous	11–34	0.78–2.08	23.0–3.0	[63, 64]
Styrene/DVB/sulfonic acid	FEP/120	γ-Radiation/simultaneous	5–52	0.39–2.27	1190.0–20.0	[77]
VBC/triethylamine	FEP/50	γ-Radiation/preirradiation	3–30	Up to 1.1	Equivalent to 10–20 mS cm^{-1}	[51, 81–83]

A number of research groups have adopted ETFE as a substrate for preparation of radiation-grafted PCMs (ETFE-g-PSSA) using preirradiation method as previously detailed, with most of the work in this area accomplished by Scherer and co-workers [85,88–91] and Horsfall and Lovell [72–75]. These authors not only established various correlations between the membrane's properties and the degree of grafting, but also evaluated their respective fuel cell performances [72–75,85,88–91]. Particularly, Horsfall and Lowell [75] studied the gas diffusion properties of these membranes and found that oxygen diffusion coefficient and permeability increased when decreasing the equivalent weight, while the proton conductivity increased an increase in water content. They also found the performance of ETFE-g-PSSA membranes to be superior to the Nafion [74] in the short term, with some membranes demonstrating stable resistivities at high current density and high power density greater than 1 A cm^{-2} [72]. However, the performance of these membranes started to deteriorate noticeably after 140 h of prolonged PEMFC test.

Scott et al. [56] tested the performance of radiation-grafted ETFE-g-PSSA membranes in comparison with Nafion in DMFC (90°C, 2M MeOH, and air). The ETFE-g-PSSA membranes showed comparable performance values with Nafion 117, especially with low current densities.

Hatanaka et al. [92] prepared and tested the performance of ETFE-g-PSSA membranes in DMFC. These membranes were prepared by preirradiation of ETFE film (120 μm) from γ-radiation followed by styrene grafting and subsequent sulfonation. The membranes with degrees of grafting exceeding 30% displayed better transport properties, i.e., higher ionic conductivity and lower methanol permeation, compared with Nafion membranes. However, the PEM fuel cell performance of these membranes was found to be inferior to that of Nafion. The authors attributed such behavior to the poor bonding of the electrodes to the membrane surface in the MEA.

Similar ETFE-g-PSSA membranes were prepared by Arico et al. [93] with electron irradiation, subsequent grafting, cross-linking, and sulfonation. The obtained membrane samples were of commercial size. The membranes showed good conductivity and low methanol crossover at a thickness of around 150 μm. MEA assemblies based on these membranes showed DMFC performance and cell resistance values comparable to Nafion 117 (210 μm). Stable electrochemical performance was recorded during 1 month of cycled operation at a temperature of 110°C.

Shen et al. [94] compared the DMFC performance of ETFE-g-PSSA membranes with that of PVDF-g-PSSA and PE-g-PSSA counterparts. These membranes were prepared by irradiating ETFE films by γ-irradiation in air to produce peroxy radicals, which were reacted with styrene by thermal decomposition to form polystyrene graft copolymers. The graft copolymers were then sulfonated and hydrolyzed in hot water. The membranes have shown interesting properties by adjusting the degree of grafting and the membrane's thickness. The DMFC performance with such membranes was superior to that of Nafion 117. The optimum thickness and degree of grafting of ETFE-g-PSSA membranes were found to be 68 μm and 27% compared with 125 μm and 17% for PE-g-PSSA and 50 μm and 36% for PVDF-g-PSSA, respectively.

Saarinen et al. [95] prepared new PCMs for DMFC by direct introduction of sulfonic acid groups to ETFE films. This was carried out by irradiation of ETFE films by means of protons followed by sulfonation. These membranes have exceptionally low water uptake, excellent dimensional stability, and 10% methanol permeability lower than Nafion 115. The performance of these membranes was tested in DMFC at 30–85°C, with a maximum power densities of 40–65% lower than the corresponding values of the Nafion 115. Chemical and mechanical stabilities of new ETFE-based membranes appeared to be promising since it was tested over 2,000 h in the DMFC without obvious performance loss.

Chen et al. [96] attempted to improve the chemical stability of ETFE-g-PSSA by grating styrene derivatives such as m,p-methylstyrene (MeSt) and p-$tert$-butylstyrene (tBuSt) of molecular structures (shown in Fig. 5.2) using the preirradiation method. These authors found that the ETFE-g-P(MeSt)SA membranes have high proton conductivity and ETFE-g-P(tBuSt)SA bear high chemical stability. Two individual cross-linkers (DVB and BVPE) and mixture of them were used to further enhance the stability and reduce the gas crossover. These authors also found that the newly obtained membranes have better chemical stability than traditionally DVB cross-linked membranes together with appropriate mechanical strength and thermal properties suitable for fuel cell. The membrane having degree of grafting of 60% showed ion exchange capacity of 2.0 mmol g^{-1} and proton conductivity of 60 mS cm^{-1} at 25°C, a six times lower methanol permeability than that of the Nafion 112 membranes. However, the stability of these membranes in the fuel cell system has yet to be proved. A summary of previous studies on radiation-grafted PCMs membranes based on ETFE films and their basic properties is shown in Table 5.4.

In their efforts to improve the membrane/electrode assembly interfacing characteristics, Scherer and co-worker [91] studied the surface properties of ETFE-g-PSSA membranes in comparison with original and grafted ETFE counterparts. The group reported an increase in total surface energy series in the sequence of: ETFE film < irradiated film < grafted film < sulfonated membrane. The group further enriched their investigation by comparing the performances of the membranes prepared from ETFE, PVDF, and FEP films. Membranes based on ETFE films exhibited better mechanical properties. This was attributed to the higher molecular weight of ETFE, better compatibility of ETFE with PS-grafted components, and the reduced extent of radiation-induced chain scission occurring on ETFE films [85,88–91].

5.3.4 Radiation-grafted PCMs Based on PTFE Films

Among fluorinated polymers, PTFE is the least polymer used for preparation of radiation-grafted membranes despite its extraordinary chemical, thermal, and mechanical stabilities. This is due to PTFE's extreme sensitivity to high-energy radiation, which produces chain scission with a very small irradiation dose [87]. Nevertheless, studies on preparation of fuel cell membranes based on PTFE films were reported in literature [27].

Table 5.4 Summary of previous studies on preparation of radiation-grafted PCMs based on ETFE films and their basic properties

Monomer and activating agent	Base polymer/ thickness (μm)	Radiation source/method	Degree of grafting (%)	Ion exchange capacity (mmol g^{-1})	Specific resistivity (Ω cm)	References
Styrene/DVB/TAC/sulfonic acid	ETFE/25	γ-Radiation/preirradiation	25	2.4	17.0	[88]
Styrene/sulfonic acid	ETFE/50	γ-Radiation/preirradiation	32–46	2.13–3.27	3.3–5.3	[73, 75]
Styrene/DVB/sulfonic acid	ETFE/100	EB/preirradiation	31 and 42	1.49 and 1.88	–	[93]
Styrene/DVB/sulfonic acid	ETFE/variable	EB/preirradiation	Variable	Variable	Up to ~6	[94]
No monomer	ETFE/35	Proton radiation/direct sulfonation	–	1.38	Equivalent to conductivity of 10 mS cm^{-1}	[95]
MeSt/tBuSt/DVB/BVPE/sulfonic acid)	ETFE/50	γ-Radiation/preirradiation	Up to 88	Up to 2.5	Equivalent to 80 mS cm^{-1}	[96]

Nasef et al. [97–103] reported the preparation of PTFE-g-PSSA membranes using the direct irradiation method with relatively low dose range (5–20 kGy), followed by sulfonation with chlorosulfonic acid. The kinetics of the graft copolymerization reactions of these membranes were established. Membranes with grafting degrees up to 36% were obtained with 60% styrene diluted in dichloromethane under N_2 atmosphere and room temperature. The obtained membranes showed a good combination of physicochemical properties and ion conductivities up to 34 mS cm^{-1} [98]. The membranes were further subjected to structural studies [101,102] and intensive evaluation of their thermal [99], chemical, and mechanical stability [97]. These authors also investigated the morphology, surface chemical properties, and chemical composition of these membranes using XPS and SEM analysis [101,103]. Interestingly, Nasef et al. [101,103] reported that these membranes have surfaces with a predominant hydrocarbon fraction, which originates from PSSA side chains despite achieving homogeneous grafting at degrees of grafting of 24% and above. The performances of these membranes were found to stand at few hundred hours in PEMFC operated at 50°C [104].

Similar PTFE-g-PSSA membranes were recently prepared by Liang et al. [105] using the same strategy with slightly different grafting conditions. Degrees of grafting up to ~35% were achieved with an irradiation dose of 20 kGy and styrene concentration of 70%. The obtained membranes showed area resistance up to 48.6 Ω cm^{-2}. However, these membranes started to suffer from substantial oxidative degradation when their chemical stability was tested at 60°C.

Recently, interest was renewed in using PTFE for preparing radiation-grafted PCMs when its radiation resistance was improved by radiation cross-linking at molten temperatures [19–24]. The cross-linked PTFE (RX-PTFE) showed remarkable improvements in radiation resistance, thermal durability, and mechanical properties, compared with those of non—cross-linked PTFE [106–111]. The use of RX-PTFE with its network structure as a base polymer could improve the gas crossover in PEMFC.

Yamaki et al. [112] prepared PCMs by radiation-induced grafting of styrene into the RX-PTFE films using the preirradiation method with γ-radiation and subsequent sulfonation. The degree of grafting was controlled in the range of 5–75% by varying grafting parameters. The resulting membranes showed ion exchange capacity reaching 2.6 mmol g^{-1}, but surprisingly their conductivity was not reported. Later, a few other groups attempted to enhance the stability of radiation-grafted membranes using the same approach, i.e., starting with RX-PTFE films [113–115]. However, these membranes suffered from the deterioration of mechanical properties and higher water uptake, compared with their counterparts, based on non—cross-linked PTFE films.

To improve the stability, Li et al. [116] suggest double cross-linking, i.e., cross-linking polystyrene grafted onto cross-linked RX-PTFE during the grafting reaction to effectively reduce gas crossover and enhance chemical stability. The group prepared RX-PTFE membranes with a thickness around 10 μm by coating the PTFE dispersion on the aluminum sheets and then cross-linking by irradiation above its melting temperature under oxygen-free atmosphere using EB. The RX-PTFE were preirradiated and grafted with a styrene/DVB mixture [116,117]. The obtained membranes properties were investigated by microscope FTIR, WAXS, and solid state [13] C NMR. Moreover, the surface chemical structure and morphology of the obtained membranes were investigated by means of XPS and SEM analysis [118]. However, reports on the chemical stability of these membranes and their perform-ance in fuel cells have not been revealed.

Chen et al. [119] attempted to prepare alternative electrolyte membranes by grafting alkyl vinyl ether monomers such as propyl vinyl ether (nPVE) and isopropyl vinyl ether (iPVE) onto RX-PTFE obtained by EB irradiation at 340°C in argon gas. The grafting reaction was initiated under direct irradiation conditions and in the presence of Lewis acid catalyst (AlCl$_3$) or at a temperature close to the boiling point of each monomer. The grafted RX-PTFE films were subsequently sulfonated by chlorosulfonic acid in dichloromethane. The structure and thermal stability of the membranes were established. Interestingly, these membranes recoded higher proton conductivity than Nafion 112 despite their lower ion exchange capacity (0.75 mmol g^{-1}).

Caro et al. [120] prepared fuel cell membranes by directly introducing sulfonic acid groups to PTFE films by EB irradiation. PTFE films were irradiated in water under atmospheric conditions and subsequently treated with fuming sulfuric acid. Evidence of the presence of sulfur-containing functional groups was presented. However, no details on the basic properties of these membranes and their fuel cell performance were reported. A summary of previous studies on radiation-grafted PCMs based on PTFE films and their properties are presented in Table 5.5.

5.3.5   Radiation-grafted PCMs Based on PFA Films

Despite its interesting properties, PFA has been less frequently used as a base polymer for radiation-grafted PCMs [27]. Nasef and co-workers [121–124] were the first to report detailed investigation into the preparation of PFA-g-PSSA membranes by styrene grafting using direct irradiation followed by sulfonation. The grafting parameters, i.e., the monomer concentration, irradiation dose and dose rate, and type of solvent were established in correlation the degree of grafting [121]. The highest degree of grafting (63%) was achieved upon using dichloromethane to dilute styrene (60 vol%) at a total dose of 30 kGy. The morphology of these membranes was found to play an important role in their chemical degradation [124]. The physicochemical properties, including ion exchange capacity, water uptake, hydration number, and ionic conductivity, were found to be functions of the degree of grafting, just as were both thermal and chemical stabilities [122]. The chemical composition, structure, and mechanical properties (tensile strength and elongation percentages) of these membranes were also characterized by FTIR, X-ray diffraction (XRD), and universal mechanical tester, respectively [121]. Nasef et al. [124] also investigated the surface properties and their chemical composition of these membranes using XPS analysis. Interestingly, PFA-g-PSSA membrane surfaces were found to be predominated by

Table 5.5 Summary of previous studies on preparation of radiation-grafted PCMs based on PTFE films and their basic properties

Film	Monomer/activating agent	Radiation source/method	Degree of grafting (%)	Ion exchange capacity (mmol g^{-1})	Ionic conductivity (mS cm^{-1})	References
PTFE	Styrene/Sulfonic acid	γ-Radiation/direct irradiation	5–36	0.36–2.20	<0.1–29	[97,98]
PTFE	Styrene/Sulfonic acid	γ-Radiation/direct irradiation	1–31	Up to 1.9	Up to 11.8	[105]
RX-PTFE (Cross-linked)	Styrene/Sulfonic acid	γ-Radiation/preirradiat-ion	61	2.6	Not given	[112]
RX-PTFE (Cross-linked)	Styrene/Sulfonic acid	γ-Radiation/preirradiat-ion	33	1.51	Not given	[113]
RX-PTFE (Cross-linked)	nPVE and iPVE/Sulfonic acid	γ-Radiation/direct irradiation	52	0.91	80	[119]

a hydrocarbon fraction originated from PSSA side chains despite achieving bulk grafting. However, the performances of these membranes were found to be limited to a few hundred hours under fuel cell conditions (H_2, O_2, 1 atm, Pt loading of 1.6 mg cm^{-2}) at 50°C [125]. In addition, XPS was used by the authors to monitor degradation of the membranes showing that the oxidation degradation after the fuel cell test is mainly caused by chemical attack at the tertiary H of α-carbon in PS side chains [126]. Membrane stability was improved by cross-linking with DVB during styrene grafting, and the obtained membranes maintained a combination of properties suitable for fuel cell applications [127].

Earlier Nezu et al. [128] reported a PEMFC test with radiation grafted PFA-*g*-PSSA membranes in comparison with counterparts based on PTFE and FEP films. However, no details on the preparation conditions and properties of these membranes were released.

Lappan et al. [115] reported preparation of new membranes by radiation-induced grafting of styrene onto the cross-linked PFA film using the preirradiation method and subsequent sulfonation. PFA was cross-linked by EB irradiation at 350°C under N_2 atmosphere. The properties of such membranes have yet to be disclosed.

5.3.6 Radiation-grafted PCMs Based on Other Fluoropolymer Films

Sundholm and co-workers [128–131] reported the preparation of PCMs by direct sulfonation of poly(vinylfluoride) (PVF) by means of irradiation with heavily charged particles (protons or electrons) prior to sulfonation. The obtained membranes (PVF-*g*-SA) displayed ionic conductivities up to 20 mS cm^{-1}. The state of water [132] and the structure [133] of these membranes were established. The water swelling properties of such membranes was lower than Nafion 117 and 112. Interestingly, PEMFC tests with these membranes showed better performance than Nafion 117 membranes, unlike Nafion 112, which performed better than PVF-*g*-SA membranes [134]. However, these membranes failed after 200 h of testing due to membrane rupture between the membrane active area and the gasket.

Nafion membranes modified by radiation grafting of vinyl phosphoric acid and 2-acrylamido-2-methyl-1-propanesulfonic acid were also tested in DMFC. The modification of Nafion membranes was found to remarkably improve the performance in terms of power density [135].

5.4 Challenges

Various research groups have used the radiation-induced graft copolymerization method to develop PCMs [27]. The obtained radiation grafted PEMs have been subjected to numerous characterization studies, which proved to possess interesting combinations of properties suitable for electrochemical applications in general and fuel cells in particular. In this regard, these membranes showed an excellent combination of properties such as proton conductivities and ion exchange capacities, gas permeabilities, and methanol transport properties compared with Nafion membranes. However, one major problem that has to be circumvented to promote the use of these membranes for practical application is chemical degradation. During the operation of PEM cells, some O_2 resulting from its incomplete reduction, diffuse across the membrane from the cathode to the anode, leading to the formation of HO˙ and HOO˙radicals [122]. These radicals attack the tertiary hydrogen at α-carbon of PSSA chains, shortening membranes lifetime in PEMFC as proved in cases such as FEP-g-PSSA [65], PVDF-g-PSSA [136], and PFA-g-PSSA [125] membranes, which were subjected to post-mortem analysis. However, the lifetimes of radiation-grafted membranes is found to vary significantly depending on their starting polymeric materials, method of grafting, and reaction parameters. This emphasizes the role of the properties of the starting base polymer in affecting the grafting behavior and the need to optimize the reaction parameters in each grafting system to achieve the desired combination of membrane compositions and properties.

To enhance the stability of radiation-grafted PCMs and meet durability demands in fuel cells, research workers used various strategies, which can be summarized as follow:

1. Cross-linking during the grafting reaction by adding small percentage (up to 10–20% of the total volume of the bulk grafting solution) of cross-linking agents (polyfunctional monomers) such as DVB, TAC, a mixture of DVB/TAC, BVPE, a mixture of DVB/BVPE, or N,N-methylene-bis-acrylamide (MBAc) to the grafting mixture [39,47,59,62,63,65,66,69,71,74,88,89,93,117,118,127,137]. However, cross-linking was found to influence membrane properties, such as mechanical properties and water swelling, together with membrane electrode assembly adhesion characteristics. Particularly, radiation-grafted membranes based on fully fluorinated films such as FEP, PTFE, and PFA films develop poor mechanical properties when cross-linked with DVB in a way that causes swelling stresses and membrane ruptures not only because of chemical degradation but also mechanical failure during fuel cell tests [66,71,88,103]. These effects would have a serious impact on the stability of these membranes in larger cells and stacks [69,85]. Such problems were avoided by replacing fully fluorinated polymers (e.g., FEP) with ETFE films, but at the expense of chemical stability, which was compromised [89,92]. Hence, cross-linking has to be carefully optimized to balance the various seemingly contradicting properties of the membranes.
2. Grafting of fluorinated styrene monomers such as α, β, β-trifluorostyrene before proceeding to sulfonation [138]. However, such fluorinated monomers require careful handing and special set-up as well as safety procedures, which eventually introduce a tremendous additional cost to the economy of these membranes.
3. Grafting of substituted styrene monomers such as, α-methylstyrene [139], chlormethylstyrene [140], m,p-methylstyrene, and p-$tert$-butylstyrene prior to sulfonation [119]. However such monomers have poor kinetics and are not easy to graft; also, careful optimization of reaction parameters has to be established.

4. Grafting of a co-monomer (e.g., acrylonitrile) at α-position of the grafted styrene monomer [137]. This approach requires careful optimization of the content of the co-monomer to retain favorable membrane fuel cell properties.
5. Using cross-linked fluorinated polymers such as PTFE [110–119], PFA [120], and FEP [120] as starting films.
6. Using a combined approach of cross-linking (double cross-linking) of fluorinated polymer backbone such as PTFE followed by cross-linking of styrene while grafting prior to sulfonation. The last two approaches (5 and 6) could help in reducing water swelling and gas crossover in fuel cells. However, the thickness and composition of the membranes have to be carefully optimized.

Another challenge associated with the MEA interfacial properties with radiation-grafted PCMs is the poor adhesion between the membrane and the electrodes caused by incompatibility rendered to the difference in the nature of these two components. Such poor interfacial properties caused unstable performance and a lengthy time to reach steady-state when a procedure with Nafion was followed that was similar to that for making MEA. The following possible solutions were suggested: (1) swell the membrane in water, and (2) coat the membrane with the Nafion solution before host pressing [71].

5.5   Future Directions

Radiation-grafted membranes have been known to be suitable for obtaining PCMs for fuel cell application for more than a decade, during which a number of studies have been conducted using various fluoropolymer films; however, few membranes reached the level of stability required for long-term fuel cell performance. Therefore, to promote the development of these membranes, further enhancement in their stability to levels greater than 5,000 h with the current technology is inevitable. One promising approach is the use of cross-linked fluoropolymers such as PTFE and FEP as starting materials while grafting substituted styrene monomers or co-grafting styrene and co-monomer. Some work has already been conducted, but more research is needed to establish the grafting parameters and correlations between composition and properties [110–120].

Developing a method to directly graft monomers containing sulfonic acid groups such as styrene sodium sulfonate (SSS) to the fluorinated backbone is yet another approach that could provide a breakthrough in improving PCM stability. This could be realized by eliminating the aggressive sulfonation reaction that causes deterioration in the crystalline structure of starting polymer films, shortening the membrane preparation time and improving the economy of the membranes. Two successful attempts to graft SSS with acrylic acid, which was grafted first onto PE films followed by SSS grafting [141,142], have been reported recently. However, COOH groups of acrylic acid and PE backbone are not desired in fuel cells because the former do not undergo complete dissociation, causing an increase in membrane resistance, whereas the latter can easily perceive chemical degradation;

therefore, more research is needed to extend SSS grafting to cross-linked and non—cross-linked fluoropolymer films.

Another promising approach is to develop a new generation of radiation-grafted membranes having sulfonic acid groups directly attached to the fluorinated polymer backbone without grafting an aromatic ring (polystyrene) host. Some attempts have been reported [95,129,130,143], but more research is needed to establish preparation procedures, the properties of the membranes, and fuel cell performance characteristics.

To enhance the use of radiation-grafted membranes for fuel cell application, more research is needed to improve the interfacial properties of MEA caused by poor adhesion between the membrane and the electrodes that eventually causes delamination during fuel cell operation. Since, radiation-grafted membranes have been developed by adopting various initial fluoropolymer films under slightly different grafting and sulfonation conditions, it is essential to properly characterize membrane surface properties to help optimize MEA-making conditions to improve fuel cell performance with each membrane. In addition, the durability of radiation-grafted membranes under dynamic fuel cell operation conditions have to be further addressed, focusing on the common mechanism of membrane degradation and failure modes.

5.6 Concluding Remarks

Radiation-induced graft copolymerization is an attractive method for preparation of PCMs for fuel cell applications. The PCMs are commonly prepared by grafting styrene or its substituents onto fluoropolymer films followed by sulfonation. These membranes have been found to possess excellent combinations of physicochemical and thermal properties that meet the requirements of fuel cells. However, chemical stability remains the main challenge, precluding the implementation of such membranes in commercial fuel cell systems despite many successful laboratory tests in the short and medium terms. Various strategies have been proposed to boost stability, including cross-linking the grafted moiety, grafting substituted styrene monomers, using cross-linked fluoropolymer films, double cross-linking (the starting film and incorporated grafts), and direct sulfonation of irradiated fluoropolymer backbone without a monomer host. However, more substantial efforts have to be undertaken to further improve the quality and lifetime of the membranes obtained from different starting materials, together with their interfacial adhesion properties with the electrodes in MEAs.

References

1. V. Mehta and J.S. Cooper, Review and analysis of PEM fuel cell design and manufacturing, *J. Power Sources* **114**, 32 (2003).
2. J. Kerres, Development of ionomer membranes for fuel cells, *J. Membr. Sci.* **185**, 3 (2001).

3. Y.S. Kim, L. Dong, M.A. Hickner, B.S. Pivovar and J.E. McGrath, Processing induced morphological development in hydrated sulfonated poly(arylene ether sulfone) copolymer membranes, *Polymer* **44**, 5729 (2003).

4. O. Savadogo, Emerging membranes for electrochemical systems: (1) Solid polymer electrolyte membranes for fuel cell systems, *J. New Mater. Electrochem. Syst.* **1**, 47 (1998).

5. B. Smitha, S. Sridhar and A.A. Khan, Solid polymer electrolyte membranes for fuel cell applications — a review, *J. Membr. Sci.* **259**, 10 (2005).

6. S. Renaud and A. Bruno, Functional fluoropolymers for fuel cell membranes, *Prog. Polym. Sci.* **30**, 644 (2005).

7. E.B. Easton, B.L. Langsdorf, J.A. Hughes, J. Sultan, Z. Qi, A. Kaufman and P.G. Pickup, Characteristics of polypyrrole/nafion composite membranes in a direct methanol fuel cell, *J. Electrochem. Soc.* **150**, C735 (2003).

8. N. Jia, M.C. Lefebvre, J. Halfyard, Z. Qi and P.G. Pickup, Modification of Nafion proton exchange membranes to reduce methanol crossover in PEM fuel cells, *Electrochem. Solid-State Lett.* **3**, 529 (2000).

9. L.J. Hobson, Y. Nakano, H. Ozu and S. Hayase, Targeting improved DMFC performance, *J. Power Sources* **104**, 79 (2002).

10. M. Walker, K.M. Baumgärtner, J. Feichtinger, M. Kaiser, E. Räuchle and J. Kerres, Barrier properties of plasma-polymerized thin films, *Surf. Coat. Technol.* **116/119**, 996 (1999).

11. N. Carretta, V. Tricoli and F. Picchioni, Ionomeric membranes based on partially sulfonated poly(styrene): Synthesis, proton conduction, and methanol permeation, *J. Membr. Sci.* **166**, 189 (2000).

12. Q. Guo, P.N. Pintauro, H. Tang and S. O'Connor, Sulfonated and cross-linked polyphosphazene-based proton-exchange membranes, *J. Membr. Sci.* **154**, 175 (1999).

13. K. Ramya, B. Vishnupriya and K.S. Dhathathreyan, Methanol permeability studies on sulphonated polyphenylene oxide membrane for direct methanol fuel cell, *J. New Mater. Electrochem. Syst.* **4**, 115 (2001).

14. F. Lufrano, I. Gatto, P. Staiti, V. Antonucci and E. Passalacqua, Sulfonated polysulfone ionomer membranes for fuel cells, *Solid State Ionics* **145**, 47 (2001).

15. F. Wang, M. Hickner, Y.S. Kim, T.A. Zawodzinski and J.E. McGrath, Direct polymerization of sulfonated poly(arylene ether sulfone) random (statistical) copolymers: Candidates for new proton exchange membranes, *J. Membr. Sci.* **197**, 231 (2002).

16. C. Manea and M. Mulder, Characterization of polymer blends of polyethersulfone/sulfonated polysulfone and polyethersulfone/sulfonated polyetheretherketone for direct methanol fuel cell applications, *J. Membr. Sci.* **206**, 443 (2002).

17. A. Schechter and R.F. Savinell, Imidazole and 1-methyl imidazole in phosphoric acid doped polybenzimidazole, electrolyte for fuel cells, *Solid State Ionics* **147**, 181 (2002).

18. Y. Woo, S.Y. Oh, Y.S. Kang and B. Jung, Synthesis and characterization of sulfonated polyimide membranes for direct methanol fuel cell, *J. Membr. Sci.* **220**, 31 (2003).

19. G.K.S. Prakash, M.C. Smart, Q.J. Wang, A. Atti, V. Pleynet, B. Yang, K. McGrath, G.A. Olah, S.R. Narayanan, W. Chun, T. Valdez and S. Surampudi, High efficiency direct methanol fuel cell based on poly(styrenesulfonic) acid (PSSA)—poly(vinylidene fluoride) (PVDF) composite membranes, *J. Fluor. Chem.* **125**, 1217 (2004).

20. X. Qiu, W. Li, S. Zhang, H. Liang and W. Zhu, The microstructure and character of the PVDF-g-PSSA membrane prepared by solution grafting, *J. Electrochem. Soc.* **150**, A917 (2003).

21. F. Finsterwalder and G. Hambitzer, Proton conductive thin films prepared by plasma polymerization, *J. Membr. Sci.* **185**, 105 (2001).

22. J.P. Shin, B.J. Chang, J.H. Kim, S.B. Lee and D.H. Suh, Sulfonated polystyrene/PTFE composite membranes, *J. Membr. Sci.* **251**, 247 (2005).

23. M.M. Nasef, H. Saidi and H.M. Nor, Radiation-induced graft copolymerization for preparation of cation exchange membranes: A review, *Nucl. Sci. J. Malaysia* **17**, 27 (1999).

24. B. Gupta and G.G. Scherer, Proton exchange membranes by radiation-induced graft copolymerization of monomers into Teflon-FEP films, *Chimia* **48**, 127 (1994).

25. T. Dargaville, G. George, D. Hill and A. Whittaker, High energy radiation grafting of fluoropolymers, *Prog. Polym. Sci.* **28**, 1355 (2003).
26. G. Ellinghorst, A. Niemoeller and D. Vierkotten, Radiation-initiated grafting of polymer films – an alternative technique to prepare membranes for various separation problems, *Radiat. Phys. Chem.* **22**, 635 (1983).
27. M.M. Nasef and E.A. Hegazy, Preparation and applications of ion exchange membranes by radiation-induced graft copolymerization of polar monomers onto non-polar films, *Prog. Polym. Sci.* **29**, 499 (2004).
28. M. Rikukawa and K. Sanui, Proton-conducting polymer electrolyte membranes based on hydrocarbon polymers, *Prog. Polym. Sci.* **25**, 1463 (2000).
29. W.W. Chen, R. Mesrobian, D. Ballantine, D. Metz and A. Glines, Graft copolymers derived by ionizing radiation, *J. Polym. Sci.* **23**, 903 (1957).
30. A. Chapiro, Radiation induced grafting, *Int. J. Radiat. Phys. Chem.* **9**, 55 (1977).
31. S. Shkolink and D. Behar, Radiation-induced grafting of sulfonates on polyethylene, *J. Appl. Polym. Sci.* **27**, 2189 (1982).
32. N. Walsby, M. Paronen, J. Juhanoja and F. Sundholm, Sulfonation of styrene-grafted poly(vinylidene fluoride) films, *J. Appl. Polym. Sci.* **81**, 1572 (2001).
33. T. Momose, H. Harada, H. Miyachi and H. Kato, *US Patent* 4,605, 685 (1986).
34. V.F. D'Agostino, J.Y. Lee and E.H. Cook, Jr., *US Patent* 4,012, 303 (1977).
35. G.G. Scherer, T. Momose and K. Tomiie, Membrane-water electrolysis cells with a fluorinated cation exchange membranes, *J. Electrochem. Soc.* **135**, 3071 (1988).
36. A. Guzman-Garcia, P. Pintauro, M. Verbrugge and E. Schneider, Analysis of radiation-grafted membranes for fuel cell electrolytes, *J. Appl. Electrochem.* **22**, 204 (1992).
37. H. Wang and G. Capuano, Behavior of Raipore radiation-grafted polymer membranes in H_2/O_2 fuel cells, *J. Electrochem. Soc.* **145**, 780 (1998).
38. S. Holmberg, T. Lehtinen, J. Naesman, D. Ostrovskii, M. Paronen, R. Serimaa, F. Sundholm, G. Sundholm, L. Torell and M. Torkkeli, Structure and properties of sulfonated poly[(vinylidene fluoride)-g-styrene] porous membranes, *J. Mater. Chem.* **6**, 1309 (1996).
39. S. Holmberg, J. Nasman and F. Sundholm, Synthesis and properties of sulfonated and cross-linked poly[(vinylidene fluoride)-*graft*-styrene] membranes, *Polym. Adv. Tech.* **9**, 121 (1998).
40. N. Walsby, M. Paronen, J. Juhanoja and F. Sundholm, Radiation grafting of styrene onto poly(vinylidene fluoride) films in propanol: The influence of solvent and synthesis conditions, *J. Polym. Sci. A: Polym. Chem.* **38**, 1512 (2000).
41. S. Hietala, S. Holmberg, M. Karjalainen, J. Näsman, M. Paronen, R. Serimaa, F. Sundholm and S. Vahvaselkä, Structural investigation of radiation grafted and sulfonated poly(vinylidene fluoride), PVDF, membranes, *J. Mater. Chem.* **7**, 721 (1997).
42. J. Ennari, S. Hietala, M. Paronen, F. Sundholm, N. Walsby, M. Karjalainen, R. Serimaa, T. Lehtinen and G. Sundholm, New polymer electrolyte membranes for low temperature fuel cells, *Macromol. Symp.* **146**, 41 (1999).
43. S. Hietala, M. Koel, E. Skou, M. Elomaa and F. Sundholm, Thermal stability of styrene grafted and sulfonated proton conducting membranes based on poly(vinylidene fluoride), *J. Mater. Chem.* **8**, 1127 (1998).
44. S. Hietala, M. Paronen, S. Holmberg, J. Näsman, J. Juhanoja, M. Karjalainen, R. Serimaa, M. Toivola, T. Lehtinen, K. Parovuori, G. Sundholm, H. Ericson, B. Mattsson, L. Torell and F. Sundholm, Phase separation and crystallinity in proton conducting membranes of styrene grafted and sulfonated poly (vinylidene fluoride), *J. Polym. Sci. A: Polym. Chem.* **37**, 1741 (1999).
45. S. Hietala, S.L. Maunu, F. Sundholm, T. Lehtinen and G. Sundholm, Water sorption and diffusion coefficients of protons and water in PVDF-*g*-PSSA polymer electrolyte membranes, *J. Polym. Sci. B: Polym. Phys.* **37**, 2893 (1999).
46. K. Jokela, R. Serimaa, M. Torkkeli, F. Sundholm, T. Kallio and G. Sundholm, Effect of the initial matrix material on the structure of radiation-grafted ion-exchange membranes:

Wide-angle and small-angle X-ray scattering studies, *J. Polym. Sci. B: Polym. Phys.* **40**, 1539 (2002).

47. M. Elomaa, S. Hietala, M. Paronen, N. Walsby, K. Jokela, R. Serimaa, M. Torkkeli, T. Lehtinen, G. Sundholm and F. Sundholm, The state of water and the nature of ion clusters in cross-linked proton conducting membranes of styrene grafted and sulfonated poly(vinylidene fluoride), *J. Mater. Chem.* **10**, 2678 (2000).

48. T. Kallio, K. Jokela, H. Ericson, R. Serimaa, G. Sundholm, P. Jacobsson and F. Sundholm, Effects of a fuel cell test on the structure of irradiation grafted ion exchange membranes based on different fluoropolymers, *J. Appl. Electrochem.* **33**, 505 (2003).

49. P. Gode, J. Ihonen, A. Strandroth, H. Ericson, G. Lindbergh, M. Paronen, F. Sundholm, G. Sundholm and N. Walsby, Membrane durability in a proton exchange membrane fuel cell studied using PVDF based radiation-grafted membranes, *Fuel Cells* **3**, 21 (2003).

50. S. Flint and R. Slade, Investigation of radiation-grafted PVDF-g-polystyrene-sulfonic-acid ion exchange membranes for use in hydrogen oxygen fuel cells, *Solid State Ionics* **97**, 299 (1997).

51. H.P. Brack and G.G. Scherer, Modification and characterization of thin polymer films for electrochemical applications, *Macromol. Symp.* **126**, 25 (1998).

52. H.P. Brack, H.G. Bührer, L. Bonorand and G.G. Scherer, Grafting of preirradiated poly(ethylene-*alt*-tetrafluoroethylene) films with styrene: Influence of base polymer film properties and processing parameters, *J. Mater. Chem.* **10**, 1795 (2000).

53. B. Soresi, E. Quartarone, P. Mustarelli, A. Magistris and G. Chiodelli, PVDF and P(VDF-HFP)-based proton exchange membranes, *J. Membr. Sci.* **166**, 383 (2004).

54. M.M. Nasef, N.A. Zubir, A.F. Ismail, K.Z.M. Dahlan, H. Saidi and M. Khayet, Preparation of radiochemically pore-filled polymer electrolyte membranes for direct methanol fuel cell, *J. Power Sources* **156**, 200 (2006).

55. M.M. Nasef, N.A. Zubir, A.F. Ismail, K.Z.M. Dahlan, H. Saidi and M. Khayet, PSSA pore-filled PVDF membranes by simultaneous electron beam irradiation: Preparation and transport characteristics of protons and methanol, *J. Membr. Sci.* **268**, 96 (2006).

56. K. Scott, W. Taama and P. Argyropoulos, Performance of the direct methanol fuel cell with radiation-grafted polymer membranes, *J. Membr. Sci.* **171**, 119 (2000).

57. T.N. Danks, R.C.T. Slade and J.R. Varcoe, Comparison of PVDF- and FEP-based radiation-grafted alkaline anion-exchange membranes for use in low temperature portable DMFCs, *J. Mater. Chem.* **12**, 3371 (2002).

58. T.N. Danks, R.C.T. Slade and J.R. Varcoe, Alkaline anion-exchange radiation-grafted membranes for possible electrochemical application in fuel cells, *J. Mater. Chem.* **13**, 712 (2003).

59. F. Büchi, B. Gupta, M. Rouilly, C. Hauser, A. Chapiro and G.G. Scherer, *27th IECEC Conference Proceedings*, Vol. 3, 1999, pp. 3419–3424.

60. M. Rouilly, E. Koetz, O. Haas, G.G. Scherer and A. Chapiro, Proton exchange membranes prepared by simultaneous radiation grafting of styrene onto Teflon-FEP films-synthesis and characterization, *J. Membr. Sci.* **81**, 89 (1993).

61. B. Gupta, F. Büchi, G.G. Scherer and A. Chapiro, Materials research aspects of organic solid proton conductors, *Solid State Ionics* **61**, 213 (1993).

62. B. Gupta, F. Büchi, G.G. Scherer and A. Chapiro, Cross-linked ion exchange membranes by radiation grafting of styrene/divinylbenzene into FEP films, *J. Membr. Sci.* **118**, 231 (1996).

63. B. Gupta, F. Büchi and G.G. Scherer, Cation exchange membranes by pre-irradiation grafting of styrene into FEP films. I. Influence of synthesis conditions, *J. Polym. Sci. A: Polym. Chem.* **32**, 1931 (1994).

64. B. Gupta, F. Büchi, M. Staub, D. Grman and G.G. Scherer, Cation exchange membranes by pre-irradiation grafting of styrene into FEP films. II. properties of copolymer membranes, *J. Polym. Sci. A: Polym. Chem.* **34**, 1873 (1996).

65. F.N. Büchi, B. Gupta, O. Haas and G.G. Scherer, Study of radiation-grafting FEP-g-polystyrene membrane as polymer electrolyte in fuel cells, *Electrochem. Acta* **40**, 345 (1995).

66. F.N. Büchi, B. Gupta, O. Haas and G.G. Scherer, Performance of differently cross-Linked, partially fluoronated proton exchange membranes in polymer electrolyte fuel cells, *Electrochem. Soc.* **142**, 3044 (1995).

67. H.P. Brack, F.N. Büchi, J. Husalge and G.G. Scherer, in: S. Gottesfeld, T.F. Fuller and G. Hapert (Eds.), *2nd Symposium on Polymer Electrolyte Fuel Cells, ESC Proceedings* Vol. 98–27, 1999.
68. J. Huslage, T. Rager, B. Schnyder and A. Tsukada, Radiation-grafted membrane/electrode assemblies with improved interface, *Electrochem. Acta* **48**, 247 (2002).
69. J. Huslage, T. Rager, J. Kiefer, L. Steuernagel and G.G. Scherer, *Electrochemical Society Meeting*, Toronto, Canada, 2000.
70. T. Rager, Pre-irradiation of styrene/divinylbenzene onto poly(tetrafluoroethylene-*co*-hexafluoropropylene) from non-solvent, *Helv. Chim. Acta* **86**, 1966 (2003).
71. L. Gubler, H. Kuhn, T.J. Schmidt, G.G. Scherer, H.P. Brack and K. Simbeck, Performance and durability of membrane electrode assemblies based on radiation-grafted FEP-g-polystyrene membranes, *Fuel Cells* **4**, 196 (2004).
72. J.A. Horsfall and K.V. Lovell, Fuel cell performance of radiation-grafted sulfonic acid membranes, *Fuel Cells* **1**, 186 (2001).
73. J.A. Horsfall and K.V. Lovell, Comparison of fuel cell performance of selected fluoropolymer and hydrocarbon based grafted copolymers incorporating acrylic acid and styrene sulfonic acid, *Polym. Adv. Technol.* **13**, 381 (2002).
74. J.A. Horsfall and K.V. Lovell, Proton exchange membranes by irradiation-induced grafting of styrene onto FEP and ETFE: Influences of the cross-linker *N,N*-methylene-bis-acrylamide, *Eur. Polym. J.* **38**, 1671 (2002).
75. C. Chuy, V.I. Basura, E. Simon, S. Holdcroft, J.A. Horsfall and K.V. Lovell, Electrochemical characterization of ethylenetetrafluoroethylene-*g*-polystyrenesulfonic acid solid polymer electrolytes, *J. Electrochem. Soc.* **147**, 4453 (2000).
76. M.M. Nasef, H. Saidi and H.M. Nor, Proton exchange membranes prepared by simultaneous radiation grafting of styrene onto FEP films.I. efffect of grafting conditions, *J. Appl. Polym. Sci.* **76**, 220 (2000).
77. M.M. Nasef, H. Saidi, H.M. Nor and M.F. Ooi, Proton exchange membranes prepared by simultaneous radiation grafting of styrene onto FEP films.II. Properties of the sulfonated membranes, *J. Appl. Polym. Sci.* **78**, 2443 (2000).
78. M.M. Nasef and H. Saidi, Thermal degradation behavior of radiation grafted FEP-*g*-polystyerne sulfonic acid membranes, *Polym. Degrad. Stab.* **70**, 497 (2000).
79. M.M. Nasef, H. Saidi and M.A. Yarmo, Surface investigations of radiation grafted FEP-*g*-polystyrene sulfonic acid membranes using XPS, *J. New Mater. Electrochem. Syst.* **3**, 309 (2000).
80. T.N. Danks, R.C.T. Slade and J.R. Varcoe, Comparison of PVDF- and FEP-based radiation-grafted alkaline anion-exchange membranes for use in low temperature portable DMFCs, *J. Mater. Chem.* **12**, 3371 (2002).
81. T.N. Danks, R.C.T. Slade and J.R. Varcoe, Alkaline anion-exchange radiation-grafted membranes for possible electrochemical application in fuel cells, *J. Mater. Chem.* **13**, 712 (2003).
82. H. Herman, R.C.T. Slade and J.R. Varcoe, The radiation-grafting of vinylbenzyl chloride onto poly(hexafluoropropylene-*co*-tetrafluoroethylene) films with subsequent conversion to alkaline anion-exchange membranes: Optimization of the experimental conditions and characterization, *J. Membr. Sci.* **147**, 218 (2003).
83. R.C.T. Slade and J.R. Varcoe, Investigations of conductivity in FEP-based radiation-grafted alkaline anion-exchange membranes, *Solid State Ionics* **176**, 585 (2005).
84. U. Lappan, U. Geißler, U. Scheler and K. Lunkwitz, Identification of new chemical structures in poly(tetrafluoroethylene-co-perfluoropropyl vinyl ether) irradiated in vacuum at different temperatures, *Radiat. Phys. Chem.* **67**, 447 (2003).
85. H.P. Brack and G.G. Scherer, Grafting of preirradiated poly(ethylene-*alt*-tetrafluoroethylene) films with styrene: Influence of base polymer film properties and processing parameters, *J. Mater. Chem.* **10**, 1795 (2000).
86. M.M. Nasef, H. Saidi and K.M. Dahlan, Electron beam irradiation effects on ethylene-tetrafluoroethylene copolymer films, *Radiat. Phys. Chem.* **68**, 875 (2003).
87. J. Forsythe and D. Hill, The radiation chemistry of fluoropolymers, *Prog. Polym. Sci.* **25**, 101 (2000).

88. H.P. Brack, F.N. Büchi, M. Rota and G.G. Scherer, Development of radiation-grafted membranes for fuel cell applications based on poly(ethylene-alt-tetrafluoroethylene), *Polym. Mater. Sci. Eng.* **77**, 368 (1997).

89. H.P. Brack, L. Bonorand, H.G. Buhrer and G.G. Scherer, Radiation grafting of ETFE and FEP films: Base polymer film effects, *Polym. Prepr. (Am. Chem. Soc. Div. Polym. Chem.)* **39**, 976 (1998).

90. H.P. Brack, L. Bonorand, H.G. Buhrer and G.G. Scherer, Radiation processing of fluoropolymer films, *Polym. Prepr. (Am. Chem. Soc., Div. Polym. Chem.)* **39**, 897 (1998).

91. H.P. Brack, M. Wyler, G. Peter and G.G. Scherer, A contact angle investigation of the surface properties of selected proton-conducting radiation-grafted membranes, *J. Membr. Sci.* **214**, 1 (2003).

92. T. Hatanka, N. Hasegawa, A. Kamiya, M. Kawasumi, Y. Morimoto and K. Kawahara, Cell performances of direct methanol fuel cells with grafted membranes, *Fuel* **81**, 2173 (2002).

93. A.S. Aricó, V. Baglio, P. Creti, A. Di Blasi, V. Antonucci, J. Brunea, A. Chapotot, A. Bozzi and J. Schoemans, Investigation of grafted ETFE-based polymer membranes as alternative electrolyte for direct methanol fuel cells, *J. Power Sources* **123**, 107 (2003).

94. M. Shen, S. Roy, J.W. Kuhlmann, K. Scott, K. Lovell and J.A. Horsfall, Grafted polymer electrolyte membrane for direct methanol fuel cells, *J. Membr. Sci.* **251**, 121 (2005).

95. V. Saarinen, T. Kallio, M. Paronen, P. Tikkanen, E. Rauhala and K. Kontturi, New ETFE-based membrane for direct methanol fuel cell, *Electrochem. Acta* **50**, 3453 (2005).

96. J. Chen, M. Asano, T. Yamaki and M. Yoshida, Preparation and characterization of chemically stable polymer electrolyte membranes by radiation-induced graft copolymerization of four monomers into ETFE films, *J. Membr. Sci.* **269**, 194 (2006).

97. M.M. Nasef, H. Saidi, A.M. Dessouki and E.M. El-Nesr, Radiation-induced grafting of styrene onto poly(tetrafluoroethylene) (PTFE) films. I. Effect of grafting conditions and properties of the grafted films, *Polym. Int.* **49**, 399 (2000).

98. M.M. Nasef, H. Saidi, H.M. Nor and O.M. Foo, Radiation-induced grafting of styrene onto poly(tetrafluoroethylene) films. II. Properties of the grafted and sulfonated membranes, *Polym. Int.* **49**, 1572 (2000).

99. M.M. Nasef, Thermal stability of radiation grafted PTFE-*g*-polystyrene sulfonic acid membranes, *Polym. Degrad. Stabil.* **68**, 231 (2000).

100. M.M. Nasef, H. Saidi, H.M. Nor and M.A. Yarmo, XPS studies of radiation grafted PTFE-*g*-polystyrene sulfonic acid membranes, *J. Appl. Polym. Sci.* **76**, 336 (2000).

101. M.M. Nasef, Structural investigation of poly(ethylene trephthalate)-*graft*-polystyrene copolymer films, *Eur. Polym. J.* **38**, 87 (2002).

102. M.M. Nasef and H. Saidi, Structure-property relationships in radiation grafted poly(tetrafluoroethylene)-*graft*-polystyrene sulfonic acid membranes, *J. Polym. Res.* **12**, 305 (2005).

103. M.M. Nasef and H. Saidi, Surface studies of sulfonic acid radiation-grafted membranes: XPS and SEM analysis, *J. Appl. Surf. Sci.* **252**, 3073 (2006).

104. M.M. Nasef, *Ph.D. Thesis*, University of Technology, Malaysia 1999.

105. G.Z. Liang, T.L. Lu, X.Y. Ma, H.X. Yan and Z.H. Gong, Synthesis and characteristics of radiation-grafted membranes for fuel cell electrolytes, *Polym. Int.* **52**, 1300 (2003).

106. A. Oshima, Y. Tabata, H. Kudoh and T. Seguchi, Radiation induced cross-linking of polytetrafluoroethylene, *Radiat. Phys. Chem.* **45**, 269 (1995).

107. Y. Tabata, A. Oshima, K. Takashika and T. Seguchi, Temperature effects on radiation induced phenomena in polymers, *Radiat. Phys. Chem.* **48**, 563 (1996).

108. A. Oshima, S. Ikeda, H. Kudoh, T. Seguchi and Y. Tabata, Temperature effects on radiation induced phenomena in polytetrafluoroetylene (PTFE)-Change of *G*-value, *Radiat. Phys. Chem.* **50**, 611 (1997).

109. Y. Tabata and A. Oshima, ESR study on free radicals trapped in cross-linked polytetrafluoroethylene (PTFE), *Macromol. Symp.* **143**, 337 (1999).

110. A. Oshima, S. Ikeda, E. Katoh and Y. Tabata, Chemical structure and physical properties of radiation-induced cross-linking of polytetrafluoroethylene, *Radiat. Phys. Chem.* **62**, 39 (2001).

111. U. Lappan, U. Geißler, L. Häußler, D. Jehnichen, G. Pompe and K. Lunkwitz, Radiation-induced branching and cross-linking of poly(tetrafluoroethylene) (PTFE), *Nucl. Instrum. Meth. Phys. Res. B* **185**, 178 (2001).

112. T. Yamaki, M. Asano, Y. Maekawa, Y. Morita, T. Suwa, J. Chen, N. Tsubokawa, K. Kobayashi, H. Kubota and M. Yoshida, Radiation grafting of styrene into cross-linked PTFE films and its sulfonation for fuel cell applications, *Radiat. Phys. Chem.* **67**, 403 (2003).

113. K. Sato, S. Ikeda, M. Iida, A. Oshima, Y. Tabata and M. Washio, Study on poly-electrolyte membrane of cross-linked PTFE by radiation grafting, *Nucl. Instrum. Meth. Phys. Res. B* **208**, 424 (2003).

114. T. Yamaki, K. Kobayashi, M. Asano, H. Kubota and M. Yoshida, Preparation of proton exchange membranes based on cross-linked polytetrafluoroethylene for fuel cell applications, *Polymer* **45**, 6569 (2004).

115. U. Lappan, U. Geißler and S. Uhlmann, Radiation-induced grafting of styrene into radiation-modified fluoropolymer films, *Nucl. Instrum. Meth. Phys. Res. B* **236**, 413 (2005).

116. J.Y. Li, K. Sato, S. Ichizuri, S. Asano, S. Ikeda, M. Iida, A. Oshima, Y. Tabata and M. Washio, Pre-irradiation induced grafting of styrene into cross-linked and non-cross-linked polytetrafluoroethylene films for polymer electrolyte fuel cell applications. I: Influence of styrene grafting conditions, *Eur. Polym. J.* **40**, 775 (2004).

117. J.Y. Li, S. Ichizuri, S. Asano, F. Mutou, S. Ikeda, M. Iida, A. Oshima, Y. Tabata and M. Washio, Pre-irradiation induced grafting of styrene into cross-linked and non-cross-linked poly(tetrafluoroethylene) films for polymer electrolyte fuel cell applications. II: Characterization of the styrene grafted films, *Eur. Polym. J.* **41**, 547 (2005).

118. J.Y. Li, S. Ichizuri, S. Asano, F. Mutou, S. Ikeda, M. Iida, T. Miura, A. Oshima, Y. Tabata and M. Washio, Surface analysis of the proton exchange membranes prepared by pre-irradiation induced grafting of styrene/divinylbenzene into cross-linked thin PTFE membranes, *Appl. Surf. Sci.* **245**, 260 (2005).

119. J. Chen, M. Asano, T. Yamaki and M. Yoshida, Preparation of sulfonated cross-linked PTFE-*graft*-poly(alkyl vinyl ether) membranes for polymer electrolyte membrane fuel cells by radiation processing, *J. Membr. Sci.* **256**, 38 (2005).

120. J.C. Caro, U. Lappan and K. Lunkwitz, Sulfonation of fluoropolymers induced by electron beam irradiation, *Nucl. Instrum. Meth. Phys. Res. B* **151**, 181 (1999).

121. M.M. Nasef, H. Saidi, M.H. Nor and K.Z.M. Dahlan, Cation exchange membranes by radiation-induced graft copolymerization of styrene onto PFA copolymer films. I. Preparation and characterization of the graft copolymer, *J. Appl. Polym. Sci.* **73**, 2095 (1999).

122. M.M. Nasef, H. Saidi, H.M. Nor and O.M. Foo, Cation exchange membranes by radiation-induced graft copolymerization of styrene onto PFA copolymer films. II. Characterization of sulfonated graft copolymer membranes, *J. Appl. Polym. Sci.* **76**, 1 (2000).

123. M.M. Nasef, H. Saidi and H.M. Nor, Cation exchange membranes by radiation-induced graft copolymerization of styrene onto PFA copolymer films, *J. Appl. Polym. Sci.* **77**, 1877 (2000).

124. M.M. Nasef, H. Saidi and M.A. Yarmo, Cation exchange membranes by radiation-induced graft copolymerization of styrene onto PFA copolymer films. IV. Morphological investigations using X-ray photoelectron spectroscopy, *J. Appl. Polym. Sci.* **77**, 2455 (2000).

125. M.M. Nasef and H. Saidi, Post-mortem analysis of radiation grafted fuel cell membrane using X-ray photoelectron spectroscopy, *J. New. Mater. Electrochem. Syst.* **5**, 183 (2002).

126. M.M. Nasef and H. Saidi, *International Exhibition on Ideas, Innovation and New Products (IENA2004)*, 28–31 October 2004, Nuremberg, Germany.

127. M.M. Nasef and H. Saidi, Preparation of cross-linked cation exchange membranes by radiation grafting of styrene/divinylbenzene mixtures onto PFA films, *J. Membr. Sci.* **216**, 27 (2003).

128. S. Nezu, H. Seko, M. Gondo and N. Ito, High performance radiation-grafted membranes and electrodes for polymer electrolyte fuel cells. In: *Fuel Cell Seminar*, 17–20 November 1996, Orlando, FL, pp. 620–627.

129. M. Paronen, F. Sundholm, E. Rauhala, T. Lehtinen and S. Hietala, Effects of irradiation on sulfonation of poly(vinyl fluoride), *J. Mater. Chem.* **7**, 2401 (1997).
130. M. Paronen, F. Sundholm, D. Ostrovskii, P. Jacobsson, G. Jeschker and E. Rauhala, Preparation of proton conducting membranes by direct sulfonation. 1. Effect of radical and radical decay on the sulfonation of polyvinylidene fluoride film, *Chem. Mater.* **15**, 4447 (2003).
131. S. Holmberg, P. Holmlund, C.E. Wilen, T. Kallio, G. Sundholm and F. Sundholm, Synthesis of proton-conducting membranes by the utilization of preirradiation grafting and atom transfer radical polymerization techniques, *J. Polym. Sci. A: Polym. Phys.* **40**, 591 (2002).
132. D. Ostrovskii, M. Paronen, F. Sundholm and L.M. Torell, State of water in sulfonated poly(vinyl fluoride) membranes: An FTIR study, *Solid State Ionics* **116**, 301 (1999).
133. M. Paronen, M. Karjalainen, K. Jokela, M. Torkkeli, R. Serimaa, J. Juhanoja, D. Ostrovskii, F. Sundholm, T. Lehtinen, G. Sundholm and L. Torell, Structure of sulfonated poly(vinyl fluoride) membranes, *J. Appl. Polym. Sci.* **73**, 1273 (1999).
134. P. Vie, M. Paronen, M. Strømgård, E. Rauhala and F. Sundholm, Fuel cell performance of proton irradiated and subsequently sulfonated poly(vinyl fluoride) membranes, *J. Membr. Sci.* **204**, 295 (2002).
135. Z. Florjanczyk, E. Wielgus-Barry and Z. Poltarzewski, Radiation-modified Nafion membranes for methanol fuel cells, *Solid State Ionics* **145**, 119 (2001).
136. B. Mattsson, H. Ericson, L.M. Torell and F. Sundholm, Degradation of a fuel cell membrane, as revealed by micro-Raman spectroscopy, *Electrochem. Acta* **45**, 1405 (2000).
137. W. Becker and G. Schmidt-Naake, Proton exchange membranes by irradiation-induced grafting of styrene onto FEP and ETFE: Influences of the cross-linker *N,N*-methylene-bis-acrylamide, *Chem. Eng. Technol.* **25**, 373 (2002).
138. T. Momose, H. Yoshioka,I. Ishigaki and J. Okamoto, Radiation grafting of α,β,β-trifluorostyrene onto poly(ethylene-tetrafluoroethylene) film by preirradiation method. I. Effects of preirradiation dose, monomer concentration, reaction temperature, and film thickness, *J. Appl. Polym. Sci.* **37**, 2817 (1989).
139. W. Becker, M. Bothe and G. Schmidt-Naake, Grafting of poly(styrene-co-acrylonitrile) onto pre-irradiated FEP and ETFE films, *Die Angew. Makromol. Chem.* **273**, 57 (1999).
140. V.I. Brunea, to Solavy S.A., *Brevet D'invention BE* 1,011,218, 1998.
141. J. Zu, M. Wu, H. Fu and S. Yao, Cation-exchange membranes by radiation-induced graft copolymerization of monomers onto HDPE, *Radiat. Phys. Chem.* **72**, 759 (2005).
142. P.R.S. Reddy, G. Agathian and A. Kumar, Preparation of strong acid cation-exchange membrane using radiation-induced graft polymerization, *Radiat. Phys. Chem.* **72**, 511 (2005).
143. V. Tricoli and N. Carreta, Polymer electrolyte membranes formed of sulfonated polyethylene, *Electrochem. Commun.* **4**, 272 (2002).

Chapter 6
Design and Development of Highly Sulfonated Polymers as Proton Exchange Membranes for High Temperature Fuel Cell Applications

Thuy D. Dang, Zongwu Bai, and Mitra Yoonessi

Abstract A series of high molecular weight, highly sulfonated poly(arylenethi oethersulfone) (SPTES) polymers were synthesized by polycondensation, which allowed controlled sulfonation of up to 100 mol %. The SPTES polymers were prepared via step growth polymerization of sulfonated aromatic difluorosulfone, aromatic difluorosulfone, and 4,4'-thiobisbenzenthiol in sulfolane solvent at the temperature up to 180°C. The composition and incorporation of the sulfonated repeat unit into the polymers were confirmed by ^{1}H nuclear magnetic resonance (NMR) and Fourier transform infrared (FTIR) spectroscopy. Solubility tests on the SPTES polymers confirmed that no cross-linking and probably no branching occurred during the polymerizations. The end-capping groups were introduced in the SPTES polymers to control the molecular weight distribution and reduce the water solubility of the polymers. Tough, ductile membranes formed via solvent-casting exhibited increased water absorption with increasing degrees of sulfona-tion. The polymerizations conducted with the introduction of end-capping groups resulted in a wide variation in polymer proton conductivity, which spanned a range of 100 –300 mS cm^{-1}, measured at 65 °C and 85 % relative humidity. The measured proton conductivities at elevated temperatures and high relative humidities are up to three times higher than that of the state-of-the-art Nafion-H proton exchange mem-brane under nearly comparable conditions. The thermal and mechanical properties of the SPTES polymers were investigated by TGA, DMA, and tensile measure-ments. The SPTES polymers show high glass transition temperatures (Tg), ~220 °C, depending on the degree of sulfonation in polymerization. SPTES-50 polymer shows a Tg of 223°C, with high tensile modulus, high tensile strengths at break and at yield as well as elongation at break. Wide angle X-ray scattering of the polymers shows two broad scattering features centered at 4.5Å and 3.3 Å, the latter peak being attributed to the presence of water molecules. The changes in the scattering features of the water in SPTES–70 membrane were examined as a function of dry-ing time during an in situ drying experiment. The in situ small angle X-ray scatter-ing from water swollen SPTES–70 membrane in a drying experiment exhibited a decrease in the water domain size morphology. AFM studies of SPTES–70 mem-brane in a humidity range (35–65 % RH) revealed an increased size of hydrophilic clusters with increasing humidity. SEM examination of cryofractured dry and swollen

S.M.J. Zaidi, T. Matsuura (eds.) *Polymer Membranes for Fuel Cells*,
doi: 10.1007/978-0-387-73532-0, © Springer Science+Business Media, LLC 2009

SPTES–70 membrane surface indicated a change from a smooth brittle fracture to a fractured surface with plastic deformation, verifying the plasticizing effects of the water molecules in the swollen membrane. Membrane electrode assemblies (MEAs), fabricated using SPTES-50 polymer as proton exchange membrane (PEM) incorporating conventional electrode application techniques, exhibit high proton mobility. The electrochemical performance of SPTES-50 membrane in the MEA was superior to that of Nafion. The SPTES polymers have been demonstrated to be promising candidates for high temperature PEM in fuel cell applications.

6.1 Introduction

6.1.1 Proton Exchange Membrane Fuel Cells

Fuel cells are attractive alternative energy devices that convert chemical energy directly into electrical energy by a series of platinum-catalyzed reactions. A PEM should minimize the mixing of the reactant gases and also serve as an electrolyte, permitting proton conduction from the anode to the cathode. The idea of using an organic cation exchange membrane as a solid electrolyte in electrochemical cells was first described for a fuel cell by Grubb in 1959. At present, the proton exchange membrane fuel cell (PEMFC) is the most promising candidate system of all fuel cell systems in terms of the mode of operation and applications.

A PEMFC consists of two electrodes and a solid polymer membrane, which acts as an electrolyte. The proton exchange membrane is sandwiched between two layers of platinum porous electrodes that are coated with a thin layer of proton conductive material. This assembly is placed between two gas diffusion layers such as carbon cloth, carbon paper enabling transfer of electrical current and humidified gas. Some single cell assemblies can be mechanically compressed across electrically conductive separators to fabricate electrochemical stacks.

In general, PEMFCs require humidified gases, hydrogen as a fuel and oxygen (or air) as an oxidant for their operation. The overall electrochemical reactions that occur at both electrodes are as follows:

$$H_2 + 1/2O_2 \rightarrow H_2O + \text{electrical energy} + \text{heat} \qquad (6.1)$$

The hydrogen molecules lose electrons at the anode demonstrated by the follo-wing reaction,

$$H_2 \rightarrow 2H^+ + 2e^- \qquad (6.2)$$

Protons, electrons, and oxygen react, producing water and heat at the cathode,

$$1/2O_2 + 2H^+ + 2e^- \rightarrow H_2O + \text{heat} \qquad (6.3)$$

In recent years, PEMFCs have been identified as promising power sources for vehicular transportation and other applications requiring clean, quiet, and portable power. Hydrogen-powered fuel cells in general have a high power density and are relatively efficient in their conversion of chemical energy to electrical energy. Exhaust from hydrogen-powered fuel cells is free of environmentally undesirable gases such as nitrogen oxides, carbon monoxide, and residual hydrocarbons that are commonly produced by internal combustion engines. Carbon dioxide, a greenhouse gas, is also absent from the exhaust of hydrogen powered fuel cells. These fuel cells are very effective for transportation applications, especially fuel cell electric vehicles (FECVs). Besides being attractive from the viewpoint of clean exhaust emissions and high energy efficiencies, they also offer an effective solution to the problem of petroleum shortage. While FCEV might provide the greatest societal benefits, its total impact would be small if only a few FCEVs are sold due to a lack of fueling infrastructure or due to high vehicle cost. The major obstacles for the commercial use of FCEV are cost of the materials and a low performance at high temperatures (over 100°C) and at low relative humidity. The first PEMFC used in an operational system was the GE-built 1 kW Gemini power plant [1]. This system was used as the primary power source for the Gemini spacecraft during the mid-1960s. The performances and lifetimes of the Gemini PEMFCs were limited due to the degradation of poly(styrene sulfonic acid) membrane employed at that time. The degradation mechanism determined by GE was generally accepted until the present time. It was postulated that HO_2 radicals attack the polymer electrolyte membrane. The second GE PEMFC unit was a 350W module that powered the Biosatellite spacecraft in 1969.

6.1.2 State-of-the-Art Proton Exchange Membranes

Proton-conducting polymers are usually polymer electrolytes with negatively charged groups on the polymer backbone. These polymer electrolytes tend to be rather rigid and are poor proton conductors unless water is absorbed. The proton conductivity of hydrated polymer electrolytes dramatically increases with water content and reaches values of $10^{-2} - 10^{-1}$ S cm^{-1}. An improved Nafion membrane manufactured by DuPont was used as the electrolyte. Figure 6.1 shows the chemical structures

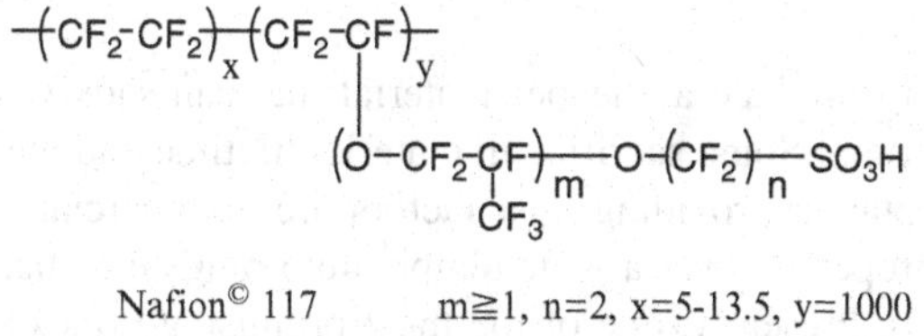

Nafion© 117   m≥1, n=2, x=5-13.5, y=1000

Fig. 6.1 Chemical structures of perfluorinated proton exchange membranes

of Nafion and other perfluorinated electrolyte membranes. The performance and lifetime of PEMFCs have significantly improved since Nafion was developed in 1968. Lifetimes of over 50,000 h have been achieved with commercial Nafion 120. Nafion 117 and 115 have equivalent repeat unit molecular weights of 1,100 and thicknesses in the dry state of 175 and 125 μm, respectively. Nafion 120 has an equivalent weight of 1,200 and a dry state thickness of 260 μm.

Perfluorinated copolymers, Nafion (DuPont), are the current fuel cell PEM materials of choice. They are chemically very stable and have good proton conductivity. Conversely, Nafion and similar perfluorinated vinylidene copolymers have several limitations, some of which include: (1) low modulus as well as modest glass-transition temperatures; (2) reduced conductivity at temperatures above 80°C; and (3) relatively high methanol permeability, which limits efficient applications for direct methanol fuel cells. They are also recognized to be currently quite expensive.

A limiting factor in PEMFCs is the membrane that serves as a structural framework to support the electrodes and transport protons from the anode to the cathode. The limitations to large-scale commercial use include poor ionic conductivities at low humidities and/or elevated temperatures, a susceptibility to chemical degradation at elevated temperatures and finally, membrane cost. These factors can adversely affect fuel cell performance and tend to limit the conditions under which a fuel cell may be operated. For example, the conductivity of Nafion reaches up to 10^{-1} S cm^{-1} in its fully hydrated state but dramatically decreases with temperature above the boiling temperature of water because of the loss of absorbed water in the membranes. Consequently, the developments of new solid polymer electrolytes, which are cheap materials and possess sufficient electrochemical properties, have become one of the most important areas for research in PEMFC.

Proton exchange membranes for high performance PEMFCs have to meet the following requirements [2]:

(1) Low-cost materials
(2) High proton conductivities over 100°C
(3) Good water uptake above 100°C
(4) Durability for 10 years

6.1.3 Proton Exchange Membranes Based on Hydrocarbon Polymers

The challenge is to produce a cheaper material that can satisfy the requirements noted in the preceding. Some sacrifice in material lifetime and mechanical properties may be acceptable, providing cost factors are commercially realistic. Good electrochemical properties over a wide temperature range may help the early marketing of PEMFCs. Presently, one of the most promising routes to high performance proton exchange membranes is the use of hydrocarbon polymers for polymer backbones. The use of hydrocarbon polymers as polymer electrolytes was abandoned in the initial stage of fuel cell development due to the low thermal and chemical stability of these materials. However, relatively cheap hydrocarbon polymers can be used for polymer electrolytes, since the lifetime of electrolytes required in FC vehicles are shorter when compared with use in space vehicles. Also, improved catalysts and fuel cell assembly technologies have brought advantages to the operational lifetimes of PEMFCs and related materials.

There are many advantages of hydrocarbon polymers that have made them particularly attractive:

- Hydrocarbon polymers are cheaper than perfluorinated ionomers, and many kinds of materials are commercially available.
- Hydrocarbon polymers containing polar groups have high water uptakes over a wide temperature range, and the absorbed water is restricted to the polar groups of polymer chains.
- Decomposition of hydrocarbon polymers can be depressed to some extent by proper molecular design.
- Hydrocarbon polymers are easily recycled by conventional methods.

Numerous works by other authors and our own research group describe the syntheses of new proton exchange membranes based on hydrocarbon polymers. The characteristics of these new materials, which determine their potential applications, are discussed in detail. A review of electrochemical properties, water uptake, and thermal stability makes possible a comprehensive understanding of the proton conduction mechanism and physical state of absorbed water in these systems.

Over the last decade, several new proton exchange membranes have been developed. The new polymers in fuel cell applications are based mostly on hydrocarbon structures for the polymer backbone. Poly(styrene sulfonic acid) is a basic material in this field. In practice, poly(styrene sulfonic acid) and the analogous polymers such as phenol sulfonic acid resin and poly(trifluorostyrene sulfonic acid), were frequently used as polymer electrolytes for PEMFCs in the 1960s. Chemically and thermally stable aromatic polymers such as poly(styrene) [3], poly(oxy-1,4-phenyleneoxy-1,4-phenylenecarbonyl-1,4-phenylene) (PEEK) [4], poly(phenylenesulfide) [5], poly(1,4-phenylene) [6, 7], poly(oxy-1,4-phenylene) [8], and other aromatic polymers [9–11], can be employed as the polymer backbone for proton conducting polymers. These chemical structures are illustrated in Fig. 6.2.

These aromatic polymers are easily sulfonated by concentrated sulfuric acid [12], by chlorosulfonic acid [13], and by pure or complexed sulfur trioxide [14]. Sulfonation with chlorosulfonic acid or fuming sulfuric acid sometimes causes chemical degradation in these polymers. According to Bishop et al. [15], the sulfonation rate of PEEK in sulfuric acid can be controlled by changing the reaction time and the acid concentration and can thereby provide a sulfonation range of 30–100% without chemical degradation or cross-linking reactions [16]. However, this direct sulfuric acid procedure cannot be used to produce truly random copolymers at sulfonation levels of less than 30%, because dissolution and sulfonation in

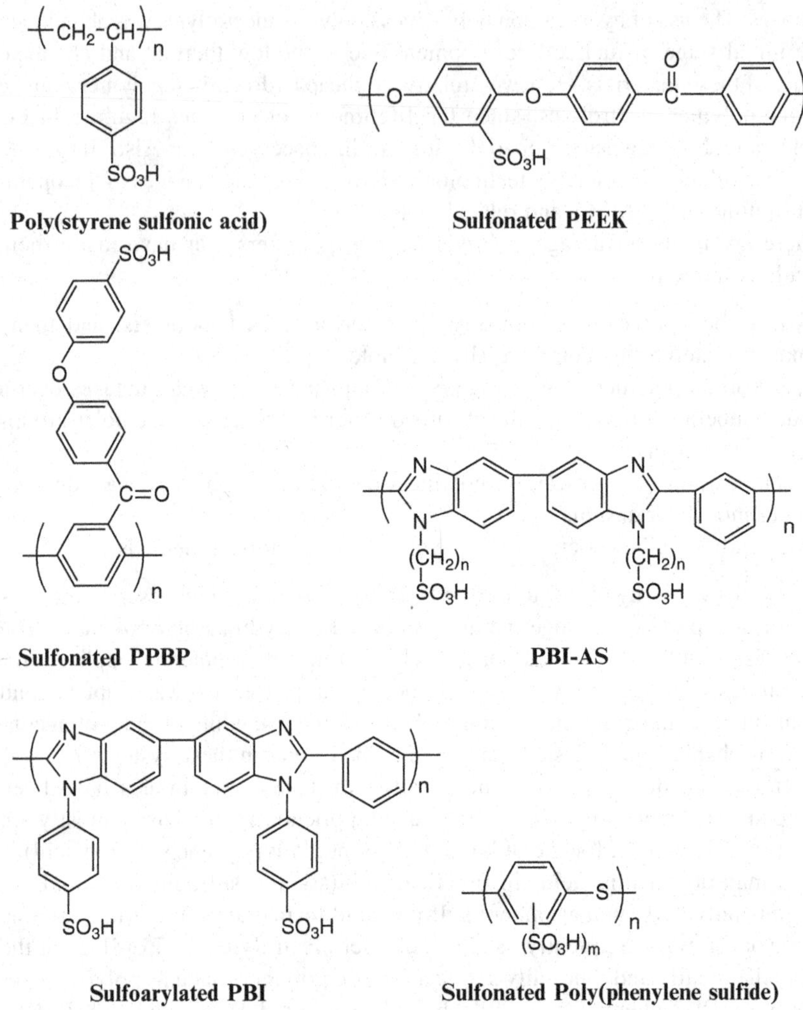

Fig. 6.2 Chemical structures of proton exchange membranes based on a hydrocarbon polymer backbone

sulfuric acid occur in a heterogeneous environment due to the increase of viscosity of reactant solutions. For this reason, the dissolution process was kept short, for less than 1 h, in order to produce a more random copolymer. All the sulfonated high performance polymers have good thermal stability and chemical resistance. The reported water uptake and proton conductivity data were promising enough for the polymers to be evaluated as PEM candidates in fuel cells.

From the application point of view, any membrane, even the one without a fluorinated aliphatic hydrocarbon backbone, but which has high temperature resistance, good mechanical strength, and high proton conductivity, would be useful. The development of new proton exchange membranes has also been necessitated by the fact that commercial Nafion membranes do not meet the requirements for high temperature (>120°C) fuel cell operation. Currently, there is an increasing interest in a variety of polymers having aryl backbones with sulfonated pendants to serve as suitable membranes for fuel cell applications.

In this article, the synthesis and characterization of highly sulfonated poly (arylenethioethersulfone) are described. The objective of the work is to develop a new class of proton exchange membrane materials with high proton conductivity by incorporating the pendant acid functionality directly onto the polymer backbone. Our approach is to utilize a wholly aromatic polymer backbone along with a high sulfonic acid content that enhances water retention for potential high temperature (>120°C) fuel cell applications. In this study, we also incorporate bulky aromatic groups to end-cap the sulfonated poly(arylenethioet hersulfone) polymer structures. The main objectives of the introduction of end-caps in the polymer structure are to increase water resistance while retaining high proton conductivity, to narrow down the molecular weight distribution, and increase the use temperature.

6.1.4 Proton Conductivity Measurements

The conductivity of the polymers was also measured using a galvanostatic four-point-probe electrochemical impedance spectroscopy technique [17]. A four-point-probe cell with two platinum foil outer current-carrying electrodes and two platinum wire inner potential-sensing electrodes was mounted on a Teflon plate. The schematic view of the cell is illustrated in Fig. 6.3. Membrane samples were cut into strips that were approximately 1.0 cm wide, 5 cm long, and 0.01 cm thick prior to mounting in the cell.

The cell was placed in a thermo-controlled humidity chamber to measure the temperature and humidity dependence of proton conductivity. In this method, a fixed AC current is passed between two outer electrodes, and the conductance of the material is calculated from the ac potential difference observed between the two inner electrodes. The method is relatively insensitive to the contact impedance at the current-carrying electrode and is therefore well suited for measuring proton conductivity. This open cell is also suited for studying the humidity dependence of conductivity for proton exchange membrane because of the low interfacial resistance.

In many cases, sealed cells have been used to investigate temperature and water content dependence of proton conductivity for samples hydrated under certain humidity prior to measurements. Conductivity measurements carried out with the sealed cell are simple and useful for conductivity measurements above 100°C and below 0°C. However, the water uptake of the hydrated sample in the sealed cell sometimes changes during measurements.

Platinum inner electrodes

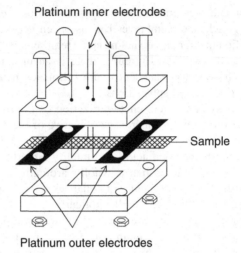

Sample

Platinum outer electrodes

Fig. 6.3 Schematic view of the cell for four-point-probe electrochemical impedance spectroscopy technique

6.2 Experimental

6.2.1 Materials

4,4′-Thiobisbenzenethiol (98%, TCI America Chemical) was recrystallized from toluene and dried at 80°C in vacuum oven prior to use. Bis (4-fluorophenyl) sulfone (99%) and Bis (4-chlorophenyl) sulfone (98%) were purchased from Sigma-Aldrich, recrystallized from isopropanol, and dried at 80°C in vacuum oven. Potassium carbonate (99%, ACS reagent, Sigma-Aldrich Co.), tetramethylene sulfone (99%, Sigma-Aldrich Co.), 4-fluorobenzophenone (97%, Sigma-Aldrich Co.), 4-chlorobenzophenone (99%, Sigma-Aldrich Co.), sulfuric acid (Fuming, 30% SO_3, Sigma-Aldrich Co.), 2-propanol (Fisher Scientific), 1-methyl-2-pyrrolidinone (NMP, Anhydrous, 99.5%, Sigma-Aldrich Co.), N,N-dimethylacetamide (DMAc, Anhydrous, 99%, Sigma-Aldrich Co.), N,N-dimethylformamide (DMF, Anhydrous, 99.9%, Sigma-Aldrich Co.), were used as received. Other chemicals were of commercially available grade and used as received unless otherwise mentioned.

6.2.2 Monomer Synthesis

3,3′-Disulfonate-4,4′-difluorodiphenylsulfone and 3,3′-disulfonate-4,4′-dichlorodiphenylsulfone were prepared by an optimized procedure following reported approaches in the literature [18]. A typical procedure (see Scheme 6.1) was as

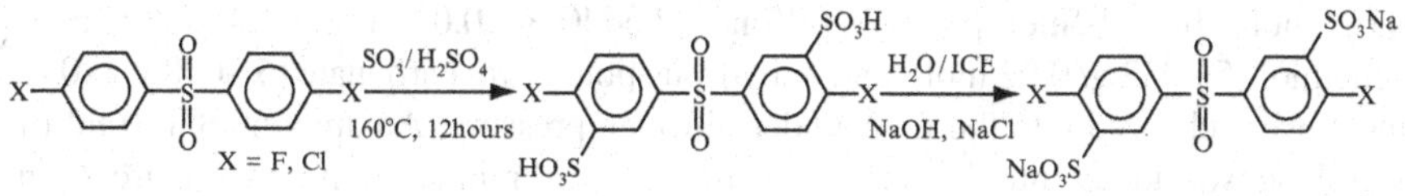

Scheme 6.1 Synthesis of disodium sulfonated monomers

follows: Bis (4-fluorophenyl) sulfone (25.4 g, 0.1 mol) was dissolved in 50 ml of 30% fuming sulfuric acid in a 100 ml, single-necked flask equipped with a magnetic stirrer and a drying tube. The solution was heated to 160°C for 12 h to produce a homogeneous solution. Then, it was poured into 300 ml of ice water under stirring. Next, a 4 M aqueous solution of sodium hydroxide was added, which reduced the pH to 6–7. An amount of 120 g of sodium chloride was added, to salt out the sodium form of the disulfonated monomer. The crude product was filtered and recrystallized from a heated mixture of isopropanol and deionized water (80/20 v/v) to produce fine, white crystals after drying overnight in a vacuum oven. The yield after recrystallization was 91%, based on bis (4-fluorophenyl) sulfone.

6.2.3 Homopolymer Synthesis

The SPTES polymerization procedures (see Scheme 6.2) with the end-capping group are as follows: A 250 ml three-neck round-bottom flask equipped with a mechanical stirrer, a nitrogen inlet, and an addition funnel was charged with 3,3'-disulfonate-4,4'-difluorodiphenylsulfone (4.5833 g, 0.01 mol), anhydrous potassium carbonate (3.0406 g, 0.022 mol), 4,4'-thiobisbenzenethiol (2.5041 g, 0.01 mol), and 120 ml of sulfolane. The mixture was stirred at room temperature for 30 min, and then heated up to100°C in an oil bath on a hot plate for 1 h under a nitrogen atmosphere. After a yellow clear solution was obtained, 4-fluorobenzophenone (1 mol% of the polymer) was then added prior to the solution being heated up to the reaction temperature at 150–180°C for 2–5 h to obtain a light brown, viscous solution. The resulting solution was cooled down to room temperature, and quenched with acetic acid. The polymer was collected by dissolving in a small amount of water, and reprecipitated in methanol. The polymer was filtered, soxhlet-extracted with methanol for 72 h, and dried in vacuum at 80°C overnight to afford a yield of 92.7%.

6.2.4 Copolymer Synthesis

A typical reaction procedure is described in the following (see Scheme 6.3). The step growth polymerization of SPTES copolymers with endcapping group was carried out in a 250 ml three-neck round-bottom flask equipped with a mechanical stirrer and a nitrogen inlet-outlet. 3,3'-Disulfonate-4,4'-difluorodiphenylsulfone (4.5833 g,

0.01 mol), bis (4-fluorophenyl) sulfone (2.5440 g, 0.01 mol), 4,4′-thiobisben-zenethiol (5.0082 g, 0.02 mol), and anhydrous potassium carbonate (5.8048 g, 0.042 mol), were charged into the flask under nitrogen pressure. An amount of 120 ml of sulfolane was added into the flask, and stirred for 30 min at room temperature, then heated to 100°C for 1 h on an oil bath using a hot plate. 4-fluorobenzophenone (1 mol% of the polymer) was then added prior to the solution being heated up to 160–180°C for 4–5 h. The viscous reaction solution was cooled down to room temperature, and quenched with acetic acid in methanol; the sulfonated copolymer was isolated in stirred methanol. The copolymer was collected, and washed several times with methanol. The precipitated copolymer was also washed several times with

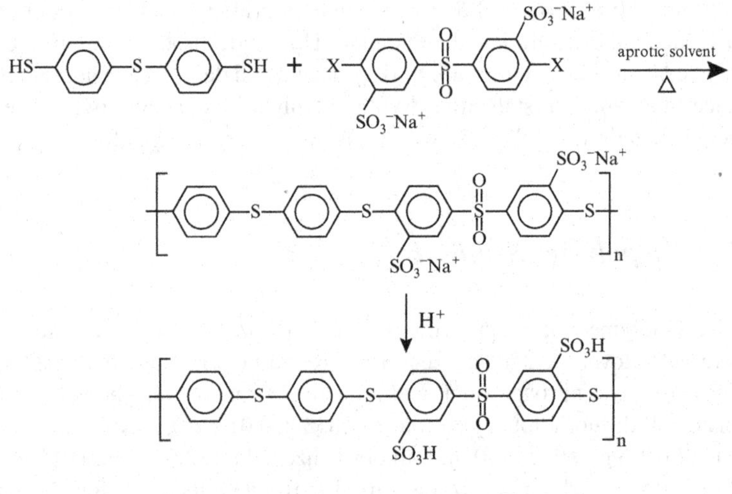

Scheme 6.2 Synthesis of sulfonated polyarylenethioether sulfone polymers

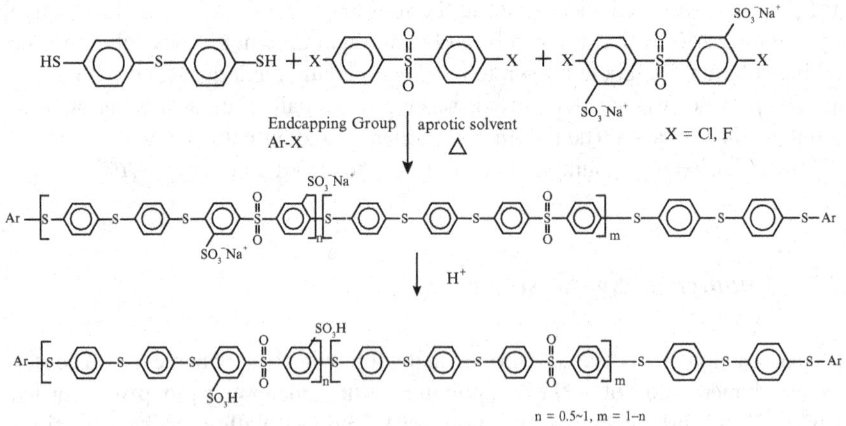

Scheme 6.3 Synthesis of sulfonated poly(arylenethioethersulfone) copolymers

deionized water to completely remove the salts and then extracted in methanol for 72 h. Finally, it was vacuum dried at 100°C for 24 h to afford a yield of 91.7%. The SPTES polymers with other compositions were prepared via similar procedures.

6.2.5   Preparation of Membranes

Membranes were prepared by dissolving the SPTES polymer salt forms in DMAc to form 5–10% clear solutions, which after filtration were taken up in a clean glass dish for casting films. The salt form membranes were carefully vacuum-dried at gradually increasing temperatures up to 100°C for 24 h and 120°C for 2 h. The drying procedure was optimized to produce flat, transparent membranes by complete removal of the solvent. Too rapid a drying rate would cause the membranes to form bubbles and be brittle. The SPTES polymer salt form membranes were converted into the acid form by the following acidification procedure. This involved the immersion of membranes for 24 h in 4 M sulfuric acid at room temperature followed by soaking for 2–4 h in deionized water, and washing with deionized water for 2 h prior to vacuum drying at 80°C for 24 h. Conversion of the membranes from the salt to the acid form was confirmed by TGA.

6.3   Characterization

6.3.1   Molecular Weight Measurement

Gel permeation chromatography (GPC, TriSEC Version 3.00) was used to determine molecular weights (MW) and molecular weight distributions, Mw/Mn, of synthesized SPTES polymers with respect to polystyrene standards (Polysciences Corporation). Molecular weight measurement was performed on GPC in 1-Methyl-2-pyrrolidinone (NMP) containing 0.5% LiBr at 70°C.

6.3.2   Viscosity Measurement

The intrinsic viscosities of SPTES polymers were determined at 30°C in 1-methyl-2-pyrrolidinone (NMP) which contained 0.5% LiBr by using an Ubbelohde Viscometer.

6.3.3   Solubility Measurement

The SPTES polymer solubility was determined at a concentration of 10% (w/v) in a number of solvents, including 1-methyl-2-pyrrolidinone (NMP), N,N-dimethylacetamide (DMAc), methanol, and water at room temperature.

6.3.4 Fourier Transform Infrared (FTIR) Spectroscopy

The powder form of the SPTES polymers were compressed into disc with KBr, and spectra were recorded for the salt forms with a Nicolet Impact 400 FTIR spectrometer to confirm the functional groups of the SPTES polymers.

6.3.5 Nuclear Magnetic Resonance (NMR) Spectroscopy

^{1}H NMR analyses were conducted on a Varian Unity 400 spectrometer. All spectra were obtained from a 10% solution (w/v) in dimethylsulfoxide-$d6$(DMSO-$d6$) at room temperature. The monomer purity and polymer compositions were analyzed via NMR spectroscopy.

6.3.6 Water Uptake Measurement

The water uptake was determined on membranes for all the SPTES polymers in both sulfonate salt and sulfonic acid forms. The membranes were first thoroughly dried at 100°C in vacuum to a constant weight, which was recorded. The dried membrane was then immersed in water at room temperature for 24 h and weighed on an analytical balance until a constant water uptake weight was obtained. Typically, the equilibrium water sorption occurred within 48 h. The water uptake of membranes is reported in weight percent as follows:

$$\text{Water uptake} = \frac{W_{\text{Wet}} - W_{\text{dry}}}{W_{\text{dry}}} \times 100 \qquad (6.4)$$

where W_{wet} and W_{dry} are the weights of the wet and dry membranes, respectively. While seemingly simple, this method has proven accurate and has been established in literature [19].

6.3.7 Ion Exchange Capacity (IEC) Measurement

The ion exchange capacity (IEC) indicates the number of milli-equivalents of ions in 1 g of the dry polymer. The IEC was calculated according to:

$$IEC = \frac{B - P \times 0.01 \times 5}{m} \qquad (6.5)$$

where IEC is the ion exchange capacity (mequiv. g^{-1}), B is the sulfuric acid used to neutralize blind sample soaked in NaOH (ml), P is the sulfuric acid used to neutralize the sulfonated membranes soaked in NaOH (ml), 0.01 is the normality of the sulfuric acid, 5 is the factor corresponding to the ratio of the amount of NaOH taken to dissolve the polymer to the amount used for titration, and m is the sample mass (g).

6.3.8   Proton Conductivity Measurement

The proton conductivity of the SPTES polymer membranes was measured using AC impedance spectroscopy and utilized a standard 4-electrode measurement setup to eliminate electrode and interfacial effects. The Teflon sample fixture was placed inside a temperature and humidity controlled oven. The fabrication of the fixture allows complete exposure of the sample to the humidified air within the chamber. The two outer electrodes were made of platinum foil and these acted to source the current in the sample. Two inner platinum wire electrodes (spaced 1 cm apart) were then used to measure the voltage drop across a known distance. By measuring the impedance of the material as a function of frequency at a set temperature and humidity, the conductivity of the membrane is obtained using the magnitude of the impedance in a region where the phase angle is effectively zero. The proton conductivities of the membranes were measured in the cell shown in Fig. 6.4.

6.3.9   Thermal Properties Measurement

Differential scanning calorimetry (DSC) was employed for the measurement of the Tg's and the detection of any other thermal transitions. For this purpose, a TA Instrument Model DSC 2910 was used with a heating rate of 10°C min^{-1} for samples weighing 5–15 mg. An Auto TGA 2950HR V5.4A thermogravimetric analyzer

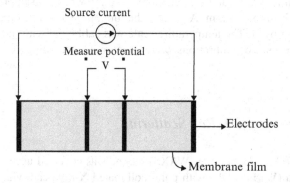

4 electrode measurement setup

Fig. 6.4 The instrumental setup for proton conductivity measurements

(TGA) was employed to study the thermal stability of both the salt form and acid-form polymers. The samples were pieces of thin films and had a weight total of 10–15 mg. The membranes were dried in vacuum for at least 12 h for the removal of absorbed water before the experiments. The samples were evaluated over the range of 30–900°C at a heating rate of 10°C min⁻¹ in air. Decomposition onset temperatures were determined by TGA.

6.3.10   Tensile Mechanical Properties Measurement

Mechanical evaluation of the tensile properties was performed on a Tinius Olsen H10KS bench top tensile tester at a speed of 5 mm min⁻¹. A stretching force was applied to one pneumatic clamp. Clamp displacement was used to determine elongation ratio and strain. Tensile stresses were calculated from the initial cross-sectional area of the sample and applied load. Young's modulus, E, were determined from the initial slope at $\lambda = 1.025$ (least squares fit, $\varepsilon = 2.5\%$). Three to five samples per mechanical measurement were used, with estimated error/uncertainty not exceeding 10% in the data. The membrane samples were cut into a rectangle shape 50 mm × 4 mm (total) and 25 mm × 4 mm (test area).

6.3.11   Dynamic Mechanical Analysis

Dynamic mechanical analysis (DMA) was performed to determine the influence of the polymer constitution on tensile modulus and mechanical relaxation behavior. For this purpose, a Perkin Elmer DMA-7 was run in tensile mode at an oscillation frequency of 1 Hz with a static stress level of 5×10^5 Pa and a superposed oscillatory stress of 4×10^5 Pa. With this stress controlled instrument, the strain and phase difference between stress and strain are the measured outputs. Typically, the resulting strain levels ranged from 0.05% to 0.2% when the sample dimensions were 8 mm × 2 mm × 0.1 mm. A gaseous helium purge and a heating rate of 3°C min⁻¹ were employed. The temperature scale was calibrated with indium, and the force and compliance calibrations were performed according to conventional methods.

6.3.12   Wide Angle X-Ray Scattering

The wide angle X-ray scattering (WAXS) regime was collected using an evacuated Statton camera (Warhus, DE) with point-collimated X-ray beam via two pinholes (0.05 mm), separated by ~15 cm at Cu Kα wavelength 1.5418 Å. In order to examine scattering from dry membranes, SPTES-50, SPTES-60, SPTES-70, and

SPTES−80 were placed in a vacuum oven at 20 mbar and 60°C for 3 days, then quickly placed in a quartz capillary and sealed.

6.3.13   Small Angle X-Ray Scattering

Small-angle X-ray scattering (SAXS) was conducted at the Advanced Polymers Beamline (X27C) of the National Synchrotron Light Source (NSLS), Brookhaven National Laboratory (BNL). The wavelength of the incident X-rays was 1.371 Å defined by a double multi-layer monochromator. The synchrotron X-rays were colli-mated to a 600 m beam size using a three pinhole collimator [20]. Data were collected using a 2D mar CCD detector at 1.8 m, calibrated using silver behenate. The DI water-swollen SPTES−70 membrane was examined in the X-ray beam while drying to inves-tigate the presence of water clusters and their change with the water content. The tensile tester used allowed simultaneous stress-strain and time resolved X-ray measurements in the X27C beam line. A tensile mode Instron with a maximum load of 500 N at room temperature and a strain rate of 1 mm min^{-1} was used. The data were corrected for background and azimuthally integrated to produce the intensity of scattering, $I(q)$, vs. wave vector, q (q is a function of scattering angle θ defined as $q = (4\pi/ \lambda) \sin (\theta/2)$).

6.3.14   Atomic Force Microscopy

Tapping mode atomic force microscopy (AFM) was used to obtain height and phase imaging data simultaneously on a Dimension 3,100 Nanoscope III Controller from Digital Instruments, CA. Etched silicon probes (Nanoprobes, Digital Instruments) were used at their fundamental resonance frequencies which typically varied from 270 to 350 Hz. The lateral scan frequency was 0.5−1 Hz. The AFM was operated in ambient condition with a double vibration isolation system. Extender electronics were used to obtain height and phase information simultaneously. The images were obtained in tapping mode. Tapping mode allows probing more delicate samples due to the lower lateral forces applied to the surface. In tapping mode, the cantilever oscillates near its resonant frequency, contacting the surface for a very short time during its oscillation cycle. Obtaining both phase images and topo-graphical information is possible in tapping mode [21].

6.3.15   Scanning Electron Microscopy (SEM)

An FEI Sirion field emission gun scanning electron microscope (5 Kv) operating in the secondary electron imaging mode was used to examine the surface of the cry-ofractured membranes. Cryofractured surfaces of the dried and swollen samples were examined.

6.3.16   Fabrication of Membrane Electrode Assembly

A painting technique was used to form electrodes on both sides of SPTES membranes, using ink slurry. This slurry consisted of platinum black, water, and Nafion solution (5 wt%) as binder. Catalyst loading of 5 mg cm^{-2} was used. The ink slurry was sonicated to break up the catalyst powder in order to obtain a homogenous distribution. Several layers of the ink were painted on both sides of the membranes. Samples were dried at 80°C under vacuum for 10 h to eliminate all solvents. Membrane electrode assembly (MEAs) with effective areas of 5 cm^2 were fabricated.

6.4   Results and Discussion

6.4.1   Monomer Synthesis

High yields of sulfonated difluorodiphenyl sulfone and sulfonated dichlorodiphenyl sulfone monomers were prepared via electrophilic aromatic substitution on bis (4-fluorophenyl) sulfone and bis(4-chlorophenyl) sulfone, respectively, in fuming sulfuric acid with the reaction conditions depicted in Scheme 6.1. In the literature [18], a precipitant was observed at high concentrations (e.g., 50%) when the monomer was synthesized at 90°C, causing lower yields for the sulfonated reaction. The higher yields may be associated with the homogeneous nature of higher temperature synthesis, and are likely concentration-dependent [22, 23]. In this research, a higher reaction temperature of 160°C was used to afford a homogeneous reaction solution, and sulfonated difluorodiphenyl sulfone and sulfonated dichlorodiphenyl sulfone monomers were produced in good yields of 91% and 89%, respectively, after recrystallization from isopropyl alcohol/water mixtures.

The higher reactivity and possible hydrolysis of the aryl C-X (F or Cl) bond in 4-halophenylsulfone required that ice-cold water be employed during the precipitation and, especially, during the sodium hydroxide neutralization steps. Hydroxide ions and, under some conditions, water may attack the activated halide to possibly produce an undesired phenolic side-product that will lower the yields of sulfonated dihalodiphenyl sulfone. The ^{1}H NMR spectrum of sulfonated diflorodiphenyl sulfone is shown in Fig. 6.5 as an example for analysis of the monomers. The resonance assignments for the aromatic protons H_1, H_2, and H_3, shown in the Fig. 6.5, are in accordance with the structure of the sulfonated monomers.

FT-IR, ^{1}H NMR, mass spectrum, and elemental analysis of sulfonated difluorodiphenyl sulfone and sulfonated dichlorodiphenyl sulfone are listed in Table 6.1. The results confirm the structure and the purity of the sulfonated monomers and agree with the literature-reported syntheses of these monomers [24].

In the synthesis reaction, the initially formed sulfonic acid was converted to the sodium sulfonate via neutralization using sodium hydroxide. The exchange was more effective than in the case of the reported procedure using excess sodium chloride.

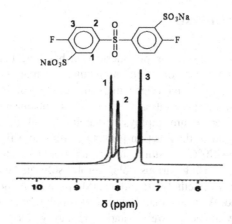

Fig. 6.5 ^{1}H NMR spectrum of sulfonated diflorodiphenyl sulfone

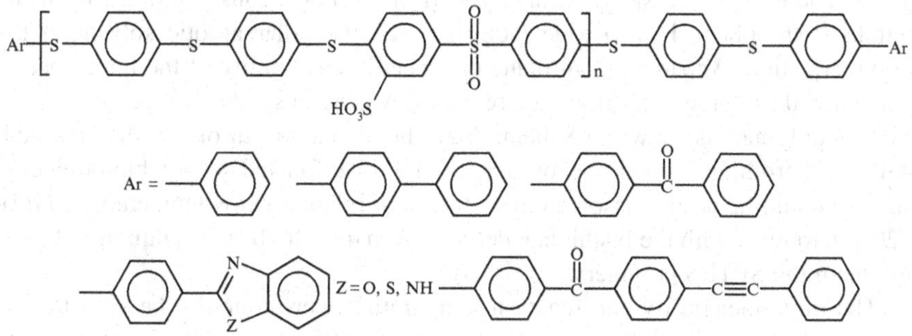

Fig. 6.6 Chemical structures of end-capping groups in the SPTES polymers

Table 6.1 FTIR, NMR, MS, elemental analysis of sulfonated monomers

Monomers	Elemental analysis		^{1}H NMR (DMSO-$d6$)	FTIR Spectra
	Calcd	**Found**		
Sulfonated Difluorodiphenyl Sulfone $C_{12}H_6O_8S_3F_2Na_2$	C, 31.44% H, 1.32% S, 20.98% Na, 10.03%	C, 30.97% H, 1.43% S, 19.96% Na, 10.4%	7.3 (t, H-1) 7.7 (m, H-2) 8.0 (dd, H-3)	1085 cm^{-1} (Ar-SO3 Na)
	Calcd	**Found**		
Sulfonated dichlorodiphenyl sulfone $C_{12}H_6O_8S_3Cl_2Na_2$	C, 29.34%. H, 1.23% S, 19.58% Na, 9.36%	C, 30.03% H, 1.21% S, 19.27% Na, 9.41%	7.4 (t, H-1) 7.9 (m, H-2) 8.1 (dd, H-3)	1085 cm^{-1} (Ar-SO3 Na)

Moreover, elemental analysis consistently showed better agreement with the theoretical composition, with the additional step of neutralization using sodium hydroxide. No melting point was observed for the ionic compound below 300°C by melting point testing.

6.4.2 Polymer Synthesis

Nucleophilic aromatic substitution polymerization has long been employed successfully in the synthesis of high molecular weight poly(aryleneether),
poly(arylenethioether), and poly(aryleneethersulfone) [25, 26]. In typical polymerization conditions, careful dehydration was required to obtain very high molecular
weight polymers. The procedure employs a polar aprotic solvent along with toluene
as an azeotroping agent in the reaction prior to heating to a final polymerization
temperature of 175–190°C. In this study, the polymerization of sulfonated
arylenethioethersulfone using various polar aprotic solvents such as dimethylsulfoxide (DMSO), N,N-dimethylacetamide (DMAc) and N-methylpyrrolidinone
(NMP) was examined. We found that the sulfonated poly(arylenethioethersulfone)s
were obtained in low molecular weight and were darkly colored, presumably due
to the solubility problems associated with the propagating polyelectrolyte species
in these solvents. However, relatively colorless, high molecular weight SPTES polymers could be successfully synthesized from homogeneous solutions, by using
tetrahydrothiophene-1, 1-dioxide (sulfolane) as the polar aprotic solvent in the
polymerization. Apparently, sulfolane enhanced the solubility of the ionic species
generated during the base-promoted reaction. A general synthesis procedure of the
SPTES polymers is shown in Scheme 6.2. The disodium salt of 3,3′-disulfonated-
4,4′ -difluorodiphenyl sulfone or that of 3,3′-disulfonated-4,4′-dichlorodiphenyl
sulfone could be easily dissolved in sulfolane at relatively low temperatures (100-
120°C), to react with the bisthiolate derived from 4,4′-thiobisbenzenthiol, at 180°C
to obtain the SPTES polymers.

The monomer purity and reaction temperature were found to be vital factors
in the polymerizations. From the viewpoint of obtaining high molecular weight
polymers, repeated recrystallization of the sulfonated monomers was carried out
in 2-propanol/water mixtures to yield polymerization grade monomers, which
were dried in vacuum at 120°C for 12 h before the polymerization.

The end-capping of the SPTES polymers was carried out by adding 1 mol% of
the monofunctional monomer in the polymerization reaction. In Fig. 6.6, is shown
the chemical structure of endcapped SPTES polymer and those of the phenyl-based
monohalide end-capping agents. The phenyl-based groups in the end-capping agent
include phenyl, biphenyl, benzophenone, and benzothiazole that are commercially
available. Endcapping was found to result in the formation of relatively colorless
SPTES polymers. Relative to the unendcapped versions, the endcapped sulfonated
polymers were also found to exhibit decreased solubility in water.

6.4.3 FTIR and NMR

FTIR and ¹H NMR were used to identify and characterize the SPTES polymers.
FTIR spectra were used for the determination of the functional groups of the

synthesized SPTES polymers. Figure 6.7 shows the comparative FTIR spectra of unsulfonated poly(arylenethioethersulfone) (PTES), the SPTES without end-capping group, (SPTES-100), and the SPTES with the end-capping group (endcapped SPTES-100) in the sodium salt form.

The absorption band at 1,501 cm^{-1} corresponds to di-substitution on the phenyl ring for the non-sulfonated polymer, whereas the new band at 1482 cm^{-1} for SPTES-100 and endcapped SPTES-100 corresponds to tri-substitution on the aromatic ring due to sulfonation in the phenyl ring. New absorption bands at 1,029 and 1,086 cm^{-1} in SPTES-100 and endcapped SPTES-100 were assigned to symmetric and asymmetric stretching vibrations of O=S=O due to the sodium sulfonate group in the polymers.

The structural determination of the polymers was made by NMR. The integration and appropriate analysis of known reference protons of the polymers enabled the relative compositions of the various polymers to be determined. The protons adjacent to the sulfonated group derived from the disulfonated dihalide monomer in the polymers were well separated from the other aromatic protons (8.25 ppm). Figure 6.8 shows ^{1}H NMR spectrum of the SPTES-100 homopolymer.

The ^{1}H NMR spectrum for the SPTES-50 polymer is in Fig. 6.9. The calculated sulfonation degree, which was provided from titration (IEC), is 45 mol% for SPTES-50. (The designed composition is 50 mol% for SPTES-50.) The integration from NMR analysis indicated that 47 mol% sulfonated monomer was incorporated into the polymer. These results indicate that the incorporation of the sulfonated groups in the polymer backbone conformed to the initial stoichiometry of the monomers used in the reaction.

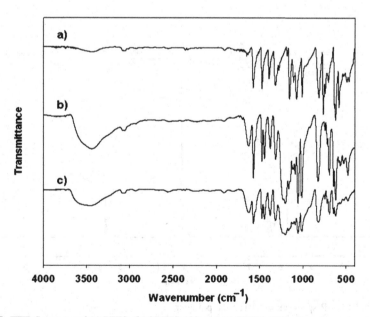

Fig. 6.7 FTIR Spectrum of (a) PTES, (b) SPTES-100, and (c) SPTES-100 with end-capping groups

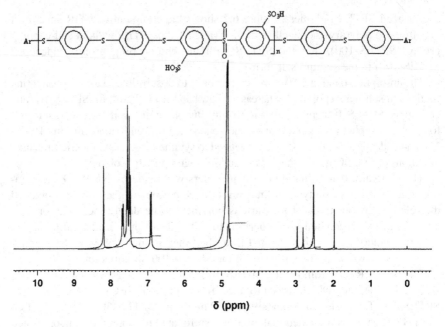

Fig. 6.8 ¹H NMR spectrum of SPTES-100 polymer

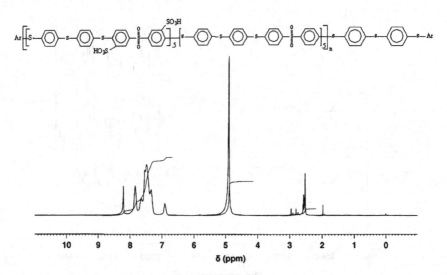

Fig. 6.9 ¹H NMR spectrum of SPTES-50 polymer

6.4.4 Physical Properties

The synthesized SPTES polymers with the endcapping group could be readily cast into membranes of both salt form and acid form. The physical properties of the SPTES polymers are listed in Table 6.2.

The solubility of the SPTES polymers and Nafion 117 was conducted at room temperature. From Table 6.2, we found that SPTES-100 polymer is soluble in water and methanol, and good solubility in polar aprotic solvents such as DMAc and NMP at room temperature; SPTES-50 and SPTES-60 were completely insoluble water, SPTES-50 polymer is insoluble in methanol, and SPTES-60 has very limited solubility in methanol while retaining their good solubility in polar aprotic solvents such as DMAc and NMP at room temperature. SPTES–80 was found to swell in water at lower temperatures, while SPTES–70 was swollen in water at higher temperatures. SPTES–70 and –80 polymers are swollen in methanol. The SPTES polymers showed increased susceptibility to water with increasing sulfonic acid content. This can be explained by the increased levels of hydration in the polymer structure as a function of increased sulfonic acid content [27].

6.4.5 Water Uptake and Ion Exchange Capacity (IEC)

The sulfonation of polymers serves to increase polymer hydrophilicity, which enhances the proton transfer and conductivity of the polymer electrolytes in the presence of water, and the water uptake for the sulfonated polymers also increases with increased degree of sulfonation. The water uptake of SPTES polymers is shown in Table 6.3.

The acid form of the SPTES membranes has a much higher water uptake than the corresponding salt form of the SPTES membranes. The water uptake increased nonlinearly from 41% for SPTES-50 membrane to 300% for SPTES-100 membrane of acid form, while the salt form of SPTES polymer membranes also shows increased water uptake (4.93–72.8%) with increased degree of sulfonation. The SPTES polymer membranes show much higher water uptake compared with Nafion

Table 6.2 Properties of sulfonated poly(arylenethioethersulfone) polymers

Polymers	Intrinsic viscosity (dL g^{-1})	Solubility in water	Solubility in methanol	Solubility in DMAc and NMP	M_w/M_n (GPC)
SPTES-100	1.89	S	S	S	93 k/41 k
SPTES-80	1.86	SW	SW	S	89 k/41 k
SPTES-70	1.65	N	SW	S	88 k/39 k
SPTES-60	1.62	N	N/SW	S	81 k/37 k
SPTES-50	1.84	N	N	S	85 k/38 k
SPTES-40	0.98	N	N	S	64 k/20 k
Nafion-117	N/A	N	N	N	N/A

N not soluble; *SW* swollen; *S* soluble

Table 6.3 Characteristics of SPTES polymer membranes

Polymers	Water uptake (%) (salt form)	Water uptake (%) (acid form)	Cal. IEC (mequiv. g^{-1})	Exp. IEC (mequiv. g^{-1})	Proton conductivity (mS cm^{-1})
SPTES-100	72.8	300	3.20	2.89	300
SPTES–80	41.5	180	2.68	2.40	215
SPTES–70	16.5	73	2.41	2.18	175
SPTES-60	10.2	62	2.12	1.94	145
SPTES-50	4.93	41	1.82	1.64	100
SPTES-40	3.3	30	1.50	1.39	60
Nafion-117	N/A	26	0.91	N/A	80

Proton conductivity was measured at 65°C, 85% relative humidity.
Water uptake was measured at room temperature.

(26%), in part due to the presence of more sulfonated groups pendent to the SPTES polymer aromatic backbone. The higher proton conductivities of the SPTES membranes can also be attributed, presumably, to the likely formation of a different microstructure for the proton transport compared with the case of Nafion. The hydrophobicity of the aromatic polymer backbone and the extreme hydrophilicity of the sulfonic acid groups could lead to a spontaneous separation of the hydrophobic/hydrophilic nanodomains [28]. Therefore, only the hydrophilic domains of the nanostructure are hydrated in the presence of water.

Usually, PEMFC performance depends on the presence of a humid environment as well as the temperature of operation. Therefore, it is important to determine the water uptake and the stability of the prepared membranes in water at various temperatures. The temperature dependence of water uptake of the SPTES membranes was determined by the following procedure. The SPTES membranes were vacuum-dried at 100°C for 24 h, weighed and immersed in deionized water at various temperatures for 1 h. Subsequently, the wet membranes were wiped dry and quickly weighed again. The water uptake of SPTES membranes is calculated according to the method described in the Experimental section to obtain weight percent of water. The results are shown in Fig. 6.10; for comparison, the Nafion-117 membrane was also tested under the same conditions.

In Fig. 6.10, the water uptake of Nafion is much lower compared with the SPTES polymers, especially those with high sulfonic acid content. While the water uptake of the SPTES polymer membranes increases steeply at higher temperatures, the water uptake of Nafion is shown to be relatively insensitive to temperature changes.

From Fig. 6.10, it also follows that the SPTES–70 and –80 membranes in acid form exhibited excessive water uptake, above 200%, already at 60°C and then lost their mechanical strength after immersion in water at 80°C for 24 h. While the SPTES-60 membrane had reasonable mechanical strength after water-immersion tests at 60°C, SPTES-50 membrane did not exhibit excessive water uptake and maintained excellent mechanical integrity even after the immersion test in water at 80°C. The tensile properties of the dry as well as the wet sulfonated copolymer membranes are discussed in Fig. 6.18.

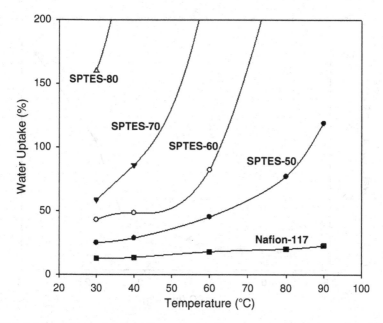

Fig. 6.10 Water uptake of the SPTES polymers

The titration was used to quantitatively determine the IECs of the SPTES polymers, and to confirm the presence of the strong acid ($-SO_3H$) in the SPTES polymeric backbone. All the experimental IEC values are listed in Table 6.3, and are in good agreement with the calculated IECs, assuming that all of the sulfonated monomer was incorporated into the polymer backbone. These results corroborate the fact that $-SO_3Na$ and proton conductive groups, $-SO_3H$, could be introduced in the polymer structure via a direct polymerization route, with no apparent side reactions, using sulfonated monomers.

The water uptake of the SPTES polymers (-100, -80, -70, -60, -50, -40) and Nafion 117 are plotted as a function of experimental ion-exchange capacity (IEC) in Fig. 6.11. The room temperature water uptake of SPTES polymers decreases from 300 wt% (SPTES-100, IEC: 2.89 mequiv. G^{-1}) to 30 wt% (SPTES-40, IEC: 1.39 mequiv. G^{-1}). The trend shown in this plot again illustrates the fact that the water uptake of the SPTES copolymers increases dramatically at higher IEC values and thus, at higher degrees of sulfonation. The steep enhancement in water uptake at higher degrees of sulfonation can also be correlated with membrane swelling (see Fig. 6.10, vide supra).

6.4.6 Proton Conductivity

The proton conductivities of the SPTES polymer membranes in the longitudinal direction were measured by AC impedance spectroscopy. All SPTES polymer membranes were initially hydrated by immersion in deionized water for at least 24 h at

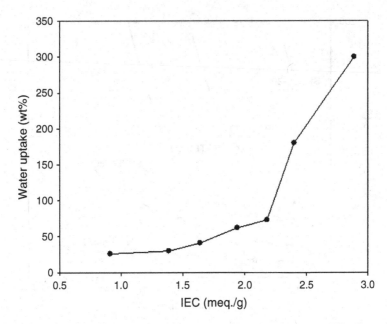

Fig. 6.11 IEC vs. water uptake of the SPTES polymers at room temperature. (The first point in the trace is due to Nafion-117.)

room temperature. The proton conductivities, listed in Table 6.3, were all measured at 65°C and at 85% relative humidity. Expectedly, the proton conductivities increased with increase in the sulfonic acid content of the polymer. The proton conductivities of the SPTES membranes were in the range of 100–300 mS cm^{-1}. These highly proton conducting PEMs are undoubtedly promising for fuel cell applications. It is noteworthy that even the sulfonated polymer with a relatively lower sulfonic acid content, i.e., SPTES-50 polymer, has a proton conductivity slightly exceeding that of Nafion-117 under the same conditions of measurement. It is also clear from Table 6.3, that the increase in proton conductivity parallels the enhancement in the water uptake of the sulfonated polymer membranes as well as their IECs.

Figure 6.12 shows the results of the proton conductivity of the SPTES polymers and Nafion-117 at 85% relative humidity at different temperatures. In all the cases, an increase of the membrane proton conductivity with increasing temperature is indicated. All the SPTES polymer membranes exhibited higher proton conductivity than Nafion-117 except in the lower temperature region for SPTES-50 polymer where a slight crossover is indicated. The proton conductivities of SPTES polymers have a linear dependence on temperature, thus showing Arrhenius behavior. The activation energy related to the proton conductivities of the SPTES polymers has been reported elsewhere [29].

Figure 6.13 shows the proton conductivity of the SPTES polymer at 65°C under different relative humidity conditions. The results indicate a significant dependence of the proton conductivity on the relative humidity. As the relative humidity was increased from 35% to 95%, the proton conductivity of the SPTES polymers was

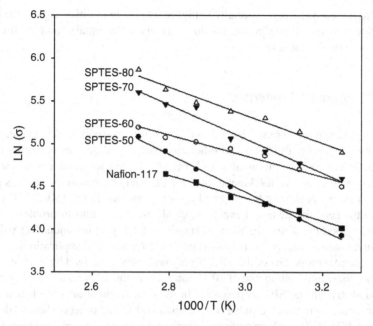

Fig. 6.12 Temperature dependence of proton conductivity of the SPTES polymers

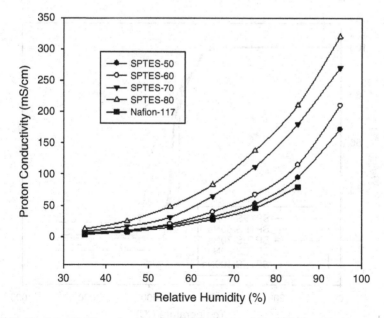

Fig. 6.13 The humidity dependence of proton conductivity of the SPTES polymers

found to increase sharply, especially at higher relative humidities. Again, the inter-relatedness of membrane proton conductivity, its water uptake, and its IEC, are fully borne out by these results.

6.4.7 Thermal Properties

Proton exchange membranes that exhibit fast proton transport at elevated temperatures are needed for PEMFCs and other electrochemical devices operating in the 100–200°C range. Operation of a PEMFC at elevated temperatures has several advantages. It increases the kinetic rates for the fuel cell reactions, reduces problems of catalyst poisoning by absorbed carbon monoxide in the 150–200°C range, reduces the need for the use of expensive catalysts, and minimizes problems due to electrode flooding. Thus, the thermal stability of the proton-conducting polymer electrolyte membranes is a very important factor for fuel cell applications.

The thermal properties of the SPTES polymers were examined by TGA under an air atmosphere at a heating rate of $10°C \ min^{-1}$. From the TGA result (see Fig. 6.14), we found that all the SPTES polymers in their salt form show excellent thermal stability; the polymers were stable at least up to 400°C (the onset of thermal decomposition). SPTES-50 (sodium sulfonate) shows a single-stage thermal decomposition. The salts of the polymers with higher sodium sulfonate contents seem to show a relatively distinct two-stage thermal decomposition behavior, attributed to the degradation of the pendant sodium sulfonate followed by that of the backbone [27].

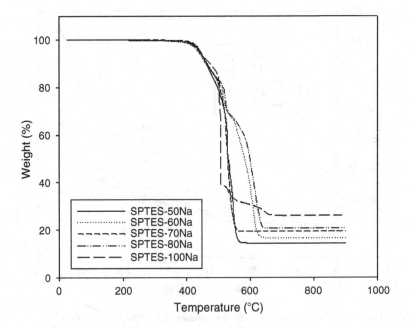

Fig. 6.14 TGA curves of the SPTES polymers (salt form)

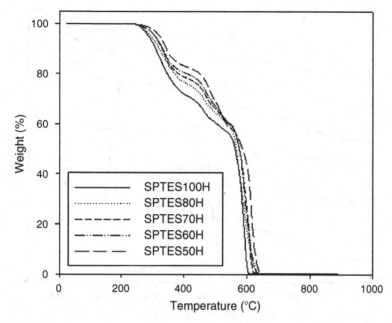

Fig. 6.15 TGA curves of the SPTES polymers

The thermal stabilities of acid form of the SPTES polymers were studied by TGA in air; the results are shown in Fig. 6.15. All the SPTES polymers in the acid form exhibited two distinct thermal degradation steps in accordance with previously made observations [30]. The first weight loss occurs at about 300°C or higher, which is associated mainly with the loss of sulfonic acid groups and the second weight loss step starts to occur at about 420°C, which is related to decomposition of the main chain.

The differential scanning calorimetry (DSC) analyses of sulfonated polymers were obtained on TA Instrument Model DSC 2910. Measurements were performed over the temperature range of 30–300°C at the heating rate of 5°C min^{-1} in hermetically sealed aluminum pans. The results (second run) are shown in Fig. 6.16. It is observed that Tg increases with decrease in the sulfonic acid content in the range of polymer compositions from SPTES-100 to SPTES-50. The Tgs of the SPTES polymers, determined by DSC to be in the range of 180 to 200°C, are shown in Fig. 6.16; the Tg values for SPTES polymer samples from DMA are shown in Fig. 6.19 (vide infra).

6.4.8 Mechanical Properties

Mechanical strength of the membrane affects manufacturing conditions of MEAs and durability of PEMFCs. The mechanical strength of the SPTES membranes and Nafion 117 were tested by a tensile tester under ambient conditions at a relative humidity (RH) of 55%; the results are shown in Fig. 6.17.

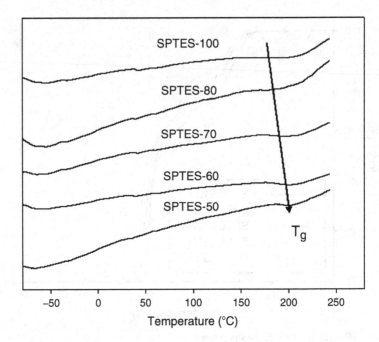

Fig. 6.16 DSC curves of the SPTES polymers (acid form)

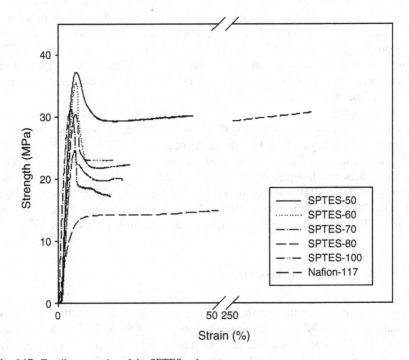

Fig. 6.17 Tensile properties of the SPTES polymers

From Fig. 6.17, we see that initial Young's modulus for the membranes of SPTES-50, -60, -70, -80, -100 are 1.23, 1.35, 1.13, 1.07, and 0.985 GPa, respectively; these values are much higher than the measured tensile modulus of Nafion-117 (0.357 GPa). Nafion-117 membrane, under our experimental conditions, shows 270% of elongation at break and the maximum stress of 29.4 MPa at break, which is in agreement with information reported by DuPont [31]. It follows from Fig. 6.17 that SPTES-50, -60, -70, -80, -100 membranes show elongation at break of 41.6, 21.4, 20.4, 25.2, 16.7%, respectively; these elongations are much lower than that of Nafion-117. As can be seen from the stress–strain curves in Fig. 6.18, SPTES-50, -60, -70, and -80 also show yield behavior with yield strengths of 38.0, 36.0, 32.0, 30.5, and 16.9 MPa, which clearly indicates the good mechanical strength and toughness of SPTES polymer membranes at 55% relative humidity. And these yield strengths are much higher than that of Nafion–117 (14.2 MPa) under comparable conditions.

It is essential for proton exchange membranes to retain their mechanical strength under humidified conditions in the light of membrane electrode assembly (MEA) to be used in fuel cells. In general, electrolyte membranes are moisture-sensitive materials. Obviously, under wet conditions, the mechanical properties of the electrolyte membranes are lowered compared with those measured under dry conditions. Figure 6.18 shows tensile strength vs. strain for dry and wet SPTES-50 membrane as well as for dry and wet Nafion-117 membrane.

In all these conditions, SPTES-50 polymer membranes exhibited greater strength than Nafion-117. It is worth mentioning that the tensile strength of wet

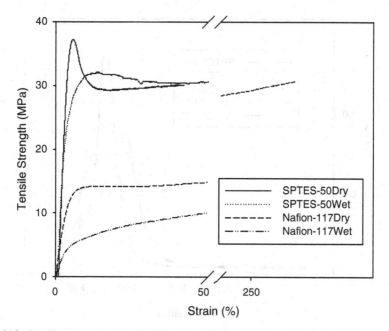

Fig. 6.18 Tensile properties of the SPTES-50 and Nafion-117 (dry and wet membranes)

SPTES-50 polymer membranes reaches a maximum at yield; the wet membrane also exhibits a higher elongation at break compared with the membrane in the dry state. It is demonstrated that hydration improves the plasticity of the SPTES-50 polymer. It appears that the maximum strength at yield of the wet SPTES-50 polymer membrane is lower than that of the dry SPTES-50 polymer membrane (38 vs. 33 MPa), while the strength at break seems marginally higher than that for the dry membrane. Upon hydration, Nafion-117 was found to lose a large proportion of its strength, whereas the corresponding loss of tensile strength for SPTES-50 polymer was relatively modest. SPTES-50 polymer membranes are also considerably stiffer than Nafion as reflected by the larger tensile modulus values. Thus, SPTES-50 membrane has been shown to be mechanically superior to Nafion-117 in both dry and wet states of the membrane.

The wet SPTES-50 polymer membrane has a higher elongation and strength at break as a function of both hydration and sulfonic acid content. Sulfonated polymers containing water of hydration would have interchain H-bonding interactions, which would be mediated by associated water molecules. However, if the sulfonic acid content is very high, as in the case of SPTES-80 polymer, the higher levels of interchain hydration would plasticize the polymer, resulting in the reduction of mechanical properties. As an extreme case, the completely sulfonated polymer (SPTES-100 polymer) is water soluble and thus has no mechanical integrity in the wet state.

Figure 6.19 shows the tan δ vs. temperature at 1 Hz frequency for the SPTES polymers. The maximum damping peaks of SPTES-100 appear at 202 and 238°C.

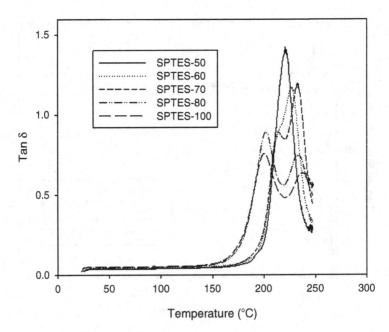

Fig. 6.19 DMA results of the SPTES polymers

The presence of two peaks is presumably, an indication of two segments in the polymer chain requiring different levels of thermal energy for initiating the chain segmental motions. The maximum damping peaks for SPTES-80 appear at 204 and 235°C, while the ones for the SPTES-70 are at 215 and 235°C. SPTES-60 shows a shoulder peak at 216 and a high intensity peak at 228°C. The Tg of the SPTES-50 is a single peak at 223°C. The lower temperature peak starts to shift toward higher temperatures with decreased sulfonic acid content in the polymer structure. Correspondingly, the higher temperature transition shifts toward lower temperatures with a concomitant increase in its intensity.

The Tg results from the peaks of the DMA loss tangents are compared with the Tg values from DSC measurements in Fig. 6.16. The Tg values obtained from DSC are in fair agreement with the Tgs based on all the lower temperature transitions obtained by DMA. Increasing the sulfonic acid content of the SPTES polymers dramatically decreases the Tg, as indicated by the observed leftward shift of the loss tangent peak with sulfonic acid content, from a value of 223°C for SPTES-50 polymer to 202°C for SPTES-100 polymer.

Typical results of a DMA experiment are displayed in Fig. 6.20, where the tensile storage modulus (E') is plotted against the sample temperature. The secondary peak in E' is due to a β transition, which is related to the side-chain motion of the polymer. Comparison of the storage moduli of the various SPTES polymers is shown in Fig. 6.20, demonstrating the influence of the sulfonated functionality on the mechanical behavior of the polymers. The fall-off in the storage modulus at temperatures near Tg, shows sensitivity to the polymer composition.

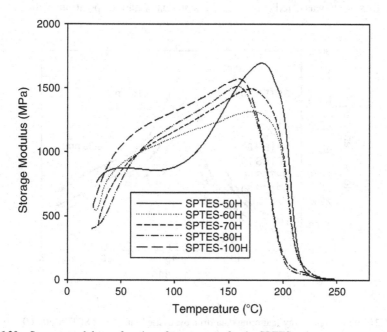

Fig. 6.20 Storage modulus as function of temperature for the SPTES polymers

6.4.9   Wide Angle X-Ray Scattering

Wide angle X-ray scattering experiments were performed on SPTES-50, -60, -70 and -80 membranes at the ambient temperature. Figure 6.21 shows the diffraction pattern from the polymers obtained by a Statton camera. A high intensity peak followed by a shoulder peak was observed for all polymers. The broad peak appearing at ~4.5 Å was attributed to the amorphous SPTES polymer. A shoulder peak appears at higher q. These peaks were deconvoluted by a Gaussian-Lorenztian fit. A second peak at ~3.3 Å was obtained for all copolymers, which was attributed to the presence of water in the polymer under ambient conditions. X-ray scans of water swollen SPTES-50, -60, -70, and -80 membranes were also obtained. The intensity of the peak appearing at   3.3 Å was very high and it was masking any scattering intensity from the polymer.

The WAXS data of vacuum-dried sealed membranes illustrate a polymer peak at ~4.5 Å and a shoulder peak at 3.3 Å. This is an indication of the presence of water even after a long exposure to vacuum. NMR studies of these polymer membranes in our laboratories have established the presence of two types of water molecules [32]. These consist of loose water molecules that are in equilibrium with the environment and tightly-bound water molecules that are difficult to remove.

A series of dynamic X-ray scattering experiments were also performed (Fig. 6.22). A swollen SPTES–70 membrane was examined by X-ray diffraction. X-ray scans were taken at a series of time intervals of 10 min in a drying experiment. The intensity and peak position was monitored at 10 min time intervals for nearly 2 h until the sample was dried (Fig. 6.22). The intensity and the position of the polymer

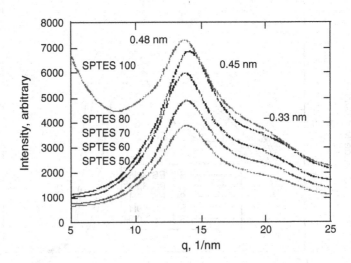

Fig. 6.21 Wide angle X-ray scattering data from oven-dried samples of SPTES-50, -60, -70, -80, and -100

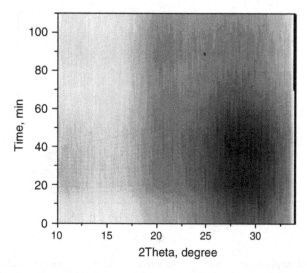

Fig. 6.22 Intensity of scattering vs. scattering angle, 2θ, for swollen SPTES-70 membrane, obtained at 10-min intervals. The intensity and peak position stayed at 4.5 Å, while the intensity of the water peak at 3.3 Å decreased with drying time

peak stayed constant during the time period of the experiment. The intensity of the peak at 3.3 Å, attributed to the presence of water, decreased with drying time. However, even after 2 h of drying, the peak at 3.3 Å did not disappear.

6.4.10 Small Angle X-Ray Scattering

Swollen SPTES-70 membrane was placed in the synchrotron X-ray beam at Brookhaven National Laboratories. Scattering data were obtained in time intervals until the sample was dry. Figure 6.23 demonstrates the scattering intensity vs. q during the drying experiment.

A change of the slope appearing at high q can be correlated to the water domains within the SPTES polymers. Neutron scattering studies at the National Institute of Standards and Technology revealed the presence of ~5 nm water domains formed in the ionomeric clusters of sulfonic groups. The change of the slope appearing at $q = 1.28$ nm^{-1} corresponding to 4.9 nm is indicative of the size of ionomeric clusters when SPTES-70 membrane is swollen. This result is in good agreement with the neutron scattering investigation of this sample [33]. The slope changes at high q shift toward higher q as water evaporates. This suggests that the size of water domains significantly become smaller and smaller as water evaporation proceeds. The slope change that started at $q = 1.28$ nm^{-1} shifted to 1.7 nm^{-1} indicating that the water domain sizes were decreased from 4.9 to 3.7 nm during the evaporation experiment. There is no X-ray scattering feature for the dry polymer confirmed by small-angle

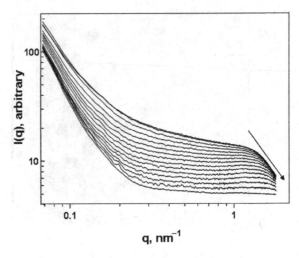

Fig. 6.23 Small angle X-ray scattering of the swollen SPTES-70 membrane, while drying in the X-ray beam. The ionic domain sizes decrease with water evaporation

neutron scattering experiments as well [21]. The upturn in the low q region is attributed to the aggregation of the ionic clusters containing water molecules.

The swollen SPTES-70 membrane was placed in the X-ray beam and stretched with a strain rate of 1 mm s^{-1} using X-27C beam line. X-ray scans were obtained every 64 s while stretching. The samples were positioned 45 degrees toward the beam direction to maximize the intensity of scattering. The maximum strain at break was 18.7%. The sample was kept moist during the stretching experiment and sprayed continuously to minimize drying effects in the X-ray scans. Figure 6.24 a shows the scattering before any stress was applied. An isotropic scattering from an SPTES swollen membrane with a broad peak appearing at ~4.5 nm is observed.

An anisotropic scattering is evident when the swollen SPTES-70 membrane was stretched to 18% before breaking (Fig. 6.24b). Azimuthally averaged intensity of scattering vs. q during the stretching process is represented in Fig. 6.25.

The orientation parameter was calculated according to the following (6.6) [33].

$$\langle \cos^2 \phi \rangle = \frac{\int_0^{2\pi} I(\phi)\cos^2 \phi \sin \phi \, d\phi}{\int_0^{2\pi} I(\phi)\sin \phi \, d\phi} \tag{6.6}$$

where I is the intensity of scattering as a function of azimuthal angle β and can be converted to $I(\phi)$ according to the (6.7). The angle between the normal to the scattering plane and the stress plane is defined by ϕ and ϕ_B is the Bragg angle.

$$\cos(\beta) \cos(\theta_B) = \cos \phi \tag{6.7}$$

Then the orientation parameter, S_d, was calculated from (6.8) by using the Herman orientation parameter.

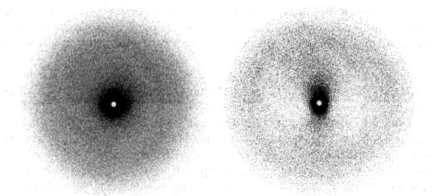

Fig. 6.24 Small angle X-ray scattering from the swollen SPTES-70 membrane. (**a**) Isotropic scattering from swollen SPTES-70 membrane with a broad peak at ~4.5 nm is evident (Left image). (**b**) Ionomeric clusters are oriented perpendicular to the stress direction (Right image)

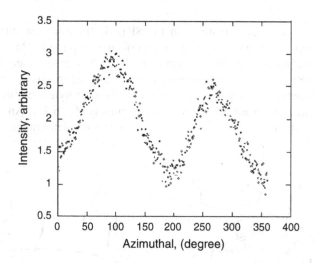

Fig. 6.25 Radially averaged intensity of scattering of the swollen SPTES-70 while stretching in the X-ray beam

$$S_d = \frac{3\langle \cos(\phi)\rangle - 1}{2} \tag{6.8}$$

The calculated orientation parameter is ~ −0.1 for the stretched DI water-swollen SPTES-70 membrane at the ultimate stress before the break. This is an indication of partial orientation of the ionic aggregates perpendicular to the stretch direction (fully perpendicular alignment results in an orientation parameter of −0.5).

The ionomeric aggregates with a cluster size in the range of ~4.1 nm are stretched and oriented perpendicular to the direction of stress (equatorial alignment). A proposed model for orientation of the ionic cluster perpendicular to the direction of stress is shown in Fig. 6.26.

The main axis of the ionic aggregates is randomly oriented with respect to the polymer chain axis in the relaxed state. The ionic clusters start to orient perpendicular to the main axis of the macromolecule during the stretching. This type of behavior upon stretching and relaxation has been observed for the semicrystalline materials [34]. In semicrystalline polymers, the orientation of the soft segment increases with increasing the draw ratio, while orientation function of the hard segments exhibits negative values initially, followed by positive values. Wang et al. [35] reported an orientation of the ionic domains containing sulfonic groups perpendicular to the stretch axis and the polymer chain axis in polytetramethylene oxide zwitterionomers for a draw ratio of 1.5.

6.4.11 Atomic Force Microscopy (AFM)

Tapping mode AFM examinations of the SPTES-70 membrane surface in the humidity range of $25–65 \pm 2\%$ RH were performed. Figure 6.27a–c shows the phase images of the SPTES-70 membrane surface at 22 ° C in the range of $25 – 50\%$ RH. The changes of the surface topology were followed in situ as a function of the relative humidity. AFM examinations of polyelectrolyte membrane surface texture like that of Nafion have been well documented [36,37]. An increase in the number and the area of cluster-like structure in a diameter range of 5–30 nm

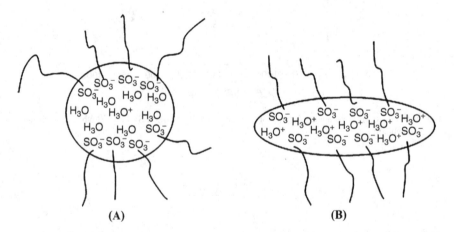

(A)   (B)

Fig. 6.26 Model indicating the effect of the stretching on the orientation of the ionic clusters with respect to the macromolecular chain. (**a**) The main axis of the nanocluster aggregates is randomly oriented relative to the macromolecular chain axis (spherical water domains). (**b**) The main axis of the ionic aggregates is perpendicular to the chain axis upon stretching

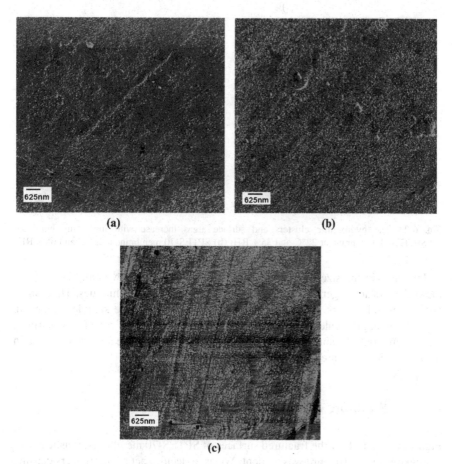

Fig. 6.27 Tapping mode AFM images of SPTES-70 membrane at 22°C and at (**a**) 25% RH, (**b**) 35% RH, and(**c**) 50% RH

have been reported by James et al. [36]. Figure 6.27a–c shows the phase images of the SPTES-70 membrane surfaces at relative humidities of 25, 35, and 50% at 22 ° C. Phase images provide information regarding polymer properties, phases, and viscoelasticity compared with height images, in which only topology information is obtained. Obtaining AFM images of the exact same position was extremely difficult due to the large expansion of the films both laterally and vertically upon water absorption. Figure 6.27a–c shows a phase separation of hydrophobic and hydrophilic clusters on the surface of the SPTES-70 membrane. The phase contrast between hydrophilic ionic sulfonated clusters and hydrophobic regions increases when water is preferentially adsorbed on to the hydrophilic regions.

Figure 6.28 is illustrative of higher resolution AFM images of SPTES-70 membrane at the beginning and end of the humidity sequence (35% and 65% relative humidity).

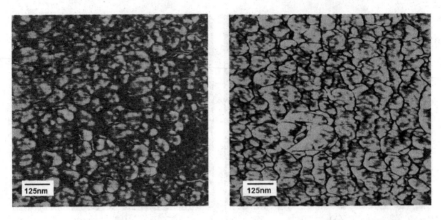

Fig. 6.28 The hydrophilic clusters and surface areas increase with increasing humidity. (**a**) SPTES-70 membrane at 22°C and 35% RH; (**b**) SPTES-70 membrane at 22°C and 65% RH

Increases in the size of surface cluster-like structures are evident. The sizes of these clusters are larger than what is observed by scattering techniques. This can be due to several factors. Scattering techniques probe the entire sample volume in centimeter length scales, while AFM can only examine the membrane surface. Also, there could be some coalescence of water clusters on the surface resulting in a larger hydrophilic area, as suggested for Nafion [37].

6.4.12   Scanning Electronic Microscopy

Figures 6.29a, b show the fractured surfaces of SPTES-70 membrane. Plastic deformation during the fracture was evident. Water molecules act as plasticizers causing chain slippage and plastic deformation of the surface during the fracture. The presence of water domains is evident. The size of the water domains is in the nanometer range. However, due to undergoing cryofracture, drying at room temperature, and exposure to vacuum, obtaining quantified information regarding the domain size was not possible.

The surface morphology of dry and cryo-swollen SPTES-70 membrane was examined. The cryofractured surface of the dry SPTES-70 membrane shows a smooth brittle fracture (Fig. 6.30). The SPTES-70 membrane was swollen in water and cryofractured in liquid nitrogen while swollen. Then the samples were dried at room temperature, lightly coated with carbon, and examined by scanning electron microscope.

6.4.13   Performances of Membrane Electrode Assembly

Membranes with painted electrodes along with a gas diffusion layer (carbon cloth) were positioned in a single cell fixture with graphite blocks as current collectors,

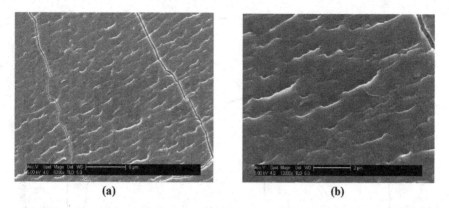

(a)   (b)

Fig. 6.29 Fractured surface of the swollen SPTES-70 membrane after drying. Surface deformation of the membrane due to the presence of the water domains is evident. The surface illustrates plastic deformation due to the plasticization effects of water molecules. (**a**) Magnification of ×5,000; (**b**) Magnification of ×12,000

Fig. 6.30 The fractured surface of dry SPTES-70 membrane indicating a smooth brittle fracture

having a serpentine flow pattern. The gases used (H_2, Air) were humidified, with the gas humidity bottles set at 80 ° C. Eighty percent fuel utilization along with a stoichiometry of 2 was used for the fuel gas and oxidant. Humidity was verified with an in-line high temperature humidity probe. Cell performance was obtained at 1 atm with cell temperatures at 80 ° C. MEAs using SPTES-50 polymer membrane and Nafion-117 cast membrane with effective areas of 5 cm^2 were fabricated. Polarization plots of Nafion and SPTES-50 based MEAs were obtained and compared; the results are shown in Fig. 6.31. Competing processes, which include

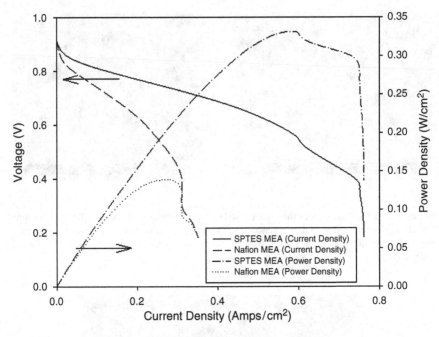

Fig. 6.31 Polarization graphs for SPTES-50 MEA and Nafion 117 MEA

catalyst activity, interfacial resistance, and diffusion processes, all occur to different extents and contribute to losses observed in the polarization curve.

Figure 6.31 shows the polarization curves of SPTES-50 polymer and Nafion-117 MEAs, providing a comparison of the relative performances of the two systems. The overall electrochemical performance of the SPTES-50 polymer MEA using conventional electrode inks with perfluorinated binders was superior to that of Nafion-117 MEA under comparable conditions (80°C, 55% relative humidity, 1 atm, stoichiometry = 2 for H_2 and air).

A higher current density of 0.52 amp cm^{-2} was obtained for the SPTES-50 polymer membrane-based MEA as compared with a current density of 0.24 amps cm^{-2} for Nafion-117 membrane-based MEA at 0.6 V potential. Similarly, a higher peak power density of 0.34 W cm^{-2} was obtained using SPTES-50 polymer membrane compared with 0.16 W cm^{-2} using Nafion-117 membrane in MEAs, measured under the same conditions. Estimates of hydrogen fuel permeability based upon measured open circuit voltage (OCV) indicate that SPTES-50 MEA and Nafion-117 exhibit similar rates of fuel crossover.

Further comparison of the electrochemical performances of the two MEA systems is also shown in Table 6.4. Overall, the electrochemical performance of the SPTES-50 MEA was superior to that of Nafion-117 MEA. The OCVs of SPTES-50 MEA and Nafion-117 are comparable. Calculated area specific resistance is lower (0.13 ± 0.03 ohm cm^{-2}) for the SPTES-50 MEA as compared with 0.23 ± 0.02 ohm cm^{-2} for the Nafion-117 MEA. The proton conductivity, as measured by the high

Table 6.4   Electrochemical performances of STPES-50 and Nafion-117 MEA

Sample	ASR (ohm cm^{-1})	Conductivity (mS cm^{-1})	R_{ct} (ohm cm^{-2})	OCV (V)
SPTES-50	0.13 ± 0.03	125	0.34	0.97
Nafion-117	0.23 ± 0.02	110	0.70	0.99

ASR area specific resistance; R_{ct} resistance; *OCV* open circuit voltage.

frequency resistance using area specific resistance (ASR) and as determined via simulation (R), is higher in the case of SPTES-50 membrane. The in situ proton conductivity of SPTES-50 membrane is 125 mS cm^{-1} compared with 110 mS cm^{-1} for Nafion-117.

6.5   Conclusions

Highly sulfonated poly(arylenethioethersulfone) (SPTES) polymers were successfully synthesized in high molecular weights via a polycondensation route. An improved method was developed for preparing sulfonated monomers in high yields. Novel sulfonated poly(arylenethioethersulfone) polymers were synthesized via the direct polymerization of the disulfonated dihalidesulfone monomers, unsulfonated dihalidesulfone monomers, and 4,4'-thiobisbenzethiol monomers. This method resulted in the random incorporation of the sulfonated monomer throughout the polymer structure. All the sulfonated polymers were soluble in polar, aprotic solvents at room temperature. The acid forms of the polymers could be fabricated into tough, flexible films from DMAc. The characterization of polymers by TGA demonstrated their thermal/thermo-oxidative stabilities. The cast polymer membranes have measured proton conductivities in the range of 100–300 mS cm^{-1}. The proton conductivity of polymers is at least three times higher than that of the state-of-the-art Nafion-H proton exchange membrane under nearly comparable conditions, indicating that these polymers are promising candidates for PEMs in fuel cells. Representative SPTES-50 polymer membrane was also successfully integrated into MEAs and their performance was compared with that of the MEA fabricated from Nafion-117 (cast membrane) as PEM.

References

1. T. A. Zawodzinski, C. Derouin, S. Radzinski, R. J. Sherman, V. T. Smith, T. E. Springer, S. Gottesfeld, Water uptake by the transport through Nafion 117 membranes, *J. Electrochem. Soc.* **140**(4), 1041–1047 (1993).
2. M. Higuchi, N. Minoura, T. Kinoshita, Photocontrol of micellar structure of an azobenzene containing amphiphilic sequential polypeptide, *Chem. Lett.* **23**, 227–230 (1994).
3. S. D. Flint, R. C. T. Slade, Investigation of radiation-grafted PVDF-g-polystyrene-sulfonic-acid ion exchange membranes for use in hydrogen oxygen fuel cells, *Solid State Ionics* **97**(1), 299–307 (1997).

4. F. Wang, J. Roovers, Functionalization of Poly(aryl ether ether ketone) (PEEK): Synthesis and properties of aldehyde and carboxylic acid substituted PEEK, *Macromolecules* **26**(20), 5295–5302 (1993).

5. Z. Qi, M. C. Lefebvre, P. G. Pickup, Electron and proton transport in gas diffusion electrodes containing electronically conductive proton-exchange polymers, *J. Electroanal. Chem.* **459**(1), 9–14 (1998).

6. A. D. Child, J. R. Reynolds, Water-soluble rigid-rod polyelectrolytes: A new self-doped, electroactive sulfonatoalkoxy-substituted poly(p-phenylene), *Macromolecules* **27**(7), 1975–1977 (1994).

7. T. Kobayashi, M. Rikukawa, K. Sanui, N. Ogata, Proton-conducting polymers derived from poly(ether-etherketone) and poly(4-phenoxybenzoyl-1,4-phenylene), *Solid State Ionics* **106**(3), 219–225 (1998).

8. A. J. Chalk, A. S. Hay, Metalation of poly(phenylene ethers), *J. Polym. Sci.: Polym. Lett. Ed.* **6**(2), 105–107 (1968).

9. X. L. Wei, Y. Z. Wang, S. M. Long, C. Bobeczko, A. J. Epstein, Synthesis and physical properties of highly sulfonated polyaniline, *J. Am. Chem. Soc.* **118**(11), 2545–2555 (1996).

10. K. Miyatake, E. Shouji, K. Yamamoto, E. Tsuchida, Synthesis and proton conductivity of highly sulfonated poly(thiophenylene), *Macromolecules* **30**(10), 2941–2946 (1997).

11. K. Miyatake, H. Iyotani, K. Yamamoto, E. Tsuchida, Synthesis of poly(phenylene sulfide sulfonic acid) via poly(sulfonium cation) as a thermostable proton-conducting polymer, *Macromolecules* **29**(21), 6969–6971 (1996).

12. X. Jin, M. T. Bishop, T. S. Ellis, F. E. Karasz, A sulfonated poly(aryl ether ketone), *Br. Polym. J.* **17**(1), 4–10 (1985).

13. J. Lee, C. S. Marvel, Poly aromatic ether-ketone sulfonamides prepared from polydiphenyl ether-ketones by chlorosulfonation and treatment with secondary amines, *J. Polym. Sci.: Polym. Chem. Ed.* **22**(2), 295–301 (1984).

14. M. I. Litter, C. S. Marvel, Polyaromatic ether-ketones and polyaromatic ether-ketone sulfoamides from 4-phenoxybenzoyl chloride and 4,4′-dichloroformyldiphenyl ether, *J. Polym. Sci.: Polym. Chem. Ed.* **23**(8), 2205–2223 (1985).

15. M. T. Bishop, F. E. Karasz, P. S. Russo, K. H. Langley, Solubility and properties of a poly(aryl ether ketone) in strong acids, *Macromolecules* **18**(1), 86–93 (1985).

16. J. Devaux, D. Delimoy, D. Daoust, R. Legras, J. P. Mercier, C. Strazielle, E. Neild, On the molecular weight determination of a poly(aryl-ether-ether-ketone) (PEEK), *Polymer* **26**(13), 1994–2000 (1985).

17. J. J. Sumner, S. E. Creger, J. J. Ma, D. D. DesMarteau, Proton conductivity in Nafion 117 and in a novel bis[(perfluoroalkyl)sulfonyl]imide ionomer membrane, *J. Electrochem. Soc.* **145**(1), 107–110 (1998).

18. M. Ueda, H. Toyota, T. Ochi, J. Sugiyama, K. Yonetake, T. Masuko, T. Teramoto, Synthesis and characterization of aromatic poly(ether Sulfone)s containing pendant sodium sulfonate groups, *J. Polym. Sci. Polym. Chem. Ed.* **31**(4), 853 (1993).

19. T. A. Zawodzinski Jr., T. E. Springer, J. Davey, R. Jestel, C. Lopez, J. Valerio, S. Gottesfeld, A comparative study of water uptake by and transport through ionomeric fuel cell membranes, *J. Electrochem. Soc.* **140**(7), 1981 (1993).

20. P. J. James, J. A. Elliot, T. J. McMaster, J. M. Newton, A. M. S. Elliot, S. Hanna, M. J. Miles, Hydration of Nafion studied by AFM and X-ray scattering, *J. Mater. Sci.* **35**(20), 5111–5120 (2000).

21. S. N. Magonov, V. Elings, M.-H. Whangbo, Phase-imaging and stiffness in tapping-mode atomic force microscopy, *Surf. Sci.* **375**(2/3), L385 (1997).

22. F. Wang, M. Hickner, J. E. McGrath, Direct polymerization of sulfonated poly(arylene ether sulfone) random (statistical) copolymers: Candidates for new proton exchange membranes, *J. Membr. Sci.* **197**(1), 231–242 (2002).

23. C. Bailly, D. J. Williams, F. E. Karasz, W. J. MacKnight, The sodium salts of sulfonated poly(aryl-ether-ether-ketone) (PEEK): Preparation and characterization, *Polymer* **28**(6), 1009–1016 (1987).

24. F. Wang, M. Hickner, Q. Ji, W. Harrison, J. Mecham, T. Zawodzinski, J. E. McGrath, Synthesis of highly sulfonated poly(arylene ether sulfone) random (statistical) copolymers via direct polymerization, *Macromol. Symp.* **175**(1), 387–396 (2001).
25. J. Kerres, W. Zhang, W. Cui, New sulfonated engineering polymers via the metalation route. II. Sulfinated/sulfonated poly(ether sulfone) PSU Udel and its crosslinking, *J. Polym. Sci. Part A: Polym. Chem.***36**(9), 1441–1448 (1998).
26. R. N. Johnson, in: *Encyclopedia of Polymer Science and Technology*, N. M. Bikales (ed.), Wiley, New York, 1969.
27. F. Wang, T. L. Chen, J. P. Xu, Sodium sulfonate-functionalized poly(ether ether ketone)s, *Macromol. Chem. Phys.* **199**(7), 1421–1426 (1998).
28. H. C. Lee, H. S. Hong, Y. M. Kim, S. H. Choi, M. Z. Hong, H. S. Lee, K. Kim, Preparation and evaluation of sulfonated-fluorinated poly(arylene ether)s membranes for a proton exchange membrane fuel cell (PEMFC), *Electrochim. Acta* **49**(14), 2315–2323 (2004).
29. M. Yoonessi, Z. Bai, T. D. Dang, M. F. Durstock, R. A. Vaia, *in preparation*.
30. M. Rikukawa, K. Sanui, Proton-conducting polymer electrolyte membranes based on hydrocarbon polymers, *Prog. Polym. Sci.* **25**(10), 1463–1502 (2000).
31. J. F. Ding, C. Chuy, S. Holdcroft, Solid polymer electrolytes based on ionic graft polymers: Effect of graft chain length on nano-structured, ionic networks, *Adv. Funct. Mater.* **12**(5), 389–394 (2002).
32. P. Mirau, NMR studies performed on SPTES membranes, Internal Air Force Research Laboratory Reports, Dec. 2004.
33. H. Koerner, Y. Luo, X. Li, C. Cohen, R. C. Hedden, C. K. Ober, Structural studies of extension-induced mesophase formation in poly(diethylsiloxane) elastomers: In situ synchrotron WAXS and SAXS, *Macromolecules***36**(6), 1975–1981 (2003).
34. F. Ye, B. S. Hsiao, B. B. Sauer, S. Michel, H. W. Siesler, In-situ studies of structure development during deformation of a segmented poly(urethane–urea) elastomer, *Macromolecules* **36**(6), 1940–1954 (2003).
35. Y. Wang, C. Pellerin, C. G. Bazuin, M. Pezolet, Molecular orientation and relaxation in uniaxially stretched segmented ptmo zwitterionomers by polarization modulation infrared linear dichroism, *Macromolecules* **38**(10), 4377–4383 (2005).
36. P.J. James, M. Antognozzi, J. Tamayo, T. J. McMaster, J. M. Newton, M. J. Miles, Interpretation of contrast in tapping mode afm and shear force microscopy. A study of nafion, *Langmuir* **17**(2), 349–360 (2001).
37. P. Krtil, A. Trojanek, Z. Samec, Kinetics of water sorption in nafionthin films – Quartz crystal microbalance study, *J. Phys. Chem. B.* **105**(33), 7979–7983 (2001).

Chapter 7
Polymer Composites for High-Temperature Proton-Exchange Membrane Fuel Cells

Xiuling Zhu, Yuxiu Liu, and Lei Zhu

Abstract Recent advances in composite proton-exchange membranes for fuel cell applications at elevated temperature and low relative humidity are briefly reviewed in this chapter. Although a majority of research has focused on new sulfonated hydrocarbon and fluorocarbon polymers and their blends to directly enhance high temperature performance, we emphasize on polymer/inorganic composite membranes with the aim of improving the mechanical strength, thermal stability, and proton conductivity, which depend on water retention at elevated temperature and low relative humidity conditions. The polymer systems include perfluoronated polymers such as Nafion, sulfonated poly(arylene ether)s, polybenzimidazoles (PBI)s, and many others. The inorganic proton conductors are silica, heteropolyacids (HPA)s, layered zirconium phosphates, and liquid phosphoric acid. Direct use of sol-gel silica requires pressurization of fuel cells to maintain 100% relative humidity for high proton conductivity above 100°C. Direct incorporation of HPAs such as phosphotungstic acid (PTA) into polyelectrolyte membranes is capable of improving both proton conductivity and fuel cell performance above 100°C; however, they tend to leach out of the membrane whenever fuel cell flooding happens. To prevent HPA leaching, amine-functionalized mesoporous silica is used to immobilize PTA in Nafion membranes, whose proton conductivity and fuel cell performance are discussed. Compared with Nafion, sulfonated poly(arylene ether)s such as sulfonated poly(arylene ether sulfone)s are cost-effective materials with excellent thermal and electrochemical stability. Their composites with HPAs show increased proton conductivity at elevated temperatures when fully hydrated. Organic/inorganic hybrid membranes from acid-doped PBIs and other polymers are also discussed.

7.1 Introduction

Recently, intense academic and industrial research efforts have focused on proton-exchange membrane fuel cells (PEMFC)s due to the promise of commercialization and mass production. In response to the rapid depletion of fossil fuels, PEMFCs use alternative and renewable energy/fuels for zero or minimal pollutant

S.M.J. Zaidi, T. Matsuura (eds.) *Polymer Membranes for Fuel Cells*,
doi: 10.1007/978-0-387-73532-0, © Springer Science+Business Media, LLC 2009

emission, significantly protecting our natural environment. Although PEMFCs have a great commercial potential in fuel cell–powered automotives, portable power supplies/batteries, and stationary power plans for buildings and residential homes, significant technical challenges, even for the PEMs alone, still await technologic resolution and improvements, such as retaining high proton conductivity at high temperatures and low relative humidity, reducing mechanical and chemical degradation, devising better water management, etc. For example, the state-of-the-art polymer electrolyte, perfluorosulfonic acid (PFSA) membrane–Nafion requires 100% hydration to achieve excellent proton conductivity at ~0.1 S cm^{-1}. Obviously, this has prevented PEMFCs from operating above 100°C at ambient pressure and requires fully hydrated fuels such as hydrogen (H_2) and oxygen (O_2) under operating conditions. Practically, the hydrogen fuels are often generated from hydrocarbon (e.g., gasoline and natural gas) reforming using water–gas shift reactions, and are often contaminated with up to 1% carbon monoxide (CO). Although CO tolerance can be achieved by using more Pt-Ru alloyed catalysts in the anode (1–2 mg cm^{-2}), it is limited to only 100–200 ppm CO levels and low current densities. Alternatively, CO tolerance can also be achieved by complicated oxygen or air bleeding at the anode. However, oxygen/air bleeding also has a limitation of low O_2 concentration at the anode side because of the low flash point threshold of H_2 (5% O_2 in H_2). Ultimately, this limits the CO tolerance up to 100 ppm.

Regarding the physisorption nature of CO onto Pt, high temperature (120–150°C) PEMFC operation, on the other hand, provides significant benefits. First, CO desorption from the Pt surfaces can be achieved at high temperatures, together with accelerated oxygen reduction kinetics at the cathode side. Second, operation at elevated temperatures enhances heat transfer and decreases weight and volume of the heat exchanger. Finally, enhanced CO tolerance helps to remove the multiple oxidation steps in the hydrocarbon reforming processes, and thus increases overall system efficiency. However, high temperature fuel cell operation seemingly has disadvantages. For example, water vapor partial pressure in the fuel stream increases, whereas the reactant (H_2 and O_2) partial pressure decreases. The relative humidity substantially drops and the membrane becomes dehydrated, accompanied by a significant decrease in proton conductivity. To ensure adequate reactant supply in the fuel stream and avoid dehydration of the membrane, pressurized fuel cell operation has to be adopted, which obviously requires extra energy and thus decreases overall system efficiency.

In this respect, modification of the polyelectrolyte membrane to permit operation at high temperature and low relative humidity seems to be a promising strategy. To pursue this goal, different approaches have been investigated, such as: (1) selecting alternative high-performance polyelectrolyte membranes based on solid-state proton conductive materials; (2) modifying PFSA membranes with hygroscopic inorganic materials to improve their water retention property at temperatures above 100°C; and (3) modifying PFSA membranes with inorganic proton conductors to obtain reasonable proton conductivity, which will be hardly dependent on the water contents.

7.2 Alternative Hydrocarbon Polymers for High Temperature PEMFC Applications

Because of good thermal and hydrolytic stability, excellent mechanical and chemical stability, low cost, and commercial availability of sulfonated aromatic hydrocarbon polymers, recent research has focused on the synthesis and development of sulfonated aromatic hydrocarbon polymers specifically for high-temperature PEMFCs. Typical examples include sulfonated poly(ether ether ketone) (SPEEK) or poly(ether ketone ketone) (SPEKK) [1,2], sulfonated poly(ether sulfone) (SPSF) [3], alkyl sulfonated polybenzimidazole (PBI), sulfonated naphthalenic polyimides (sNPI) [4–6], sulfonated poly(phenylene sulfide) [7,8]. Both post- and pre-sulfonation methods have been used in the past. Other than the post-sulfonation modification of aromatic polymers, recently, efforts have been dedicated to direct polycondensation from sulfonic acid containing monomers to synthesize sulfonated polymers [9]. The latter approach, namely pre-sulfonation, is widely applied because of the ease of controlling sulfonation degree and deactivated sites in the arylene backbones, which further avoid side reactions such as decomposition and hydrolysis of polymers resulted from the post-sulfonation method.

SPEEK is a widely studied candidate for high-temperature fuel cell membranes. However, how to balance the achievement of high proton conductivity and minimization of membrane swelling behavior at high degrees of sulfonation is still under investigation. Possible solutions are: (1) partial cross-linking of the membranes; and (2) blending or compounding with other nonproton conductive, high-performance polymers. Compared with Nafion, SPEEK shows less hydrophobic/hydrophilic separation in morphology; therefore, the proton transport coefficient decreases greatly with decreasing the water content. For post-sulfonated PEEK membranes, several disadvantages exist. For example, chain-scission or cross-linking occur under severe sulfonation conditions, or the main-chain ether linkages are susceptible to hydrolytic degradation under chemical and/or electrochemical conditions.

McGrath et al. synthesized novel biphenol-based aromatic poly(arylene ether sulfones) (e.g., PBPSH, see Scheme 7.1) with the deactivated sulfonic acid substituents on the polymer backbone [10,11]. The copolymers having a high degree of

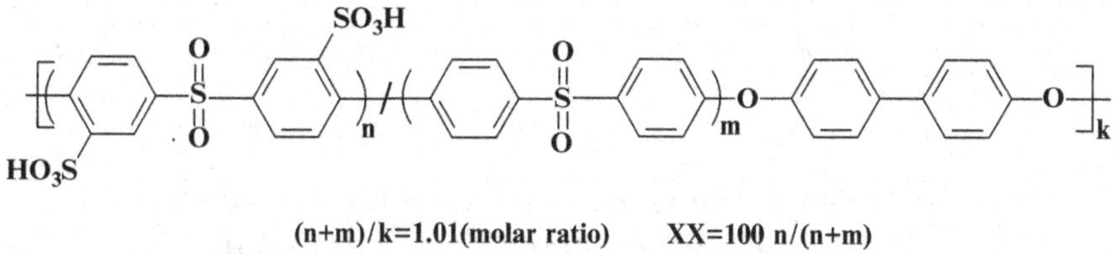

(n+m)/k=1.01(molar ratio)   XX=100 n/(n+m)

PBPSH-XX or BPSH

Scheme 7.1 Chemical structure of randomly sulfonated poly(arylene ether sulfones), where XX represents the sulfonation level

sulfonation, such as PBPSH-60, demonstrated a significant increase in water uptake than those with lower contents of sulfonic acid groups (see Fig. 7.1). Tapping-mode (TM) atomic force microscopy (AFM) studies showed that at low sulfonation degrees (PBPSH-40 and below), the hydrophilic ionic domains were isolated, which might be responsible for low water uptake and low proton conductivity. At high sulfonation

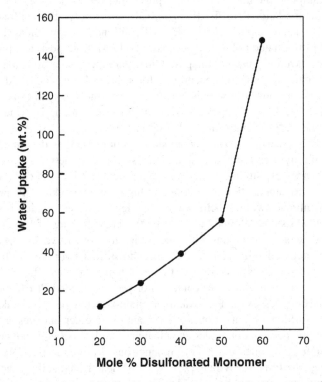

Fig. 7.1 Influence of the degree of sulfonation on the water uptake of sulfonated poly(arylene ether sulfone) copolymers (PBPSH). Reprinted with permission from Ref. [10]: F. Wang, et al., Direct polymerization of sulfonated poly(arylene ether sulfone) random (statistical) copolymers: Candidates for new proton exchange membranes. *J. Membr. Sci.* **197**, 231–242 (2002). Copyright Elsevier

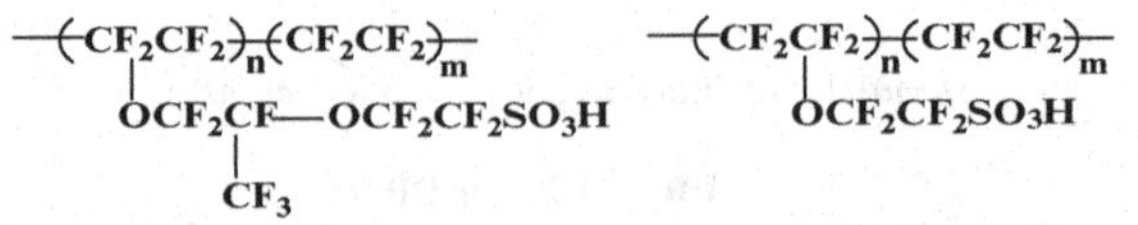

Scheme 7.2 Chemical structures of Nafion and Dow Hyflon membranes

levels (PBPSH-50 and PBPSH-60), the hydrophilic ionic domains percolated to form continuous microdomains. It was speculated that the percolation threshold was obtained from a PBPSH-50 sample. The percolated hydrophilic domain morphology was reminiscent of that in Nafion at ambient humidity [10]. However, for both PBPSH and Nafion, proton conductivity showed a significant dependence on the relative humidity, i.e., an exponential decay with decreasing relative humidity.

The process of acid treatment in the last step of synthesis, converting the sodium form PBPS to acid form PBPSH, showed a significant influence on the performance of the membranes. Two methods were used. Method 1 referred to the sulfuric acid treatment of membranes at 30°C, and method 2 referred to the process at 100°C. Generally, the fully hydrated membranes treated using method 2 achieved higher proton conductivity, which depended less on temperature than the membranes from method 1 in a temperature range of 70–120°C (see Fig. 7.2) [11]. The BPSH-40 membranes had proton conductivity higher than 0.1 S cm^{-1}. The AFM studies further confirmed that the membranes made by method 2 possessed larger hydrophilic ionic domains with more connectivity than the membranes made by method 1.

Considerable research work has been dedicated to the improvement of sulfonated polyamide (SPI) for fuel cell applications because of their excellent chemical, thermal, and mechanical stability, as well as their reasonably good electrochemical properties [4–6]. Okamoto et al. prepared anhydride-terminated SPNI oligomers from 1,4,5,8-naphathlenetetracarboxylic dianhydride and sulfonated diamine, and 1,3,5-tris(4-aminophenoxy)benzene, serving as a cross-linking reactant, was subsequently added to afford the branching of SNPI [4]. The obtained membranes exhibited good proton conductivity of 0.02–0.25 S cm^{-1} at 50–100% relative humidity, together with low membrane swelling and good

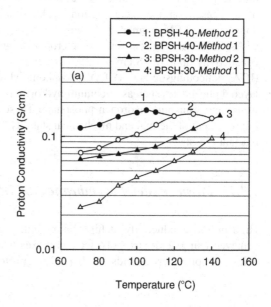

Fig. 7.2 Influence of temperature and H$_2$SO$_4$ treatment methods on the proton conductivity of PBPSH-30 and PBPSH-40. Reprinted with permission from Ref. [11]: Y. S. Kim, et al., Effect of acidification treatment and morphological stability of sulfonated poly(arylene ether sulfone) copolymer proton-exchange membranes for fuel-cell use above 100°C, *J. Polym. Sci.: Part B: Polym. Phys.* **41**, 2816–2828 (2003). Copyright Wiley-VCH

mechanical integrity. However, SNPI was shown to exhibit problems in long-term (>5,000 h) hydrolytic stability, which should be considered for real-world fuel cell applications.

Granados-Focil and Litt et al. synthesized poly(p-phenylene sulfonic acids) (SPPP) and its copolymers with high molecular weights through copper catalyzed coupling of dibromo-aromatic sulfonates [12,13]. The membranes had good chemical stability up to 270°C. Biphenylsulfone-containing copolymers were also synthesized and were promising candidates to achieve thermally cross-linked, highly conductive, and dimensionally stable membranes [12].

7.3 Polymer Composites for High Temperature PEMFCs

Although synthesis of new high performance proton conductive polymers for PEMFC applications is promising, many molecular and structural parameters are difficult to manipulate and obtain high fuel cell performance. An alternative approach to achieve high proton conductivity in PEMs at reduced relative humidity is polymer composite membranes. The composite membranes may be divided into two categories: organic/organic polymer blend membrane and organic/inorganic composite membranes with the polymer as the host matrix. The first method is associated with the blend of sulfonic acid–containing polymers and the basic polymers such as amine-containing polymers. Sometimes, sulfonic acid–containing polymers are also impregnated with imidazole or PBI with phosphoric acid, where imidazole and phosphoric acid are used to replace water as proton conductors. Incorporation of hygroscopic inorganic materials such as silica into PEM membranes may retain water in the membranes, and thus keep high proton conductivity at high temperatures. Most recently, approaches have focused on the incorporation of hydrophilic inorganic proton–conducting dopants into PEM membranes, such as heteropolyacids (HPAs) and isopolyacids (IPAs). HPAs are the acid form of heteropolyanions and have extremely high proton conductivity up to 0.2 S cm^{-1} at 25°C. Examples of HPAs are phosphotungstic acid (PTA), silicotungstic acid (STA), phosphomolybdic acid, silicomolybdic acid, and layered phosphates such as zirconium hydrogen phosphate, tungsten hydrogen phosphate, and vanadium hydrogen phosphate. These solid acids can also assist in the improvement of thermal and mechanical stability, water absorption, and resistance to reactant crossover.

7.3.1 Polymer Blend Membranes for High-Temperature PEMFCs

High proton conductivity at high temperature and low relative humidity can be achieved using acid-base polymer complexes between basic polymers and strong acids or polymeric acids [1,14]. The proton-accepting polymers include poly(ethylene oxide) (PEO), poly(vinyl acetate) (PVAc), polyacrylamide (PAAM), and poly(ethylene imine) (PEI). Imidazole and pyrazole were used to substitute the solvating water in the proton-conducting polymeric matrix, and proton conduction could take place at high temperatures in the absence of water [15]. Although high proton conductivity comparable to Nafion (~0.1 S cm^{-1} at 160–180°C) was obtained at high temperatures, imidazole-impregnated membranes poisoned the catalysts and no current could be produced from the fuel cell [16].

Kerres investigated novel acid-base polymer blend membranes composed of acid polymers such as SPEEK or sulfonated poly(ethersulfone) (SPSU) and basic polymers such as poly(4-vinylpyridine) (P4VP), poly(ethyleneimine) (PEI), and PBI [14]. The ionic complexation between the basic and acidic polymer compounds via the specific interaction of the $-SO_3H$ groups and the basic $- N =$ groups are formed by proton transfer from the acidic to the basic groups. These blend membranes displayed good performance in H_2/O_2 PEMFC, which was slightly higher than that of Nafion 112 with decomposition temperatures up to 270 and 350°C, depending on different types of acid-basic polymer blends.

Yamada and Honma have investigated poly(vinylphosphonic acid) (PVPA)/ organic basic heterocycle, such as imidazole (Im) or pyrazole (Py), as well as 1-methylimidazole (MeIm), acid-base composites [17]. The study suggested that the basicity and molecular structure of heterocycle in the acid-base polymer complexes are important factors to obtain the proton conductivity at elevated temperature (see Table 7.1). It revealed that pyrazole molecule with a low basicity did not behave as a proton acceptor in the composite material, since the free proton from the PVPA polymer could not strongly interact with the non-protonated nitrogen group in the pyrazole ring. Imidazole molecules have been reported to form molecular clusters, consisting of approximately 20 molecules [18], through intermolecular hydrogen bonding. Therefore, PVPA-Im composites might possess fast proton transfer between heterocycle molecules in the composite material [17]. The fuel cell test of PVPA-Im composite material using dry H_2/O_2 showed a power density of approximately 10 mW·cm^{-2} at 80°C.

Table 7.1 Maximum proton conductivities of PVPA-heterocycle composites and the pKa values of various heterocycle molecules. Reprinted from Ref. [17]: Yamada and Honma, Anhydrous proton conducting polymer electrolytes based on poly(vinylphosphonic acid)-heterocycle composite material, *Polymer* **46**, 2986–2992 (2005), Copyright Elsevier.

PVPA-heterocycle	Maximum conductivity,[a] S·cm^{-1}	PKa value of heterocycles [18]	
		PKa$_1$	pKa$_2$
PVPA-Im	7×10^{-3}	7.2	14.5
PVPA-Py	8×10^{-4}	2.5	14
PVPA-MeIm	1×10^{-3}	7.4	—

[a]Maximum proton conductivity at 150°C under dry conditions.

7.3.2 Organic/Inorganic Composite Membranes for High-Temperature PEMFCs

Since Malhotra and Datta proposed; the incorporation of inorganic solid acids in conventional polymer electrolyte membranes, such as Nafion with a goal of improving water retention and providing additional acidic sites in 1997 [19], this applicable approach attracted much attention because of their enhanced proton conductivity at elevated temperature and low relative humidity operating conditions. In these organic/inorganic hybrid membranes, the polymer matrices usually are Nafion or other PFSAs, PBI, sulfonated polysulfone (SPSF), and SPEEK, etc. The solid inorganic fillers, which can provide the composites with enhanced mechanical and thermal stability, usually include inorganic proton conductors such as heteropolyacids or zirconium phosphates with high proton conductivity and inorganic solid oxides such as amorphous silica (SiO_2) with high internal surface areas to retain water [20,21]. The HPAs are highly conductive and thermally stable in their crystalline forms [22,23]. For example, PTA ($H_3PW_{12}O_{40} \cdot nH_2O$, where n is up to 29) exhibits a room temperature conductivity of 0.19 $S \cdot cm^{-1}$, and its sodium form has a conductivity of $\sim 10^{-2}$ S cm^{-1} [24]; therefore, it is the most widely used in research work. Kreuer suggested that the PTA acted as a Brönsted acid toward hydration and water was loosely bound in the PTA structure, resulting in high proton conductivity [25].

7.3.2.1 Nafion-Based Composite Membranes

Watanabe et al. [26] and Antonucci et al. [27] investigated the reduction of water above 100°C to maintain proton conductivity by incorporating hydrophilic micron-sized metal oxide particles such as SiO_2 and TiO_2 into Nafion with limited success. Peak power densities of 250 and 150 $mW \cdot cm^{-2}$ in oxygen and air, respectively, were reported. Mauritz et al. used an in situ sol-gel technique to introduce a polymeric form of SiO_2 into Nafion to form composite membranes [28]. Recently, Adjemian et al. utilized Mauritz synthetic procedure to fabricate Nafion/SiO_2 membranes and demonstrated that water management within Nafion improved at elevated temperatures in a 3-atm pressurized H_2/O_2 PEMFC [29]. Various PFSAs, including Nafion and Aciplex, were studied as pure and in the SiO_2 composite membranes for operation in H_2/O_2 PEMFCs from 80 to 140°C. These cells demonstrated acceptable current densities, for instance, Nafion-117/SiO_2 and Nafion-112/SiO_2 achieved 850 and 1280 $mA \cdot cm^{-2}$, respectively, at 0.4 V up to at least 130°C, 3 atm pressure, and 100% relative humidity.

One of the early studies for the high-temperature operation of PEMFCs was the impregnation of Nafion 117 with HPAs [19]. The composite membranes showed promising results for short-term operation of PEMFCs at 110–115°C and 1 bar. Unfortunately, these membranes tended to lose HPAs by dissolution in the water present in the fuel cell environment, e.g., in the case of flooding. Efforts have been pursued recently to impregnate Nafion membranes with a highly hygroscopic but insoluble solid such as zirconium phosphate (ZrP) or zirconium hydrogen phosphate (ZrHP) to minimize the dissolution/leaching problem [30]. It was demonstrated that 4–6 times better fuel cell performance was achieved at 130°C at 3 bar pressure and 100% relative humidity, although only a limited increase in proton conductivity was achieved, as compared with pure Nafion. α-ZrP was also dispersed in SPEEK and its proton conductivity improved three times, with a loading of 10 wt% inorganic fillers [31]. Si et al. studied ZrHP/Nafion composite membranes for high-temperature fuel cells [32]. The composite membrane showed lower resistance (0.3 $\Omega \cdot cm^2$) than that (0.6 $\Omega \cdot cm^2$) of Nafion at 120°C and 31% relative humidity.

Jalani et al. [33] and Thampan et al. [34] reported the synthesis of Nafion-MO$_2$ (M = Zr, Si, Ti) nanocomposite membranes for high-temperature PEMFCs. An in situ sol-gel approach was applied and completely transparent membranes were obtained. A Nafion-ZrO$_2$ sol-gel nanocomposite showed slightly higher conductivity than pure Nafion at the same activity of water vapor, as shown in Fig. 7.3. This was possibly due to its improved water uptake, strong acid sites, and a higher ratio of bulk to surface water in the presence of sol-gel ZrO$_2$. At 120°C, Nafion-SiO$_2$ sol-gel nanocomposite exhibited lower conductivity than Nafion, as opposed to 8–10% higher conductivity for the ZrO$_2$ composite membranes. Figure 7.4 showed a

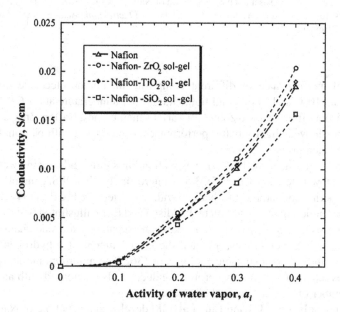

Fig. 7.3 Conductivity dependence on the activity of water vapor for nanocomposite Nafion-MO$_2$ (M = Zr, Si, Ti) and Nafion membrane at 120°C. Reprinted with permission from Ref. [33]: N. H. Jalani, et al., Synthesis and characterization of Nafion®-MO$_2$ (M = Zr, Si, Ti) nanocomposite membranes for higher temperature PEM fuel cells, *Electrochim. Acta* **51**, 553–560 (2005). Copyright Elsevier

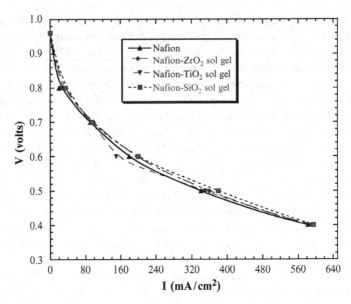

Fig. 7.4 The cell performance of Nafion 112 membrane electrode assembly (MEA) and Nafion-MO2 sol-gel composite MEAs. O2 and H2 at 2.0 and 1.3 times stoichiometry flows, respectively, pressure at 1.0 atm, humidifier at 80C, and cell at 110C. Reprinted with permission from Ref. [33]: N. H. Jalani, et al., Synthesis and characterization of Nafion®-MO2 (M=Zr, Si, Ti) nanocomposite membranes for higher temperature PEM fuel cells, *Electrochim. Acta* **51**, 553–560 (2005). Copyright Elsevier

single cell performance of different composite membrane electrode assemblies (MEAs) at 110°C and 1 atm conditions (humidifier temperature at 80°C). Although no significant difference was found for various membranes, Nafion-ZrO$_2$ composite membrane showed slightly better performance as compared with Nafion and other composite membranes.

A PFSA with the same backbone as Nafion but shorter side chains was synthesized by Dow Chemical Co. [35]. Most interestingly, this PFSA membrane possessed superior conductivity to Nafion with a shorter -OCF$_2$CF$_2$SO$_3$H side chain. Molecular modeling investigation by Paddison and Elliot illustrated that the number of CF$_2$ groups (n = 5, 7, and 9) in the backbone separating the side chains in short side chain PFSA membranes, affected the establishment of a hydrogen bonding network of the terminal sulfonic acid groups [36]. The minimum number of water molecules required for effective proton conduction also increased with an increase in the number of CF$_2$ groups in the backbone.

Most recently, Liu [37] and Liu et al. [38] developed a new type of Nafion/SiO$_2$ composite membranes with immobilized PTA via both in situ and ex situ sol-gel procedures. The silica was first functionalized with aminopropyl triethoxylsilane (APTES) to form amine-containing silica materials. PTA was immobilized onto silica by ionic complexation with the amine groups in modified silica. The in situ sol-gel reaction was carried out in the presence of Nafion, while the ex situ sol-gel reaction was performed without Nafion, which was blended in a second step. Although the in situ strategy was able to immobilize PTA in the membrane, the obtained composite membranes were brittle at SiO_2 contents exceeding 10 wt%. It was speculated that competitive ionic complexion existed between Nafion and PTA with the amine groups on functionalized silica. The in situ Nafion/APTES/PTA composite membrane showed much lower conductivity and performance than the Nafion/PTA composite membrane at the same condition at 80–120°C and 1 atm pressure. The reason for low fuel cell performance of the in situ Nafion/APTES/PTA composite membranes was possibly due to the residual-NH_2 poisoning of Pt catalysts in the MEA.

An ex situ method was developed in order to overcome these problems; a mesoporous silica material (SBA-15) was prepared according to the literature [39], followed by functionalization using APTES [40]. PTA was immobilized in the nanopores by ionic complexation with monolayer amine groups at the internal pore surfaces. The mesoporous silica morphology was observed using field emission scanning electron microscope (FE-SEM). A low magnification micrograph in Fig. 7.5 shows wheat-like particles with an average size of 1~2 μm, while a high magnification FESEM image in Fig. 7.6 reveals the hexagonally packed pore microstructure.

Composite membranes were obtained by solution-casting of Nafion with PTA-immobilized SBA-15 at different weight percentages, and were denoted as

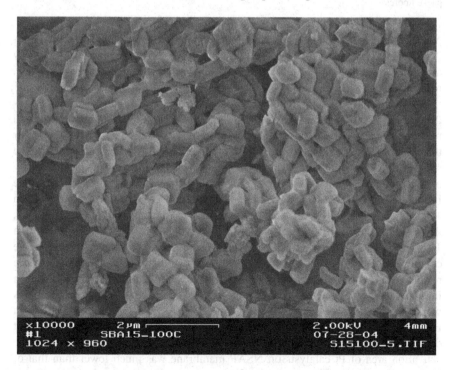

Fig. 7.5 A low-magnification FE-SEM micrograph of PTA-loaded SBA-15 composite

Fig. 7.6 A high-magnification FE-SEM micrograph of PTA-loaded SBA-15, showing hexagonally packed pores

NSAP-xx (xx is the weight percentage of the inorganic particles). The proton conductivity was studied and showed that the NSAP-2.5 had proton conductivity around 0.16 S·cm^{-1} and 0.01 S·cm^{-1} at fully hydrated and 50% relative humidity, respectively, at 120°C.

NSAP composite membranes, especially NSAP-2.5, exhibited promising conductivity at 120°C and a fully hydrated condition. A further investigation of the cell performance showed that the cell performance of NSAP-2.5 membrane was lower than that of Nafion 112 and the results at 80°C and 120°C under 1 atm pressure are shown in Fig. 7.7. For example, at 80°C, the voltage was 0.57 V at a current density of 400 mA·cm^{-2}, while it was 0.74 V for Nafion 112. However, the cell resistance for NSAP-2.5 (~0.05 Ω·cm^2) was slightly lower than that (0.08 Ω·cm^2) of Nafion 112. For NSAP-2.5 at 120°C, the cell performance was even lower. Surprisingly, the cell resistance (~0.16 Ω·cm^2) was still lower than that (0.22 Ω·cm^2) of Nafion 112. It seemed that the lower cell performance could not be explained by membrane conductivity. We speculated that the Pt catalysts could also be contaminated by the residual amount of free amines in the membrane, although most APTES had been immobilized to the internal surfaces of SBA-15. Cyclic voltametry experiments were performed on both NSAP-2.5 and Nafion MEAs, and the results are shown in Fig. 7.8. For the Nafion 112 MEA, much larger oxidation/reduction current densities were obtained than those for the NSAP-2.5 MEA. This indicated that the activation area of Pt catalysts on NSAP membrane was much lower than that of Nafion 112 under the same conditions. It indirectly supported our speculation that the

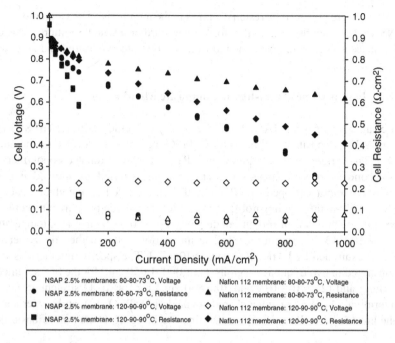

Fig. 7.7 Polarization curves and cell resistance of NSAP and Nafion-112 MEAs at various fuel cell conditions. Anode: H_2 with 33% utilization, cathode: O_2 with 25% utilization, cathode Pt loading: 0.3 mg cm^{-2}

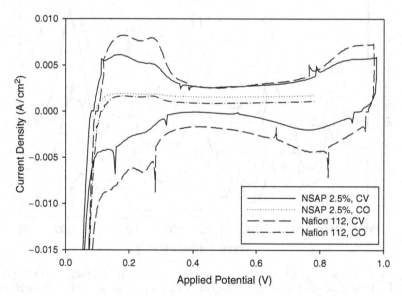

Fig. 7.8 Cyclic voltammetry (CV) curves and crossover (CO) curves of NSAP-2.5 and Nafion-112 MEAs. Temperature: 30°C, anode: H_2 and cathode: N_2, cathode Pt loading: 0.3 mg/cm^2, scan rates: CV at 30 mVs^{-1} and CO at 4 mVs^{-1}

Pt catalysts were poisoned by residual amines in the membrane. The H_2 crossover rate for NSAP-2.5 is slightly higher that for Nafion. Further research to improve the cell performance at high temperature and low relative humidity is desired and underway.

7.3.2.2 Poly(arylene ether)-Based Composite Membranes

Kim et al. investigated PTA/sulfonated poly(arylene ether sulfone) (BPSH) (see Scheme 7.1) composite membranes for PEMFCs operated at elevated temperatures [22,41]. The influence of incorporation of PTA into the sulfonated copolymers on proton conductivity at various temperatures was studied. It showed that the composite membranes doped with PTA significantly reduced the membrane swelling problem, without influencing proton conductivity at room temperature. Furthermore, a pronounced increase in proton conductivity up to 0.15 S·cm^{-1} at a temperature range of 100–130°C was observed. The interaction between the copolymer and PTA was examined by FTIR, as shown in Fig. 7.9. The specific interactions were demonstrated in two frequency regions: (1) 1,000–1,050 cm^{-1} for the SO$_3$ symmetric stretching; and (2) 870–1,010 and 700–850 cm^{-1} for the W–O stretching, i.e., W with terminal O (W = O$_t$), edge-shared octahedral O (W-O$_e$-W), and corner-shared octahedral O (W-O$_c$-W). It was known that the bridging O was more basic than

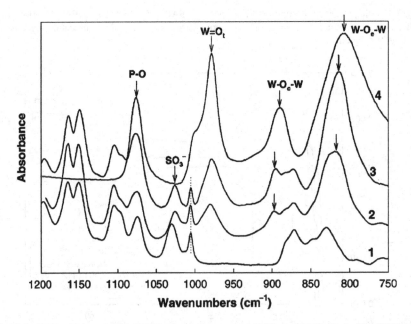

Fig. 7.9 FTIR spectra of PTA/BPSH-40 membranes at 140°C. 1: BPSH-40; 2: PTA/BPSH-40 (3:7); 3: PTA/BPSH-40 (6:4); 4: H$_3$PW$_{12}$O$_{40}$·6H$_2$O. Reprinted with permission from Ref. [22]: Y. S. Kim, et al., Fabrication and characterization of heteropolyacid (H$_3$PW$_{12}$O$_{40}$)/directly polymerized sulfonated poly(arylene ether sulfone) copolymer composite membranes for higher temperature fuel cell applications, *J. Membr. Sci.* **212**, 263–282 (2003). Copyright Elsevier

terminal O, and was thus more capable of forming hydrogen bonding with protonated H_2O molecules. First, after blending with PTA, the symmetric SO_3 stretching at 1030 cm^{-1} in pure BPSH-40 shifted to 1026 cm^{-1}, indicative of sulfonic acid group interaction with PTA. Second, the W = O_t absorption band blue-shifted from 1,007 to 980 cm^{-1}, the W-O_c-W absorption red-shifted from 887 to 897 cm^{-1}, and the W-O_e-W band red-shifted from 795 to 816 cm^{-1} for the composite membranes, suggesting that both the terminal and bridging O in PTA formed hydrogen bonding with the sulfonic acid groups in BPSH-40. On the basis of these results, it was speculated that an intermolecular hydrogen bonding interaction between the BPSH copolymer and the PTA existed, with sulfonic acid groups on the polymer backbone interacting with both bridging and terminal oxygen atoms in the PTA (see Scheme 7.3).

The extraction of PTA from the composite membranes, HPA/BPSH-40, was carefully examined using tapping mode AFM, as shown in Fig. 7.10. PTA/Nafion 117 composite membrane was used as a control experiment. After immersion in liquid water, PTA/Nafion 117 showed irregular holes (0.2 μm in diameter) on the membrane surface, which was supposed to be traces of PTA extraction (Fig. 10a). In contrast, HPA/BPSH-40 composite membrane did not show any holes after liquid water treatment, indicative of a good retention of PTA in the composite, as shown in Fig. 7.10b. This could be partly attributed to the strong hydrogen bonding interaction between BPSH and PTA shown in Scheme 7.3.

Polysulfone (PSF) is a commercially available polymer, less expensive and having a rigid backbone. It showed considerably good resistance toward oxidative and reducing agents during rigorous modification conditions [23,42]. The composite membranes of HPA/PSF and HPA/SPSF were obtained from solution blending [23]. They exhibited improved mechanical strength and lower water uptake than the neat membranes. Compared with Nafion 117, the composite membranes, consisting of 40 wt% PTA and 60 wt% SPSF (with a sulfonation degree of 40%), exhibited enhanced proton conductivity at elevated temperatures. The conductivity increased linearly from 0.089 S cm^{-1} at room temperature to 0.14 S cm^{-1} at 120°C. This PTA/SPSF composite membrane also showed an increase in the glass transition temperature (Tg = 219°C) as compared with the neat SPSF (185°C), suggesting a strong interaction between PTA and SPSF.

Membranes based on aromatic SPEEK are promising for fuel cell applications, since they possess good mechanical properties, thermal stability, as well as reasonably good proton conductivity, depending on their degree of sulfonation [43]. Water absorption of SPEEK membranes increased as the number of sulfonic acid groups per repeating unit increased. In highly sulfonated PEEK, the high density of -SO_3H

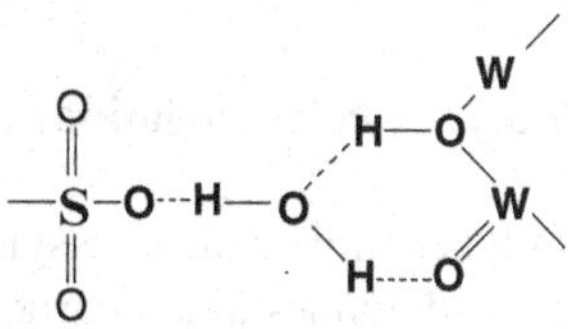

Scheme 7.3 Proposed hydrogen-bonding structure in the PTA/BPSH composite membranes Reprinted with permission from Ref. [22]

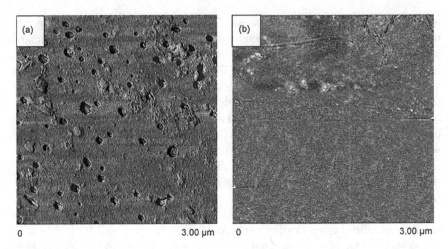

0   3.00 μm   0   3.00 μm

Fig. 7.10 Tapping mode AFM images after immersion of composite membranes in liquid water at 30°C for 48 h. (**a**) PTA/Nafion 117 (4:6) and (**b**) PTA/BPSH-40 (3:7). Reprinted with permission from Ref. [22]: Y. S. Kim, et al., Fabrication and characterization of heteropolyacid $(H_3PW_{12}O_{40})$/ directly polymerized sulfonated poly(arylene ether sulfone) copolymer composite membranes for higher temperature fuel cell applications, *J. Membr. Sci.* **212**, 263–282 (2003). Copyright Elsevier

groups might induce ionic clustering for better water absorption [44]. A series of composite membranes have been prepared by incorporation of PTA, its disodium salt (Na-PTA), and MPA into SPEEK [24]. These membranes (SPEEK sulfonation degree of 70%) showed good conductivity of 3×10^{-3} S·cm^{-1} at ambient temperature and up to a maximum of about 0.02 S·cm^{-1} for SPEEK/PTA composite above 100°C, as shown in Fig. 7.11. From Fig. 7.11 it is seen that the proton conductivity increased significantly for all the composite membranes in comparison with the neat SPEEK membrane. Particularly, the SPEEK/PTA membrane achieved the highest conductivity for SPEEK with different sulfonation degrees.

A fluorinated DF-F polymer was synthesized by condensation polymerization of decafluorobiphenyl (DF) and 4,4'-(hexafluoroisopropylidene)diphenol (F). Post-sulfonation attached -SOH$_3$ groups to the F monomers to yield sDF-F, as shown in Scheme 7.4. Composite membranes, sDF-F/SiO$_2$, were obtained by mixing sDF-F with an inorganic material, SiO$_2$, synthesized by the sol-gel process [45]. Compared with Nafion 117, sDF-F/SiO$_2$ composite membranes showed higher water uptake and higher tensile strength. It was speculated that the SiOH group in silica was capable of retaining water, resulting in high water uptake and good conductivity at elevated temperature. Therefore, better fuel cell performance of the composite membranes was observed.

7.3.2.3 Polybenzimidazole-Based Composite Membranes

Polybenzimidazoles are synthesized from condensation polymerization of aromatic bis-*o*-diamines and dicarboxylates (acids, esters, amides) in the molten state or in solution. Among these, commercially available poly[2,2'-(*m*-phenylene)-5,5'-

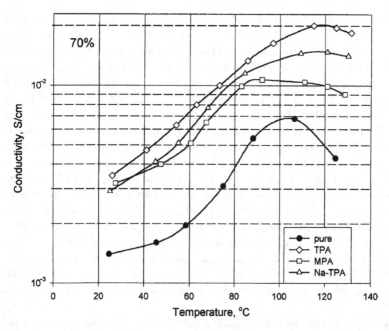

Fig. 7.11 Effect of different HPAs on the proton conductivity of composite membranes as a function of temperature. The sulfonation degree for the SPEEK was 70%. Reprinted with permission from Ref. [24]: S. M. J. Zaidi, et al., Proton conducting composite membranes from polyether ether ketone and heteropolyacids for fuel cell applications, *J. Membr. Sci.* **173,**17–34 (2000). Copyright Elsevier

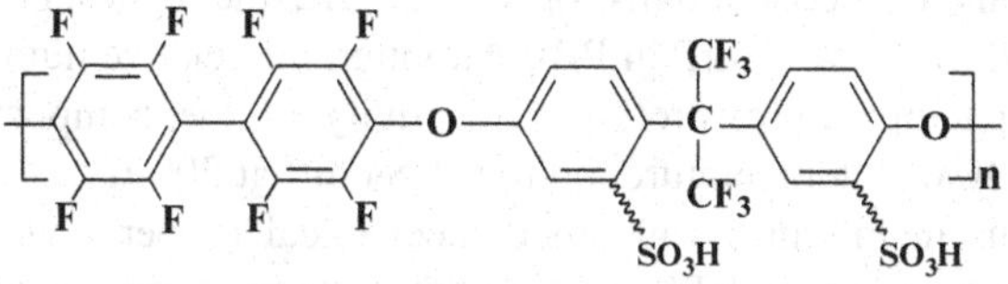

sDF-F

Scheme 7.4 Chemical structure of fluorinated poly(arylene ether) sDF-F

bibenzimidazole] has attracted much attention because of its excellent chemical and thermal stability in the fuel cell application [46]. It is reported that Tg is around 420°C. As a hydrocarbon polymer, it is more cost effective than Nafion; however, it has much lower proton conductivity at room temperature. For example, neat PBI had proton conductivity in the range of 2×10^{-4}– 8×10^{-4} S·cm^{-1} when relative humidities ranged from 0% to 100% [47]. Accordingly, it would not be appropriate for direct uses in fuel cells. Due to its basic characteristic (pK_a = −5.5), PBI could complex with inorganic and organic acids [46]. Phosphoric acid doped PBI membranes thus were suggested for fuel cell applications [48,49]. Since then, acid-doping has been attempted with various inorganic acids such as phosphoric, sulfuric, hydrochloric, perchloric acid, and so on. The doping acid concentration for the basic PBI allowed up to ca. 50 wt%. Depending on the weight percentage of the acid, the PBI/H$_3$PO$_4$ membranes could have a proton conductivity reaching 3.5×10^{-2} S·cm^{-1} at 190°C [49]. The mechanism of proton conduction in PBI/acid membranes was proposed. For the proton migration, the hydrogen bonding interaction immobilized the anions of the acid and formed a network to facilitate proton transfer by the Grotthuss mechanism. The rate of proton transfer involving H$_2$O was faster, leading to higher proton conductivity with an increase in relative humidity [50]. Phosphoric acid-doped PBI membrane fuel cells could operate up to 200°C with little gas humidification, and CO tolerance could be up to a few percent [51].

Composite membranes based on PTA-impregnated SiO$_2$ and PBI had been prepared and their physicochemical properties were studied. The membranes with a high tensile strength and a thickness of <30 μm were prepared by solution casting. They were chemically stable in boiling water and thermally stable in air up to 400°C [52]. The presence of silica in the composite and 100% relative humidity allowed the membranes to maintain stable proton conductivity at temperatures up to 130°C. For example, the proton conductivity measured at 130°C was 10^{-3} S·cm^{-1}. In these membranes, the PBI formed a network to keep the PTA supported on silica.

Bjerrum reported the PBI composite membranes contain inorganic proton conductors including ZrP, PTA, and STA [21]. Layered ZrP is a surface proton conductor, and its surface area and capacity for surface adsorption have a significant influence on proton conductivity. For the acid-doped ZrP/PBI membranes, there were a variety of chemical species forming hydrogen bonds around the ZrP particle, such as the H$_3$O$^+$, PO$_4^-$, H$_2$O, P-O$^-$, and P-OH groups [53]. The fine ZrP particles in the membrane therefore attracted the protonated active ions or groups to enhance proton transfer and thus conductivity. From Fig. 7.12, the highest proton conductivity was obtained for 15 wt% ZrP in PBI. Although the relative humidity decreased with increasing temperature, proton conductivity of the composite membranes increased linearly with temperature, up to 0.1 S·cm^{-1} at 200°C.

Homogeneous membranes with good mechanical property were obtained by doping PBI membranes with PTA and STA. The conductivity was higher than or at least comparable to that of the H$_3$PO$_4$-doped PBI membranes up to 110°C [21]. Above 110°C, however, their conductivity became lower. Figure 7.13 shows the conductivity of PTA/PBI and STA/PBI membranes as a function of relative humidity. The conductivities of the composite membranes containing 20 and 30 wt% PTA at 140°C and 20 and 30 wt% STA at 200°C were all lower than those of pure PBI membranes with 4.4 and 5.1 H$_3$PO$_4$ doping levels under the same conditions.

Xiao et al. reported main-chain pyridine-containing PBI (PPBI, see Scheme 7.5) membranes prepared by direct polymerization during a solution-casting process [54]. The PPBI membranes had high H$_3$PO$_4$-doping levels from 15–25 mole of H$_3$PO$_4$ per PBI repeating unit, and good proton conductivity of 0.1–0.2 S·cm^{-1} at 160°C. These composite membranes achieved good thermal stability and mechanical integrity even at high H$_3$PO$_4$ doping levels. Figure 7.14 demonstrates that the conductivity of 2,5-PPBI with 20.4 mole H$_3$PO$_4$-doped membrane increased with

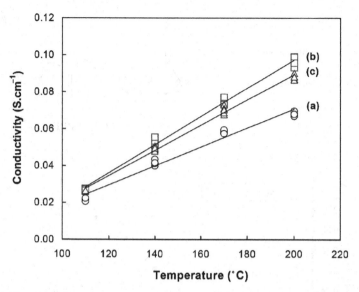

Fig. 7.12 Conductivity of PBI and ZrP/PBI composite membranes at different temperatures with a H$_3$PO$_4$ doping level of 5.6. The relative humidity was 20% for 110 and 140°C, 10% for 170°C, and 5% for 200°C. (**a**) PBI; (**b**) 15 wt% of ZrP in PBI; and (**c**) 20 wt% of ZrP in PBI. Reprinted with permission from Ref. [21]: R. He, et al., Proton conductivity of phosphoric acid doped poly-benzimidazole and its composites with inorganic proton conductors, *J. Memb. Sci.* **226**, 169–184 (2003). Copyright Elsevier

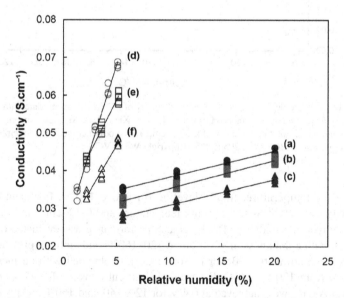

Fig. 7.13 Conductivity as a function of relative humidity for PBI and PTA/PBI and STA/PBI composite membranes. (**a**) PBI, H$_3$PO$_4$ doping level of 4.4 at 140°C; (**b**) 20% PTA in PBI, H$_3$PO$_4$ doping level of 4.4 at 140° C; (**c**) 30% PTA in PBI, H$_3$PO$_4$ doping level of 4.4 at 140° C; (**d**) PBI, H$_3$PO$_4$ doping level of 5.1 at 200° C; (**e**) 20% STA in PBI, H$_3$PO$_4$ doping level of 5.1 at 200°C; (**f**) 30% STA in PBI, H$_3$PO$_4$ doping level of 5.1 at 200° C. Reprinted with permission from Ref. [21]: R. He, et al., Proton conductivity of phosphoric acid doped polybenzimidazole and its composites with inorganic proton conductors, J. Membr. Sci. 226, 169–184 (2003). Copyright from Elsevier

Scheme 7.5 Chemical structure of 2,5-PPBI

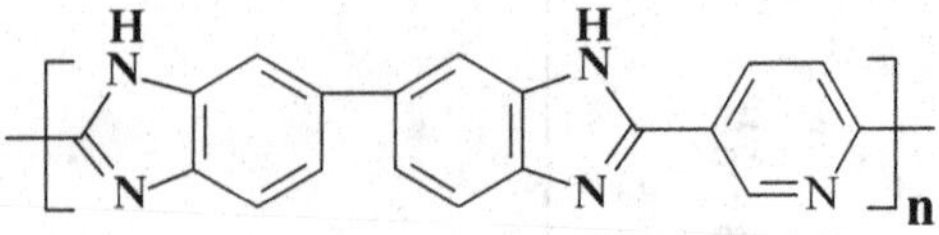

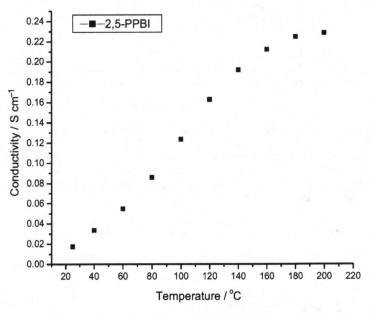

2, 5 −PPBI

Fig. 7.14 Temperature dependence of proton conductivity of a 2,5-PPBI membrane with 20.4 mol H_3PO_4 doping. Reprinted with permission from Ref. [54]: L. Xiao, et al., Synthesis and characterization of pyridine-based polybenzimidazole for high temperature polymer electrolyte membrane fuel cell application, *Fuel Cell*, **5**, 287–295 (2005). Copyright Wiley-VCH

increasing the temperature, finally leveling off above 180°C. For example, the conductivity was 0.018 S·cm^{-1} at room temperature and 0.2 S·cm^{-1} at 160–200°C under an anhydrous condition. The H_2/O_2 cell performance was evaluated on a 2,5-PPBI MEA with a thickness of ~200 μm at 120–160°C and ambient pressure without external humidification. Better performance was obtained with an increase in temperature from 120 to 160°C. For example, current density of 0.95 A·cm^{-2}, 1.2 A·cm^{-2}, 1.5 A·cm^{-2} were achieved at 0.4 V for 120, 140, and 160°C, respectively.

Poly(2,5-benzimidazole) (ABPBI, see Scheme 7.6) has recently been reviewed for fuel cell applications [55]. Compared with PBI, higher molecular weight ABPBI was achieved, resulting in better mechanical properties. Proton conductivity of H_3PO_4-doped ABPBI membranes ranged from 0.01 to 0.05 S·cm^{-1} above 100°C under dry conditions, and increased with the doping level.

Scheme 7.6   Chemical structure of ABPBI

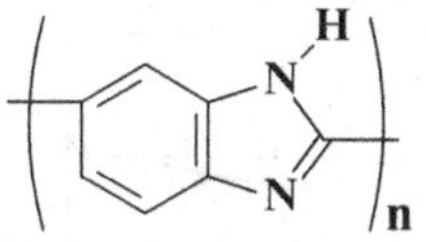

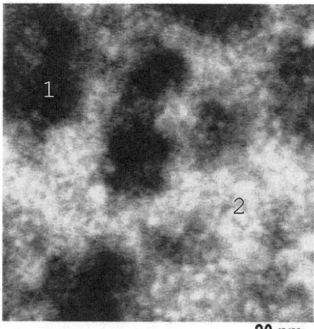

Fig. 7.15 TEM micrograph of a PTA impregnated organic/inorganic membrane. Nano-phase separation of the macromolecules is observed between organic phase (region 1: *dark*) and bicontinuous inorganic phase (region 2: *bright*). Reprinted with permission from Ref. [56]: I. Honma, et al., Organic/inorganic nano-composites for high temperature proton conducting polymer electrolytes, *Solid State Ionics* **162–163**, 237–245 (2003). Copyright Elsevier

7.3.2.4   Other Types of Polymer-Based Composite Membranes

Besides the above polymer-based composite membranes for high temperature PEMFCs, other types of polymer-based composite membranes were also reported. Honma investigated organic/inorganic nanocomposite membranes based on a family of bridged polysilsesquioxanes [56–58]. These membranes were prepared by a sol-gel process of macromonomers that contained a telechelic oligomer with two ethoxysilane groups. With incorporation of PTA, transparent nanocomposite membranes were obtained. The TEM photograph showed nano-phase separation on a length scale of 20 to 50 nm (Fig. 7.15). The dark region represented the organic

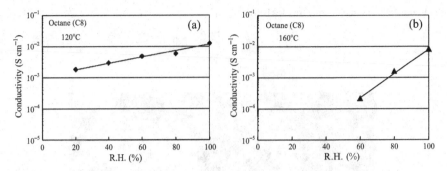

Fig. 7.16 Dependence of proton conductivity on relative humidity of a hybrid octane membrane at (**a**) 120°C and (**b**) 160°C. Reprinted with permission from Ref. [56]: I. Honma, et al., Organic/inorganic nano-composites for high temperature proton conducting polymer electrolytes, *Solid State Ionics* **162–163**, 237–245 (2003). Copyright Elsevier

phases, and the bright region represented continuous inorganic phases [56]. It was considered that the PTA was anchored to inorganic silicate moieties by strong electrostatic interaction to form ionic clusters. The nano-sized channels of the hybrid material might facilitate fast proton transport through the membranes. The nano-phase separated hybrid membranes doped with PTA exhibited high proton conductivity at a high temperature up to 160°C. The conductivity dependence on relative humidity for an octane hybrid membrane at 120 and 160°C was studied and is shown in Fig. 7.16. The conductivity at 120°C showed less relative humidity dependence than that at 160°C, and the conductivity exceeded 10^{-3} S·cm^{-1} even at 20% relative humidity.

7.4 Conclusions

PEMFCs operating at high temperature are being pursued by researchers worldwide, because they have the advantage of enhanced tolerance to CO, accelerated reaction kinetics on both the anode and cathode sides, and significant improvement of the overall cell efficiency. This chapter briefly reviewed advances in polymer composite membranes for PEMFCs operating at high temperature and low relative humidity. The aims of polymer composite membranes are to improve mechanical strength, thermal stability, and particularly high temperature proton conductivity, which would not largely depend on water retention at dehydrated conditions. Different approaches have be investigated, including modifying PEMs with hygroscopic inorganic materials to improve their water retention above 100°C and modifying PEMs with inorganic proton conductors to obtain reasonable conductivity independent of water content. Most popular polymers are perfluoronated Nafion, sulfonated poly(arylene ethers), sulfonated polyimides, and polybenzimidazoles. Inorganic proton conductors are silica, HPAs, layered ZrPs, and liquid phosphoric acid. Successful high-temperature composite membranes have been achieved using mostly HPAs and layered ZrPs, or their combination with silica. However, only a few reports have addressed the prevention of HPAs from leaching out of the composite membranes. Although Nafion is electrochemically stable, it may not be applicable for PEMFCs above 130°C due to its limited Tg at 130°C. Hydrocarbon high-performance polyelectrolytes such as sulfonated poly(arylene ether sulfones) and PBI/acid complexes are thus promising candidates for high-temperature PEMFCs. Nevertheless, challenges in high-temperature PEMs still remain to improve proton conductivity, and thermal, electrochemical, and mechanical stability.

Acknowledgments   The authors appreciate the financial support of this work by U.S. Army Phase II portable fuel cell program through Connecticut Global Fuel Cell Center at the University of Connecticut. We also appreciate the partial support of this work from National Science Foundation of China (grant no. 50373005). The authors are grateful to Dr. Ruichun Jiang (Chemical Engineering Department at University of Connecticut) for the assistance in the fuel cell performance tests and cyclic voltametry study. Helpful discussions with Prof. James M. Fenton (Central Florida University) and Prof. H. Russell Kunz (University of Connecticut) are also acknowledged.

References

1. K. D. Kreuer, On the development of proton conducting polymer membranes for hydrogen and methanol fuel cells, *J. Membr. Sci.* **185**, 29–39 (2001).
2. G. Alberti, M. Casciola, L. Massinelli and B. Bauer, Polymeric proton conducting membranes for medium temperature fuel cells (110–160°C), *J. Membr. Sci.* **185**, 73–81 (2001).
3. L. E. Karlsson and P. Jannasch, Sulfone ionomers for proton-conducting fuel cell membranes 2. Sulfophenylated polysulfones and polyphenylsulfones, *Electrochim. Acta* **50**, 1939–1946 (2005).
4. Y. Yin, S. Hayashi, O. Yamada, H. Kita and K. -I. Okamoto, Branched/crosslinked sulfonated polyimide membranes for polymer electrolyte fuel cells, *Macromol. Rapid Commun.* **26**, 696–700 (2005).
5. T. Watari, J. Fang, K. Tanaka, H. Kita, K. -I. Okamoto and T. Hirano, Synthesis, water stability and proton conductivity of novel sulfonated polyimides from 4,4-bis(4-aminophenoxy) biphenyl-3,3′-disulfonic acid, *J. Membr. Sci.* **230**, 111–120 (2004).
6. S. Sundar, W. Jang, C. Lee, Y. Shul and H. Han, Crosslinked sulfonated polyimide networks as polymer electrolyte membranes in fuel cells, *J. Polym. Sci. Part B: Polym. Phys.* **43**, 2370–2379 (2005).
7. K. Miyatake, H. Iyotani, K. Yamamoto and E. Tsuchida, Synthesis of poly(phenylene sulfide sulfonic acid) via poly(sulfonium cation) as a thermostable proton-conducting polymer, *Macromolecules* **29**, 6969–6971 (1996).
8. K. Miyatake, E. Shouji, K. Yamamoto and E. Tsuchida, Synthesis and proton conductivity of highly sulfonated poly(thiophenylene), *Macromolecules* **30**, 2941–2946 (1997).
9. M. A. Hickner, H. Ghassemi, Y. S. Kim, B. R. Einsla and J. E. McGrath, Alternative polymer systems for proton exchange membranes (PEM)s, *Chem. Rev.* **104** 4587–4612 (2004).
10. F. Wang, M. Hickner, Y. S. Kim, T. A. Zawodzinski and J. E. McGrath, Direct polymerization of sulfonated poly(arylene ether sulfone) random (statistical) copolymers: Candidates for new proton exchange membranes, *J. Membr. Sci.* **197**, 231–242 (2002).
11. Y. S. Kim, F. Wang, M. Hickner, S. Mccartney, Y. T. Hong, W. Harrison, T. A. Zawodzinski and J. E. McGrath, Effect of acidification treatment and morphological stability of sulfonated poly(arylene ether sulfone) copolymer proton-exchange membranes for fuel-cell use above 100°C, *J. Polym. Sci. Part B: Polym. Phys.* **41**, 2816–2828 (2003).

12. S. Granados-Focil and M. H. Litt, A new class of polyelectrolytes, polyphenylene sulfonic acid and its copolymers, as proton exchange membranes for PEMFC's, *Prepr. Symp. (Div. Fuel Chem., Am. Chem. Soc.)* **49**(2), 528–529 (2004).

13. S. Granados-Focil and M. H. Litt, Novel highly conductive poly(phenylene sulfonic acid)s and its evaluation as proton exchange membranes for fuel cells, *PMSE Prepr. (Div. Polym. Mater. Sci. Eng., Am. Chem. Soc.)* **89**, 438–439 (2003).

14. J. Kerres, A. Ullrich, F. Meier and T. Haring, Synthesis and characterization of novel acid–base polymer blends for application in membrane fuel cells, *Solid State Ionics* **125**, 243–249 (1999).

15. K. D. Kreuer, A. Fuchs, M. Ise, M. Spaeth and J. Maier, Imidazole and pyrazole-based proton conducting polymers and liquids, *Electrochim. Acta* **43**, 1281–1288 (1998).

16. C. Yang, P. Costamagna, S. Srinivasan, J. Benziger and A. B. Bocarsly, Approaches and technical challenges to high temperature operation of proton exchange membrane fuel cells, *J. Power Sources* **103**, 1–9 (2001).

17. M. Yamada and I. Honma, Anhydrous proton conducting polymer electrolytes based on poly(vinylphosphonic acid)-heterocycle composite material, *Polymer* **46**, 2986–2992 (2005).

18. R. M. Acheson, An Introduction to the Chemistry of Heterocyclic Compounds, 3rd edn., Wiley, New York, (1976).

19. S. Malhotra and R. Datta, Membrane-supported nonvolatile acidic electrolytes allow higher temperature operation of proton-exchange membrane fuel cells, *J. Electrochem. Soc.* **144**, L23–L26 (1997).

20. Q. Li, R. He, J. O. Jensen and N. J. Bjerrum, PBI based polymer membrane for high temperature fuel cells-preparation, characterization and fuel cell demonstration, *Fuel Cells* **4**(3), 147–159 (2004).

21. R. He, Q. Li, G. Xiao and N. J. Bjerrum, Proton conductivity of phosphoric acid doped polybenzimidazole and its composites with inorganic proton conductors, *J. Membr. Sci.* **226**, 169–184 (2003).

22. Y. S. Kim, et al., Fabrication and characterization of heteropolyacid ($H_3PW_{12}O_{40}$)/directly polymerized sulfonated poly(arylene ether sulfone) copolymer composite membranes for higher temperature fuel cell applications, *J. Membr. Sci.* **212**, 263–282 (2003). Copyright Elsevier

23. B. Smith, S. Sridhar and A. A. Khan, Proton conducting composite membranes from polysulfone and heteropolyacid for fuel cell applications, *J. Polym. Sci. Part B: Polym. Phys.* **43**, 1538–1547 (2005).

24. S. M. J. Zaidi, S. D. Mikhailenko, G. P. Robertson, M. D. Guiver and S. Kaliaguine, Proton conducting composite membranes from polyether ether ketone and heteropolyacids for fuel cell applications, *J. Membr. Sci.* **173**, 17–34 (2000).

25. K. D. Kreuer, Proton conductivity: materials and applications, *Chem. Mater.* **8**, 610–641 (1996).

26. M. Watanabe, H. Uchida, Y. Seki, M. Emori and P. Stonehart, Self-humidifying polymer electrolyte membranes for fuel cells, *J. Electrochem. Soc.* **143**, 3847–3852 (1996).

27. P. L. Antonucci, A. S. Arico, P. Creti, E. Ramunni and V. Antonucci, Investigation of a direct methanol fuel cell based on a composite Nafion®-silica electrolyte for high temperature operation, *Solid State Ionics* **125**, 431–437 (1999).

28. K. A. Mauritz, I. D. Stefanithis, S. V. Davis, R. W. Scheez, R. K. Pope, G. L. Wilkes and H. H. Huang, Microstructural evolution of a silicon-oxide phase in a perfluorosulfonic acid ionomer by an in-situ sol-gel reaction, *J. Appl. Polym. Sci.* **55**, 181–190 (1995).

29. K. T. Adjemian, S. Srinivasan, J. Benziger and A. B. Bocarsly, Investigation of PEMFC operation above 100°C employing perfluorosulfonic acid silicon oxide composite membranes, *J. Power Sources* **109**, 356–364 (2002).

30. P. Costamagna, C. Yang, A. B. Bocarsly and S. Srinivasan, Nafion® 115/zirconium phosphate coposite membranes for operation of PEMFC above 100°C, *Electrochim. Acta* **47**, 1023–1033 (2002).

31. B. Ruffmann, H. Silva, B. Schulte and S. P. Nunes, Organic/inorganic composite membranes for application in DMFC, *Solic State Ionics* **162–163**, 269–275 (2003).

32. Y. Si, H. R. Kunz and J. M. Fenton, Nafion-Teflon-Zr(HPO$_4$)$_2$ composite membranes for high-temperature PEMFCs, *J. Electrochem. Soc.* **151**, A623–A631 (2004).

33. N. H. Jalani, K. Dunn and R. Datta, Synthesis and characterization of Nafion®-MO$_2$ (M = Zr, Si, Ti) nanocomposite membranes for higher temperature PEM fuel cells, *Electrochim. Acta* **51**, 553–560 (2005).

34. T. Thampan, N. H. Jalani, P. Choi and R. Datta, Systematic approach to design higher temperature composite PEMs, *J. Electrochem. Soc.* **152**, A316–A325 (2005).

35. B. R. Ezzell, W. P. Carl and W. A. Mod, Preparation of vinyl ethers, *U.S. Patent* **4**, 358–412 (1982).

36. S. J. Paddison and J. A. Elliott, Molecular modeling of the short-side-chain perfluorosulfonic acid membrane, *J. Phys. Chem. A* **109**, 7583–7593 (2005).

37. Y. Liu, Organic/inorganic composite membranes for high temperature proton exchange membrane fuel cells, M.S. Dissertation, University of Connecticut, USA, 2005.

38. Y. Liu, H. R. Kunz, J. M. Fenton and L. Zhu, Development of nafion/SiO$_2$/phosphotunstic acid nanocomposite membranes for high temperature proton exchange membrane fuel cells, *PMSE Prepr. (Div. Polym. Mater. Sci. Eng., Am. Chem. Soc.)* **93**, 703–704 (2005).

39. D. Zhao, Q. Huo, J. Feng, B. F. Chmelka and G. D. Stucky, Nonionic triblock and star diblock copolymer and oligomeric surfactant syntheses of highly ordered, hydrothermally stable, mesoporous silica structures, *J. Am. Chem. Soc.* **120**, 6024–6036 (1998).

40. X. Feng, G. E. Fryxell, L. Q. Wang, A. Y. Kim, J. Liu and K. M. Kemner, Functionalized monolayers on ordered mesoporous supports, *Science* **276**, 923–926 (1997).

41. Y. S. Kim, F. Wang, M. Hickner, T. A. Zawodzinski and J. E. McGrath, Heteropolyacid/sulfonated poly(arylene ether sulfone) composites for proton exchange membranes fuel cells, *PMSE (Div. Polym. Mater. Sci. Eng., Am. Chem. Soc.)* **85**, 520–521 (2001).

42. P. Genova-Dimitrova, B. Baradie, D. Foscallo, C. Poinsignon and J. Y. Sanchez, Ionomeric membranes for proton exchange membrane fuel cell (PEMFC): sulfonated polysulfone associated with phosphatoantimonic acid, *J. Membr. Sci.* **185**, 59–71 (2001).

43. T. Kobayashi, M. Rikukawa, K. Sanui and N. Ogata, Proton conducting polymers derived from poly(ether-ether ketone) and poly(4-phenoxybenzoyl-,4-phenylene), *Solid State Ionics* **106**, 219 (1998).

44. C. Bailly, D. J. Williams, F. E. Karasz and W. J. McKnight, The sodium salts of sulfonated poly(aryl ether-ether ketone) (PEEK): preparation and characterization, *Polymer* **28**, 1009 (1987).

45. Y. M. Kim, S. H. Choi, H. C. Lee, M. Z. Hong, K. Kim and H.-I. Lee, Organic-inorganic composite membranes as addition of SiO$_2$ for high temperature-operation in polymer electrolyte membrane fuel cells (PEMFC)s, *Electrochim. Acta* **49**, 4787–4796 (2004).

46. D. J. Jones and J. Rozière, Recent advances in the functionalisation of polybenzimidazole and polyetherketone for fuel cell applications, *J. Membr. Sci.* **185**, 41–58 (2001).

47. D. Hoel and E. Grunwald, High protonic conduction of polybenzimidazole films, *J. Phys. Chem.* **81**, 2135–2136 (1977).

48. J.-T. Wang, R. F. Savinell, J. Wainright, M. Litt and H. Yu, A H$_2$/O$_2$ fuel cell using acid doped polybenzimidazole as polymer electrolyte, *Electrochim. Acta* **41**, 193–197 (1996).

49. J. S. Wainright, J.-T. Wang, D. Weng, R. F. Savinell and M. Litt, Acid-doped polybenzimidazoles: a new polymer electrolyte, *J. Electrochem. Soc.* **142**, L121–L123 (1995).

50. Y.-L. Ma, J. S. Wainright, M. H. Litt and R. F. Savinell, Conductivity of PBI membranes for high-temperature polymer electrolyte fuel cells, *J. Electrochem. Soc.* **151**, A8–A16 (2004).

51. Q. Li, R. He, J. O. Jensen and N. J. Bherrum, PBI based polymer membrane for high tem fuel cells-preparation, characterization and fuel cell demonstration, *Fuel Cells* **4**, 147–159 (2004).

52. P. Staiti, M. Minutoli and S. Hocevar, Membranes based on phosphotungstic acid and polybenzimidazole for fuel cell application, *J. Power Sources* **90**, 231–235 (2000).

53. A. Clearfield, Structural concepts in inorganic proton conductors, *Solid State Ionics* **46**, 35–43 (1991).

54. L. Xiao, H. Zhang, T. Jana, E. Scanlon, R. Chen, E.-W. Choe, L. S. Ramanathan, S. Yu and B. C. Benicewicz, Synthesis and characterization of pyridine-based polybenzimidazole for

high temperature polymer electrolyte membrane fuel cell application, *Fuel Cell* **5**, 287–295 (2005).

55. J. A. Asensio and P. Gomez-Romero, Recent developments on proton conducting poly(2,5-benzimidazole) (ABPBI) membranes for high temperature polymer electrolyte membrane fuel cells, *Fuel Cells* **5**, 336–343 (2005).

56 I. Honma, H. Nakajima, O. Nishikawa, T. Sugimoto and S. Nomur, Organic/inorganic nanocomposites for high temperature proton conducting polymer electrolytes, *Solid State Ionics* **162–163**, 237–245 (2003).

57. I. Honma, H. Nakajima and S. Nomura, High temperature proton conducting hybrid polymer electrolyte membranes, *Solid State Ionics* **154–155**, 707–712 (2002).

58. I. Honma, S. Nomura and H. Nakajim, Protonic conducting organic/inorganic nanocomposites for polymer electrolyte membrane, *J. Membr. Sci.* **185**, 83–94 (2001).

Chapter 8
Blend Concepts for Fuel Cell Membranes*

Jochen Kerres

Abstract Differently cross-linked blend membranes were prepared from commercial arylene main-chain polymers from the classes of poly(ether-ketones) and poly(ethersulfones) modified with sulfonate groups, sulfinate cross-linking groups and basic N-groups. The following membrane types have been prepared: (a) van-der Waals/dipole-dipole blends by mixing a polysulfonate with unmodified PSU. This membrane type showed a heterogeneous morphology, leading to extreme swelling and even dissolution of the sulfonated component at elevated temperatures. (b) Hydrogen bridge blends by mixing a polysulfonate with a polyamide or polyetherimide. This membrane type showed a partially heterogeneous morphology, also leading to extreme swelling/dissolution of the sulfonated blend component at elevated temperatures. (c) Acid-base blends by mixing a polysulfonate with a polymeric N-base (self-developed/commercial). With this membrane type, we could reach a wide variability of properties by variation of different parameters. Membranes showing excellent stability and good fuel cell performance up to 100°C (PEFC) and 130°C (DMFC) were obtained. (d) Covalently cross-linked (blend) membranes by either mixing of a polysulfonate with a polysulfinate or by preparation of a polysulfinatesulfonate, followed by reaction of the sulfinate groups in solution with a dihalogeno compound under S-alkylation. Membranes were prepared that showed effective suppression of swelling without H^+-conductivity loss. The membranes showed good PEFC (up to 100°C) and DMFC (up to 130°C) performance. (e) Covalent-ionically cross-linked blend membranes by mixing polysulfonates with polysulfinates and polybases or by mixing a polysulfonate with a polymer carrying both sulfinate and basic N-groups. The covalent-ionically cross-linked membranes were tested in DMFC up to 110°C and showed a good performance. (f) Differently cross-linked organic-inorganic blend composite membranes via different procedures. The best results were obtained with blend membranes having a layered zirconium phosphate "ZrP" phase: They were transparent, and showed good H^+-conductivity and stability. Application of one of these composite membranes to a PEFC yielded good performance up to T = 115°C.

---

*This book chapter has been published as a review article in the journal Fuel Cells (J. A. Kerres, Fuel Cells 5, No. 2, 230-247 (2005)). Kind permission for complete publication as a book chapter has been given by VCH-Wiley Rights and Licenses, Weinheim, Germany

8.1   Overview: State of the Art in Fuel Cell Membrane Development

Fuel cell research and development is one of the key topics in material science and engineering, because fuel cells could help to solve the problems connected with consumption of the global energy carrier reserves, and with environmental problems connected with the use of fossil fuels in transport systems and energy production. Fuel cells are obviously a building block in developing environment-friendly economies, because they can be used for energy supply in transport applications (cars, buses, trucks, railway engines) as well as stationary applications (decentralized power stations, home energy supply) and mobile applications (laptop computers, cell phones, handhelds). For these reasons, fuel cell component research and development efforts have increased considerably during the last decade. The developed membrane systems can be roughly separated into the following material classes [1]:

1. Perfluorinated ionomer membranes of the Nafion, Flemion, Dow Membrane type.
2. Partially fluorinated ionomer membranes: Among this material class is the BAM3G membrane type composed of sulfonated or phosphonated poly(α,β,β-trifluorostyrene) and its copolymers [2–4] and the different types of grafted membranes based on partially fluorinated polymer foils, as developed by Scherer [5], Sundholm [6], and others.
3. Nonfluorinated ionomer membranes: Numerous different types of nonfluorinated ionomer membranes, among them ionomer membranes based on styrene polymers and copolymers containing polystyrene units [7], arylene main-chain polymers of different poly(phenylene) [8], poly(ethersulfone) [9–11], poly(etherketone) [12–15], poly(phenylene oxide) [16,17], poly(phenylene sulfide) [18] types, and such membranes based on an inorganic backbone like poly(phosphazenes) [19,20], poly(siloxane)s [21], have been developed in the past years
4. Composite membranes: The composite materials can be roughly subdivided into the following material types:

   • Ionomers filling the pores of a porous support material (fleeces, nonwovens, textiles, porous teflon foils of the GoreTex type ("GoreSelect" [22,23])
   • High-molecular/low-molecular composites such as blends of poly (benzimidazoles) with phosphoric acid as the proton-conducting component or phosphoric acid blended into other organopolymers [24–27], or blends of a sulfonated polymer with amphoterics such as imidazoles or pyrazoles or imidazole-containing oligomers or polymers alone [28,29], or blends of sulfonated polymers with heteropolyacids [30–32].
   • Organic/inorganic microcomposites or nanocomposites such as an (proton-conducting) organopolymer filled with an inorganic oxide (SiO_2 [33–35], TiO_2, ZrO_2 [36]), hydroxide, or salt (layered zirconium phosphates or zirconium

sulfophenylphosphonates [37,38], in which the inorganic or inorganic-organic component is also capable to contribute to proton transport, etc.).

5. Blend membranes from different organopolymers, particularly where interactions exist between the proton-conducting components and the other polymer(s).

The shortcomings of the present membrane types with respect to their application in fuel cells are given in the following:

1. Application in H_2 membrane fuel cells (H_2-PEFC): The commercial perfluorinated ionomer membranes such as Nafion are too expensive (US\$500–800 m^{-2}). Moreover, at T > 100°C the membranes show strong drying out, leading to a conductivity drop by several orders of magnitude [39,40]. One general problem, especially of nonfluorinated ionomer materials, is that they show too high water uptake when having a proton conductivity sufficient for fuel cell operation.
2. Application in direct methanol fuel cells (DMFC): The Nafion-type membranes are too expensive, hindering their broad application. The perfluorinated membranes also show too high methanol permeability [41,42], leading to poisoning of the cathode catalyst and therefore to strong reduction in power density. Some ionomer membrane types are unstable in methanol solutions, leading in extreme cases (particularly high temperature) to dissolution of the polymer.

From these shortcomings, the tasks for the designation of improved fuel cell membranes, compared with the state of the art, can be defined. The property profile of improved ionomer fuel cell membranes includes high H$^+$-conductivity, low water/methanol uptake, low methanol (and other liquid fuel) permeability, and fuel cell-applicability also at T > 100°C, because the higher the fuel cell operation temperature, the higher the fuel utilization, and applicability also in other (electro)membrane processes. Last but not least, the membranes should have a low price.

The following section describes in detail our scientific-technological approach(es) for development of novel ionomer membranes fulfilling the preceding property profile.

8.2   Review of Membrane Development

The preceding property profile of ionomer membranes for use in fuel cells had led to the development of the approaches listed in Table 8.1.

The water uptake of ionomer membranes can be reduced by introduction of specific interactions (Fig. 8.1) between macromolecular chains.

In polymers different types of interactions are always present simultaneously. For example, van der Waals interactions between macromolecules are present in every polymer, and electrostatic interactions are always connected with hydrogen bonding and dipole—dipole interactions. In any event, introduction of chemical bonds between the macromolecules has the strongest impact on the polymer

Table 8.1 Approaches for polymer and membrane development for fuel cells

Requirements	Approaches
High H⁺-conductivity	Use of sulfonated polymers as the proton-conductive component in the fuel cell membranes at T < 100°C
Low water/methanol uptake	Use of nonfluorinated ionomers physical and/or chemical cross-linking of the fuel cell membranes
Low methanol (and other liquid fuel) permeability	Use of nonfluorinated ionomers physical and/or chemical cross-linking of the fuel cell membranes
Fuel cell-applicability also at T > 100°C	Development of organic-inorganic composite membranes, based on our cross-linked ionomer membrane systems, in which the inorganic membrane component serves as water storage or even contributes to H⁺-conduction
Low price	Use of commercially available polymers for chemical modification and membrane formation, which avoids expensive development of novel polymers

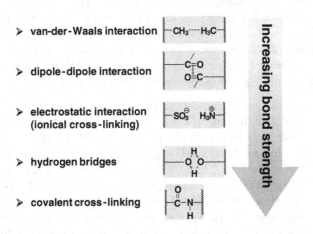

Fig. 8.1 Specific interactions between macromolecules

structure, because covalent cross-links are "fixing" the polymer morphology, while physical interactions between the macromolecules can be detached, e.g., hydrogen bondings can be dissociated by temperature increase.

8.2.1 Interaction-Blend Membrane Types

Ionomer membrane types have been developed that show the different types of interaction forces between the blend components. In Table 8.2, an overview is given of the different ionomer membrane types developed. In the following, the developed (blend) membrane types are described in more detail.

Table 8.2 Overview over the different ionomer membrane types developed

Interaction type	Ionomer systems	Remarks
An der Waals/dipole-dipole interaction	Blends from sulfonated PSU and unmodified PSU	Inhomogeneous morphology, too high swelling
H-bondings	Blends from sulfonated PEEK and PEI Ultem® or PA Trogamid P	Partially inhomogeneous morphology, too high swelling
Electrostatic interaction (ionical cross-linking)/hydrogen bridges	Blends of sulfonated poly(etherketone)s/sulfonated poly(ethersulfone)s and basic polymers (commercial and self-developed)	In most cases homogeneous morphology, partially too high swelling at elevated T
Covalent cross-linking	Blends from sulfonated arylene main-chain polymers and polysulfinates, polyarylenesulfonate-sulfinates, cross-linked with dihalogeno compounds	Homogeneous morphology, swelling effectively reduced

8.2.1.1 Van der Waals/Dipole–Dipole Interaction Blends

The basic idea for this membrane type was to "reinforce" the sulfonated ionomer membrane with unmodified polymer, due to the fact that polymers lose much of their mechanical strength by sulfonation or other modification reactions. As basic polymer, we selected the commercial PSU Udel, because this polymer is relatively inexpensive, shows good chemical and mechanical stability, and can be easily sulfonated by different methods. First trials of mixed unmodified and sulfonated PSU showed that the mixing of the unmodified and sulfonated PSU in a dipolar-aprotic solvent (N-methylpyrrolidinone NMP, N,N-dimethylacetamide DMAc or dimethylsulfoxide DMSO) led to inhomogeneous solutions and, after solvent evaporation, to inhomogeneous blend membranes, which showed very poor mechanical stability. We concluded that the sulfonic acid group is responsible for the incompatibility of the polymers, due to its hydrophilicity and ionogenity. Moreover, the van der Waals and dipole—dipole forces between the two polymers are obviously too weak to lead to blend component compatibility. Therefore, in a second approach the unmodified PSU was mixed with a nonionic precursor of the polymeric sulfonic acid to improve compatibility of the polymers forming the blend. For this purpose, we transformed the PSU-SO$_3$H into PSU-SO$_2$Cl, PSU-SO$_2$OCH$_3$, and PSU-SO$_2$NHC$_3$H$_7$ [43]. However, the blend membranes of unmodified PSU with these modified sulfopolymers had unsatisfying properties. The PSU/PSU-SO$_2$Cl and PSU/PSU-SO$_2$OCH$_3$ blend membranes still had, after hydrolysis of the nonionic sulfogroups to SO$_3$H, an inhomogeneous morphology (although the solutions of PSU with PSU-SO$_2$Cl and PSU-SO$_2$OCH$_3$ in tetrahydrofuran (THF) were homogeneous), leading to high membrane swelling in water and therefore to mechanical instability. Moreover, considerable leaching out of PSU-SO$_3$H was observed due to insufficient entanglement of PSU and PSU-SO$_3$H because of polymer incompatibility. In the PSU/PSU-SO$_2$NHC$_3$H$_7$ blend membranes, the SO$_2$NHC$_3$H$_7$ groups could not be hydrolyzed into SO_3H groups. For these reasons, this van der Waals/ dipole-dipole interaction blend membrane approach was not further considered by us.

8.2.1.2 Hydrogen-Bonding Interaction Blend Membranes

Many polymers that are capable forming hydrogen bonds and show good chemical and mechanical stability are commercially available. Therefore, we concluded that blending of a sulfonated ionomer with a hydrogen-bond-forming polymer should be a cost-effective way for reduction of swelling of the sulfonated ionomer. Therefore, blend membranes were prepared from sulfonated poly(etheretherketone) SPEEK and the polyamide PA Trogamid P (Hüls) and the poly(etherimide) PEI Ultem [44]. The resulting membranes showed good proton conductivity. The hydrogen bridge interaction between the PEEK-SO_3H polymer and PA or PEI polymer in the blend membrane was indicated by an increase of the glass transition temperature (T_g) of the blend by 5–15K, compared with pure PEEK-SO_3H. However, the swelling value of the hydrogen-bonded blend membranes at elevated temperatures was too high, in some cases even leading to dissolution of the membrane at $T = 90°C$. Moreover, some phase separation occurred in the blend membranes, and there are concerns that the PA amide bonds and PEI imide bonds show insufficient hydrolysis stability in an acidic environment, which is present during fuel cell operation. Therefore, the work with these blend membrane types was stopped.

8.2.1.3 Ionically Cross-Linked Acid-Base Blends and Acid-Base Ionomer Membranes

Since membrane types 1 and 2 showed unsatisfying properties, we searched for blend membrane types in which the blend membrane components show stronger interactions. We discovered acid-base blend membranes accessible by mixing polysulfonates and polybases and showing good mechanical and thermal stabilities, which are even better than the mechanical and thermal stability of the sulfonated polymers alone, and very good performance in fuel cells [45,46]. The structure of acid-base blend membranes is depicted schematically in Fig. 8.2.

The interaction forces between the acidic and basic blend component include electrostatic and hydrogen bridge interaction. The sulfonated poly(ethersulfones) and poly(etherketones) were combined both with commercially available basic polymers (e.g., polybenzimidazole Celazole (Celanese), poly(4-vinylpyridine), poly(ethylene imine)), and with self-developed basic polymers derived from poly(ethersulfones) [47] and poly(etherketones), including polymers that carry both sulfonic and basic groups onto the same backbone [48]. A wide variety of acid-base blend membranes with a broad property range were obtained. The most important characterization results of the ionically cross-linked ionomer membranes are

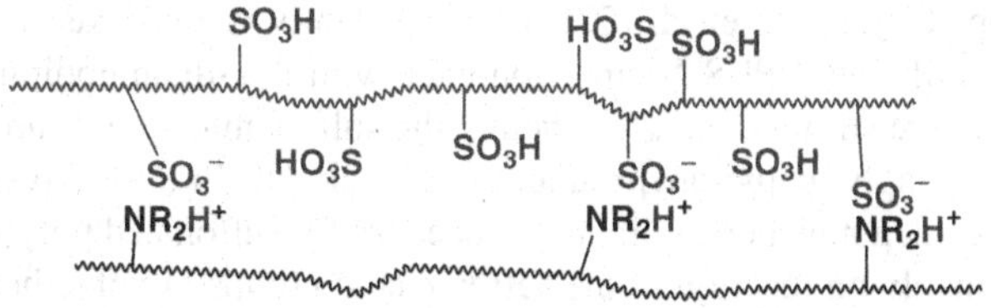

Fig. 8.2 Scheme of ionically cross-linked acid-base blend membranes

discussed in the section Membrane Characterization and Fuel Cell Results. Acid-base blend membranes also have been prepared from sulfonated polysulfone PSU Udel and polybenzimidazole PBI Celazol, and the properties of these membranes have been compared with those of Nafion [49]. They confirmed the previous finding [50,51] that ionomer-PBI blends show a marked reduction in methanol permeability both ex situ [50] and in situ [51], compared with Nafion. The synthesis and characterization of acid-base blend membranes from sulfonated poly(2,6-dimethyl-1,4-phenylene oxide) and polybenzimidazole PBI Celazole [52,53] has been reported recently. Interestingly, it was found that these membranes show good oxidative stability: after immersing the membrane samples for 72 h in 80°C hot 3%H_2O_2 aqueous solution containing ferrous ions, no weight decrease of the membranes was observed. These results confirm thermogravimetric analysis results with acid-base blend membranes, which also indicated an excellent thermal stability of acid-base blends, particularly blend membranes with PBI [54]. Moreover, other groups have also started investigating acid-base ionomer blends with PBI as the base component, such as blend membranes of SPEEK, PBI, and PAN [55]. Hasiotis et al. [56] report blends from sulfonated PSU and PBI, which were doped with phosphoric acid. They found that the ternary blend membranes showed better H^+-conductivity and mechanical stability than binary blends of PBI/H_3PO_4. The membranes were investigated in a PEFC up to T = 190°C.

One disadvantage of the ionically cross-linked (blend) membranes from polysulfonates and polybases is that the hydrogen bridges and electrostatic interactions break in aqueous environment when the temperature is raised to T > 70–90°C, leading to unacceptable swelling in water and therefore to mechanical instability, which could lead to destruction of the membrane in the fuel cell. To overcome this instability, covalently cross-linked (blend) membranes also have been developed.

8.2.1.4 Covalently Cross-Linked (Blend) Membranes

The need for an effective reduction of ionomer membrane swelling led us to search for covalent cross-linking procedures that show good chemical stability in the aqueous acidic environment present in the fuel cell. In the literature, only a limited number of covalently cross-linked ionomer membrane types is found, one approach being developed by Nolte et al. [57], who covalently cross-linked a partially N-imidazolized sulfonated PES Victrex ionomer with 4,4'-diaminodiphenylsulfone. However, there are strong concerns whether the sulfonamide bonds are sufficiently stable in the strongly acidic environment of a fuel cell. In a very recent paper of Mikhailenko et al. a novel cross-linking procedure for sulfonated poly(etherketone) is described, involving reaction of the sulfonic acid groups of polyetheretherketone with oligoalcohols such as ethylene glycol, glycerine, and meso erythrite under condensation (ester formation) [58]. However, the stability of these covalently cross-linked membrane systems in acidic environment was not investigated. Some years ago, we have discovered a novel cross-linking process in which sulfinate groups $SO_2Me(Me = Li, Na...)$ are involved. It consists of a nucleophilic substitution (S-alkylation) of the sulfinate group with di- or oligohalogenealkanes or – arylenes [59]:

Polymer - SO_2 Li + Hal - R - Hal + LiO_2 S - Polymer → Polymer - $S(O)_2$ - R - $S(O)_2$ - Polymer Preferred halogen alkanes were α,ω-dibromo- or α, ω -diiodoalkanes Br(I)-$(CH_2)_x$-Br(I) with x = 3–12, preferred dihalogenoarylenes were bis(4-fluorophenyl)sulfone, bis(3-nitro-4-fluorophenyl)sulfone, bis(4-fluorophenylphenyl)phosphinoxide, decafluorobenzophenone, and decafluorobiphenyl. The cross-links created by S-alkylation are stable in aqueous environment, both under alkaline, neutral, and acidic conditions, and in hot dipolar-aprotic solvents such as NMP or DMAc. Moreover, the membranes showed good thermal stabilities. Two different membrane types (Fig. 8.3) have been developed:

1. Cross-linked blend membranes by mixing sulfonated polymers with sulfinated polymers and the cross-linker in a dipolar-aprotic solvent (mostly NMP)
2. Cross-linked blend membranes by mixing a polymer carrying both sulfonate and sulfinate groups onto the same backbone in NMP with the cross-linker

The advantage of the membrane type 1 is that it can be prepared very easily: Both polymers are dissolved in the same solvent, and the cross-linker is added. A further advantage of the type 1 ionomer blend membranes is that a very broad

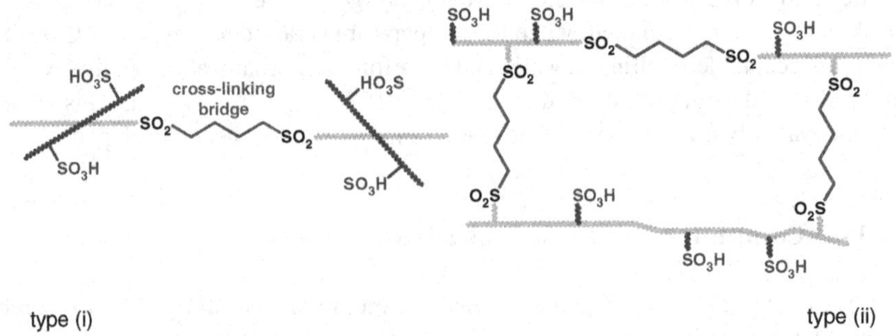

type (i)   type (ii)

Fig. 8.3 Scheme of covalently cross-linked membranes; type (i): ionomer blend; type (ii): ionomer

property range can be obtained by variation of the mass relation of the sulfinated and sulfonated blend component, the ion exchange capacity of both blend components, the backbone type of the blend components, and the cross-linker (different chain length of the cross-linkers, use of aliphatic or aromatic cross-linkers, use of mixtures of cross-linkers, etc.). The disadvantage of this type is that the polysulfonate macromolecules can diffuse out of the blend membrane, because they are only entangled in the covalent network built up by the sulfinated polymer and cross-linker. However, this problem can be minimized by increasing the cross-linking density of the network. The advantage of the membrane type (2) is that all macromolecules are taking part in the network; therefore, no bleeding-out of the sulfonated component can take place. The disadvantage is that the effort for preparation of mixed sulfonated/sulfinated polymers is higher than that for the preparation of 100% sulfonated or 100% sulfinated polymers. The preparation of the starting polymers for the cross-linked membranes is described in the section Polymer Modification for Blend Membranes, and principal characterization results of a selection of the covalently cross-linked membranes are given in the section Membrane Characterization and Fuel Cell Results.

8.2.1.5 Covalent-Ionically Cross-Linked (Blend) Membranes

Both ionically cross-linked membranes (splitting-off of the ionic bonds at T = 70–90°C) and covalently cross-linked membranes (bleeding-out of sulfonated macromolecules from covalently cross-linked blend membranes, brittleness of dry membranes) show disadvantages. To overcome these disadvantages, we started the development of covalent-ionically cross-linked membranes [60]:

1. Blending of a polysulfonate with a polysulfinate and a polybase, under addition of a dihalogeno cross-linker, which is capable to react with both sulfinate groups and tertiary amino groups under alkylation and therefore cross-linking (Fig. 8.4)

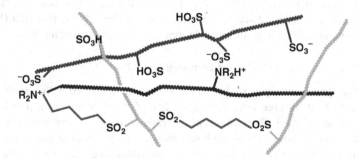

Fig. 8.4 Scheme of covalent-ionically cross-linked membranes

2. Blending of a polymer carrying both tertiary amino and sulfinate groups with a polysulfonate under addition of a dihalogeno cross-linker that alkylates both sulfinate and tertiary amino groups   •
3. Blending of a polymer carrying both sulfinate and sulfonate groups with a polybase under addition of a dihalogeno cross-linker

We found that the disadvantage of the membrane type 1 is incompatibility between the polysulfinate and polyamine, leading to phase-separated membranes that show unsatisfying mechanical stability and insufficient suppressing of water uptake at elevated temperatures. Although not yet clear, it may be speculated that the incompatibility of sulfinate and base polymers is due to the repulsion of the base group lone electron pair and the sulfinate group lone electron pair. This disadvantage of type 1 can be avoided with the type 2 membrane, in which the incompatible functional groups are bound to the same backbone in statistical distribution, and with type 3 membranes, in which the repulsion of the basic polymer with the sulfinate groups of the second polymer can be balanced by hydrogen bridges and/or dipole—dipole interaction of the base groups with the sulfonate groups of the second polymer. The membrane types 2 and 3 are transparent to visible light, indicating a homogeneous membrane morphology. One could also think of preparation of a polymer carrying sulfonate groups as well as sulfinate groups and basic groups onto the same backbone, e.g., by reaction of a lithiated polymer with the three electrophiles SO_2 (for sulfinate groups), SO_2Cl_2 (for sulfochloride $\rightarrow$ sulfonic acid groups), and an aromatic carbonyl base (for basic groups). This polymer would inherently form a morphologically homogeneous membrane. However, such a polymer would be very expensive due to the need for careful dosage of the electrophile mixture, possible reaction between the different electrophiles, and reaction at low temperatures and under protective atmosphere, so that it probably would not be suitable for mass production.

8.2.1.6 Composite Blend Membranes

Due to the fact that pure organic sulfonated ionomer membranes progressively lose their H^+-conductivity when raising the temperature above 100°C, if the fuel cell is not pressurized, due to drying out (evaporation of the water that is a "vehicle" for proton transport), we have started combining our differently cross-linked (blend) ionomer membranes with different inorganics, as schematically presented in Fig. 8.5.

We have prepared the following composite membrane types:

1. Covalently or ionically cross-linked blend membranes, filled with μ-sized oxide (SiO_2, TiO_2) or layered zirconium phosphate ZrP, introduced as a powder into the polymer solution. The problem of this membrane type is that inorganic oxide or salt powders tend toward agglomeration in the polymer solution and, after solvent evaporation, in the membrane, which reduces the active surface for water adsorption and possible proton transport dramatically. Therefore, we applied

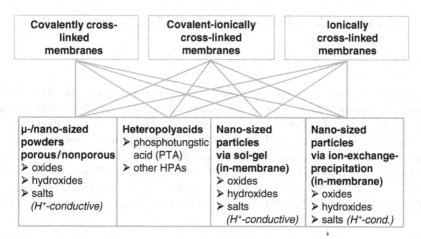

Fig. 8.5 ICVT strategies for ionomer composite membranes

literature-known procedures for formation of in-membrane nanoparticles to our membrane types (particularly, Jones, Rozière, Bauer et al. performed pioneering work in this field) [61–65].

2. Covalently or ionically cross-linked blend membranes, filled with layered ZrP by ion-exchange precipitation:

    (a) Ion-exchange of the SO_3H form of the membranes in a $ZrOCl_2$ solution to yield the membranes in the $(SO_3)_2ZrO$-form
    (b) Immersion of the ion-exchanged membranes in H_3PO_4; during immersion, the H_3PO_4 diffuses into the membrane, and by reaction with the ZrO^{2+} ions, the layered ZrP is precipitating in the membrane matrix

3. Ionically cross-linked blend membranes from sulfonated poly(etherketone)s, PBI, and the heteropolyacid tungstenic phosphoric acid (TPA): Heteropolyacids (HPA) are strong Broenstedt acids [66], and are interesting candidates for blending with ionomer membranes to increase their H^+-conductivity. However, HPAs are water-soluble, which leads to concerns that they diffuse out of the ionomer membranes with time, and different authors have reported that indeed partial leaching-out of the HPAs from the membrane takes place [30]. A marked reduction of the leaching-out rate of the HPA molecules could be realized by synthesis of novel HPAs, as reported by Ponce et al. [31]. Our motivation for this development of ionomer-HPA blend membranes was to answer the question whether the ionical cross-links in ionically cross-linked blend membranes and resulting interactions between HPA molecules and ionomer blend membrane components (ionical cross-linking, H bridges, dipole–dipole interaction) could prevent leaching-out of the HPA.

8.2.2   Polymer Modification for Blend Membranes

8.2.2.1   Synthesis of Sulfonated Polymers

For the synthesis of sulfonated poly(etherketones), well-known procedures were applied that involve sulfonation in 96% sulfonic acid [12,67] or oleum [14,68]. For the synthesis of sulfochlorinated poly(etherketone)s, the method described in [12] was used. PSU Udel and PPSU Radel R were sulfonated with n-butyllithium, as first described by Guiver [69]. The procedure involves reaction of the lithiated poly(ethersulfone) with SO_2, yielding PSU/PPSU-sulfinate, followed by oxidation with NaOCl to the PSU-sulfonate [11], or reaction of the lithiated poly(ethersulfone) with SO_2Cl_2, yielding PSU-SO_2Cl [70], followed by hydrolysis to the PSU/PPSU sulfonic acid.

8.2.2.2   Synthesis of the Sulfinated Polymers

For the synthesis of the PSU/PPSU-sulfinates, the discussed reaction of lithiated polymers with SO_2 was used. For the synthesis of poly(ethersulfones) carrying both sulfinate and sulfonate groups, we partially oxidized the PSU/PPSU-sulfinates with subportions of NaOCl [71]. Poly(etheretherketone sulfinates) were prepared by reduction of poly(etheretherketone sulfochlorides) with aqueous Na_2SO_3 [72].

8.2.2.3   Synthesis of Basic Polymers

Amino groups were introduced into poly(ethersulfone)s by the following methods:

1. Introduction of the NH_2 group *ortho* to the sulfone bridge of the poly(ethersulfone) was performed via a method developed by Guiver et al. [73,74]. The poly(ether sulfone amine) was then stepwise alkylated to the secondary and the tertiary polymeric amine by sequential addition of n-BuLi and CH_3I [43].
2. Introduction of the NH_2 group *ortho* to the ether bridge of poly(ethersulfone)s or *ortho* to the ether bridge of poly(etheretherketone) was done according to [75,76]. The poly(ethersulfone) primary amines were then alkylated to the secondary amines by deprotonation with LDA, followed by alkylation with CH_3I. The poly(ethersulfone) secondary amines were alkylated to the tertiary amines by reaction with KOH/CH_3I in DMSO. The poly(etheretherketone) primary amines were alkylated to the tertiary amines by reaction with KOH/CH_3I in one step.
3. Poly(ethersulfone)s were modified with tertiary amines by reaction of their lithiated form with basic aromatic aldehydes and ketones [47,77,78].
4. Poly(ethersulfones) were modified with tertiary amines also by reaction of their lithiated species with basic aromatic carboxylic acid esters, as shown in [47].

Interestingly, the created keto bridges of the product polymers are not attacked by residual PSU-Li sites to cross-link the polymer [47,77].

8.2.2.4 Synthesis of Polymers Containing Both Basic and Sulfinate Groups

Poly(ethersulfone)s carrying both sulfinate and basic groups were prepared starting from lithiated poly(ethersulfone) [60,79,80].

8.2.2.5 Synthesis of Polymers Containing Both Basic and Sulfonate Groups

Poly(etheretherketone) carrying both primary amino and sulfonate groups was prepared by nitration-reduction procedure [48].

8.2.3 Membrane Characterization and Fuel Cell Results

We review here the dependence of the most important characterization and fuel cell application results of the different membrane types we have developed in our lab on different parameters.

8.2.3.1 Ionically Cross-Linked Blend Membranes

1. Dependence of the membrane properties on size of the repeating unit of the polybase. To ensure complete ionical cross-linking, the repeating units of the acidic and basic blend component should have comparable size and density of functional groups. We compared the extent of formation of ionical cross-links using the same polymeric sulfonic acid (sulfonated PSU, ion-exchange capacity (IEC) = 1.6 mequiv. g^{-1}), and the two polybases polybenzimidazole (PBI, base capacity 6.5 mequiv.g^{-1}) and polyethylenimine (PEI, base capacity 23.2 mequiv. g^{-1}) [81]. When the calculated (from the molar relation acid-base) and experimentally obtained IECs of the two acid-base blend membranes were compared, we found that for the SPSU/PBI blends the calculated and experimental IECs were similar, whereas for the SPSU/PEI blends the experimental IECs were much higher than calculated. This led to the conclusion that of the SPSU/PEI blends, from the sterical point of view, it is not possible that every amino group finds its acidic counterpart, due to the extreme difference in functional group densities of SPSU and PEI, respectively. On the contrary, of the SPSU/PBI blends, functional group densities are not that strongly different, as is the case of the SPSU/PEI blends, allowing that every acidic group finds its basic counterpart, taking into account the excess in acidic groups in the SPSU/PBI blends and chain segment mobilities of the blend components.

2. Dependence of the membrane properties from the base strength of the base. We investigated to what extent the strength of the polybase influences the percentage of formation of acid/base cross-links. Calculated and experimental acid/base blend ion exchange capacities were compared using the relatively strong polybases PBI (calculated p_k^a of its protonated form 5.6), poly(4-vinylpyridine) (P4VP, calculated p_k^a of its protonated form 6), and PSU-ortho-ether-diamine (POSAII, calculated p_k^a of its protonated form 4) with the IECs of acid-base blends using the very weak polybase PSU-ortho-sulfone-diamine (POSAI, calculated p_k^a of its protonated form – 1.5). The results showed very clearly that the calculated and experimental IECs were very close to each other for the PBI, P4VP-, and POSAII-containing acid-base blends [54,75,81], whereas the experimental IECs were much higher than the calculated ones for the POSAI acid-base blends [75,81], assuming that at p_k^a of the conjugated acids to the respective bases of < 0–2 the formation of acid-base bonds, or, in other words, the protonation of the base groups is incomplete. Therefore, for an effective suppression of swelling by ionic cross-links, the chosen base should be strong enough to ensure complete protonation when forming acid-base blends. A very good polybase candidate on that score is PBI, due to its high base strength and excellent chemical stability [82].
3. Dependence of the membrane properties from the ionical cross-linking density. We recently [68] compared the properties of membranes, composed of sulfonated poly(etherketoneetherketoneketone) SPEKEKK Ultrapek and PBI, having the same IEC of 1.35 mequiv. SO_3Hg^{-1}, but different ionical cross-linking densities. The different cross-linking density was obtained by use of sulfonated SPEKEKKs having different IECs ranging from 2.5 to 4.1 mequiv SO_3Hg^{-1}. As expected, the water uptake of the membranes decreased with increasing cross-linking density: The membrane having the lowest cross-linking density in this series had a swelling value of 260% at 90°C, whereas the 90°C swelling degree of the membrane with the highest cross-linking density amounted to 90%. Interestingly, in the same membrane series an increase in electrical resistance with increasing cross-linking density was observed (membrane with the lowest cross-linking density: R_{sp}^{H+}=5 Ωcm; membrane with the highest cross-linking density: R_{sp}^{H+}=11 Ωcm). An explanation for this finding could be that the ionic cross-links hinder the H^+ transport through the H^+-conducting channels of the membranes by repulsion forces [68]. The thermal stability of all membranes up to $\approx$240°C from this series is comparable. In Fig. 8.6, the TGA traces of the membranes with the highest and lowest cross-linking density in this series are shown. When the PBI portion of acid-base blend membranes, containing the same sulfonated polymer having the same IEC, is increased, a mass relation is reached in which the amount of acidic groups balance those of basic groups, in other words, where no excess free SO_3H groups are available. Consequently, such membranes are no longer H^+ conductive [54].
4. Dependence of the membrane properties from the membrane preparation method. In the case of very weak polybases, such as POSAI, it was possible to mix the polybase with the polymeric sulfonic acid in the same (dipolar-aprotic)

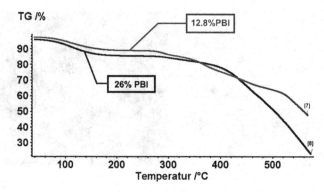

Fig. 8.6 TGA of SPEKEKK/PBI blend membranes with a PBI content of 12.8% and of 26%, respectively

solvent without precipitation of a polyelectrolyte complex. This phenomenon is due to the very low extent of formation of ionical cross-links at such blends [54]. In the case of stronger bases, such as PBI or P4VP, the salt form of the polymeric sulfonate must be mixed with the polybase to avoid polyelectrolyte complex precipitation, which would take place when the polymeric sulfonic acid is mixed with the polymeric base. Two procedures have been applied:

(a) The polymeric sulfonic acid was dissolved in a dipolar-aprotic solvent, followed by neutralization of the SO_3H groups with a base such as n-propylamine or triethylamine. Then the polybase solution in a dipolar-aprotic solvent was added, followed by membrane casting and solvent evaporation. During solvent evaporation, a part of the ammonium sulfonate groups decomposed, leaving back the sulfonic acid group, which immediately reacted with a basic group present in its environment. In other words, with this procedure a part of the ionical cross-links is created during membrane formation.

(b) The polymeric metal sulfonate (cations: Li, Na) was dissolved in a dipolar-aprotic solvent, followed by polybase addition, and then the solvent was evaporated. The resulting membrane had to be treated with mineral acid to protonate the sulfonate groups and therefore to create the ionical cross-links. The membranes prepared via methods (1) and (2) showed different morphology: the (1) membranes showed a homogeneous morphology, whereas at the membranes (2) a partially phase-separated morphology occurred in some cases, such as that reported in [79]. Two membranes, composed of sulfonated PEK (IEC = 1.8 mequiv. g^{-1}) and the two bases PBI and PSU-ortho-sulfone-C(OH)(4-diethylaminophenyl)$_2$ were prepared, the first prepared from SPEK in the SO_3H form, neutralized with n-propylamine (membrane 504H), and the second prepared from SPEK in the SO_3Li form (membrane 504Li). The morphology of both membranes was determined with transmission electron microscopy (TEM). The TEMs of both membranes

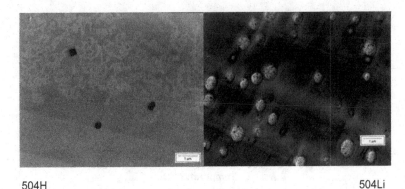

504H   504Li

Fig. 8.7 TEM of membrane 504H and 504Li, magnification 6,600×

are shown in Fig. 8.7. The 504H membrane showed a homogeneous morphology, whereas at 504Li some phase separation could be observed. These findings can be explained as follows. In membrane 504H, during membrane formation a part of the ionical cross-links between the acidic and basic polymers is formed, leading to compatibility of the blend components. In membrane 504Li, only weak van der Waals and dipole–dipole interaction is present between the acidic and basic blend components—the insufficient interactions between the blend components lead to partial incompatibility. The partial incompatibility also influences the membrane properties [79]. Due to incomplete ionical cross-link formation at the 504Li membrane, the IEC of the 504Li membrane has a value of 1.34 mequiv g^{-1} (calculated from acid/base molar relation, IEC = 1.11 mequiv g^{-1}), whereas the experimental IEC of 504H is 1.12 mequiv. g^{-1}. Moreover, the swelling value of the 504Li membrane, particularly at higher temperatures in water, is higher than that of the 504H membrane. At 90°C, the 504Li membrane shows a swelling value of 134%, whereas the swelling value of 504H at 90°C amounts to 74%. In the DMFC experiment, the 504Li membrane shows a slightly better performance than the 504H membrane, which is due to their higher free SO_3H group concentration.

5. Influence of the addition of a radical scavenger. It is known that in fuel cell membrane degradation processes, radicals are involved that are generated during fuel cell operation [83]. Therefore, it was investigated whether the addition of substances that are capable of acting as radical scavengers improves the chemical stability of the ionomer membranes. For this purpose, blend membranes from PEKSO$_3$Li, PBI, and the radical scavenger poly(N,N'-bis-(2,2,6,6-tetramethyl-4-piperidinyl)-1,6-diamino-hexane-co-2,4-dichloro-6-morpholino-1,3,5-triazine) were prepared [84]. The membranes were investigated via thermogravimetry in 65% O$_2$ atmosphere to check their thermal stability. However, no positive effect

of the radical scavenger onto the thermal stability could be detected from TGA. All the membranes—SPEK/PBI/scavenger, SPEK/PBI, and SPEK/scavenger— had nearly identical thermal stabilities. For all three membranes, the splitting-off of the sulfonic acid group, which is the first step of membrane decomposition, started at around 230°C. The membrane decomposition process in the TGA was investigated by a TGA-FTIR coupling setup, which allows the FTIR-analysis of the TGA decomposition gases [84]. On the other hand, direct investigation of these membrane types in a fuel cell, placed in the resonator of an electron resonance spectrometer, showed very low radical concentration in the membranes, which is comparable to the radical concentration observed in Nafion membranes placed in the same setup [85], suggesting good radical stability of these acid-base blend membranes.

6. Dependence of the DMFC performance on DMFC operation parameters. Different ternary acid-base blend membranes, composed of SPEK, PBI, and PSU-C(OH)(4-diethylaminophenyl)$_2$, were tested in a DMFC up to temperatures of 130°C. Good performance was detected at 110°C, which was comparable to Nafion 112, reaching a peak power density of 0.25 W cm^{-2} in air operation [86]. However, the catalyst loading was 12 mg noble metal cm^{-2} membrane, which is much too high. One possibility to reduce noble metal loading of the electrodes is to increase the DMFC operation temperature, because higher operation temperatures lead to improved catalyst utilization. Another possibility is to reduce the membranes' methanol permeability. Therefore, the following DMFC operation conditions were applied [84]: the DMFC temperature was increased to 130°C, and the noble metal loading was reduced by a factor of 4. Indeed, by application of these measures the power density of the fuel cell could even be increased, compared with the higher loading and lower operation temperature. Moreover, by a reduction of the air flow from 4 to 0.3 Lmin^{-1}, the methanol permeability of the membranes could be drastically reduced. In Fig. 8.8, the current density/peak power density curves of the 565 acid-base blend membrane in the DMFC under these conditions are shown.

7. Dependence of the PEFC performance from operation temperature. One problem of ionomer membranes requiring water for proton transport, like the sulfonated ionomer blend membranes, is that these membranes suffer from progressive drying out at fuel cell operation temperatures approaching the boiling point of water, leading to a dramatic drop in proton conductivity and consequent loss in power density. Therefore, we investigated the temperatures to which our acid-base blend membranes can be used in PEFC. We again selected the previously mentioned ternary membrane type composed of SPEK/PBI/PSU-C(OH) (4-diethylaminophenyl)$_2$ for the test. The catalyst loading was 1 mg cm^{-2} per electrode. The operation temperature was raised from 85 to 100°C. The result of the PEFC experiment was that the membrane operated well up to 100°C. Even an increase of power density of the membrane from 85 to 100°C could be detected [80]. At 85°C, the maximum power density was 0.26 W cm^{-2} (@0.68 A cm^{-2}), at 90°C, 0.29 W cm^{-2} (@0.7 A cm^{-2}), and at 100°C 0.33 W cm^{-2} (@0.8 A cm^{-2}).

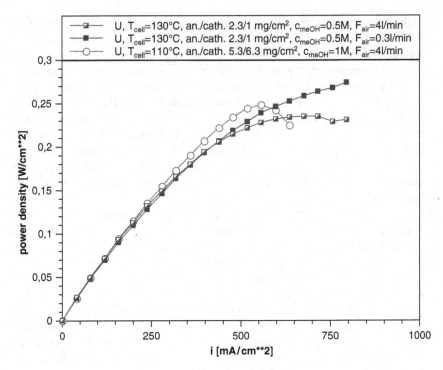

Fig. 8.8 Current density/power density curves of membrane 565 under different applied conditions

8.2.3.2 Covalently Cross-Linked Blend Membranes

1. Dependence of the membrane properties from the type of cross-linker. It was investigated whether the use of an aromatic cross-linker for sulfinate-S-alkylation in blend membranes of sulfonated poly(etherketone)s with sulfinated poly(ethersulfones) leads to better thermal stability of the membrane. Therefore, blend membranes were prepared from sulfochlorinated poly(etherketone) SPEK (IEC = 3.5 mequiv. g^{-1}), sulfinated PSU (1 group per RU), and cross-linkers 1,4-diiodobutane or bis(3-nitro-4-fluorophenyl)sulfone, respectively [87]. After hydrolysis of the SO_2Cl groups to SO_3H groups of the membranes, the following results were obtained. The proton conductivity of the two membranes was nearly identical, the water uptake characteristics were similar, and the thermal stability of the two membranes was nearly the same. To summarize, the use of an aromatic cross-linker does not improve the thermal stability of the membranes.

2. Dependence of the membrane properties from type of cross-linker (different aromatic). A number of aromatic cross-linkers (Fig. 8.9) have been tested for their suitability for the sulfinate S-alkylation reaction. When determining the properties of the membranes, we could see that all cross-linkers were capable of

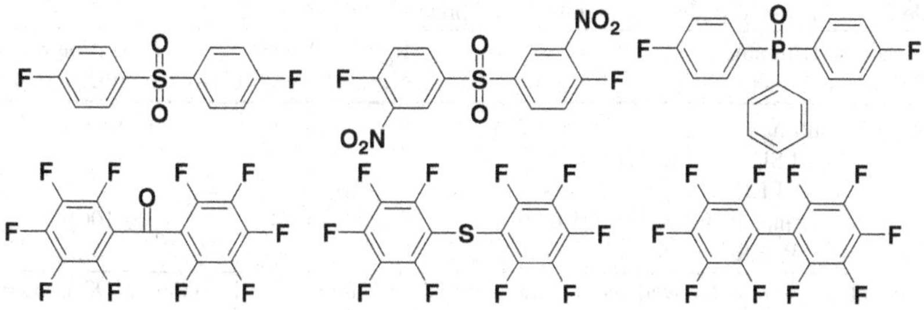

Fig. 8.9 Tested cross-linkers for sulfinate S-alkylation

Table 8.3 Covalently cross-linked blend membranes with different cross-linking density

Membrane (no.)	IEC of sulfonated PEK	IEC exp. (theo.) (mequiv. g^{-1})	R$_{sp}^{H+}$ (Ω*cm)	Cross-linking density (mmol CL g^{-1})	SW90°C/ SW25°C^a (–)
1,025	1.8 (water-insoluble)	1.07 (1.23)	9.8	0.296	2.94
1,251	3.5 (water-soluble)	1.21 (1.23)	7.7	0.607	1.43

^{a}Quotient between water uptake at 90°C and water uptake at 25°C.

S-alkylating the sulfinate groups of the blend membrane. The properties of all prepared membranes were comparable [87,88], their specific resistance being in the range 8–12 cm, which is a value comparable to Nafion. Moreover, their swelling at elevated temperatures (90°C) was limited to 40–60%, which indicates a high degree of cross-linking. The cross-linking was also confirmed by extraction experiments of the membranes in 90°C hot DMAc. The extraction residue was in good accordance with the mass share of the sulfinated and cross-linked blend membrane component. The thermal stability of all investigated blend membranes was in the same range.

3. Dependence of the membrane properties from the covalent cross-linking density. The water uptake (swelling) of ionomer membranes in fuel cells should be reduced to maintain good mechanical stability, and this is possible without reduction in H$^+$-conductivity. We prepared covalently cross-linked membranes that had the same calculated IEC but different cross-linking density. In Table 8.3, the composition and some properties of the two membranes are listed. Cross-linking polymer was sulfinated PSU Udel, one sulfinate group per repeat unit, and the cross-linker was 1,4-diiodobutane. From Table 8.3 it follows that the H$^+$-conductivity of both membranes is comparable, being in the Nafion range. Interestingly, the swelling value (SW) of the 1,251 membrane is by a factor of 2 lower than that of the 1,025 membrane, as shown by the SW90°C/SW25°C quotient (Table 8.3). From the results it can be concluded that the swelling values of the covalently cross-linked membranes can be varied independently from the proton conductivity of the membranes. The thermal stabilities of both membranes were nearly similar, as determined by TGA.

Table 8.4 Comparison: ionomer membrane/ionomer blend membrane

Membrane (no.)	Membrane polymers	IEC exp. (theo.) (mequiv. G^{-1})	R$_{sp}^{H+}$ (Ω*cm)	SW90°C/ SW25°C (–)	Extraction residue exp(theo)[a] (%)
1,030	Sulfinated PSU + sulfonated PEEK	1.14(1.23)	9.5	2.05	91.2(54.5)
Zh31	Sulfinated-sulfonated PEEK	0.84(1.26)	9.6	1.38	100(100)

[a]The membranes were dry-weighed, followed by an 48-h immersion in DMAc at 90°C. During this time, all non-cross-linked polymers dissolved in DMAc.

4. Dependence of the membrane properties from the type of sulfonated poly(etherketone). The three different sulfonated poly(etherketones) SPEK, SPEEK, and SPEKEKK have been used in covalently cross-linked blend membranes as the H+-conductive component. It was found that the properties of the different membranes were very close to one another [87].

5. Dependence of the membrane properties from the type of sulfonated poly(ethersulfone). The two different poly(ethersulfones), PSU and PPSU, were used as the cross-linking component in the covalently cross-linked blend membranes. The membranes prepared from these polymers showed comparable properties. There are some indications that the thermal and mechanical stability of the membranes from sulfonated PPSU is slightly better than the thermal and mechanical stability of membranes using sulfonated PSU [88].

6. Dependence of the membrane properties from membrane type (blend/not blend). Covalently cross-linked blend membranes such as those described herein have been compared with covalently cross-linked membranes prepared from sulfochlorinated PEEK, which has been partially reduced using Na$_2$SO$_3$ to yield sulfonated-sulfonated PEEK [72] (the cross-linker was 1,4-diiodobutane). The properties of the two membranes are gathered in Table 8.4. Comparison of the properties yields two points of interest: the SW90°C/SW25°C quotient of the completely cross-linked membrane was markedly lower than the SW quotient of the blend membrane, and the extraction residue of the Zh31 membrane is 100%, indicating that all macromolecules of the Zh31 are integrated in the covalent network. Obviously it is an advantage when all macromolecules are integrated in the covalent network, because this efficiently limits the water uptake of the membranes, leading to improved mechanical stability. TGA investigations of both membranes indicated an improved thermal stability of the Zh31 membrane [89].

7. Dependence of the membrane properties from membrane type (nonfluorinated and partially fluorinated ionomer). We have developed partially fluorinated covalently cross-linked membranes by reaction of disulfinated poly (ethersulfones) with pentafluorobenzene sulfochloride and different cross-linkers [90]. The scheme for the preparation of such partially fluorinated covalent ionomer networks is given in Fig. 8.10. The obtained membranes showed high H+-conductivities and moderate SW. In Table 8.5, some of the properties of one

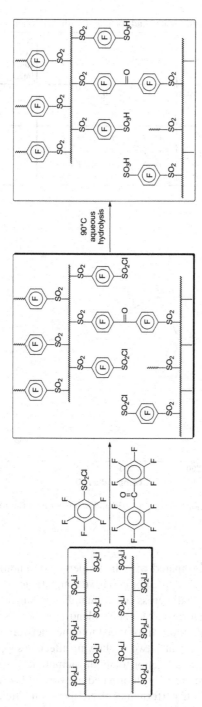

Fig. 8.10 Partially fluorinated ionomer network preparation

Table 8.5 Properties of nonfluorinated and partially fluorinated ionomer membranes

Membrane [no.]	Membrane polymers	Cross-linker	IEC exp. (theo.) [meq/g^{-1}]	R_{sp}^{H+} [Ω*cm]	SW90°C/ SW25°C (–)
1030	monosulfinated PSU + sulfonated PEEK	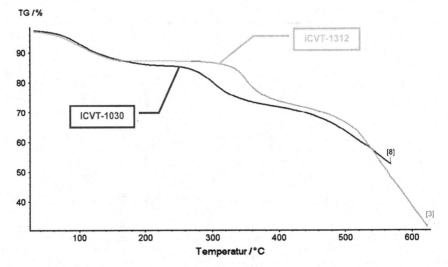	1.14 (1.23)	9.5	2.05
1312	disulfinated Radel R		0.81	13.1	1.46

Fig. 8.11 TGA traces of the nonfluorinated 1,030 and of the partially fluorinated 1,312 membrane, respectively

novel membrane are compared with the properties of a nonfluorinated ionomer blend membrane [88]. Both membranes have comparable H⁺-conductivities. The SW quotient of the partially cross-linked membrane is markedly reduced, compared with the non-fluorinated ionomer blend membrane. The reason for this finding is that at the partially fluorinated ionomer network all macromolecules are taking part in the covalent network, allowing effective suppression of swelling. A comparative TGA investigation of both membranes suggests an improved thermal stability of the partially fluorinated network membrane (Fig. 8.11).

8. Dependence of the PEFC performance from operation temperature Some of the covalently cross-linked blend membranes were tested in a PEFC at different temperatures. The i/U polarization curves of one of these membranes, the 1,251

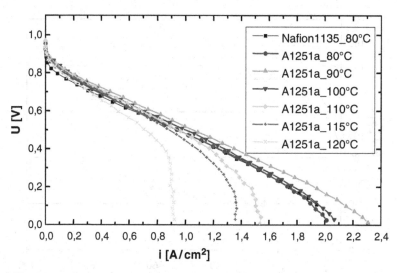

Fig. 8.12 i/U polarization curves of the 1,251 membrane in a PEFC at T = 80–120°C, O$_2$/H$_2$ operation, catalyst loading 2 mg cm^{-2}, pressure 2 bar

membrane, are shown in Fig. 8.12. It is obvious that the performance of the membrane is comparable to Nafion up to temperatures of 100°C. Above this temperature, however progressive drying-out of the membrane takes place, leading to dramatic decrease of performance. The results suggest that it is required to improve the water-storage capacity of sulfonated ionomer membranes to ensure fuel cell-applicability in the mid-temperature range 100–150°C.

8.2.3.3 Covalent-Ionically Cross-Linked Membranes

1. Comparison of the morphology of binary and ternary covalent-ionically cross-linked membranes As mentioned, the ternary covalent-ionically cross-linked blend membranes show phase separation, due to incompatibility of the sulfinate and the basic blend component. This problem was overcome by preparation of binary covalent-ionically cross-linked blend membranes: The polysulfonate was mixed with a polymer carrying both sulfinate and basic groups in statistical distribution onto the same backbone. The TEM micrographs of the binary blend membrane clearly showed a homogeneous morphology of this membrane [80].
2. Comparison of the properties of binary and ternary covalent-ionically cross-linked membranes. The properties of a ternary covalent-ionically cross-linked blend membrane (ICVT-1028) were compared with the properties of a binary covalent-ionically cross-linked blend membrane (ICVT-WZ-054), both membranes showing comparable IEC and proton conductivity [79,80]. The main difference in properties between the two membranes was their different swelling

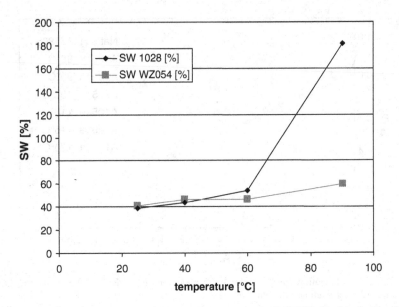

Fig. 8.13 Water uptake of ICVT-1028 (ternary blend) and ICVT-WZ-54 (binary blend) in dependence of T

behavior. In Fig. 8.13, the water uptake (swelling) of both membranes in dependence of T is shown. The findings can be explained as follows: The membrane WZ 054 has a homogeneous morphology, the covalent network spreads all over the membrane matrix. Therefore, the water uptake of this membrane can be effectively limited. In contrast, the membrane 1,028 is phase separated, whereas the covalently cross-linked membrane phase is the discontinuous (disperse) membrane phase. Therefore, only the water uptake of the disperse phase is limited, whereas the water uptake of the continuous membrane phase, in which most of the sulfonated macromolecules are placed, is not suppressed, leading to extreme membrane swelling at T > 60°C where the ionic cross-links are split off [86]. It is advantageous to ensure that covalent-ionically cross-linked ionomer membranes show a homogeneous morphology in order to efficiently suppress the water uptake of the membranes.

8.2.3.4 Composite Blend Membranes

1. Composite membranes by introduction of inorganic powders. Our first approach for the preparation of hybrid membranes was the addition of inorganic particles to the polymer solutions, followed by solvent evaporation. The following inorganics have been added as μ-particles: SiO_2 (Aerosil 380, Degussa) [79], Rutil TiO_2 (Aldrich 22,422-7) [80], and layered ZrP powder (Prof. Linkov, Univ. of the Western Cape, Cape Town, South Africa) [89]. Unfortunately, agglomeration

of the inorganic particles in the membrane matrix took place, leading to large inorganic particles within the membrane morphology that were not effective in adsorption of water, particularly at elevated temperatures of >80°C. When applied to DMFC, a strong reduction of MeOH permeability could be realized by addition of SiO_2 particles, but also the DMFC performance was reduced. Consequently, the fuel cell performance of these membranes was not improved, compared with the pure organomembranes [80].

2. Comparison of organo blend membranes with ZrP hybrid blend membranes. By the application of the $ZrOCl_2$-H_3PO_4 procedure to the covalently cross-linked blend membranes, transparent hybrid membranes were obtained, indicating that the size of particles was well below the wavelength of visible light [88]. In Table 8.6, some of the properties of a covalently cross-linked blend membrane (ICVT-1228) and ZrP hybrid membrane based onto this membrane (ICVT-1228ZrP) are listed. Interestingly, the room temperature proton conductivity of the composite membrane, compared with the organomembrane, is reduced by a factor of two. This finding can be explained by a stricture of the ion-conducting channels by the growth of the inorganic phase, which preferably takes place in the vicinity of the SO_3H ion-aggregates. In Fig. 8.14, TEM micrographs of the 1228 and 1228ZrP membrane are presented. The growth of the inorganic phase can be seen very clearly. The even distribution of the ZrP microphase within the membrane matrix was proven by SEM-EDX mapping [88]. One of the ZrP hybrid blend ionomer membranes was tested in a PEFC up to temperatures of 115°C and compared with a pure organomembrane. While the pure organomembrane showed a peak power density of 0.35 W cm^{-2} at 80°C, the ZrP hybrid

Table 8.6 Some Properties of ICVT-1228 and ICVT-1228ZrP

Membrane (no.)	IEC exp(theo) (mequiv. g^{-1})	R_{sp}^{H+} ($\Omega*$cm)	SW90°C/SW25°C (–)
1,228	1.24 (1.23)	9.5	1.33
1,228ZrP	1,21	19.6	1.24

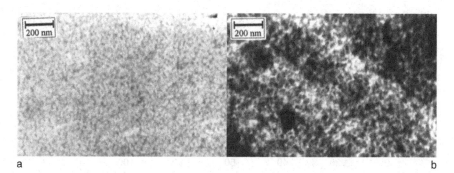

a   b

Fig. 8.14 TEM micrographs of ICVT-1228 (**a**) and ICVT-1228ZrP (**b**), sulfonate groups exchanged with Pb^{2+} for a better contrast, magnification 52,000×

membrane had a peak power density of 0.6 W cm^{-2} at 115°C, indicating that the ZrP phase had a positive impact on PEFC performance. The improvement of PEFC performance by the ZrP phase could be due to both improvement of water storage ability and the contribution of the ZrP phase to the proton transport at elevated temperatures, as suggested in the literature [63,65,38]. However, further work has to be done for in-depth clarification of the influence of the ZrP phase to PEFC performance of the hybrid membranes.

3. Properties of composite membranes, repeatedly treated with $ZrOCl_2$-H_3PO_4. Blend membranes from sulfonated PEKEKK Ultrapek (IEC = 3 mequiv g^{-1}) and sulfinated PSU Udel (1 group per RU), cross-linked with 1,4-diiodobutane (ICVT-1202), were repeatedly treated with $ZrOCl_2$-H_3PO_4. The change in membrane properties with the number of treatments of the ICVT-1202 membrane was monitored. In Fig. 8.15, the IEC and H$^+$-conductivities of the membranes are presented. In Fig. 8.16, the water uptake of the neat and the $ZrOCl_2$-H_3PO_4–

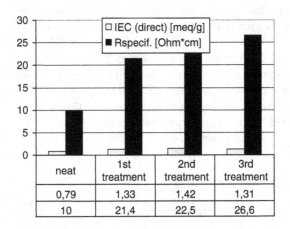

	neat	1st treatment	2nd treatment	3rd treatment
IEC (direct) [meq/g]	0,79	1,33	1,42	1,31
Rspecif. [Ohm*cm]	10	21,4	22,5	26,6

Fig. 8.15 IEC and H$^+$-conductivities of the 1,202 membranes

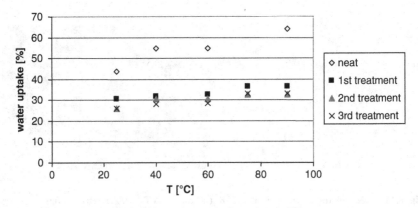

Fig. 8.16 Water uptake of the 1,202 membranes in dependence of T

Table 8.7 TPA Loss in SPEKEKK-PBI-TPA membranes

Content TPA (%)	Calc. residuals (dehydrated TPA) (%)
0	0
10	8.56
20	17.12
40	34.24
100	85.6

treated membranes are shown. From Figs. 8.15 and 8.16, the following can be observed: by the ZrP treatment, the H$^+$ resistance is increased, supporting the hypothesis that the deposit of ZrP in the ion-conducting channels narrows them, leading to hindrance of H$^+$-transport. Moreover, the water uptake is reduced, which can be explained with (hydrogen bridge, dipole–dipole) interactions between the ZrP phase, which contains $Zr(HPO_4)_2$ groups [91], and the SO_3H groups of the organo-ionomer. Repeated treatment of ionically cross-linked blend membranes with $ZrOCl_2/H_3PO_4$ lead to similar results: the H$^+$-resistance of the membranes was increased, and their water uptake was reduced [91].

4. Properties of ionically cross-linked membranes containing tungstenic phosphoric acid. Blend membranes from sulfonated PEKEKK Ultrapek, PBI Celazole, and tungstenic phosphoric acid (TPA) were prepared [68]. Indeed, by introduction of TPA the proton conductivity of the membranes could be enhanced. We investigated the possible leaching out of HPA molecules from the membrane matrix by post-treatment of the pure organomembrane and the composite membranes, initially containing 10, 20, and 40 wt% TPA, in (a) 10%HCl at 90°C for 48 h and (b) water at 60°C for 48 h, followed by investigation of the posttreated membranes in TGA up to 600°C. TPA remains as a residual when the organic membrane part is thermally removed. The pure organo blend membrane was completely decomposed after TGA, whereas the TGA of the composite membranes showed that only a part of the HPA was present in the membrane after the post-treatment (Table 8.7). The interactions present in the SPEKEKK-PBI-TPA membranes are not strong enough to prevent leaching out of the TPA ions, and methods must be found for immobilization of the heteropolyacids in the ionomer blend membrane matrix, which could be achieved by generation of chemical bonds between the TPA molecules and the organomembrane.

8.2.4 Comparison of the Properties of the Different Ionomer Membrane Systems

To be able to assess the suitability of the differently cross-linked blend membrane types for the fuel cell application, we compared representative membranes of the different types having comparable calculated IEC with each other.

8.2.4.1 Comparison of the Properties of Differently Cross-Linked Ionomer Membrane Types Having Comparable Calculated IEC

Differently cross-linked membranes having a calculated IEC of 1.2–1.3 mequiv g^{-1} were compared. The composition and characterization results are presented in Table 8.8.

From Table 8.8 it can be concluded that the covalently cross-linked membrane 1,398 clearly showed the best properties in this series: This membrane has the lowest resistance and is the only one which does not dissolve when immersed in 90°C hot water. In the other membranes of this series, the interactions between the acidic blend component and other components are not strong enough to prevent dissolution in 90°C hot water. Particularly, the membrane 1,397 shows a low IEC and a high H^+-resistance, indicating that already a considerable amount of sulfonated macromolecules had diffused out from the blend membrane matrix. Membrane 1,397 is not transparent, indicating a microphase-separated morphology, which facilitates leaching out of the sulfonated blend membrane component.

The dissolution of the three membranes also reflects the undesirable property of SPEEK, having an IEC of 1.8 mequiv g^{-1} this ionomer itself dissolves in water at T = 90°C. In the morphologically homogeneous-covalently cross-linked membrane, obviously the strong entanglement of the SPEEK macromolecules in the

Table 8.8 Composition and characterization results of differently cross-linked membranes having comparable calculated IEC (comparison: pure $PEEKSO_3H$, IEC = 1.8 mequiv. g^{-1})

Membrane (no.)	Type cross-linking	Composition	IEC_{calc}(IEC_{exp}) (mequiv. g^{-1})	R_{sp}^{H+} (HCl)[a] (Ω cm)	SW^b 25–40–60–90 (°C)
SPEEK	—	SPEEK (1.8)	1.8(1.54)	21.5	31–37–50[e]
1,397	Van der Waals/ dipole-dipole	SPEEKCl (1.8)[c] PSU^d	1.23(0.82)	69.2	25–35–42[e]
1,392	H-bridge	SPEEK (1.8) PEI Ultem[f]	1.23(1.01)	10.3	30–46–59[e]
1,389	Ionically	SPEEK (1.8) PBI[VII]	1.3(1.12)	28.4	25–33–34[e]
1,398	Covalently	SPEEKCl (1.8)[c] PPSU-SO_2Li[h] B4FPhPhPO[i]	1.23(1.18)	6.7	32–50–56–108

[a] Measured in 0.5N HCl.

[b] Measured in water of the respective T.

[c] Hydrolysis after membrane formation.

[d] Unmodified PSU Udel (Solvay).

[e] Dissolved at 90°C.

[f] Product of General Electric.

[g] PBI Celazole (Celanese).

[h] Sulfinated PPSU Radel R, 1 group per RU.

[i] Cross-linker bis(4-fluorophenyl)-phenylphosphinoxide.

covalent network prevents their leaching out. Interestingly, use of SPEK (IEC = 1.8 mequiv g⁻¹) instead of SPEEK (IEC = 1.8 mequiv g⁻¹) in ionically cross-linked membranes with PBI, having the same IEC as membrane 1,389, prevents dissolution in water at T = 90°C [84]. Obviously, in SPEK-PBI blends stronger interactions between the blend components are present than in SPEEK-PBI blends. The reason for this finding is not yet clear. However, due to the higher concentration of carbonyl groups in SPEK, compared with SPEEK, one can speculate that the carbonyl group markedly contributes to the interaction between the macromolecular chains by dipole–dipole forces and H-bridges.

The thermal stability of ionomer membranes is of great importance for the application in fuel cells also. Therefore, the membranes have also been investigated by TGA-FTIR coupling. In Fig. 8.17, the TGA traces of the mentioned membranes are presented. The decomposition gases of the TGA were investigated by TGA-FTIR coupling to determine the temperature at which splitting-off of the SO_3H groups starts, which is the first step in membrane degradation [84]. The starting temperatures of SO_3H splitting-off at the different membranes are listed in Table 8.9.

The ionically cross-linked membrane showed the best thermal stability, being similar to the thermal stability of pure SPEEK, followed by the covalently cross-linked membrane. The thermal stabilities of the membranes 1,392 and 1,397 are

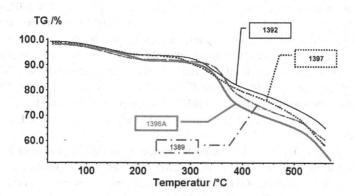

Fig. 8.17 TGA traces of the membranes 1,397, 1,392, 1,389, and 1,398A

Table 8.9 Starting temperatures of SO_3H splitting-off at the different blend membranes (comparison with Pure PEEKSO$_3$H, having IEC of 1.8 mequiv. g⁻¹)

Membrane (no.)	Type cross-linking	Start-T SO_3H splitting-off (°C)
PEEK-SO$_3$H	—	240
1,397	Van der Waals/dipole-dipole	224.1
1,392	H-bridge	228.1
1,389	Ionically	239.2
1,398	Covalently	232.6

markedly worse. The reason for this finding is not yet clear. In the literature it is reported that in polymer blends in some cases one blend component facilitates the degradation of the other blend component, as was shown for PBI/polyarylate blends [92]. From the obtained results it can be concluded that the covalently and ionically cross-linked blend membrane are interesting candidates for fuel cell application.

8.2.4.2   Comparison of the Properties of Covalently, Ionically, and Covalent-Ionically Cross-Linked Ionomer Membranes Having the Same IEC

To find out which of the ionomer blend membranes are the most promising for the application in membrane fuel cells, if any, we prepared a covalently, covalent-ionically, and ionically cross-linked membrane having comparable IEC [86]. Membranes were obtained that showed nearly identical H^+-conductivities and thermal stabilities. However, the membranes showed marked differences in swelling behavior: The covalently cross-linked membrane (ICVT-1025) had a water uptake value of 87% at 90°C, whereas the water uptake value of the ionically cross-linked membrane (ICVT-1029) amounted to 190%, and that of the covalent-ionically cross-linked membrane (ICVT-1028) was 181%. As pointed out, the unexpected high swelling of ICVT-1028 can be explained by ineffective covalent cross-linking taking place in the disperse blend membrane phase of the inhomogeneous ICVT-1028 blend membrane, whereas the high swelling value of ICVT-1029 finds its explanation in the splitting-off of the ionical cross-links at elevated temperatures in water [86]. In 2M meOH solution, the swelling behavior of the three membranes is comparable to that in H_2O [93]: at 90°C, the ICVT-1025 swells to 95%, whereas the ICVT-1028 has a swelling value of 150%, and the ICVT-1029 increases in weight by 195%. All three membranes were applied to a DMFC at temperatures from 25 to 110°C [86]. The following current density/power density curves were obtained at 110°C (Fig. 8.18).

From Fig. 8.18 it results that the performance of the three membranes is comparable, only a bit inferior to that of the Nafion 105 membrane. Therefore, from the DMFC test, no advantage of one of the three membrane types over the others could be observed, which is reflected by the fact that the H^+-resistance of the membranes, measured in situ via impedance spectroscopy, was nearly the same. The ICVT-1029 had an area resistance of 0.13 Ω cm^2, whereas the ICVT-1028 had an area resistance of 0.16 Ωcm^2 and the ICVT-1025 an area resistance of 0.17 Ωcm^2. The MeOH permeability of our membranes was lower by a factor 2–3 than that of Nafion 105 [86]. From this finding, one would expect that the performance of the cross-linked blend membranes is better than that of Nafion 105 under the same conditions. The reason why this is not the case lies in the bad connection between the polyaryl membranes and catalytic electrodes that contain Nafion as the binder and H^+-conductive component. After the DMFC test, partial delamination between membranes and electrodes was observed [86]. Substitution of Nafion as a component by a polyaryl ionomer in the electrode has not led to satisfying results.

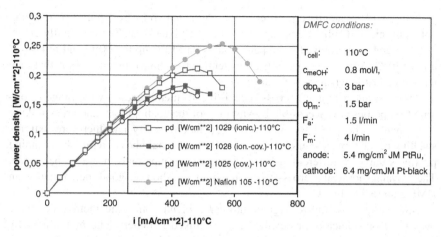

Fig. 8.18 Current density/power density curves of the membranes ICVT-1025, ICVT-1028, and ICVT-1029 at 110°C, compare with Nafion 105)

8.3 Conclusions

We have used commercially available arylene main-chain polymers for modification with proton-conductive groups. Blend concepts for low-temperature (<100°C) fuel cell ionomer membranes were developed. We blended differently modified polymers using cross-linking procedures (physical and chemical cross-linking) because of the convenience of mixing different polymers in a joint solvent, swelling reduction by cross-linking, and the possibility of tailoring the membrane morphology and therefore tailoring the membrane properties by selective choice of blend component type.

The developed membrane types had the following properties:

The van der Waals/dipole-dipole blends and hydrogen bridge blends showed a heterogeneous morphology, leading to extreme swelling at elevated temperatures. Therefore, these membrane types were regarded as not suitable for fuel cell application.

A wide variability in properties could be reached in acid-base blends by variation of the materials used for blending. The membranes showed good performance in PEFC up to 100°C and in DMFC up to 130°C, reaching peak power densities approaching 0.3 W cm^{-2} at reduced noble metal loadings of 3 mg cm^{-2}. A disadvantage of the ionically cross-linked membranes is that the ionic cross-links split off at T > 70–80°C, leading to extreme swelling above this temperature range in liquid water.

At the covalently cross-linked (blend) membranes, also a wide variability in properties could be obtained. This membrane type showed good H$^+$-conductivities and stabilities. Leaching-out of the sulfonated component could be avoided by preparation of covalently cross-linked ionomeric networks from polymers carrying both sulfonic and sulfinate groups onto the same backbone. The covalently cross-linked membranes showed good performance in PEFC up to 100°C and in DMFC up to 130°C. A disadvantage of the covalently cross-linked membranes is that they tend to be brittle when they are drying out.

At the covalent-ionically cross-linked (blend) membranes, a wide range of different properties was accessible by variation of the acidic, basic, and cross-linking components. The problem of morphologic heterogeneity and therefore extreme swelling of the ternary membranes at T > 60–80°C could be overcome by blending polymers carrying different types of functional groups onto the same backbone. The covalent-ionically cross-linked membranes showed good performance in DMFC up to T = 110°C.

Differently cross-linked organic-inorganic blend composite membranes were prepared from ionically or covalently cross-linked ionomer blend membranes by adding μ-sized inorganic powders to the solution of the organopolymers or heteropolyacids to the organopolymer solution, and by ion-exchange of the formed organomembrane SO_3H protons with metal cations such as ZrO^{2+} and TiO^{2+} followed by immersion of the membrane in phosphoric acid, leading to precipitation of metal phosphates or metal hydrogenphosphates within the membrane matrix.

Blend membranes by mixing inorganic powders into the polymer solutions ended up in composite membranes containing large, agglomerated μ-sized particles. Therefore, no positive effect of the inorganic phase to the H^+-conductivity of the membranes could be found. Ionically cross-linked blend membranes prepared by addition of heteropolyacid powders to the polymer solution showed significant heteropolyacid leaching-out during the membrane post-treatment, which disqualifies this membrane type for application in fuel cells. Blend membranes having a layered zirconium phosphate ZrP phase were transparent, indicating nanosized inorganic particles, and showed good H^+-conductivity, and mechanical and thermal stability. Application of one of these composite membranes to a PEFC yielded good performance up to T = 115°C.

In the future, we will focus onto the following membrane systems.

We will further develop membranes from arylene main-chain polymers carrying both sulfonate and sulfinate groups onto the same backbone. Moreover, partially fluorinated covalent ionomer networks, which show an improved chemical and thermal stability, compared with non-fluorinated systems, will be prepared. We will concentrate on the further development of blend membranes of sulfonated poly(etherketone)s with stable and strong polybases such as poly(benzimidazole)s, due to their high durability. The development of morphologically homogeneous ionomer membranes by preparation and further development of polyarylenes carrying several types of functional groups on the same backbone to avoid phase separation will be continued. Finally, we will concentrate onto the further development of differently cross-linked hybrid membrane systems having ZrP-type inorganic phase. Our objective is to obtain membranes whose inorganic phase is both proton conductive and serves as water storage to allow fuel cell operation in the temperature range 100–150°C.

References

1. J. Kerres, Development of ionomer membranes for fuel cells, *J. Membr. Sci.* **185**, 3–27 (2001)
2. J. Wei, C. Stone, A. E. Steck, US 5,773,480; June 30, 1998
3. C. Stone, C. L. Q. Hu, T. Daynard, C. Mah, Lecture, 3rd International Symposium on New Materials for Electrochemical Systems, 4th to 8th July 1999, Montreal, Canada
4. A. E. Steck, C. Stone, in: "Proceedings of the Second International Symposium on New Materials for Fuel Cell and Modern Battery Systems", Eds. O. Savadogo, P. R. Roberge, Montreal, Canada, July 6–10, 1997, p. 792
5. H. P. Brack, D. Ruegg, H. Bührer, M. Slaski, S. Alkan, G. G. Scherer, Differential scanning calorimetry and thermogravimetric analysis investigation of the thermal properties and degradation of some radiation-grafted films and membranes, *J. Polym. Sci. Part B: Polym. Phys.* **42**, 2612–2624 (2004)
6. P. Gode, J. Ihonen, A. Strandroth, H. Ericson, G. Lindbergh, M. Paronen, F. Sundholm, G. Sundholm, N. Walsby, Membrane durability in a PEM fuel cell studied using PVDF based radiation grafted membranes, *Fuel Cells* **3**, 21–27 (2004)
7. G. D. Alelio, US 2,366,007; December 12,1944
8. N. Ogato, M. Rikukawa, US 5,403,675; April 4, 1995
9. A. Noshay, L. M. Robeson, Sulfonated polysulfone, *J. Appl. Polym. Sci.* **20**, 1885–1903 (1976)
10. B. C. Johnson, I. Yilgör, C. Tran, M. Iqbal, J. P. Wightman, D. R. Lloyd, J. E. McGrath, Synthesis and characterization of sulfonated poly(acrylene ether sulfones), *J. Polym. Sci. Polym. Chem. Ed.* **22**, 721–737 (1984)
11. J. Kerres, W. Cui, S. Reichle, New sulfonated engineering polymers via the metalation route. I. Sulfonated poly(ethersulfone) PSU Udel ® via metalation-sulfination-oxidation, *J. Polym. Sci. Part A: Polym. Chem.* **34**, 2421–2438 (1996)
12. F. Helmer-Metzmann, F. Osan, A. Schneller, H. Ritter, K. Ledjeff, R. Nolte, R. Thorwirth, EP 0574 791 B1, 22.12.1999
13. J. Rozière, D. J. Jones, Non-fluorinated polymer materials for proton exchange membrane fuel cells, *Annu. Rev. Mater. Res.* **33**, 503–555 (2003)
14. H. H. Ulrich, G. Rafler, Sulfonated Poly(aryl ether ketone)s, *Angew. Makromol. Chem.* **263**, 71–78 (1998)
15. P. Xing, G. P. Robertson, M. D. Guiver, S. D. Mikhailenko, K. Wang, S. Kaliaguine, Synthesis and characterization of sulfonated poly(ether ether ketone) for proton exchange membranes, *J. Membr. Sci.* **229**, 95–106 (2004)
16. J. Schauer, P. Lopour, J. Vacik, The preparation of ultrafiltration membranes from a moderately sulfonated poly(oxy(2,6-Dimethyl-1,4-Phenylene), *J. Membr. Sci.* **29**, 169–175 (1992)
17. S. Percec, G. Li, Chemical modification of poly(2,6-Dimethyl 1,4-Phenylene Oxide) and properties of the resulting copolymers, *Polym. Prepr. (Am. Chem. Soc., Div. Polym. Chem.)* **27**(2), 19–20 (1986)
18. K. Miyatake, H. Iyotani, K. Yamamoto, E. Tsuchida, Synthesis of poly(phenylene sulfide sulfonic acid) via poly(sulfonium cation) as a thermostable proton-conducting polymer, *Macromolecules.* **29**, 6969–6971 (1996)
19. Q. Guo, P. N. Pintauro, H. Tang, S. O Connor, Sulfonated and crosslinked polyphosphazene-based proton-exchange membranes, *J. Membr. Sci.* **154**, 175–181 (1999)
20. H. R. Allcock, M. A. Hofmann, C. M. Ambler, S. N. Lvov, X. Y. Zhou, E. Chalkova, J. Weston, Phenyl phosphonic acid functionalized poly[aryloxyphosphazenes] as proton-conducting membranes for direct methanol fuel cells, *J. Membr. Sci.* **201**, 47–54 (2002)
21. I. Gautier-Luneau, A. Denoyelle, J. Y. Sanchez, C. Poinsignon, Organicinorganic protonic polymer electrolytes as membrane for low-temperature fuel cell, *Electrochim. Acta* **37**, 1615–1618 (1992)
22. K. Tabata, F. Fujibayashi, M. Aimu, US 6,723,464; April 20, 2004

23. M. Murthy, M. Esayian, A. Hobson, S. MacKenzie, W. Lee, J. W. van Zee, Performance of a polymer electrolyte membrane fuel cell exposed to transient CO concentrations, *J. Electrochem. Soc.* **148**, A1141–A1147 (2001)

24. R. F. Savinell, M. H. Litt, US 5,525,436; June 11, 1996

25. J. S. Wainright, J.-T. Wang, D. Weng, R. F. Savinell, M. H. Litt, Acid-doped polybenzimidazoles: a new polymer electrolyte, *J. Electrochem. Soc.* **142**, L121–L123 (1995)

26. G. Calundann, M. Sansone, B. Benicewicz, E. W. Choe, Oe. Uensal, J. Kiefer, DE 10246459 A1, 2004

27. Y. L. Ma, J. S. Wainright, M. H. Litt, R. F. Savinell, Conductivity of PBI membranes for high-temperature polymer electrolyte fuel cells, *J. Electrochem. Soc.* **151**(1) A8–A16 (2004)

28. K. D. Kreuer, On the development of proton conducting polymer membranes for hydrogen and methanol fuel cells, *J. Membr. Science* **185**, 29–39 (2001)

29. M. Schuster, *Protonenleitung in imidazolhaltigen Materialien als Modellsysteme für wasserfreie Brennstoffzellenmembranen*, Ph.D. Dissertation, University of Mainz 2002; available via Internet: http://archimed.uni-mainz.de/pub/2002/0017/diss.pdf

30. S. M. J. Zaidi, S. D. Mikhailenko, G. P. Robertson, M. D. Guiver, S. Kaliaguine, Proton conducting composite membranes from polyether ether ketone and heteropolyacids for fuel cell applications, *J. Membr. Sci.* **173**, 17–34 (2000)

31. M. L. Ponce, L. Prado, B. Ruffmann, K. Richau, R. Mohr, S. P. Nunes, Reduction of methanol permeability in polyetherketone-heteropolyacid membranes, *J. Membr. Sci.* **217**, 5–15 (2003)

32. Y. S. Kim, F. Wang, M. Hickner, T. A. Zawodzinski, T. A. J. E. McGrath, Fabrication and characterization of heteropolyacid ($H_3PW_{12}O_{40}$)/directly polymerized sulfonated poly(arylene ether sulfone) copolymer composite membranes for higher temperature fuel cell applications, *J. Membr. Sci.* **212**, 263–282 (2002)

33. A. S. Aricò, P. Cretì, P. L. Antonucci, V. Antonucci, Comparison of ethanol and methanol oxidation in a liquid-feed solid polymer electrolyte fuel cell at high temperature, *Electrochem. Solid-State Lett.* **1**(2), 66–68 (1998)

34. K. T. Adjemian, S. J. Lee, S. Srinivasan, J. Benziger, A. B. Bocarsly, Silicon oxide Nafion® composite membranes for proton-exchange membrane fuel cell operation at 80–140°C, *J. Electrochem. Soc.* **149**(3), A256–A261 (2002)

35. I. Honma, H. Nakajima, O. Nishikawa, T. Sugimoto, S. Nomura, Amphiphilic organic/inorganic nanohybrid macromolecules for intermediate-temperature proton conducting electrolyte membranes, *J. Electrochem. Soc.* **149**(10), A1389–A1392 (2002)

36. K. A. Mauritz, Organic-inorganic hybrid materials: perfluorinated ionomers as sol-gel polymerization templates for inorganic alkoxides, *Mat. Sci. Eng.* **C6**, 121–133 (1998)

37. C. Yang, S. Srinivasan, A. S. Aricò, P. Creti, V. Baglio, V. Antonucci, Composite Nafion®/zirconium phosphate membranes for direct methanol fuel cell operation at high temperature, *Electrochem. Solid-State Lett.* **4**(4), A31–A34 (2001)

38. G. Alberti, M. Casciola, Composite membranes for medium-temperature PEM fuel cells, *Annu. Rev. Mater. Res.* **33**(1), 129–154 (2003)

39. K. T. Adjemian, S. Srinivasan, J. Benziger, A. B. Bocarsly, Investigation of PEMFC operation above 100°C employing perfluorosulfonic acid silicon oxide composite membranes, *J. Power Sources* **109**(2), 356–364 (2002)

40. S. C. Yeo, A. Eisenberg, Physical properties and supermolecular structure of perfluorinated ion-containing (Nafion®) polymers, *J. Appl. Polym. Sci.* **21**, 875–898 (1977)

41. X. Ren, T. E. Springer, S. Gottesfeld, Water and methanol uptakes in Nafion® membranes and membrane effects on direct methanol cell performance, *J. Electrochem. Soc.* **147**, 92–98 (2000)

42. V. M. Barragán, C. Ruiz-Bauzá, J. P. G. Villaluenga, B. Seoane, Transport of methanol and water through Nafion® membranes, *J. Power Sources* **130**(1–2), 22–29 (2004)

43. W. Zhang, C. -M. Tang, J. Kerres, Development and characterization of sulfonated-unmodified and sulfonated-aminated PSU Udel® blend membranes, *Sep. Purif. Technol.* **22**, 209–221 (2001)
44. W. Cui, J. Kerres, G. Eigenberger, Development and characterization of ion-exchange polymer blend membranes, *Sep. Purif. Technol.* **14**, 145–154 (1998)
45. J. Kerres, W. Cui, US 6,194,474; 27 Feb 2001 J. Kerres, W. Cui, US 6,300,381; 9 October, 2001 J. Kerres, W. Cui, EP 1,073,690, 14 January, 2004
46. J. Kerres, A. Ullrich, T. Häring, EP 1,076,676, 28 January, 2004 J. Kerres, A. Ullrich, T. Häring, US 6,723,757, 20 April, 2004
47. J. Kerres, A. Ullrich, T. Häring, US 6,590,067, 8 July, 2003
48. W. Cui, J. Kerres, DE Appl. 198 13 613.7; 27 March, 1998
49. C. Manea, M. Mulder, New polymeric electrolyte membranes based on proton donor-proton acceptor properties for direct methanol fuel cells, *Desalination* **147**, 179–182 (2002)
50. M. Walker, K.-M. Baumgärtner, M. Kaiser, J. Kerres, A. Ullrich, E. Räuchle, Proton conducting polymers with reduced methanol permeation, *J. Appl. Polym. Sci.* **74**, 67–73 (1999)
51. L. Jörissen, V. Gogel, J. Kerres, J. Garche, New membranes for direct methanol fuel cells, *J. Power Sources* **105**, 267–273 (2002)
52. B. Kosmala, J. Schauer, Ion-exchange membranes prepared by blending sulfonated poly(2,6-dimethyl-1,4-phenylene oxide) with polybenzimidazole, *J. Appl. Polym. Sci.* **85**, 1118–1127 (2002)
53. K. Bouzek, S. Moravcova, Z. Samec, J. Schauer, H⁺ and Na⁺ Ion transport properties of sulfonated poly(2,6-dimethyl-1,4-phenyleneoxide) membranes, *J. Electrochem. Soc.* **150**(6) E329–E336 (2003)
54. J. Kerres, A. Ullrich, T. H. Häring, M. Baldauf, U. Gebhardt, W. Preidel, Preparation, characerization, and fuel cell application of new acid-base blend membranes, *J. New Mater. Electrochem. Syst.* **3**, 229–239 (2000)
55. S. Motupally, V. Lightner, U. S. Department of Energy, Energy Efficiency and Renewable Energy, Hydrogen, Fuel Cells and Infrastructure Technologies Program, FY 2003 Progress Report, p. 7
56. C. Hasiotis, Q. Li, V. Deimede, J. K. Kallitsis, C. G. Kontoyannis, N. J. Bjerrum, Development and characterization of acid-doped polybenzimidazole/sulfonated polysulfone blend polymer electrolytes for fuel cells, *J. Electrochem. Soc.* **148**(5) A513–A519 (2001)
57. R. Nolte, K. Ledjeff, M. Bauer, R. Mülhaupt, Partially sulfonated poly(arylene ether sulfone) - a versatile proton conducting membrane material for modern energy conversion technologies, *J. Membr. Sci.* **83**, 211–220 (1993)
58. S. D. Mikhailenko, K. Wang, S. Kaliaguine, P. Xing, G. P. Robertson, M. D. Guiver, Proton conducting membranes based on cross-linked sulfonated poly(ether ether ketone) (SPEEK), *J. Membr. Sci.* **233**, 93–99 (2004)
59. J. Kerres, W. Cui, W. Schnurnberger, DE 196 22 237.7; 12 March, 1998, US 6,221,923; 24 April, 2001, US 6,552,135; 22 April, 2003
60. J. Kerres, W. Zhang, C.-M. Tang, DE Appl. 10024576;19 May, 2000, Int. Appl. PCT/EP200105644 from 17th May, 2001
61. D. J. Jones, M. El Haddad, B. Mula, J. Rozière, *New proton conductors for fuel cell applications*, Environmental Research Forum "Chemistry and Energy", C. A. C. Sequeira, Ed., Transtec, 1–2, pp. 115 – 126 (1996)
62. J. Rozière, D. J. Jones, *Inorganic-organic Composite Membranes for PEM Fuel Cells*, Handbook of Fuel Cell Technology, W. Vielstich, A. Lamm, H. Gasteiger, Eds., Wiley, Vol. 3, pp. 447–455 (2003)
63. B. Bauer, D. J. Jones, J. Rozière, L. Tchicaya, G. Alberti, L. Massinelli, M. Casciola, A. Peraio, and E. Ramunni, Hybrid organic-inorganic membranes for a medium temperature fuel cell, *J. New Mater. Electrochem. Appl.* **3**, 87–92 (2000)
64. D. J. Jones, J. Roziere, Recent advances in the functionalisation of polybenzimidazole and polyetherketone for fuel cell applications, *J. Membr. Sci.* **185**, 41–58 (2001)

65. L. Tchicaya-Bouckary, D. J. Jones, J. Rozière, Hybrid polyaryletherketone membranes for fuel cell applications, *Fuel Cells* **2**, 40–45 (2002)
66. K. D. Kreuer, Proton conductivity: materials and applications, *Chem. Materials* **8**, 610–641 (1996)
67. S. Kaliuguine, S. D. Mikhailenko, K. P. Wang, P. Xing, G. P. Robertson, M. D. Guiver, Properties of SPEEK based PEMs for fuel cell application, *Catalysis Today* **82**, 213–222 (2003)
68. J. Kerres, C.-M. Tang, C. Graf, Improvement of properties of poly(ether ketone) ionomer membranes by blending and cross-linking, *Ind. Eng. Chem. Res.* **43**(16), 4571–4579 (2004)
69. M. D. Guiver, *Aromatic Polysulfones Containing Functional Groups by Synthesis and Chemical Modification*, Ph.D. Dissertation, Carletown University 1987
70. J. A. Kerres, A. J. van Zyl, Development of new ionomer blend membranes, their characterization and their application in the perstractive separation of alkenes from alkene-alkane mixtures. 1. Polymer modification, ionomer blend membrane preparation and characterization, *J. Appl. Polym. Sci.* **74**, 428–438 (1999)
71. J. Kerres, W. Zhang, W. Cui, New sulfonated engineering polymers via the metalation route. 2. Sulfinated-Sulfonated Poly(ethersulfone) PSU Udel® and its crosslinking, *J. Polym. Sci.: Part A: Polym. Chem.* **36**, 1441–1448 (1998)
72. J. Kerres, W. Zhang, T. Häring, DE Appl. 102 09 784.4 from 28th Febr., 2002, Int. Appl. PCT/DE02/03260 from 2nd Sept., 2002 W. Zhang, V. Gogel, K. A. Friedrich, J. Kerres, Novel covalently cross-linked poly(etheretherketone) ionomer membranes, *J. Power Sources* **155**(1), 3–12 (2006)
73. M. D. Guiver, G. P. Robertson, Chemical modification of polysulfones: a facile method of preparing azide derivatives from lithiated polysulfone intermediates, *Macromolecules.* **28**, 294–301 (1995)
74. M. D. Guiver, G. P. Robertson, S. Foley, Chemical modification of Polysulfones II: an efficient method for introducing primary amine groups onto the aromatic chain, *Macromolecules* **28**, 7612–7621 (1995)
75. C. M. Tang, W. Zhang, J. Kerres, Preparation and characterization of ionically cross-linked proton-conducting membranes, *J. New Mat. Electrochem. Syst.* **7**, 287–298 (2004)
76. H. A. Naik, I. W. Parsons, P. T. McGrail, P. D. MacKenzie, Chemical modification of polyarylene ether/sulphone polymers: preparation and properties of materials aminated on the main chain, *Polymer* **32**(1), 140–145 (1991)
77. J. Kerres, A. Ullrich, M. Hein, Preparation and characterization of novel basic polysulfone polymers, *J. Polym. Sci.: Part A: Polym. Chem.* **39**, 2874–2888 (2001)
78. J. Kerres, A. Ullrich, Synthesis of novel engineering polymers containing basic side groups and their application in acid-base polymer blend membranes, *Sep. Purif. Technol.* **22**, 1–15 (2001)
79. J. Kerres, W. Zhang, A. Ullrich, C.-M. Tang, M. Hein, V. Gogel, T. Frey, L. Jörissen, Synthesis and characterization of polyaryl blend membranes having different composition, different covalent and/or ionical cross-linking density, and their application to DMFC, *Desalination* **147**, 173–178 (2002)
80. J. Kerres, M. Hein, W. Zhang, N. Nicoloso, S. Graf, Development of new blend membranes for polymer electrolyte fuel cell applications, *J. New Mat. Electrochem. Syst.* **6**(4), 223–229 (2003)
81. J. Kerres, A. Ullrich, F. Meier, Th. Häring, Synthesis and characterization of novel acid-base polymer blends for application in membrane fuel cells, *Solid State Ionics* **125**, 243–249 (1999)
82. D. R. Coffin, G. A. Serad, H. L. Hicks, R. T. Montgomery, Properties and applications of celanese PBI-Polybenzimidazole fiber, *Textile Res. J.* **52**(7), 466–472 (1982)
83. A. Panchenko, H. Dilger, J. Kerres, M. Hein, A. Ullrich, T. Kaz, E. Roduner, In-situ spin trap electron paramagnetic resonance study of fuel cell processes, *Phys. Chem. Chem. Phys.* **6**, 2891–2894 (2004)

84. J. Kerres, A. Ullrich, M. Hein, V. Gogel, K. A. Friedrich, L. Jörissen, Cross-linked polyaryl blend membranes for polymer electrolyte fuel cells,*Fuel Cells* **4**, 105–112 (2004)
85. E. Roduner, A. Panchenko, J. Kerres, unpublished results
86. J. Kerres, W. Zhang, L. Jörissen, V. Gogel, Application of different types of polyaryl-blend-membranes in DMFC, *J. New Mat. Electrochem. Syst.* **5**, 97–107 (2002)
87. J. Kerres, M. Hein, W. Zhang, N. Nicoloso, S. Graf, S., Invited lecture (J. Kerres), 5th International Symposium "New Materials for Electrochemical Systems", 6. to 11. July 2003, Montreal, Canada J. Kerres, W. Zhang, T. Häring, Covalently cross-linked ionomer (blend) membranes for fuel cells, *J. New Mat. Electrochem. Syst.* **7**, 299–309 (2004)
88. J. Kerres, Invited Lecture, 204th Meeting of The Electrochemical Society, October 12-October 16, 2003, Orlando, Florida, proceedings submitted
89. J. Kerres, Invited lecture, Symposium "Advances in Materials for Proton Exchange Membrane Fuel Cell Systems", February 23–27, 2003, Asilomar, California.
90. J. Kerres, DE Appl. 103 08 462.2 from 19th February, 2003
91. J. Kerres, W. Zhang, unpublished results
92. T. S. Chung, P. N. Chen, Polybenzimidazole (PBI) and polyarylate blends, *J. Appl. Polym. Sci.* **40**, 1209–1222 (1990)
93. F. Meier, J. Kerres, G. Eigenberger, Characterization of polyaryl-blend-membranes for DMFC application *J. New Mat. Electrochem. Syst.* **5**, 91–96 (2002)

Chapter 9
Organic-Inorganic Membranes for Fuel Cell Application

Suzana Pereira Nunes

Abstract Organic-inorganic membranes are worldwide under investigation with the purpose of achieving reduced methanol crossover for direct methanol fuel cell (DMFC) and increasing the proton conductivity at temperatures higher than 100°C in hydrogen fuel cells. The advantages and disadvantages of these membranes are discussed here with membrane examples containing aerosol, layered silicates, and modified silica as passive fillers as well as zirconium phosphate and heteropolyacids as conductive fillers.

9.1 Requirements for Fuel Cell Membranes

The establishment of fuel cells as a competitive technology for energy conversion still depends on the development of materials with better performance. A large challenge is the improvement of the currently available membranes.

The requirements for good membranes are high proton conductivity, low permeability to the reactants (fuel and oxygen), and high chemical stability. While the bulk of the membrane should not allow electron transport, electron conductivity is recommended in the membrane-catalyst-electrode interface region to stimulate the electrochemical reaction.

9.2 Key Issues for Fuel Cell Membrane Development

A key issue is the preparation of new membranes able to effectively operate above 100°C and under external low humidification. At this temperature most of the available membranes start to dehydrate, requiring more complex operating conditions to compensate the consequent conductivity decrease. Reviews on materials under investigation to overcome this problem have been published by Alberti and Casciola [1], Li et al. [2], and Hogarth et al. [3]. The motivations for operating at high temperatures are: (1) improved reaction kinetics, (2) minimization of catalyst poisoning by CO, and (3) simplification of the heat and water (humidification) management in the cell.

S.M.J. Zaidi, T. Matsuura (eds.) *Polymer Membranes for Fuel Cells*,
doi: 10.1007/978-0-387-73532-0, © Springer Science+Business Media, LLC 2009

The combination of organic and inorganic materials is under investigation by different groups in an attempt to minimize the membrane drawbacks at high temperature. Inorganic additives can help to keep water in the membrane at higher temperatures or contribute for the proton conductivity themselves.

Another fundamental issue is the fuel crossover [4,5], which is much more relevant for direct methanol fuel cells. Methanol crossing the membrane from the anode to the cathode will react in the presence of the catalyst, competing with oxygen reduction. This competition leads to a loss of cell performance. A frequently used approach to minimize methanol crossover has been the introduction of an inorganic phase. Besides the methanol crossover, the water transport in the membrane is at least as important. Most of the available membranes are based on sulfonated polymers. A reasonably high degree of sulfonation is required to reach enough proton conductivity. This also increases membrane hydrophilicity and stimulates water transport. However, if an excessive amount of water reaches the cathode, the active catalyst sites will be protected from the access to oxygen, hindering the cathode reaction and again reducing cell performance. The introduction of an inorganic phase in the membrane can also reduce water transport and minimize the cathode flooding problem.

9.3   Different Strategies for Organic-Inorganic Materials

The development of organic-inorganic materials has increased dramatically in the last 15 years, with applications that include not only membranes, but also catalysis, sensors, protective coatings, devices for non-linear optics, mechanics, and electronics. Reviews of different approaches to incorporating inorganic building blocks into organic polymers recently have been published by Kickelbick [6] and Sanchez et al. [7]. The use of organic-inorganic materials for membranes in applications other than fuel cell (gas separation, pervaporation) [8–15] has been explored by our group for a long time. Figure 9.1 summarizes many of the alternative approaches for organic-inorganic materials for membranes. The simplest approach (Fig. 9.1a) is to add isotropic nanosized inorganic particles such as fumed silica, TiO_2, or ZrO_2. The effect can be understood using the conventional Maxwell equation [16]:

$$P_c = P_p(1-\phi)/(1+\phi/2) \qquad (9.1)$$

Where
P_c = permeability of the composite
P_p = permeability of the plain membrane
ϕ = filler volume fraction

The permeability decreases with the increase of the filler volume fraction. These particles might be organically modified, mainly with silanes to incorporate groups which strengthen the interaction with the polymer matrix or even covalently bind to the matrix polymer chains (Fig. 9.1b). Inorganic particles with higher aspect ratios (L/W), like flakes, might contribute with a more effective barrier

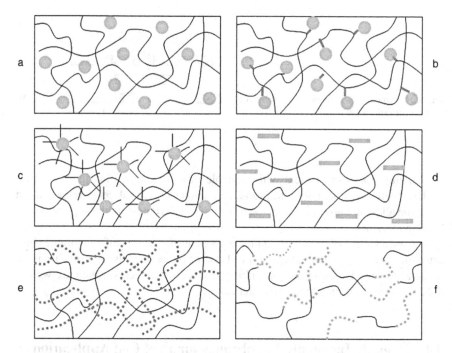

Fig. 9.1 Strategies for preparation of organic-inorganic membranes

(lower permeability) effect to the membrane, as shown in Figs. 9.1d and 9.2 [17–19] and described by Equation (9.2).

$$P_c = P_p(1-\phi)/[1+(L/2W)\phi] \tag{9.2}$$

More versatile hybrid inorganic materials could be developed by making use of soft inorganic chemistry processes. With the sol-gel process, using alkoxy precursors, inorganic networks can be generated in situ into polymer matrices (Fig. 9.1e), one of the simplest precursors being tetraethoxysilane (TEOS). In the presence of acid or basic catalysts, TEOS follows a hydrolysis and condensation reaction to form a silica network. The use of organically modified silanes as precursors gives rise to networks, which are linked through stronger chemical bonds (covalent, iono-covalent or Lewis acid-base bonds) to the matrix. Hybrid polymers are prepared from a combination of organic and inorganic monomers, which react to form a common chain or network, giving rise to membranes with tailored selectivity, as previously explored for gas separation [11] (Fig. 9.1f). Organic polymers can be also functionalized with pending alkoxy groups, which may be further polymerized to form a cross-linked organic-inorganic network. This approach has been used for nanofiltration or ultrafiltration membranes [20]. New emerging routes for organic-inorganic materials are for instance the use of self-assembly templates [21] or organic structure directing agents and the incorporation

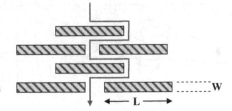

Fig. 9.2 Permeability reduction by fillers with high aspect ratio (L/W)

of building blocks or clusters with tuneable functionalities. Mauritz et al. [21] achieved self-assembled organic-inorganic membranes via sol-gel polymerization of silicon alkoxides around sulfonated blocks of polystyrene-soft block-polystyrene block copolymers.

The introduction of active inorganic components, having specific favorable characteristics for the preferential transport of one of the permeants are also discussed in the following. Examples are phosphates and heteropolyacids, which are well in the development of membranes for fuel cells.

9.4 Organic-Inorganic Membranes for Fuel Cell Application

Early reports and patent applications of Stonehart and Watanabe [22], Antonucci et al. [23], and Antonucci and Arico [24] claim the advantage of the introduction of small amounts of silica particles to Nafion to increase the retention of water and improve the membrane performance above 100°C. The effect is believed to be a result of water adsorption on the oxide surface. As a consequence the back-diffusion of the cathode-produced water is enhanced and the water electro-osmotic drag from anode to cathode is reduced [3]. A recent report of the group of Arico et al. [25] confirms the effect of water retention with the inclusion of oxide particles in Nafion and the importance of the acidity of the particle surface. An increase in both strength and amount of acid surface functional groups in the fillers enhances the water retention in the membrane: SiO_2-PWA (modified with phosphotungstic acid) > SiO_2 > neutral-Al_2O_3 > basic-Al_2O_3 > ZrO_2.

Instead of adding fumed silica or other particles in the membrane, the inorganic phase can be generated in situ by sol-gel chemistry. The in situ generation of a silica phase in Nafion has been reported by Mauritz et al. [26], impregnating a manufactured membrane with silanes and inducing their hydrolysis. By introducing silanes to a Nafion solution and casting films it is possible to change the morphology of the ionic clusters [9,27]. The investigation of Nafion or SPEEK in situ generated SiO_2 or ZrO_2 membranes for fuel cell has been reported [28–30]. The reduction of methanol and water permeability of sulfonated poly (ether ether ketone membranes) with the generation of ZrO_2 in the membrane casting solution was investigated by Silva et al. [31,32]. However, a remarkable decrease in proton conductivity was detected for filler contents higher than 10 wt% in the membrane. The best DMFC performance was obtained with filler content near 5 wt%. Jung et al. [33] generated silica in a preformed Nafion membrane, by impregnation of TEOS. More than 10 wt% TEOS impregnation led to an excessive decrease of conductivity.

As mentioned, inorganic fillers with high aspect ratios are expected to lead to more effective permeability reduction in membranes [19]. According to Fig. 9.2, fillers in the form of flakes would be much more effective in reducing the diffusion and therefore the permeability of membranes than spherical particles.

In Fig. 9.3, the effect of inorganic fillers with different aspect ratios on the membrane permeability to methanol and proton conductivity is compared [34]. For the same filler concentration, the layered silicate was more effective in reducing methanol permeability. Clays and layered silicates such as laponite have been used by different groups for the development of fuel cell membranes. Examples are Nafion/mordenite [35] and Nafion/montmorillonite [36], Nafion/sulfonated montmorillonite [37], SPEEK/montmorillonite [38,39], SSEBS/montmorillonite [40].

However, the surface modification of the inorganic filler is at least as important as the aspect ratio.

Figure 9.4 shows how the methanol permeability decreases when the inorganic phase, independent of form or aspect, is treated with organo-modified silanes. In the samples of Fig. 9.4, the fillers were modified with silanes containing basic groups such as amine or imidazole. The main function of the basic surface is to improve the compatibilization to the acid polymer matrix and inhibit the formation of cavities or defects between filler and matrix, which would otherwise act as a path

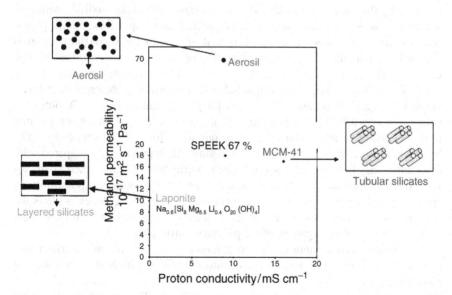

Fig. 9.3 Methanol permeability and proton conductivity of membranes with fillers of different aspect ratios

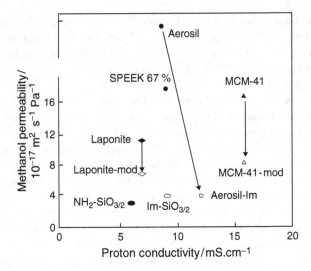

Fig. 9.4 Influence of the filler surface modification on the methanol permeability and proton conductivity of the membranes. Adapted from [34]

for water and methanol transport. The methanol permeability was dramatically reduced with the surface modification. Figure 9.4 also shows the examples of amino modified silica network (polysilsesquioxane) (NH_2-$SiO_{3/2}$ and Im-$SiO_{3/2}$) prepared by the sol-gel process, as well as analogous networks modified with imidazole groups. In both cases, the methanol permeability is quite low, but the proton conductivity is also low. The reason is that an amino (or imidazole) group is attached to each silicon atom. The interaction of the basic groups to the sulfonic groups of the polymer matrix is strong and partially reduces the conductivity. The importance of the surface modification has been also discussed before for the development of organic-inorganic membranes for gas separation [10,41]. Zeolites are being used as a filler for the development of gas separation membranes for a long time [41] and other reports have been also published for fuel cell application [42], however, for fuel cells no considerable additional improvement compared with other fillers has been so far observed. The modification of silica or ZrO_2 particles with silanes, containing organic functionalized groups, open the possibility of covalently linking the particles to the main chain (Fig. 9.1 b). An example is the treatment with amino silanes and further with carbodiimidazole, which then reacts with part of the sulfonic groups of the polymer matrix [30].

A Nafion/silicon composite has been reported recently for miniaturized fuel cells, using silicon in the form of porous substrate filled with Nafion to work as a membrane [43].

Silica and silicates are usually rather passive fillers. They may help in keeping water or reducing the methanol permeability, but they do not contribute themselves for the proton conduction. Zirconium phosphate and heteropolyacids are inorganic fillers, which are able to conduct protons. Zirconium phosphates and phosphonates with α- and γ-layered structures have been extensively investigated by Alberti [1,44–45]. He has shown that their properties could be improved by introducing a suitable choice of organic groups. Zirconium phosphonates containing carboxylic and sulfonic as well as amino functions were investigated [46]. Zirconium sulfophenylphosphate has been shown by his group to be particularly effective for proton conduction. For instance, the α-type $Zr(O_3PR')_{1.3}(O_3PR)_{0.7}$, with R= $-C_6H_4-$ SO_3H and R'= $-CH_2OH$ has a proton conductivity of 8 mS cm^{-1} at 100°C and 60% relative humidity. With R= $-(CH_2)_2 - C_6H_4 - SO_3H$ and R'= $- C_6H_5$, the proton conductivity is 1 mS cm^{-1}. The phosphonates are stable at least at temperatures as high as 180–190°C, for which the sulfonic groups starts to decompose. The acid strength of the metal sulfophosphonates changes with the organic radical to which the $-SO_3H$ group is bonded, influencing the proton conductivity.

The introduction of phosphates in polymers like sPEEK is reported to improve the proton conductivity at higher temperatures [1,47]. Zirconium phosphate has been used in combination with Nafion by DuPont [48,49] and later by other groups [50,51]. Yang et al. [51] prepared the composites by immersion Nafion membranes in zirconyl chloride and further treated with phosphoric acid. They suggest that the zirconium phosphate forms an internal rigid scaffold within the membrane that permits increased water uptake by the membrane. An analogous method was applied by Bauer and Willert-Porada [52]. We have tested zirconium phosphate/ SPEEK membranes for direct methanol fuel cells [30,53]. Phosphate was prepared by treating very fine zirconium oxide particles previously obtained from zirconyl chloride with phosphoric acid. The high hydrophilic phosphate in this case was not effective in decreasing the methanol crossover. A first attempt to reach a balance between proton conductivity and low methanol crossover was performed by additionally creating a network of zirconium oxide in the polymer matrix. Later a treatment of the phosphate particles with alkyl ammonium and further with polyimidazole promoted the phosphate exfoliation and simultaneously increased the compatibilization between the phosphate particles and sulfonated polymer matrix (Fig. 9.5), decreasing the crossover [54]. However, the amount of filler must be carefully controlled. Excessive contents of phosphate/ polybenzimidazole can lead to decrease of proton conductivity and low performance in DMFC. Composites of zirconium phosphates and polybenzimidazole have been tested to increase the performance of the polymer in operation up to 200°C [55].

Composites of zirconium tricarboxybutylphosphonate and polybenzimidazole were prepared by Jang and Yamaguchi [56]. The membranes were thermally stable and the conductivities measured for composites with 50 wt% phosphate were about 3.8 mS cm^{-1} at 200°C. Zaidi [57] and Zaidi et al. [58] introduced boron phosphate to improve the conductivity of SPEEK membranes. By adding polybenzimidazole again the compatibility between the phosphate and the polymer matrix could be improved. Up to 40% phosphate was added to the polymer matrix. The boron phosphate was synthesized from orthophosphoric and boric acid and added as a solid powder to blends of SPEEK and PBI.

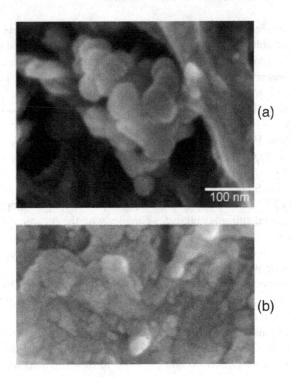

Fig. 9.5 Zirconium phosphate/SPEEK composites (**a**) without and (**b**) after treatment with polybenzimidazole. Adapted from [54]

More recently [59] boron phosphate was prepared in situ by reacting tripropyl borate and phosphoric acid. The resulting network is claimed to be less soluble than the previous inorganic phase.

Suzuki et al. [60] proposed a series of proton conducting boronsiloxane solid electrolytes containing $-SO_3H$ groups as proton sources, as well as the preparation of their composites with Nafion. The presence of boron as a Lewis acid should enhance the $-SO_3H$ dissociation and alkyl groups should overcome the tendency of the materials towards deliquescence at high temperature and high humidity.

Heteropolyacids are known for their high proton conductivity and have been considered by different groups as filler for polymeric membranes for fuel cell [61–66]. Zaidi et al. [62] reported promising results for membranes based on SPEEK and heteropolyacids for hydrogen fuel cell. Honma et al. [67] prepared hybrids of SiO_2 and poly(ethylene oxide) (PEO) and incorporated the heteropolyacids in the PEO domains. However, there are limiting problems for the application of heteropolyacids such as phosphotungstic or molibdophosphonic acid as fillers for membranes, particularly if DMFC is the aimed application. They are soluble and can be easily bled out from the membrane. To overcome the problem of the electrolyte dissolution,

Fig. 9.6 Functionalized heteropolyacid

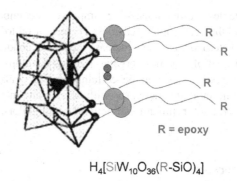

$$R = epoxy$$

$$H_4[SiW_{10}O_{36}(R\text{-}SiO)_4]$$

different approaches have been tried. Ahmed et al. proposed the use of heteropoly-acids entrapped in zeolites [68]. An approach proposed by our group was the in situ generation of an oxide network by the sol-gel process from alkoxysilanes and the modification of the anion structure of the heteropolyacid [69,70]. In the catalyst literature the entrapping of the heteropolyacid in the host material, mostly silica oxide networks has been reported [71–73]. Usually the heteropolyacid is fixed through covalent bonds or columbic interactions, which involve acid sites of the polymer matrix. More recently [70] our group investigated an alternative way to avoid the consumption of the acid sites. Organosilyl derivatives of the heteropolyacid, for instance a divacant tungstosilicate $[\gamma\text{-}SiW_{10}O_{36}]^{8-}$, were prepared using 3-glycidox-ypropyltrimethoxysilane (GPTS) (Fig. 9.6). The introduction of GPTS in the anion structure of the heteropolyacid enables its attachment to a host material, by an epoxy ring opening reaction with appropriate functional groups present in the host surface. For instance, the modified heteropolyacid was introduced into a sulfonated polymeric matrix, containing insoluble inorganic filler modified with amino groups. The amino groups react with the epoxy groups of the heteropolyacid molecules and fix them without reducing their acidity.

A more recent approach has been the covalent bonding of sulfonated polymer segments to inorganic insoluble particles, allowing the incorporation of highly functionalized insoluble fillers.

9.5   Conclusion

Two kinds of fillers are mainly used for the modification of proton conductive membranes for fuel cell. Passive fillers such as aerosol and layered silicates are effective to reduce the methanol crossover of the membrane used in DMFC and their efficiency depends on their aspect ratio. Fillers with a higher aspect ratio are more effective. However, the surface chemistry of the filler is much more important than the aspect ratio. A first reason is the adhesion between matrix and filler.

Suitable modifications, which improve the compatibility between filler and matrix, avoid the generation of cavities. The use of proton conductive fillers like phosphates and heteropolyacids can increase the proton conductivity of low conductive matrices. The field of organic-inorganic hybrids develops fast and an increasing number of new possibilities will be certainly explored also for fuel cell applications, like organic-inorganic networks with organic and inorganic segments in the same chain and ordered tridimentional networks with tailor-made ionic channels.

References

1. G. Alberti and M. Casciola, *Annu. Rev. Mater. Res.* **33**, 129–154 (2003).
2. Q. Li, R. He, J. O. Jensen, and N. J. Bjerrum, *Chem. Mater.* **15**, 4896–4915 (2003).
3. W. H. J. Hogarth, J. C. Diniz da Costa, and G. Q. Lu, *J. Power Sources* **142**, 223–237 (2005).
4. A. Heinzel and V. M. Barragan, *J. Power Source* **84**, 70–74 (1999).
5. A. S. Arico, S. Srinivasan, and V. Antonucci, *Fuel Cells* **1**, 133–161 (2001).
6. G. Kickelbick, *Prog. Polym. Sci.* **28**, 83–114 (2003).
7. C. Sanchez, G. J. A. A. Soler-Illia, F. Ribot, T. Lalot, C. R. Mayer, and V. Cabui, *Chem. Mater.* **13**, 3061–3083 (2001).
8. S. P. Nunes, J. Schulz, and K. V. Peinemann. *J. Mater. Sci. Lett.* **15**, 1139–1141 (1996).
9. R. A. Zoppi, I. V. Yoshida, and S. P. Nunes. *Polymer* **30**, 1309–1315 (1998).
10. S. P. Nunes, K. V. Peinemann, K. Ohlrogge, A. Alpers, M. Keller, and A. T. N. Pires. *J. Membr. Sci.* **157**, 219–226 (1999).
11. M. L. Sforça, I. V. Yoshida, and S. P. Nunes. *J. Membr. Sci.* **159**, 197–207 (1999).
12. R. A. Zoppi, I. V. P. Yoshida, and S. P. Nunes, *Polymer* **39**, 1309–1315 (1998).
13. R. A. Zoppi and S. P. Nunes. *J. Electroanal. Chem.* **44**, 39–34 (1998).
14. M. L. Sforca, I. V. P. Yoshida, C. P. Borges, and S. P. Nunes. *J. Appl. Polym. Sci.* **82**, 178–185 (2001).
15. D. Gomes, S. P. Nunes, and K. V. Peinemann. *J. Membr. Sci.* **246**, 13–25 (2005).
16. J. C. Maxwell, *Treatise on Electricity and Magnetism*, Vol. 1, Clanderon Press, London, 1881.
17. L. E. Nielsen, *J. Macromol. Sci.* A1 (1967) 929.
18. S. P. Nunes and K. V. Peinemann, *Membrane Technology in the Chemical Industry*, Wiley, New York, 2001.
19. W. R. Falla, M. Mulski, and E. L. Cussler, *J. Membr. Sci.* **119**, 129–138 (1996).
20. S. Zhang, T. Xu, and C. Wu, *J. Membr. Sci.* **269**, 142–151 (2006).
21. K. A. Mauritz, D. A. Mountz, D. A. Reuschle, and R. I. Blackwell, *Electrochim. Acta* **50**, 565–569 (2004).
22. P. Stonehart and M. Watanabe US Patent 5,523,181, June 4, 1996.
23. P. L. Antonucci, A. S. Arico, P. Creti, E. Rammunni, and V. Antonucci. *Solid States Ionics* **125**, 431–437 (1999).
24. V. Antonucci and A. Arico EP 0926754A1 199.
25. A. S. Arico, V. Baglio, A. Di Blasi, E. Modica, P. L. Antonucci, and V. Antonucci, *J. Power Sources* **128**, 113–118 (2003).
26. K. A. Mauritz *Mater. Sci. Eng. C-Biomimetic and Supramolecular Syst.* **6**, 121–133 (1998).
27. R. A. Zoppi and S. P. Nunes. *J. Electroanal. Chem.* **445**, 39–45 (1998).
28. S. P. Nunes, E. Rikowski, A. Dyck, D. Fritsch, M. Schossig-Tiedemann, and K. V. Peinemann. *Euromembrane* 2000, Jerusalem, Proceedings, pp. 279–280.
29. D. J. Jones and J. Roziere. in *Handbook of Fuel Cells*, W. Vielstich, A. Lamm, and H. A. Gasteiger, editors. Wiley, New York 2003, Vol. 3, p. 447.

30. S. P. Nunes, B. Ruffmann, E. Rikowski, S. Vetter, and K. Richau. *J. Membrane Sci.* **203**, 215–225 (2002).
31. V. S. Silva, B. Ruffmann, H. Silva, Y. A. Gallego, A. Mendes, L. M. Madeirab, and S. P. Nunes, *J. Power Sources* **140**, 34–40 (2005).
32. V. S. Silva, J. Schirmer, R. Reissner, B. Ruffmann, H. Silva,A. Mendes, L. M. Madeira, and S. P. Nunes, *J. Power Sources* **140**, 41–49 (2005).
33. D. H. Jung, S. Y. Cho, D. H. Peck, D. R. Shin, and J. S. Kim, *J. Power Sources* **106**, 173–177 (2002).
34. C. S. Karthikeyan et al. *J. Membr. Sci.* **254**, 139–146 (2005).
35. S. -H. Kwak, T. -H. Yang, C. -S. Kim, and K. H. Yoon, *Solid State Ionics* **160**, 309–315 (2003).
36. D. H. Jung, S. Y. Cho, D. H. Peck, D. R. Shin, and J. S. Kim, *J. Power Sources* **118**, 205–211 (2003).
37. C. H. Rhee, H. K. Kim, H. Chang, and J. S. Lee, *Chem. Mater.* **17**, 1691–1697 (2005).
38. C. S. Karthikeyan, S. P. Nunes, and K. Schulte, *Eur. Polym. J.* **41**, 1350–1356 (2005).
39. Z. Gaowen and Z. Zhentao, *J. Membr. Sci.* **261**, 107–113 (2005).
40. J. Won and Y. S. Kang, *Macromol. Symp.* **204**, 79–91 (2003).
41. T. T. Moore and W. J. Koros, *J. Mol. Structure* **739**, 87–98 (2005).
42. V. Tricoli and F. Nannetti, *Electrochim. Acta* **48**, 2625–2633 (2003).
43. Tristan Pichonat, Bernard Gauthier-Manuel, and Daniel Hauden, *Fuel Cell Bulletin*, August 2004, p. 11
44. G. Alberti, M. Casciola, U. Costantino, and R. Vivani, *Adv. Mat.* **8**, 291–303 (1996).
45. G. Alberti, M. Casciola, L. Massineli, and B. Bauer, *J. Membr. Sci.* **185**, 73–81 (2001).
46. G. Alberti, M. Casciola, and R. Palombari, *J. Membr. Sci.* **172**, 233–239 (2000).
47. L. Tchicaya-Bouckary, D. Jones, and J. Roziere, *Fuel Cells* **2**, 40 (2002).
48. W. G. Grot and G. Rajendran, US Patent 5919583, July 6, 1999.
49. C. Yang, P. Costamagna, S. Srinivasan, J. Benziger, and A. B. Bocarsly, *J. Power Sources* **103**, 1–9 (2001).
50. F. Damay and L. C. Klein. *Solid State Ionics* **162–163**, 261–267 (2003).
51. C. Yang, S. Srinivasan, A. B. Bocarsly, S. Tulyani, and J. B. Benziger, *J. Membr. Sci.* **237**, 145–161 (2004).
52. F. Bauer and M. Willert-Porada. *J. Membr. Sci.* **233**, 141–149 (2004).
53. B. Ruffmann and S. P. Nunes. *Solid State Ionics* **162–163**, 269–275 (2003).
54. L. A. S. A. Prado, H. Wittich, K. Schulte, G. Goerigk, V. M. Garamus, R. Willumeit, S. Vetter, B. Ruffmann, and S. P. Nunes. *J. Polym. Sci.-Phys.* **42**, 567–575 (2003).
55. R. He, Q. Li, G. Xiao, and N. J. Bierrum. *J. Membr. Sci.* **226**, 169–184 (2003).
56. M. Y. Jang and Y. Yamazaki *Solid State Ionics* **167**, 107–112 (2004).
57. S. M. J. Zaidi. *Electrochim. Acta* **50**, 4771–4777 (2005).
58. S. M. J. Zaidi, S. D. Mikhailenko, and S. Kaliaguine, *J. Polym. Sci. B: Polym. Phys.* **38**, 1386 (2000).
59. K. Palanichamy, J. S. Park, T. H. Yang, Y. G. Yoon, W. Y. Lee, and C. S. Kim, *Proceedings of the ICOM 2005*, Seoul, Korea, August 21–26, p. 98.
60. H. Suzuki, Y. Yoshida, M. A. Mehta, M. Watanabe, and T. Fujinami, *Fuel Cells* **2**, 46–51 (2002).
61. B. Tazi and O. Savadogo *Electrochem. Acta* **45**, 4329–4339 (2000).
62. S. Zaidi, S. Mikhailenko, G. Robertson, M. Guiver, and S. Kaliaguine, *J. Membr. Sci.*, **173**, 17–34 (2000).
63. P. Staiti, M. Minutoli, and S. Hocevar, *J. Power Sources* **90**, 231–235 (2000).
64. P. Staiti and M. Minutoli *J. Power Sources* **94**, 9–13 (2001).
65. P. Staiti *J. New. Mat. Electrochem. Syst.* **4**, 181–186 (2001).
66. Y. S. Kim, F. Wang, M. Hickner, T. Zawodzinski, and J. Mc Grath. *J. Membr. Sci.* **212**, 263–282 (2003).
67. I. Honma, Y. Takeda, and J. M. Bae, *Solid State Ionics* **120**, 255–264 (1999).

68. I. Ahmed, S. M. J. Zaidi, and S. U. Rahman, *Proceedings of the ICOM 2005*, Seoul, Korea, August 21–26, p. 102.
69. M. L. Ponce, L. Prado, B. Ruffmann, K. Richau, R. Mohr, and S. P. Nunes. *J. Membr. Sci.* **217**, 5–15 (2003).
70. M. L. Ponce, L. A. S. A. Prado, V. Silva, and S. P. Nunes. *Desalination* **162**, 383–391 (2004).
71. M. Misono, *Cat. Rev.- Sci. Eng.* **29**, 269–321 (1987).
72. I. V. Kozhevnikov, *Chem. Rev.* **98**, 171–198 (1998).
73. N. Mizuno and M. Misono, *Chem. Rev.* **98**, 199–218 (1998).

Chapter 10
Thermal and Mechanical Properties of Fuel Cell Polymeric Membranes: Structure–Property Relationships

Ibnelwaleed A. Hussein and S.M. Javaid Zaidi

Abstract Fuel cells provide pollution-free clean energy and are extremely efficient. The most important component of fuel cell is the proton conductive membrane which transports proton from the anode to the cathode of the fuel cell, and also separates the fuel and the oxidant. Therefore, desirable properties of proton exchange membrane (PEM) include high proton conductivity, high resistance to electrons, impermeability to fuel and oxidants to avoid diffusion and leakage, long-term chemical and thermal stability, and good mechanical properties. In this chapter, effort has been made to highlight the response of thermal and mechanical properties with variation of different parameters characteristic of a typical fuel cell environment. These parameters include: water, solvent, temperature, degree of sulfonation, and filler. A detailed literature review has also been made regarding the studies conducted worldwide related to novel membrane development and property associated with these new materials. This review highlights the structure-property relationships in polymeric membranes.

10.1 Background on Thermal and Mechanical Analyses

Thermal analysis technique represents a group of methods for determining changes in physical and chemical properties of a material with variation in temperature in a controlled atmosphere. Thermal analysis is mainly used to measure

- Phase transition temperatures and associated heat exchange
- Thermal decomposition of solids and liquids
- Identification of impurities in materials

Using this method one can characterize polymers, organic or inorganic chemicals, metals and other types of materials. The principal techniques of thermal analysis are differential scanning calorimetry (DSC) and thermo gravimetric analysis (TGA).

Another important tool is dynamic mechanical thermal analysis (DMTA) by which viscoelastic properties of the polymer can be studied at different temperatures below the melting point of the material. In this technique, an oscillating force is applied to the sample and the displacement of the sample is measured and vice

S.M.J. Zaidi, T. Matsuura (eds.) *Polymer Membranes for Fuel Cells*,
doi: 10.1007/978-0-387-73532-0, © Springer Science+Business Media, LLC 2009

versa. The stiffness and modulus of the material can be determined. The time lag in the displacement and the applied force can be used to determine the damping properties of the material (storage and loss moduli). At temperatures below the glass transition, all polymers behave like a glass (high elastic modulus and brittle). On the other hand, at high temperatures (above T_g or T_m) the modulus decreases at different rates depending on polymer structure. Therefore, by changing the temperature during DMTA testing the glass transition and the α-transition as well as other relaxations can be observed. The DSC can also measure these transitions and relaxations, but DMTA signals are stronger.

Uniaxial tensile strength can be measured in an Instron test machines equipped with a static load cell of suitable capacity (see Fig. 10.1). To measure the tensile properties of the membranes while immersed in a fluid, an assembly consisting of a testing chamber and tensile grips are fitted for the Instron machine. DMTA technique has the option of performing mechanical testing in immersed fluids.

The testing chamber allows the recirculation of the liquid when coupled with a circulator bath. The experiments can be conducted under several constant temperature conditions. The extension of the specimen could be followed by an optical extensometer. The values of L, L_o, L_1, W and W_o (see Fig. 10.1) depend on the ASTM method. The length between two marks on the specimen–the gauge length– is approximately 20% of the field view.

Before actual testing, each specimen should be left in the testing environment once it has been attached to the grips. But there is an exception to this. We know that membranes markedly swell when they come in contact with warm water, which causes dimensional change in the already clamped specimen, causing jaw breaks at the very beginning. Therefore, samples should be soaked in water at the testing temperature for a specific time in a separate container and then clamped and tested in the testing container for a few minutes. With this we can measure the water absorption rates of SPEEK and other membranes, the apparent elastic modulus, E, the yield stress, σ_y, the yield strain, ε_y, the stress at break, σ_b, etc. The load elongation curves obtained can be converted to stress-strain curves. The apparent tensile modulus is usually computed as the ratio of stress to strain of the initial slope of the stress-strain curve. The speed of elongation can be used to estimate the strain rate. Stress-strain curves can be used to determine yield stresses and strains, as well as stresses and strains at break.

10.2   Introduction

Fuel cells are devices that convert the chemical energy stored in the fuel directly into electricity. They provide pollution-free clean energy and are extremely efficient. The most important component of fuel cell is the proton conductive membrane which transports proton from the anode to the cathode of the fuel cell, and also separates the fuel and the oxidant. Therefore, desirable properties of proton exchange membrane (PEM) include high proton conductivity, high resistance to

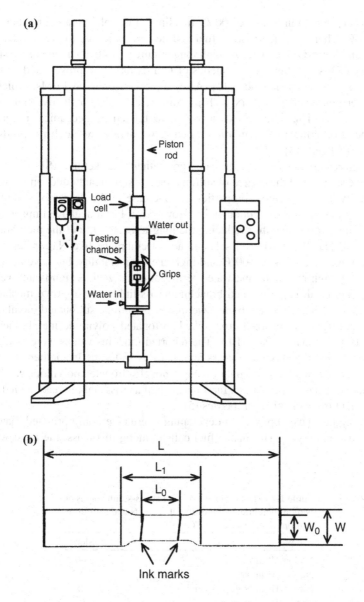

Fig. 10.1 (a) Instron tensile analyzer (b) Tensile test specimen dimensions (in mm): overall length (L); length of calibrated section (L_1); gauge length (L_0); width of narrow section (W_0); overall width (W)

electrons, impermeability to fuel and oxidants to avoid diffusion and leakage, long-term chemical and thermal stability, and good mechanical properties [1].

Most widely used perfluorinated membranes are Nafion, Flemion, etc. There are some drawbacks related to these membranes. They are mechanically unstable for operation at temperatures above 100°C; this limits fuel cell operating temperatures up to 80°C. There is a decrease of fuel cell performance due to slower electrode kinetics and low CO tolerance. At high temperatures (>140°C) anode catalyst poisoning by CO is less important, kinetics of fuel oxidation improves, and cell efficiency is significantly enhanced [2]. Also, these conventional PEMs, which are mainly perfluorinated ionomers such as Nafion, etc., are conductive only when soaked in water. This in turn adds to problems related to mechanical properties. Thus, the development of membranes is an active area of research for producing economic fuel cells [3].

Characterization studies of sulfonated poly(ether ether ketone) (SPEEK) membrane have shown that the degree of sulfonation has significant effect on the proton conductivity and mechanical properties. An increase in the degree of sulfonation improves water uptake and conductivity. However, it has a negative impact on the mechanical performance, thus reducing the long-term stability of the membrane in the fuel cell [1]. Perfluorinated sulfonic acid membranes like Nafion have high proton conductivities below 100°C and high chemical stability, but have the drawback of very high methanol and water permeability as well as significant osmotic drag [4]. Yin et al. [5] attempted branching/cross-linking to improve mechanical properties and lessen the membrane swelling in polyimide sulfonated membranes. They successfully synthesized crosslinked sulfonated polyimide membranes and the results are shown in Table 10.1. The reported modulus values were as high as 2.4 GPa, whereas the maximum stress reported was 130 MPa. In another study, Yin et al. [6] investigated the potential of using a novel sulfonated polyimide for use as a polymer conducting membrane. Incorporation of a cross-linking reactant led to an improvement of mechanical properties [5].

Many studies have appeared in developing alternative nonfluorinated materials for low cost and high performance fuel cells. Among this class, the most widely

Table 10.1 Initial slope (MPa) of stress-strain curves of Nafion membrane with different solvents and salt forms at room temperature (~27°C)

Sample	Slope
Nafion-H⁺ as received	2.00
Nafion-H⁺ in water 24 h	0.95
Nafion-H⁺ in boiling water 1 h	1.28
Nafion-H⁺ in methanol 24 h	1.12
Nafion-H⁺ in ethanol 24 h	0.50
Nafion-H⁺ in ethanol/water 24 h	1.35
Nafion-Li⁺	1.62
Nafion-Na⁺	2.49
Nafion-K⁺	3.46
Nafion-Rb⁺	5.58
Nafion-Cs⁺	5.33

Reference [8], reprinted with permission of the Brazilian Polymer Association.

investigated PEMs are the aromatic thermoplastics such as polyethersulfone, poly-benzimidazole, polyimides, and poly(ether ether ketone), poly(arylene ether sulfone), and poly(phenylene oxide). In most cases, electrophilic sulfonation is used to induce proton conductivity in these materials making them suitable for fuel cell applications. The introduction of sulfonic acid groups was achieved either by direct sulfonation of the parent polymers or polymerization of sulfonated monomers. On the other hand, biopolymers were also examined for use in fuel cell membranes [4,7]. Smitha et al. [4] observed enhancement in tensile strength of Chitosan when blended with sodium alginate with marginal change in elongation at break. These modifications were attributed to ionic cross-linking of polymer membrane; hence, increased chain stiffness. Therefore, the development of economically competitive novel PEMs with good mechanical characteristics and high proton conductivity at high temperature remains an important challenge to the realization of practical PEM fuel cells.

Kawano et al. [8] studied the stress-strain curves of Nafion membrane in acid and salt forms, using DMTA. It was found that the initial slope of the stress-strain curves decreases with increasing water or solvent content in the Nafion membrane. The initial slope decreases also with increasing temperatures. It was also tested for different cations and found that the initial slope increases when the cation is replaced, in the following order: Li^+, Na^+, K^+, Cs^+, and Rb^+. Nafion in salt form has higher value of initial slope from room temperature up to 90°C. Nafion consists of polytetrafluoroethylene (PTFE) backbone, the side chains terminated by sulfonic groups $-SO_3^-H^+$ (acid form) that can be substituted by $-SO_3^-$ Me^+ (salt form). The PTFE segment in these membranes represents the hydrophobic region, whereas hydrophilic region consists of side chains terminated by ion-exchange groups. The ionic species aggregate to form clusters that are interconnected by channels and distributed through the polymer matrix [8]. The water/solvent sorbed by the membrane is incorporated into the clusters and channels. They swell according to nature and content of the solvent, changing the mechanical and thermal properties of the membranes.

Degradation of mechanical properties in polymeric membranes can occur in many forms like cracks, tears, punctures, or pinhole blisters. Hence, adequate care must be taken during membrane preparation to prevent nonuniform pressure between membrane electrode assembly and bipolar plates during operation. Lack of water can make a membrane brittle and fragile. Several steps can be taken to slow or prevent the degradation process. These methods include the following [9]:

1. Reduce metal contaminants from all possible sources.
2. Decrease gas permeability across the membrane.
3. Optimize the membrane water content (increasing water content lowers the concentration of H_2O_2, however, it allows increased gas crossover at the same time).
4. Deposit radical inhibitors or peroxide-decomposition catalysts within membrane to decrease the severity of radical attacks.
5. Include in situ temperature mapping to detect hotspots and identify problem areas which could lead to pinholes/perforations.
6. Include humidity mapping.

In general, to find solutions for the membrane degradation we need to collect more in situ information from a working fuel cell to improve our understanding of the membrane's degradation process.

Chang et al. [10] developed a hybrid proton conductive membrane that was more flexible and had good thermal stability. The material was hybrid organic-inorganic composites, the properties of which depend on the chemical nature of the organic and the inorganic constituents, and the interface between them. The composite membrane consisted of PEG end-capped with 3-isocyanatopropyltriethoxy silane and monophenyl triethoxysilane (MPh). The membrane doped with 4-dodecylbenzene sulfonic acid (DSBA) moieties had good proton conductivity. MPh was added to improve the chemical stability and structural flexibility of the membrane. The important observation made is that the initial decomposition temperature (IDT) decreases continuously with increasing DSBA doping, whereas in the absence of DSBA, IDT increases with increasing MPh. Integrity of the polymer network is likely to be destroyed by excess amounts of DBSA leading to marked drop in storage modulus.

Watanabe and coworkers [11] reported that perfluorinated ionomers are highly proton conductive, but their performance is lowered at temperatures higher than their glass transition temperature, T_g (100°C). Effort has been made to produce proton conductive nonfluorinated aromatic ionomers with good stability for fuel cell application. Many of these new ionomers have sulfonic groups that improve hydrolytic stability of the ionomer membranes. The PEFC membrane should have a proton conductivity > 0.1 S cm^{-1}. Their ionomer showed high proton conductivity (0.2 S cm^{-1}) and excellent hydrolytic stability under harsh hydrolytic conditions (140°C and 100% relative humidity for 700 h). Investigation on the properties of poly(arylene ether) ionomer containing sulfofluorenyl group was done. The authors claim that the properties of poly(arylene ether) ionomers are uppermost among the hydrocarbon based ionomers for fuel cell applications. In a separate study, the same group investigated the influence of introducing aliphatic groups in the main chain or side chains to improve the mechanical performance of SPI containing fluorenyl groups [12].

10.2.1   Effect of Water, Solvents, and Cations on Stress-strain Relationships

Understanding the mechanical properties of membranes is essential for producing quality membranes with good durability under fuel cell operating conditions [13]. Mechanical properties of Nafion are influenced by hydration and metal ions. High conductivities require the presence of moisture. On the other hand, when Nafion is soaked in water or any solvent, its Young's modulus decreases. It was observed [8] that soaking the membrane in water had a direct impact on the stress-strain relationship. The high water content swells the membrane, reduces the intermolecular forces, increasing the degree of elongation making the Nafion membrane more ductile (see Table 10.1 for the effect of water on the modulus). We need to have thinner membranes to enhance permeability of protons. The problem associated with low Young's modulus and yield strength is that it makes the membrane susceptible to permanent deformation, gradual weakening, and eventual failure in fuel cell when it is exposed to pressure gradients and pressure pulses. Therefore, a robust membrane with high mechanical strength, under fuel cell condition is desired. Some studies have shown that membranes prepared in a film of polyimide frames have shown improvement in yield strength of membranes under hydrated conditions [13].

Kawano et al. [8] obtained the stress-strain relationships for Nafion membranes with different solvents (see Fig. 10.2). The initial slope of the curve decreases in the order: as received Nafion, Nafion soaked in methanol, ethanol/water, and ethanol. Nafion soaked in ethanol/water mixture and ethanol has a higher degree of elongation with a modulus compared with the membrane soaked in water indicating a lower molecular interaction, particularly ionic interaction in membrane. The difference in the stress-strain curves with different solvents is in the degree of swelling. For short exposure times (24 h), the severity of the solvents on the membrane increases in the direction: water-methanol-ethanol [8].

With different cations, such as Li+, Na+, K+, Rb+, and Cs+, that substituted the side chains, it was observed that the Young's modulus of membranes increases in the order of Li+, Na+, K+, Cs+, and Rb+, and the degree of elongation is lower than as-received Nafion. The reason for this trend is that water content in Nafion membrane decreases with increasing atomic radius of metal cations. Therefore, the membrane becomes hard and brittle with the modulus increasing in the order Li+, Na+, K+, Cs+, and Rb+ [8]. In general, cross-liking and crystallization make the membrane hard

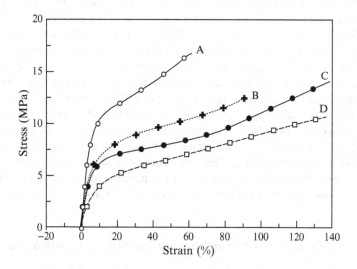

Fig. 10.2 Stress-strain curves of Nafion membranes with different solvents (A) (—o—) as received; (B) (— + —) soaked in methanol for 24 h; (C) (— ● —) soaked in ethanol/water for 24 h; and (D) (— □ —) soaked in ethanol for 24 h. Reference [8], reprinted with permission of the Brazilian Polymer Association

Table 10.2 Effect of degree of sulfonation, temperature, and moisture conditions on mechanical properties of SPEEK membranes

DS (%)	Test temperature (°C)	Moisture conditions	E (at 1% ε) MPa	ε_y (%)	σ_y (MPa)	ε_b (%)	σ_b (MPa)
Pure PEEK	23	30%RH	3,034	—	—	6	94.2
	23	30%RH	2,404	5	62	50	51.5
63	23	In water	716	3	15.1	60	16.6
	40	In water	607	3	10.4	130	17.4
	23	30%RH	2,258	6	61.9	90	55.8
69	23	In water	781	3	14.4	120	23.6
	40	In water	449	3	8.5	180	19.5
	23	30%RH	2,599	5	66.8	80	59.2
83	23	In water	607	2	9.2	140	
	40	In water	144	4	3.9	200	11.3

Reference [1], reprinted with permission of John Wiley & Son, Inc.

and brittle. The ionic cluster in Nafion works as a physical cross-linking. So, when ionic interaction is high chain mobility is lowered and the polymer becomes hard.

The membranes exchanged with different cations were also tested by Kundu et al. [13]. Their results showed that with an increase in radius, the Young's modulus and yield strength increased. It is known that the contaminant ions reduce the ability of the membrane to conduct protons, but on the other hand it improves the mechanical properties of the membranes. With ions such as Na^+, K^+, Mg^+, Cu^{2+}, Ni^{2+} inside the polymer a larger force is required to cause elastic deformation. Also, the increased yield strength serves to keep the membrane from plastic deformation.

In DMFC, operating temperatures of 140°C are highly desirable for enhancing the methanol oxidation kinetics. At such high temperatures, the Nafion membranes tend to dehydrate, which will impact its mechanical properties as shown in Table 10.2 and Fig. 10.2. Recent efforts to increase the water retention of the Nafion membrane at elevated temperatures to increase proton conductivity have focused on the incorporation of hydrophilic silica particles. Ladewig et al. [14] used a sol gel technique to homogenously disperse silica nanoparticles in the nanosized (~5 nm) hydrophilic domains of Nafion membranes. Dehydration of the water from composite membranes on increasing temperature is much less pronounced than that of bare Nafion. Water is needed for proton conductivity; however, it has negative impact on the mechanical properties. Therefore, water content of polymeric membranes should be optimized.

10.2.2 Influence of Temperature on Mechanical Properties

Figure 10.3 below shows the effect of increasing temperature on the Young's modulus and the degree of elongation of as-received Nafion-H^+ membranes. As the temperature rises we expect the polymer to show more flexible behavior. At around

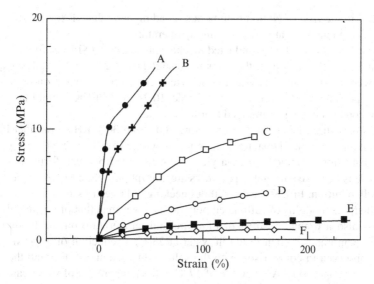

Fig. 10.3 Stress-strain curves of as received Nafion membranes at different temperatures (A) (— • —) room temperature; (B) (— + —) 60°C; (C) (— □ —) 90°C; (D) (— ○ —) 120°C; (E) (— ■ —) 150°C; and (F) (— ◊ —) 180°C. Reference [8], reprinted with permission of the Brazilian Polymer Association

60–90°C the behavior is similar to that of hard and brittle plastic. Curves A and B are similar in behavior since the membrane is below its glass transition (i.e., in the glassy state). Still the modulus (slope of the curve) at 60°C is lower than that at room temperature.

Above 90°C, the molecular motion becomes very fast, and Young's modulus decreases drastically like a plastic deformation behavior. It should be noted that the glass transition temperature (Tg) of the Nafion membrane used in this study is 90°C. At temperatures above 150°C, the Nafion shows very low modulus and high degree of elongation. The test on Nafion-Na$^+$ [8] showed values of Young's modulus at $T < Tg$ that are similar to those obtained at room temperature. They suggested this to be due to a small increase in crystallinity in the interphase between the amorphous and crystalline regions. No evidence was shown to support this assumption. Our explanation for their observation is that in the glassy state the modulus vs. temperature curve reaches a plateau and the major drop takes place above Tg [15]. However, in this case we also see from Fig. 10.3 that the cluster is destroyed at 120–150°C. Hence a very low initial slope was obtained. Stress-strain curves for Nafion membrane substituted by Cs$^+$ cation had a slope greater than Na$^+$ substituted Nafion membranes at corresponding temperatures. This is because of the high ionic interaction in the cluster is the reason for this observation. However, we believe that this observation is likely due to the increase in ionic radius which results in a decrease in water content.

Kawano et al. [8] compared the performance of PTFE film at different temperatures with Nafion membrane. As PTFE has almost 100% crystallinity as compared with 25% in the case of Nafion membrane, it had higher value of Young's modulus than Nafion membrane at corresponding temperatures.

Reyna-Valencia et al. [1] conducted mechanical testing on SPEEK films in air at 23°C and 30% RH. In water, the films were stretched at 23 and 40°C. The experiments were carried out at constant crosshead speed of 5 mm min^{-1} on dog-bone–shaped specimens. Their results (see Table 10.2) on SPEEK membrane under varying conditions are summarized below:

The stress-strain behavior at room temperature and 30% RH shows that PEEK sample exhibits brittle behavior as compared with SPEEK. The SPEEK gave a ductile behavior characterized by a yield point followed by neck formation. This difference is because sulfonated polymers are amorphous while PEEK is semicrystalline. In addition, PEEK breaks without necking at large stresses.

Further, the modulus of sulfonated polymer is lower than that of virgin polymer. There is also a dramatic effect of moisture and temperature on the modulus of SPEEK membranes. When tested in water at 23°C, a reduction of approximately 70% is observed in comparison with the values obtained at 30%RH and the same temperature conditions. An increase of 17°C in the temperature of water caused an additional modulus drop, in such a way that membranes lost 75–94% of their original stiffness depending on their degree of sulfonation. When SPEEK is exposed to water at room temperature, the amorphous part absorbs water and swells. Solvent molecules act as plasticizers for polymeric chains, increasing their segmental mobility. Consequently, a decrease in the modulus and yield stress and increase in strain at break will follow.

10.2.3 Effect of Filler on the Tensile Properties of the Membranes

Kaneko et al. (2003) [34] reported improved mechanical properties of PBI-BS without decrease in proton conductivity (10^{-2} S cm^{-1}). The thermal decomposition temperature was about 400°C while desulfonation occurred above 400°C. High conductivity was attributed to the aggregation of the sulfonic acid groups and increase of flexibility of the alkyl side chain. Tensile strength of PBI-BS/apatite hybrid membranes showed sufficient tensile strength of 30 MPa. Tensile strength of fully hydrated PBI-BS membrane was about 15 MPa, insufficient for fuel cell application. On the other hand, the increase in filler concentration induces a transition from ductile to brittle behavior (see Fig. 10.4). The incorporation of the second phase deteriorates the initial modulus of the matrix. The enhanced chain orientation causes strain hardening and increases the strain at break. Incorporation of foreign particles in the polymer cause T_g to be depressed. For two components the glass transition temperature, T_g, is given by:

$$\frac{1}{T_g} = \frac{W_A}{T_{gA}} + \frac{W_B}{T_{gB}}$$

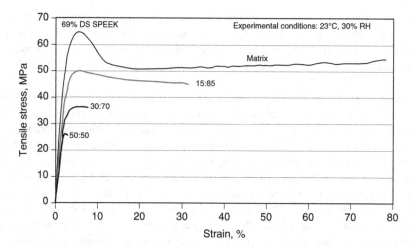

Fig. 10.4 Stress-strain curves of BPO$_4$/SPEEK membranes, illustrating the effect of the filler to matrix ratio on the tensile properties of the membranes. Tests were performed at 23°C and 30% RH. Reference [1], reprinted with permission of John Wiley & Sons, Inc

where A and B represent the polymer and the plasticizer, and W_A and W_B are the weight fractions of the polymer and the plasticizer, respectively. It should be noted that there are many mixing rules other than the above relationship and polymer textbooks could be consulted in this regard [15]. The reader is advised to perform a few experiments to examine which relationship fits his or her system.

Properties that are preferred or required for DMFC are as follows: high proton conductivity of at least about 5×10^{-2} S cm^{-1} in order to reduce Ohmic losses; good mechanical resistance of films of 100-μm thickness; low permeation of reactants and products of the electrochemical combustion; high chemical and electrochemical stability in cell operating conditions; and cost compatible with commercial requirements. The main problems associated with the traditional membranes used in fuel cells are high temperature application and high cost. Research is going on in this area, and a number of ideas have been put forward to improve one property or the other; however, Nafion continues to be the most acceptable proton-conducting membrane. One of the main reasons for this is that researchers have tried to look into the commercial membranes and their problem with regard to one property or the other; but overall properties (mechanical, thermal, conducting properties) together with economic feasibility have often been overlooked. For example, there is lack of research to find out whether a particular membrane will have good thermal and mechanical properties while maintaining good proton conducting properties.

The mentioned disadvantages of PFI membranes induced many efforts to synthesize PEM based on hydrocarbon-type polymers and brought about the emergence of partially fluorinated and fluorine-free ionomer membranes as alternatives to Nafion membranes. Among them the membranes based on aromatic PEEK were shown to be promising for fuel cell application, as they possess good mechanical properties, thermal stability, toughness, and some conductivity, depending on the degree of sulfonation [16]. Nevertheless, the proton conductivity of SPEEK is yet to reach a level sufficient to enable adequate performance in a fuel cell. A proposed improvement in this connection could be the introduction of heteropolyacids or Y-zeolite.

Yen et al. [17] used SPEEK as the basic material for fuel cell membranes. Their materials had high proton conductivity, and were made from inexpensive, readily available materials. According to that invention, proton-conducting membranes are formed based on a sulfonic acid-containing polymer. One preferred material is PEEK. This was an inexpensive starting material, and it was suggested to enhance protection against fuel crossover. In a particular embodiment, PEEK was sulfonated with H_2SO_4 to give H-SPEEK, a polymer that is soluble in an organic solvent and water mixture. It was found that although sulfonic acid increased the proton-conducting performance of PEEK (the sulfonate groups are responsible for proton conductivity), it degraded the physical structure of the resulting membrane. Therefore, they developed a tradeoff between the amount of sulfonation and appropriate physical structure by sulfonating less than one third of the benzene rings. Yen et al. [17] showed methods of modifying the morphology of the processed polymers to limit the transport of methanol across the membrane (to reduce the free volume) by using zeolites tin mordenite or the like; but such solids did not impart high enough proton conductivity to the composite membrane. Hence, heteropolyacids can be suggested to improve the proton conductivity of the polymer. Their invention was not limited to SPEEK. Other polymer matrices were suggested in accordance. Examples thereof include polysulfones, polystyrenes, polyether imides, polyphenylenes, poly (α-olefins), polycarbonates, and mixtures thereof.

Kurano et al. [18] proposed a covalent cross-linking of ion-conducting materials via sulfonic acid groups to be applied to various low-cost electrolyte membrane base materials for improved fuel cell performance metrics relative to such base material. Many aromatic and aliphatic polymer materials have significant potential as proton exchange membranes if a modification can increase their physical and chemical stabilities without sacrificing electrochemical performance or significantly increasing material and production costs. They incorporate a cross-linking agent into the ion-conducting base material through hydroxyl and sulfonic acid condensation or through amine and sulfonic acid condensation. The incorporation takes place in a nonaqueous environment. The ion conducting base material is an organically or inorganically based material, or a composition thereof. Xiao et al. [19] showed that pyridine-based polybenzimidazole membranes displayed thermal stability and maintained mechanical integrity.

Cross-linking agents improve properties such as water uptake; reduce methanol crossover; and increase thermal stability and mechanical strength without significantly decreasing the base material's positive attributes. The incorporation of functionalized cross-linking additives offers a promising technology for material modification for fuel cell applications [18]. Their invention provided performance-enhancing cross-linking agents, a method for their incorporation into ion-conducting materials, and their incorporation into a fuel cell as a high-performance membrane material. The cross-linking components improved the tensile strength of the base material in both wet and dry conditions. This is primarily due to the increase in molecular weight and decrease in chain mobility. The inherent nature of the cross-linking agent imparts either hydrophobic or hydrophilic regions affecting the overall hydrodynamic nature of the resulting material. Base ion-conducting materials may have inherent water uptake properties that may be either too high or too low for their objective purposes. Very high water uptake may cause the weakening of the material's physical properties, whereas very low water uptake may limit the ion-conducting material's ability to conduct protons at high efficiency. Therefore, water balance of the membrane is critical for high fuel cell performance. To tailor the material for usage, covalent cross-linking can act as an ideal modifier for a polymer electrolyte membrane.

Qing et al. [20] prepared and characterized a series of sulfonated polybenzimidazoles (sPBI-IS) with controlled sulfonation degrees. The thermal stability of these membranes, investigated by TGA under nitrogen atmosphere, increased with the increase in the degree of sulfonation. However, the acidic forms of these membranes were less thermally stable than nonsulfonated sample. The glass transition temperature of the acidic form was 196°C. The sulfonated membranes had higher storage and loss moduli than nonsulfonated ones (see Fig. 10.5). The storage moduli of all the prepared membranes were high due to their rigid backbones. The membranes had higher hygroscopicity and showed potential for high-temperature fuel cell applications.

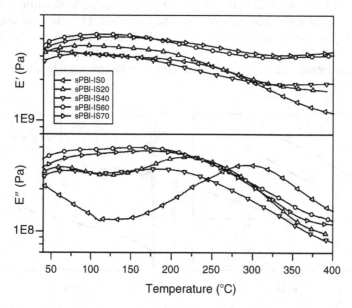

Fig. 10.5 Temperature dependence of the storage (E′) and loss (E″) modulus of sPBI-IS0 (nonsulfonated) and the acidic form sPBI-IS. The numbers 0–70 represent the degree of sulfonation. Reference Qing et al. (2005), reprinted with permission of Elsevier

Xue and Yin [21] modified SPEEK membranes with chemically in situ polymerized polypyrrole (PPy). It was found that the solution uptake and the swelling ratio of
the SPEEK/PPy decreased upon incorporation of PPy. At temperatures >60°C, the
membrane in 10 vol% methanol began to lose its mechanical stability. XRD analysis
showed that the structure was almost amorphous. The DSC and TGA measurements
showed that the microstructure change of SPEEK/PPy membranes cannot be
removed, even after drying the membranes. The incorporation of PPy into the SPEEK
matrix decreased methanol permeation and proton conductivity. It was suggested that
this is likely due to PPy particles blocking the methanol and protons.

Hill et al. [22] developed a series of zirconium hydrogen phosphate/disulfonated
poly(arylene ether sulfone) copolymer composite membranes for PEM fuel cells.
Disulfonated poly(arylene ether sulfone) copolymers (BPSH) were also tested separately. It was found that both pure BPSH and BPSH/ZrP composite membranes
had better mechanical properties when hydrated due to plasticization (see Fig.
10.6). Moderate amounts of ZrP (17%) had a positive effect on the mechanical
properties, whereas high amounts (36%) resulted in brittle membranes that are not
desirable for fuel cell applications.

Smitha et al. [23] synthesized membranes made from chitosan (CS) and
poly(vinyl pyrrolidone) for application in a DMFC. Chitosan has high hydrophilicity,
good chemical and thermal resistance properties. It has both hydroxyl and amino
groups. It is blended with hydrophilic polymers to overcome the disadvantage of
the loss in mechanical strength in the wet state. Poly(vinyl pyrrolidone) is a tough
hydrophilic polymer that gets dispersed into the CS matrix when doped. This is
cross-linked with glutaraldehyde followed by cross-linking with sulfuric acid to
improve resistance to excessive swelling. The cross-linked blends were suggested
to be safe for use in pervaporation applications at temperatures up to 150°C.
Valencia et al. [1] measured the stress-strain curves for SPEEK, and their results
showed a downtrend in both yield and break stresses with growing filler concentration.

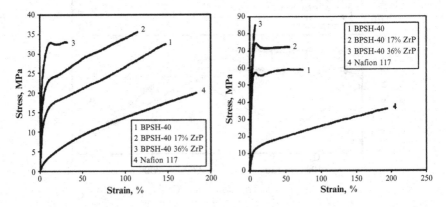

Fig. 10.6 Stress-strain curves for BPSH and BPSH/ZrP composite membranes show the reinforc-
ing effect of the inorganic component under both wet (*left*) and dry (*right*) conditions. Reference
[22] reprinted with permission of Elsevier

With no filler the break strain was 90% and break stress 55.8 MPa. With 30% filler, the break strain was 4% and break stress 24.8 MPa at 23°C and 30% RH.

Tian et al. [24] investigated the properties of novel sulfonated poly(phthalazinone ether ketone) (SPPEK) –based PEM. DSC measurements in the range 30–300°C did not show glass transition temperature before thermal decomposition. The rigid nature and ionomer characteristics of the SPPEK molecule gave high thermal stability resulting in absence of glass temperature below 250°C.

Loyens et al. [25] used $NaClO_4$ containing poly(ethylene oxide) PEO clay nanocomposite for polymer electrolyte membrane. Incorporating $NaClO_4$ containing PEO/Cloisite 30B nanocomposites displayed improvement of mechanical stiffness. At lower salt content it revealed an increased conductivity attributed to exfoliated structure. That effect was minimized at higher $NaClO_4$ content due to interaction between Na^+ and silicate layers. At temperatures above T_m the nanocomposites displayed slightly lower conductivities compared with corresponding $PEO/NaClO_4$ complex due to presence of clay platelets hindering ion transport.

Nam et al. [26] used organic-inorganic nanocomposite material like Nafion/ poly(phenylmethyl silsequioxane, PPSQ). Incorporation of PPSQ improved initial degradation temperature of Nafion membrane and increased the crystallinity of the recast composite membrane. The membrane was reported to have lower methanol permeability as compared with bare Nafion due to interruption of organic filler.

Su et al. [27] used silica nanopartilces for preparation of sulfonated poly (phthalazinone ether ketone) membranes. The nanoparticles were found homogeneously dispersed in matrix. Hybrid membrane exhibited improved swelling behavior, thermal stability, and mechanical properties, with better cell performance than conventional Nafion 117 [32,33]. In addition, Jung et al. [28] used Nafion/montmorillonite and reported higher thermal resistance of the composite membranes than pure copolymer resin due to the presence of MMT. Membranes with 3 and 5 wt% showed improved performance with temperatures up to 125°C. The performance of the Nafion/MMT nanocomposite was higher than commercial Nafion at high operating temperatures.

Thermal stability up to 150°C is clearly vital to the successful operation of DMFC, since thermal effect facilitates an oxidation reaction of methanol fuel at the anode and suppresses the electrode flooding and methanol diffusion processes. One possible approach to realizing such proton conductive membranes is to combine a highly proton-conductive nanomaterial with a thermostable polymer-composite proton conductive membranes [29]. The authors reported a relatively high proton conductivity of 3.82×10^{-3} S cm^{-1} in a fully humid condition at 200°C, which is attributed to hydrogen bonding. Post-sulfonation thermal treatment, which has a great effect on the ligand structure of PBI, gives a marked increase in the conductivity of the membrane by a factor of 2 in the same condition. This effect is mainly attributed to the proton transport via sulfonic acid groups bonded to the PBI unit [29].

Chen et al. [30] reported fabrication of Nafion/zeolite nanocomposite membranes by hydrothermal growth of acid-functionalized zeolite nanocrystals into commercial Nafion membranes for DMFCs. The presence of zeolite nanocrystals in the composite is confirmed by X-ray diffraction techniques. The tensile strength and water uptake were studied and the performance was compared with that of the

Nafion membrane. The membrane was reported to be stable during fuel cell operation up to 120°C. Further, Yu et al. [31] tested polystyrene sulfonic acid (PSSA) and Nafion101 composite membranes and were found to be stable after 835 h.

There is a tremendous scope for improvements in fuel cell membranes. The properties desired and costs involved have to be dealt with simultaneously and an optimum solution is required. Then, we can think of the bulk use of membranes to replace the currently used Nafion.

10.3 Conclusion

This present chapter covers an extensive study of the physical and mechanical properties related to fuel cell membranes. Currently, Nafion membranes have been extensively used as the polymer electrolyte in fuel cell membranes. The development of membranes is an active area of research for producing economic fuel cells. We have seen problems associated with that membrane in regard to methanol permeability (in DMFCs), high temperature operations, and conductivity. Therefore, it is critical for researchers to look into the solution of these problems, and try to develop novel membranes that would provide better performance and efficiency. Studies have shown that composite and nanocomposite membranes would give better response to the mechanical and thermal stresses. The latest in development is the use of SPEEK-incorporating heteropolyacids, HPAs, zeolites, or MCM-41. However, there is also emphasis on the suppression of permeability of methanol and enhancement of proton conductivity. These have to be tested in high-temperature applications and at high mechanical stresses and if not too successful they can be tested after incorporation of nanocomposites or other forms of materials. Therefore, it is worth mentioning that the development of membranes with excellent thermal and mechanical properties as well as low cost is crucial for the use of fuel cells. The influence of water, solvent, temperature, degree of sulfonation, and filler are some of the factors that influence the thermal and mechanical properties of the membranes.

References

1. Valencia, A. R., S. Kaliaguine, and M. Bousmina, "Tensile mechanical properties of sulfonated poly(ether ether ketone) (SPEEK) and BPO_4/SPEEK membranes", *J. Appl. Poly. Sci.* 98 (**2005**) 2380–2393.
2. Savadogo, O., "Emerging membranes for electrochemical systems Part II. High temperature composite membranes for polymer electrolyte fuel cell (PEFC) applications", *J. Power Sources*, 127 (**2004**) 135–161.
3. Jalani, N. H., K. Dunn, and R. Datta, "Synthesis and characterization of Nafion – MO_2 (M = Zr, Si, Ti) nanocomposite membranes for higher temperature PEM fuel cells", *Electrochim. Acta*, 51 (**2005**) 553–560.
4. Smitha, B., S. Sridhar, and A. A. Khan, "Chitosan-sodium alginate polyion complexes as fuel cell membranes", *Eur. Polym. J.*, 41 (**2005**) 1859–1866.

5. Yin, Y., S. Hayashi, O. Yamada, H. Kita, and K. Okamoto, "Branched/crosslinked sulfonated polyimide membranes for polymer electrolyte fuel cells", *Macromol. Rapid Commun.*, 26, **2005**, 696–700.

6. Yin, Y., J. Fang, Y. Cui, K. Tanaka, H. Kita, and K. Okamoto, "Synthesis, proton conductivity and methanol permeability of a novel sulfonated polyimide from 3-(2',4'-diaminophenoxy) propane sulfonic acid", *Polymer*, 44 (**2003**) 4509–4518.

7. Won, J., S. K. Chae, J. H. Kim, H. H. Park, Y. S. Kang, and H. S. Kim, "Self-assembled DNA composite membranes", *J. Membr Sci.*, 249 (**2005**) 113–117.

8. Kawano, Y., Y. Wang, R. A. Palmer, and S. R. Aubuchon, "Stress-strain curves of Nafion membranes in acid and salt forms", *Polímeros*, 12(2) (**2002**) 96–101.

9. Collier, A., H. Wang, X. Z. Yuan, J. Zhang, and D. P. Wilkinson, "Degradation of polymer electrolyte membranes", *Int. J. Hydrogen Energy*, 31 (**2006**) 1838–1854.

10. Chang, H. Y., R. Thangamuthu, and C. W. Lin, "Structure-property relationships in PEG-SiO$_2$ based proton conducting hybrid membranes – A ^{29}Si CP/MAS solid state NMR study", *J. Membr. Sci.*, 228 (**2004**) 217–226.

11. Chikashige, Y., Y. Chikyu, K. Miyatake, and M. Watanabe, "Poly(arylene ether) ionomers containing sulfofluorenyl groups for fuel cell applications", *Macromolecules*, 38 (**2005**) 7121–7126.

12. Yasuda, T. K., M. M. Hirai, M. Nanasawa, and M. Watanabe, "Synthesis and properties of polyimide ionomers containing sulfoalkoxy and fluorenyl groups, *J. Polym Sci. A: Polym. Chem*, 43 (**2005**) 4439–4445.

13. Kundu, S., L. C. Simon, M. Fowler, and S. Grot, "Mechanical properties of Nafion™ electrolytic membranes under hydrated conditions", *Polymer* 46 (**2005**) 11707–11715.

14. Ladewig, B. P., D. J. Martin, J. C. Diniz da Costa, and G. Q. Lu, "Nanocomposite Nafion/silica membranes for high power density direct methanol fuel cells", *5th International Membrane Science and Technology Conference (IMSTEC 03)* (ed. Technology, U. C. f. M. S. a.) paper 115 pp. 1–5 (Sydney NSW, Australia, 2003).

15. Rodriguez, F., *"Principles of Polymer Systems"*, 4th Ed., Philadelphia, PA, Taylor & Francis, 1989.

16. Kaliaguine, S., Mikhailenko, S., and S. M. J. Zaidi, "Composite electrolyte membranes for fuel cells and methods of making same", *US Patents* 6, (**2004**) 548–716.

17. Yen, S. S., S. R. Narayanan, G. Halpert, E. Graham, and A. Yavrouian, "Polymer material for electrolytic membranes in fuel cells", *US Patents* 5, (**1998**) 496–795.

18. Kurano, M. R., P. G. M. Kannan, A. Nadar, and T. K. Milton, III, "Composite electrolyte with crosslinking agents", *US Patents* 6, (**2005**) 959–962.

19. Xiao, L., H. Zhang, T. Jana, E. Scanlon, R. Chen, E. W. Choe, L. S. Ramanathan, S. Yu, and B. C. Benicewicz, "Synthesis and characterization of pyridine-based of polybenzimidazoles for high temperature polymer electrolyte membrane fuel cell applications", *Fuel Cells* 5(2) (**2005**) 287–295.

20. Qing, S., W. Huang, and D. Yan, "Synthesis and characterization of thermally stable sulfonated polybenzimidazoles", *Eur. Polym. J.*, 41 (**2005**) 1589–1595.

21. Xue, S. and G. Yin, "Proton exchange membranes based on modified sulfonated poly(ether ether ketone) membranes with chemically in situ polymerized polypyrrole", *Electrochem. Acta*, 52 (**2006**) 847–853.

22. Hill, M. L., Y. S. Kim, B. R. Einsla, and J. E. McGrath, "Zirconium hydrogen phosphate / disulfonated poly(arylene ether sulfone) copolymer composite membranes for proton exchange membrane fuel cells", *J. Membr. Sci.*, 283 (**2006**) 102–108.

23. Smitha, B., S. Sridhar, and A. A. Khan, "Chitosan — poly(vinyl pyrrolidone) blends as membranes for direct methanol fuel cell applications", *J. Power Sources*, 159 (**2006**) 846–854.

24. Tian, S. H., D. Shu, Y. L. Chen, M. Xiao, and Y. Z. Meng, "Preparation and properties of novel sulfonated poly(phthalazinone ether ketone) based PEM for PEM fuel cell application", *J. Power Sources*, 158 (**2006**) 88–93.

25. Loyens, W., F. H. J. Maurer, and P. Jannasch, "Melt-compounded salt-containing poly(ethylene oxide)/clay nanocomposites for polymer electrolyte membrane", *Polymer*, 46 (**2005**) 7334–7345.

26. Nam, S. E., S. A. Song, S. G. Kim, S. M. Park, Y. Kang, J. W. Lee, and K. H. Lee, "Preparation of organic-inorganic nanocomposite membranes as proton exchange membranes for direct dimethyl ether fuel cell application", *Desalination*, 200 (**2006**) 584–585.

27. Su, Y. H., Y. L. Liu, Y. M. Sun, J. Y. Lai, M. D. Guiver, and Y. Gao, "Using silica nanoparticles for modifying sulfonated poly(phthalazinone ether ketone) membrane for direct methanol fuel cell: A significant improvement on cell performance", *J. Power Sources*, 155 (**2006**) 111–117.

28. Jung, D. H., S. Y. Cho, D. H. Peck, D. R. Shin, and J. S. Kim, "Preparation and performance of a Nafion/montmorillonite nanocomposite membrane for direct methanol fuel cell", *J. Power Sources*, 118 (**2006**) 205–211.

29. Yamazaki, Y., M. Y. Jang, and T. Taniyama, "Proton conductivity of zirconium tricarboxy-butylphosphonate/PBI nanocomposite membrane", *Sci. Tech. Adv. Mat.*, 5 (**2004**) 455–459.

30. Chen, S. L., A. B. Bokarsly, and J. Benziger, "Nafion-layered sulfonated polysulfone fuel cell membranes", *J. Power Sources*, 152 (**2005**) 27–33.

31. Yu, J., B. Yi, D. Xing, F. Liu, Z. Shao, Y. Fu, and H. Zhang, "Degradation mechanism of polysterene sulfonic acid membrane and application of its composite membranes in fuel cells", *Phys. Chem. Chem. Phys.*, 5 (**2003**) 611–615.

32. Silva, V. S., B. Ruffmann, H. Silva, A. Gallego, A. Mendes, M. Madeira, and S. P. Nunes, "Proton electrolyte membrane properties and direct methanol fuel cell performance. Characterization of hybrid sulfonated poly(ether ether ketone)/zirconium oxide membranes", *J. Power Sources*, 140 (**2005**) 34–40.

33. Zhang, X., L. P. Filho, C. Torras, and R. Gracia-Valls, "Experimental and computational study of proton and methanol permeabilities through composite membranes", *J. Power Sources*, 145 (**2005**) 223–230.

34. Kaneko, K., Y. Takeoka, M. Aizawa, M. Rikukawa, "Fabrication and electrical properties of proton conducting polymer hybridized with apatite compounds", *Synthetic Metals*, 135–136 (**2003**) 73–74.

Chapter 11
Membrane and MEA Development in Polymer Electrolyte Fuel Cells

Panagiotis Trogadas and Vijay Ramani

Abstract The polymer electrolyte fuel cell (PEFC) is based on Nafion® polymer membranes operating at a temperature of 80°C. The main characteristics (structure and properties) and problems of Nafion®-based PEFC technology are discussed. The primary drawbacks of Nafion® membranes are poor conductivity at low relative humidities (and consequently at temperatures >100°C and ambient pressure) and large crossover of methanol in direct methanol fuel cell (DMFC) applications. These drawbacks have prompted an extensive effort to improve the properties of Nafion® and identify alternate materials to replace Nafion®. Polymer electrolyte membranes (PEMs) are classified in modified Nafion®, membranes based on functionalized non-fluorinated backbones and acid-base polymer systems. Perhaps the most widely employed approach is the addition of inorganic additives to Nafion® membranes to yield organic/inorganic composite membranes. Four major types of inorganic additives that have been studied (zirconium phosphates, heteropolyacids, metal hydrogen sulfates, and metal oxides) are reviewed in the following. DMFC and H_2/O_2 (air) cells based on modified Nafion® membranes have been successfully operated at temperatures up to 120°C under ambient pressure and up to 150°C under 3–5 atm. Membranes based on functionalized non-fluorinated backbones are potentially promising for high-temperature operation. High conductivities have been obtained at temperatures up to 180°C. The final category of polymeric PEMs comprises non-functionalized polymers with basic character doped with proton-conducting acids such as phosphoric acid. The advanced features include high CO tolerance and thermal management. The advances made in the fabrication of electrodes for PEM fuel cells from the PTFE-bound catalyst layers of almost 20 years ago to the present technology are briefly discussed. There are two widely employed electrode designs: (1) PTFE-bound, and (2) thin-film electrodes. Emerging methods include those featuring catalyst layers formed with electrodeposition and vacuum deposition (sputtering). The thin-film electrodes have significantly increased performance and reduced the level of platinum loading required. Thin sputtered layers have shown promise for low catalyst loading with adequate performance. Electrodeposition methods are briefly discussed. Finally, the relationship between MEA processing and the durability of the membrane/electrode interface and hence the fuel cell as a whole is presented.

S.M.J. Zaidi, T. Matsuura (eds.) *Polymer Membranes for Fuel Cells*,
doi: 10.1007/978-0-387-73532-0, © Springer Science+Business Media, LLC 2009

11.1 Introduction

Fuel cells are primarily classified into five categories (Table 11.1) according to the electrolyte material used. Since the choice of electrolyte material plays an important role in determining the operating temperature of the fuel cell, this parameter may be used interchangeably in fuel cell classification. Table 11.1 presents the different kinds of fuel cells, respective electrolytes used, and operating temperature range of the fuel cell.

11.1.1 Polymer Electrolyte Fuel Cells

Polymer electrolyte fuel cells (PEFCs) are unique in that they are the only variety of low-temperature fuel cell to utilize a solid electrolyte. The most common polymer electrolyte used in PEFCs is Nafion®, produced by DuPont, a perfluorosulfonic ionomer that is commercially available in films of thicknesses varying from 25 to 175 μm. This material has a fluorocarbon polytetrafluoroethylene (PTFE)–kbone with side chains ending in pendant sulfonic acid moieties. The presence of sulfonic acid promotes water uptake, enabling the membrane to be a good protonic conductor, and thereby facilitating proton transport through the cell. This chapter reviews PEFC development, structure, and properties and presents an overview of PEM technology to date.

11.1.2 Background

One of the first attempts to use a polymeric ion exchange membrane as a solid electrolyte for fuel cells was described by Thomas Grubb of GE in 1959. Initially, between 1959 and 1961, polysulfuric sulfonic acid (PSSA) membranes were used. The early versions of the PEFC, as used in the NASA Gemini spacecraft, had a lifetime of only about 500 h because of membrane degradation, but that was sufficient for those limited early missions. The development program continued

Table 11.1 Fuel cell classification

Fuel cell type	Electrolyte	Typical operating temperature
Polymer electrolyte	Polymer membrane	60–80°C
Alkaline	Potassium hydroxide	70–120°C
Phosphoric acid	Phosphoric acid	160–200°C
Molten carbonate	Lithium/potassium carbonate	650°C
Solid oxide	Yttrium stabilized Zirconia	1,000°C

with the incorporation of a new polymer membrane in 1967 called Nafion®, synthesized by DuPont. This type of membrane became standard for the PEFC and remains today.

However, the problem of degradation and water management in the electrolyte was judged too difficult to manage reliably, and, for the Apollo vehicles, NASA selected the alkaline fuel cell [1]. General Electric also chose not to pursue commercial development of the PEFC, probably because the costs were seen as higher than other fuel cells, such as the phosphoric acid fuel cell then being developed. At that time catalyst technology was such that 28 mg of platinum was needed for each square centimeter of electrode, compared with 0.2 mg cm^{-2} or less now.

The development of proton exchange membrane (PEM) cells was held in abeyance in the 1970s and early 1980s. However, in the latter half of the 1980s and early 1990s, there was a renaissance of interest in this type of cell [2].

The developments over recent years have brought the current densities up to around 1 A cm^{-2} or more, while at the same time reducing the use of platinum by a factor of >100. These improvements have led to huge reduction in cost per kilowatt of power, and much improved power density.

PEFCs are now being actively developed for use in cars and buses, as well as for a very wide range of portable applications, and also for combined heat and power (CHP) systems. A sign of the dominance of this type of cell is that they are again the pReference:red option for NASA and the new space shuttle Orbiter will use PEM cells. It could be argued that PEFCs exceed all other electrical energy–generating technologies with respect to the scope of their possible applications. They are a possible power source at a few watts for powering mobile phones and other electronic equipment such as computers, right through to a few kilowatts for boats and domestic systems, to tens of kilowatts for cars and hundreds of kilowatts for buses and industrial CHP systems. Further bibliographic information in this technology is available in the literature [3].

11.2 Structure

The structure of PEMs and their concomitant relationship to important properties such as proton conductivity is best described by considering a model perfluorosulfonic acid ionomer such as Nafion®, a perfluorinated polymer with a polytetrafluoroethylene (PTFE)–like backbone that contains small proportions of sulfonic or carboxylic groups dangling from regularly spaced side chains.

Nafion® was introduced by DuPont in the mid- to late 1960s. Nafion® films are produced by both extrusion and solution casting processes. Additionally, dispersions of Nafion® in selected solvents in concentrations ranging from 5% to 20% are also available. These dispersions may be employed to form recast Nafion® membranes. Since their introduction, Nafion® membranes have been studied by numerous researchers to determine their physicochemical properties and applicability to fuel cells. The water uptake of Nafion® increases with the temperature of the liquid water to which the membrane is exposed. The effect of thermal pretreatment on the water uptake of Nafion® membranes was studied in detail [4,5]. The water uptake decreases as alkali metal cations of increasing size replace the proton in the sulfonic acid group. The water uptake of Nafion® when exposed to water vapor has also been extensively studied [4–6]. The water uptake from the vapor phase is lower than that from the liquid phase, even at the same water activity—an anomaly termed Schroeder's paradox. This has been explained by the hydrophobic nature of the membrane surface when exposed to water vapor, which has been demonstrated by contact angle measurements. The proton conductivity of Nafion® as a function of humidity was estimated by Anantaraman and Gardner [7], who reported a lowering of conductivity with humidity. Zawodzinski and co-workers [8] also reported a decrease in the proton conductivity of Nafion® with decreasing water content and estimated the water diffusion coefficients in Nafion® membranes [8]. Proton conductivity as a function of both temperature and humidity was reported by Sone and co-workers [9]. The dependence of conductivity on pressure has also been reported for Nafion® membranes [10]. In addition to studies on water uptake and conductivity, researchers have investigated the microstructure and properties of the Nafion® membrane using FTIR spectroscopy [11], Raman spectroscopy, small angle X-ray scattering (SAXS) [12], and thermogravimetric analysis (TGA) [13]. Advances in perfluorinated ionomer membranes have been summarized in the literature [14].

The general structure of Nafion® is shown in Fig. 11.1 [15]. It is generally accepted that the ionic groups tend to form clusters within the polymeric matrix, and several models have been proposed to explain such clustering. Eisenberg [16] proposed perhaps one of the first theories treating clustering of ions in polymers. Other models include the Gierke cluster network model [17,18], the Yeager three-phase model, and the Eisenberg–Hird–Moore (EHM) model [19]. The Gierke cluster network model proposes that the ionomer microstructure is divided into a hydrophobic domain (the fluorinated backbone in Nafion®) and a hydrophilic domain consisting of ionic clusters comprising the sulfonic acid moiety and absorbed water. These clusters are interconnected by a network of short and narrow channels, approximately 1 nm in diameter. The size of the cluster is strongly dependent on the system water content, and ranges from 2 to 5 nm. Evidence for such cluster formation has been obtained by small angle X-ray scattering experiments. The Yeager "three-phase" model introduces a third region—the so-called interfacial region—that contains side chains, water, and some of the ionic groups that are not part of the clusters. The development in the understanding of Nafion® structure is perhaps best summarized in a recent review by Mauritz and Moore [20].

Finally, the structure and properties of recast Nafion® membranes are strongly dependent on the solvent used for casting and the temperature at which the casting is performed. This has been illustrated by Moore and Martin [21], who have also conducted studies to explain the chemical and morphological properties of recast Nafion® membranes prepared under different conditions [22]. These studies have revealed that casting the membranes at high temperatures (>130°C) in the presence of a high boiling solvent such as dimethyl formamide (DMF) enhances the crystallinity

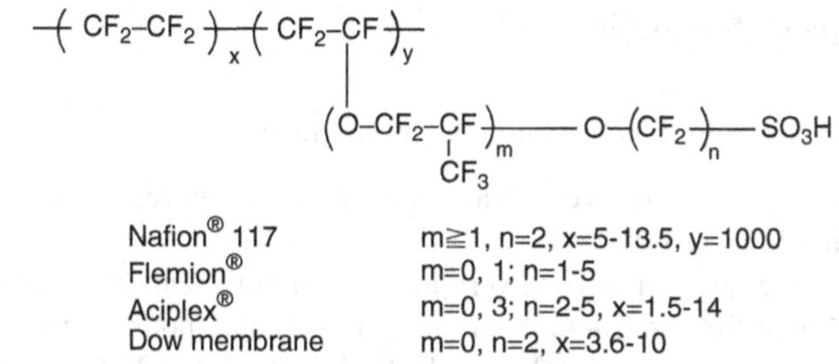

$$-\left(CF_2-CF_2\right)_x\left(CF_2-CF\right)_y$$

$$\left(O-CF_2-CF\right)_m \!\!\!\! -O\!-\!\left(CF_2\right)_n\!\!-SO_3H$$
$$\quad\quad\quad CF_3$$

Nafion® 117	m≧1, n=2, x=5-13.5, y=1000
Flemion®	m=0, 1; n=1-5
Aciplex®	m=0, 3; n=2-5, x=1.5-14
Dow membrane	m=0, n=2, x=3.6-10

Fig. 11.1 General structure of perfluorosulfonic acid (PFSA) ionomers

Fig. 11.2 Qualitative sketch of a-ZrP layer, and the orientation of the phosphate groups with respect to the plane of Zr(IV) ions

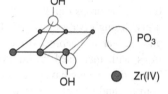

Fig. 11.3 Molecular structure of the Keggin-type $[PW_{12}O_{40}]^{3-}$ heteropolyanion

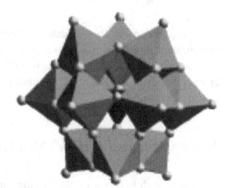

of the membrane and prevents dissolution in common solvents. This corroborated the conclusions of Gebel and co-workers [23], who earlier reported increasing crystallinity when Nafion® films with added high boiling solvent were recast and annealed. More recently, the density and solubility of recast and commercial Nafion® membranes have been reported [24]. The application of Nafion® membranes in PEFCs has been recently reviewed [25].

11.3   Desired Properties

A good PEM should possess the following properties:

1. High proton conductivity over a wide range of temperatures ($-40°-150°C$) and relative humidities (10–100%)
2. Excellent barrier properties (i.e., very low gas and methanol permeabilities)
3. Sufficient durability for 10–40,000 h of cyclic or continuous operation
4. High thermal stability for membrane electrode assembly (MEA) processing
5. Compatibility with the electrode framework to ensure proton conductivity at the electrode
6. Low cost to facilitate fuel cell commercialization

Nafion®, which is one of the best PEMs available today, satisfies requirements b (for gases only) d and e, and perhaps even requirement c. However, the conductivity of Nafion® drops rapidly with decreasing relative humidity (RH), methanol permeability through Nafion® is very high (a disadvantage for direct methanol fuel cells), and the material is presently extremely expensive (~$700 m^{-2}), although cost reductions with increased production rates have been projected by the manufacturer. These drawbacks have prompted an extensive effort to improve the properties of Nafion® and identify alternate materials to replace Nafion®.

11.4   Progress Made to Date

For the purposes of this discussion, PEMs are classified in modified Nafion®, membranes based on functionalized non-fluorinated backbones and acid-base polymer systems. The prior approaches adopted by researchers and some characteristics of the resultant membranes under each classification are presented in this section.

11.4.1   Modified Nafion® Membranes

The primary drawbacks of the Nafion® membranes are poor conductivity at low relative humidities (and consequently at temperatures >100°C and ambient pressure) and large crossover of methanol in direct methanol fuel cell (DMFC) applications. As a result, considerable efforts have been made in recent years to overcome these drawbacks. Perhaps the most widely employed approach is the addition of inorganic additives to Nafion® membranes to yield organic/inorganic composite membranes. Three major types of inorganic additives that have been studied (zirconium phosphates, heteropolyacids, metal hydrogen sulfates and metal oxides) are reviewed in the following.

11.4.1.1 Zirconium Phosphates

Zirconium phosphates can be expressed in the form of $M^{IV} (RXO_3)_2 \cdot nS$ [26], where M is a tetravalent metal such as Zr, Ti, Ce, Th, or Sn; R an inorganic or organic group -H, -OH, -CH$_3$OH, or $(CH_2)_n$; X is P or As; and S a solvent, i.e., H$_2$O. They form two types of layered structures, named α and γ [27–29], which are a group of inorganic polymers where the O$_3$POH groups of the α-type Zr(O$_3$POH)$_2 \cdot n$H$_2$O and the O$_2$P(OH)$_2$ groups of γ-type ZrPO$_4 \cdot$ O$_3$P(OH)$_2 \cdot n$H$_2$O are replaced with O$_3$POR or O$_2$PR′R-groups [26] (Fig. 11.2). When the organic moieties R contain a proton-generating function such as -COOH, -PO$_3$H, -SO$_3$H, or NH$_3^+$, these compounds become proton conductors and exhibit very good conductivity in a temperature range up to 200°C. Some of the mixed zirconium alkyl-sulfophenyl phosphates have proton conductivity of 5×10^{-2} S cm^{-1} at 100°C with good thermal stability at temperatures up to 200°C (in the form of glassy plates or films), a room-temperature conductivity of 10^{-2} S cm^{-1} has been reported [30]. The proton conductivity can be enhanced by composite formation with SiO$_2$ or Al$_2$O$_3$ [31,32]. The possible uses of these membranes for medium-temperature sensors and fuel cells have been recently reviewed [33].

11.4.1.2 Heteropolyacids

Heteropolyacids exist in a series of hydrated phases. Their basic structure unit is the $[PM_{12}O_{40}]^{3+}$ cluster, the so-called Keggin unit (Fig. 11.3). Typical compounds include H$_3$PW$_{12}$O$_{40} \cdot n$H$_2$O (PWA), H$_3$PMo$_{12}$O$_{40} \cdot n$H$_2$O (PMoA), and H$_4$SiW$_{12}$O$_{40} \cdot n$H$_2$O (SiWA) [26]. In their crystalline forms, hydrated with 29 water molecules, these acids exhibit high proton conductivities [34]; for example, 0.18 S cm^{-1} for H$_3$PW$_{12}$O$_{40} \cdot$ 29H$_2$O and 0.17 S cm^{-1} for H$_3$-PMo$_{12}$O$_{40} \cdot$ 29H$_2$O. Aqueous solutions of heteropolyacids, e.g., SiWA and PWA [35–37], have been explored as fuel cell electrolytes, showing fast electrode kinetics and less CO poisoning at the electrode–electrolyte interface. Solid electrolyte fuel cells have also been proposed [36]. Furthermore, phosphotungstic acid (PTA) has also been added to Nafion® to prepare composite membranes. These additives were initially proposed as liquid electrolytes (similar to phosphoric acid) for fuel cells [35], but were then adapted in the solid form within a Nafion matrix to permit retention of the advantages of a solid electrolyte. Malhotra and Datta [38] first proposed the use of Nafion®/PTA composite membranes for operation >100°C at lower relative humidities. Nafion®/HPA membranes also have been investigated by Savodago and Tazi [39,40]. Savadogo and Tazi [39] developed Nafion®/silicotungstic acid with and without thiophene (named NASTATH and NASTA, respectively) composite membranes, and the obtained results are available in the literature. Finally, investigators have employed PTA supported on metal oxides such as silicon dioxide as additives to Nafion to promote stability of the PTA within the membrane matrix [41,42].

11.4.1.3 Hydrogen Sulfates

Hydrogen sulfates can be expressed in the form of $MHXO_4$, where M is large alkali species NH_4^+, Cs, Rb, and X is Se, S, As, or P [26]. The most known compound is $CsHSO_4$, which at temperatures >141°C exhibits a high proton conductivity of 10^{-2} S cm^{-1} [43], due to its dynamically disordered network of hydrogen bonds. Also, this compound does not contain water molecules in its structure; as a result it has high thermal and electrochemical stability. Its conductivity does not depend on atmospheric humidity and can be enhanced by using high surface area metal oxides (inorganic composites of $CsHSO_4$ with SiO_2 [44] and TiO2 [45]. During recent years, new solid compounds $CsHSO_4 \cdot CsH_2PO_4$ and $2CsHSO_4 \cdot CsH_2PO_4$ have been prepared and shown to have lower transition temperatures and high proton conductivity [46,47], but they suffer from poor mechanical properties, water solubility, as well as extreme ductility and volume expansion at raised temperatures. $CsHSO_4$ has been used as the electrolyte of an H_2/O_2 fuel cell [47]. The development of composite membranes using $CsHSO_4$ [48] and direct methanol fuel cell applications of $CsHSO_4$ membranes [49] have also been reported.

11.4.1.4 Metal Oxides

Nafion®-MO_2 (M = Zr, Ti) nanocomposite membranes are used to increase the proton conductivity and water retention at higher temperatures and lower relative humidities (120°C, 40% RHs) as well as to improve the thermomechanical properties. Usually, the sol-gel approach is utilized to incorporate inorganic oxide nanoparticles within the pores of Nafion® membrane. The membranes synthesized by this approach are reported to be completely transparent and homogeneous as compared with membranes prepared by alternate casting methods, which are cloudy due to the larger particle size [50]. At 90 and 120°C, all Nafion®-MO_2 sol-gel composites exhibit higher water sorption than Nafion® membrane, but, at these conditions the conductivity is enhanced in only Nafion®-ZrO_2 sol-gel composites, with a 10% enhancement at 40% RH over Nafion®. This can be attributed to the increase in acidity of Zirconia-based sol-gel membranes shown by a decrease in equivalent weight in comparison to other nanocomposites based on Ti and Si. In addition, the TGA and DMA analyses showed improvement in degradation and glass transition temperature for nanocomposite membranes over Nafion®.

Furthermore, Watanabe and co-workers [51] modified Nafion® PEMs by the incorporation of nano-sized particles of SiO_2, TiO_2, Pt, Pt–SiO_2, and Pt–TiO_2 to decrease the humidification requirements of PEMs. When operated at 80°C under low humidification PEMFC, the modified PEMs showed lower resistance than Nafion®. This improvement was attributed to the suppression of H_2 crossover by in situ Pt along with the subsequent sorption of the water produced on the incorporated oxides.

Also, Adjemian and co-workers [52] introduced nanosized SiO_2 into Nafion® pores and tested various thickness and EW membranes. The benefit of these composite membranes appears to be in more stable operation versus conventional

Nafion® at a cell temperature of 130°C due to higher rigidity, when both were tested under fully humidified conditions. The investigators noted that the unmodified PEMs showed thermal degradation, whereas the SiO_2 modified PEMs did not show such damage. Adjemian and co-workers [52] have also synthesized Nafion® PEMs containing silicon oxide, as well as zirconium phosphate particles. They found that silicon oxide modified membranes exhibited better robustness and water retention and better performance.

Based on this work accomplished on higher-temperature membranes, it can be concluded that the approach of synthesizing nanocomposite membranes either by casting a bulk mixture of powder or colloidal state of inorganics with a polymer solution, or in situ formation of inorganic particles utilizing the membrane as template [50], is very promising. The advantage of the in situ method is that the particle size can be controlled by the concentration of precursors because the size and dispersion of these solid particles are of utmost importance in final performance of fuel cells. Mostly, the in situ methods are based on sol-gel reactions between the organometallic compound as the precursor and water within the pores of the membrane. Although these membranes show better water sorption and proton conduction properties, better mechanical properties with higher fuel cell performance and long-term stability are yet to be established.

11.4.1.4.1 Silica

Nafion®-SiO_2 nanocomposite membranes are used to increase the proton conductivity and water retention at higher temperatures as well as improve the thermomechanical properties. Mauritz [53] investigated the preparation of Nafion/SiO_2 composites by infiltration of Nafion membranes with silicon alkoxides. They exploited the fact that the pendant $SO_3^-H^+$ group clusters catalyze the sol-gel reaction. In their procedure, a Nafion membrane is swelled in an alcohol/water solution. A mixture of tetraethylorthosilicate (TEOS) and alcohol is added to the swelling solution containing the Nafion membrane, where TEOS molecules presumably migrate to the clusters. After the sol-gel reactions, the in situ inorganic phase is cured. This is accomplished by placing the membrane in a vacuum oven at 100°C, where the solvents, alcohol, and water evaporate and the condensation of SiOH group proceeds.

In Nafion® 117 (1,100 equiv. wt) the weight uptake of the dried samples increases linearly with immersion time [54,55]. This suggests that as immersion time increases the SiO_2 begins to percolate through the Nafion®. However, a significant amount of SiO_2 deposits on the surface, making it necessary to clean the surfaces of the membranes with alcohol after the immersion process to avoid forming surface layers of silica. Chemical analysis shows a profile across the membrane thickness with greater SiO_2 concentration near the surface, decreasing to a minimum in the middle [56]. This gradient may be reduced using an acid catalyzed, pre-hydrolyzed silicon alkoxide solution in alcohol, so that partially hydrolyzed species, such as $(RO)_{4-x}Si(OH)_x$ molecules, migrate to the polar clusters of Nafion® at the same time that the membrane is swelled. A slight concentration gradient is hard to avoid because of the difficulty of diffusion in the narrow channels of Nafion® membrane [56].

Moreover, Nafion®/SiO$_2$ composites showed a higher water uptake at room temperature (w20%) compared with unfilled Nafion (w15%) [57]. Presumably, the sol-gel silica provides a large number of hydroxyl groups that tie up water molecules. The hydrophobicity of the fluorocarbon backbone and pendant side chains appears to be reduced by the incorporation of SiO$_2$.

Furthermore, Nafion®/SiO$_2$ membranes can also be prepared using Nafion® solutions and silicon alkoxides in a mutual solvent. The mixed solution is cast into shallow containers and dried at ambient temperature. This procedure allows the incorporation of continuously increasing amounts of silica. With high silica contents (>50%), the films are transparent but brittle. Hence, the composites are formed from a mixed solution. Instead of infiltrating a Nafion film, the silica phase is not restricted to the clusters of Nafion®. When part of the TEOS is substituted by 1,1,3,3-tetramethyl-1,3-diethoxy disiloxane (TMDES), phase separation, higher flexibility, and higher ionic conductivity are observed [58,59].

Finally, particles of Nafion®/SiO$_2$ nanocomposites have been obtained by capturing Nafion® solutions in silicon alkoxides [60,61]. The Nafion® phase (20–60 nm) is trapped within a porous silica network. By controlling the variables of the process (composition, pH, drying), the resulting material can have a wide range of surface area (85–560 m^2 g^{-1}) and pore diameter (25–2.1 nm), and highly dispersed Nafion® is readily accessible to reactants.

11.4.2 PEMs with Functionalized Non-fluorinated Backbones

The second category of PEMs studied, broadly involve membranes containing non-fluorinated backbones—i.e., hydrocarbon based polymers. Since most polymers do not possess inherent proton-conducting properties, their backbones are functionalized by a sulfonation process to introduce sulfonic acid groups into the structure. These SO$_3$H groups enhance the water uptake of the membrane and improve proton conductivity. There are various ways to prepare partially perfluorinated and non-perfluorinated ionomer membranes, based on chemical modification of the polymers [62,63] or monomer sulfonation and subsequent polymerization [64]. Several polymeric systems have been studied in this regard, with sulfonated polystyrene being one of the earliest. Sulfonated polystyrene has been employed in different capacities, ranging from direct application as PEMs to application as proton-conducting grafts on fluoropolymer backbones. The primary demerit of sulfonated polystyrene as a PEM is the rapid degradation by the free radical mechanism due to peroxide attack, especially given that peroxide is produced by incomplete oxygen reduction at the cathode and by reduction of oxygen crossing over through the membrane to the anode. The mechanism of degradation has been studied and reported [65]. Other extensively studied functionalized hydrocarbon polymers include sulfonated polysulfones [66], sulfonated poly ether sulfones [67–72], sulfonated poly ether ether ketones

[26,73,74], sulfonated poly ether ketone ketones [75], sulfonated polyimides [76–78], sulfonated poly phthalazinones [79], sulfonated polybenzimidazoles [31,32,80–82], and sulfonated polyphosphazines [83]. Developments in this field have been numerous, and have been summarized in several reviews [80,84–86]. In addition to the work reviewed in the preceding, several studies report the preparation and characterization of sulfonated hydrocarbon/inorganic additive composite membranes, with heteropolyacids being the most common inorganic additive reported [87–89]. In the following sections are described analytically the categories of PEMs with functionalized non-fluorinated backbones.

11.4.2.1   Sulfonated Polysulfones

Polysulfone (PSU, Fig. 11.4) is an interesting polymer for its low cost, commercial processability and availability. Two different procedures for the sulfonation of polysulfone are reported. In the first, a sodium-sulfonated group is introduced in the base polysulfone via the metalation-sulfination-oxidation process [90], and in the other, the trimethylsilyl chlorosulfonate is used as the sulfonating agent. However, following either of these two procedures, the result remains the same: The proton conductivity of the sulfonated PSU membrane is lower than Nafion® membrane, because the C-F chains in Nafion® have high hydrophobicity and large-phase separation.

11.4.2.2   Poly(ether ether ketone)

Poly(ether ether ketone) (PEEK) (Fig. 11.5) is a thermostable polymer with an aromatic, non-fluorinated backbone, in which 1,4-disubstituted phenyl groups are separated by ether (–O–) and carbonyl (–CO–) linkages. PEEK can be functionalized by sulfonation and the degree of sulfonation (DS) can be controlled by reaction time and temperature.

The sulfonation can be carried out directly in concentrated sulfuric acid or oleum, [64,90], although it has been reported that this method is not appropriate for the preparation of polymers with a low degree of sulfonation (<30%) because the sulfonation reaction takes place at the same time as polymer dissolution; the resulting sulfonation is heterogeneous and the polymer microstructure difficult to reproduce [64]. The degradation of the polymer does not occur <400°C, but this temperature

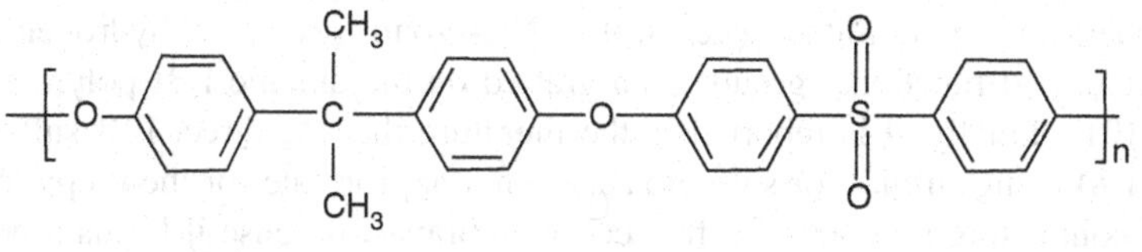

Fig. 11.4   Structure of polysulfone (PSU)

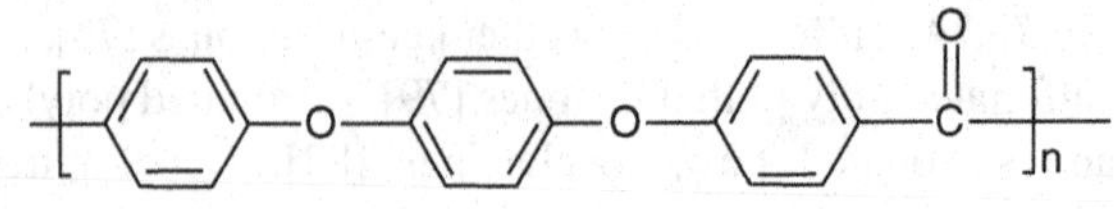

Fig. 11.5 Polymer structure of
poly(ether ether ketone) (PEEK)

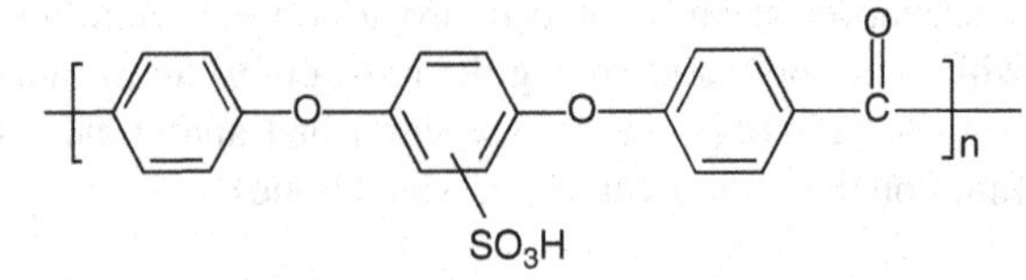

Fig. 11.6 Polymer structure of
sulfonated poly (ether ether ketone)
(SPEEK)

range is limited by the temperature at which the sulfonic acid group degrades, around 270°C.

Sulfonated poly(ether ether ketone) (SPEEK) (Fig. 11.6) is very promising for hydrogen fuel cells application, as it has good thermal stability, mechanical strength, and adequate conductivity [33,91–95]. However, the chemical durability and proton conductivity at low relative humidity (RH) are still to be resolved.

11.4.2.3 Poly (Benzimidazole) PBI

Polybenzimidazoles are synthesized from aromatic bis-o-diamines and decarboxylates (acids, esters, amides), in the molten state or in solution. Their thermal properties depend on the nature of the component tetramine, and dicarboxylic acids and have been reported in the literature [96].

The commercially available polybenzimidazole is poly-[2,20-(m-phenylene)-5,50-bibenzimidazole] (so-called PBI; Fig. 11.7). This is synthesized from diphenyl-iso-phthalate and tetra-aminobiphenyl [97]. It has excellent mechanical and thermal stability. Reports of the PBI proton conductivity are conflicting. Even though values in the range of 2×10^{-4}–$8 \cdot 10^{-4}$ S cm^{-1} at relative humidities between 0% and 100% were published [98], other authors [99,100] observed proton conductivities of two to three orders of magnitude lower. These latter values are those generally accepted for non-modified PBI, and are too small for any use of PBI membranes in fuel cell applications. As a result, two principal routes have been developed to improve proton conduction properties, and these are based on the particular reactivity of PBI, which is twofold, and comes from the $-N =$ and $-NH$-groups of the imidazole ring. Furthermore, PBI complexes with inorganic and organic acids due to its basic character. In addition, however, the $-NH$-group is reactive; hydrogen can be abstracted, and functional groups then grafted on the anionic PBI polymer backbone [101]. Finally, it is reported in the literature that the direct PBI sulfonation (Fig. 11.8) using sulfuric or sulfonic acid is not appropriate for the preparation of proton-conducting polymers for fuel cell membranes, because it forms a polymer of low degree of sulfonation and increased brittleness [86].

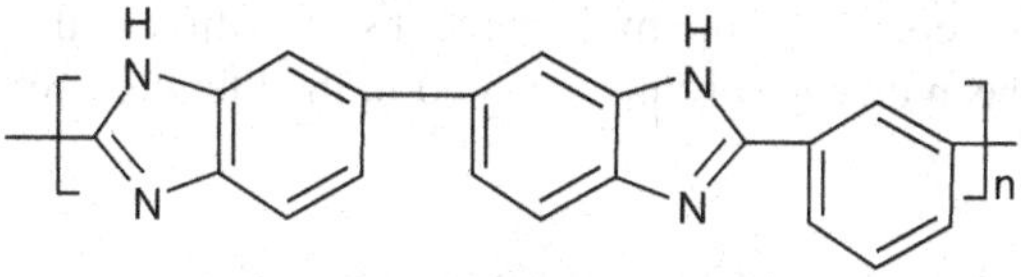

Fig. 11.7 Polymer structure of poly (benzimidazole) (PBI)

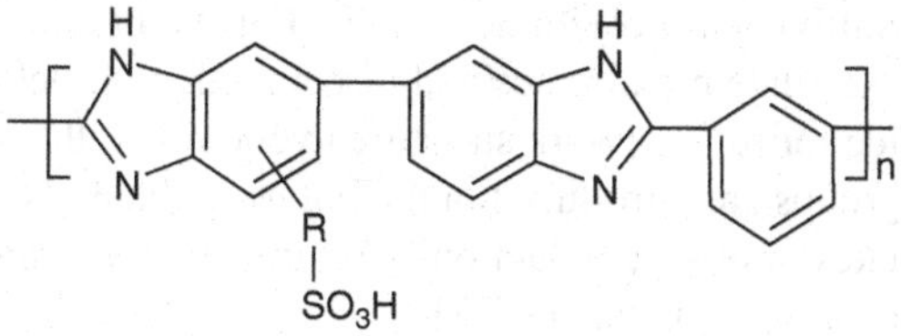

Fig. 11.8 Polymer structure of sulfonated poly (benzimidazole) (SPBI)

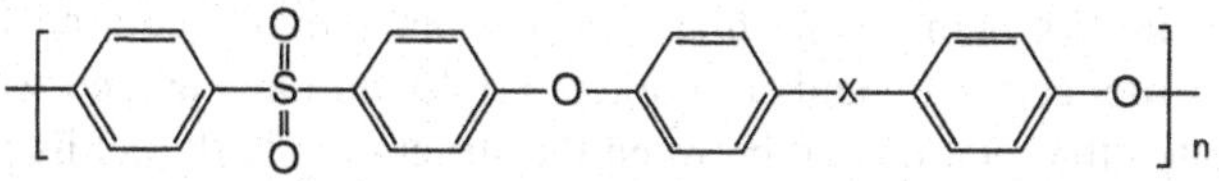

Fig. 11.9 Polymer structure of poly(arylene ether sulfone) (PES)

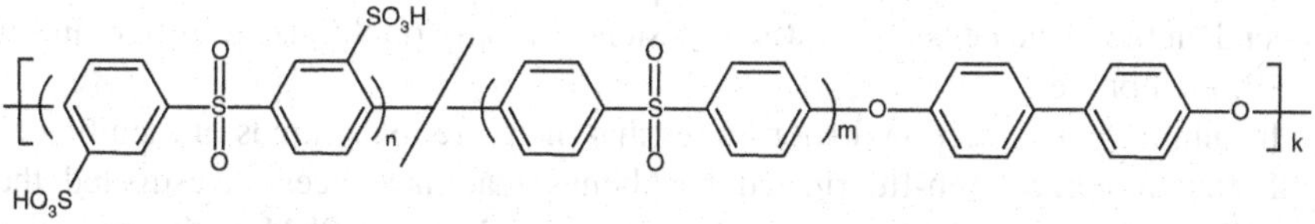

Fig. 11.10 Polymer structure of sulfonated poly(arylene ether sulfone) (SPES)

11.4.2.4 Poly (Ether Sulfone)

Poly(arylene ether sulfones) (Fig. 11.9) are well-known engineering thermoplastics characterized by excellent thermal and mechanical properties, as well as resistance to oxidation and acid catalyzed hydrolysis [25].

In order to prepare advanced molecules of poly(arylene ether sulfones) for fuel cell applications without sacrificing their excellent physical properties, Noshay and Robeson developed a mild sulfonation procedure for the commercially available bisphenol-A-based poly(ether sulfone) [62,63]. The sulfonation agents that have been used for this polymer modification are chlorosulfonic acid and a sulfur trioxide–triethyl phosphate complex. Recently, Kerres and co-workers [102] reported an alternative sulfonation process of commercial polysulfone based on a series of steps, including metalation–sulfination–oxidation reactions.

The proton conductivity of the sulfonated PES is above 0.08 S cm^{-1}, which is in the range needed for high-performance fuel cell proton exchange membranes and it depends on the mole percent sulfonation of the membranes. These materials have been extensively investigated in the literature [68–72].

11.4.2.5 Sulfonated Polyimide (SPI)

Sulfonated polyimides (SPI, Fig. 11.11) are potential candidates for proton exchange membranes and direct methanol fuel cells (DMFC) because the polyimide forms a network structure to control methanol permeability, and if sulfonic acid groups are introduced in the polymer chain, then it becomes hydrophilic and facilitates proton conduction. Polyimide is mechanically and thermally stable and chemically resistant [77].

Various structures of SPI materials were recently studied [103,104]. For instance, Genies and co-workers [78] synthesized SPI membranes with random and sequence polymer, but did not achieve the high proton conductivity required for practical application of fuel cell. Guo and co-workers [103] and Fang and co-workers [104] also investigated the relationship between the structure of SPI and its properties in detail. It was revealed that the more flexible the main chain, the more the membrane's hydrolysis stability was improved. Finally, Woo and co-workers [77] found that proton conductivity and methanol permeability of SPI membranes do not depend on the distance between water clusters, but only on the size of water cluster in SPI membrane.

In summary, this section demonstrates that there are many kinds of membranes with functionalized non-fluorinated backbones that have been investigated the recent years in order to substitute the expensive Nafion® PEM with properties (especially proton conductivity, methanol permeability, and mechanical stability) superior to those of Nafion. The last category of polymeric PEMs is the acid-base polymer systems, which are presented in the following section.

11.4.3 Acid–Base Polymer Systems

The final category of polymeric PEMs to be reviewed in this section comprises non-functionalized polymers with basic character doped with proton conducting acids such as phosphoric acid. Pioneering work was performed in this area by

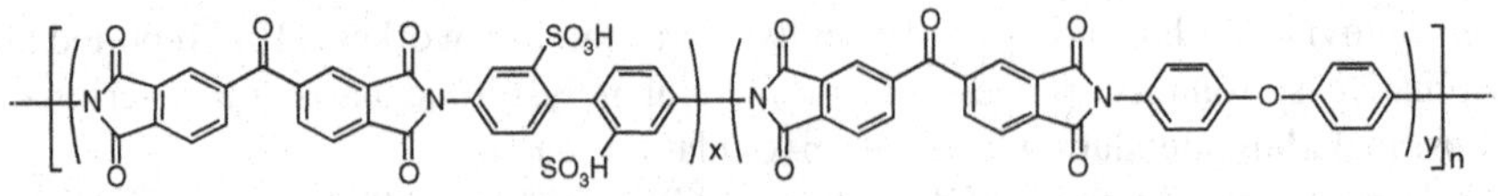

Fig. 11.11 Polymer structure of sulfonated polyimide (SPI)

Savinell and co-workers, [105–107], who have principally studied phosphoric acid doped polybenzimidazole (PBI). Similar systems have been reported by He and co-workers [108], who in addition report the conductivities of PBI based membranes doped with PTA and zirconium hydrogen phosphate [108]. Acid-base interactions in entirely polymeric systems have been reported by Kerres and co-workers [102], who prepared and studied several membranes prepared by blending polymers with acidic (sulfonated–PEEK, sulfonated polyethersulfone) and basic (polybenzimidazole, poly-vinylpyridine) characteristics. Selected acid-base polymer systems are discussed in the following.

11.4.3.1 Acid-PBI

The basic character of the PBI polymer allows doping levels of up to ca. 50 wt%. Two routes to the complexation of H_3PO_4 by PBI have been reported: (1) PBI films are immersed in an acid solution of molarity M for time t [107,109,110], and (2) films are cast directly from a solution of the polymer and phosphoric acid in a suitable solvent [111]. As a result, the doped films are produced directly, and the preparation time is reduced [112,113]. The conductivity of such systems is between 5×10^{-3} and 2×10^{-2} S cm^{-1} at room temperature [111], and even $3{:}5 \cdot 10^{-2}$ S cm^{-1} at 190°C [110], depending on the quantity of acid in the complex PBI/H_3PO_4. The nature of the acid influences the conductivity of doped PBI, and after contact with acid of high concentration (11 mol dm^{-3}) the conductivity follows the order H_2SO_4 > H_3PO_4 > HNO_3 > $HClO_4$ > HCl [114]. Recently, the dependence of the room temperature conductivity of PBI/H_3PO_4 and PBI/H_2SO_4 films has been investigated as a function of the concentration of acid solution and their immersion time in acid solution. According to Jones and Rozière [80], two types of behavior are observed. The first type of membrane, prepared at shorter doping times, displays conductivity in the range 10^{-5}–10^{-4} S cm^{-1}, whereas the second, with conductivity $>10^{-3}$ S cm^{-1}, is formed after more prolonged immersion. There is a switch-over from one state to the other which occurs after 10^-11 h in H_3PO_4 and after 2^-3 h in H_2SO_4. Most notably, the crossover to more highly conducting system is independent of the concentration of the acid solution [100].

11.4.3.2 Poly-(Diallyldimethylammonium-Dihydrogenphosphate) (PAMA$^+$- $H_2PO_4^-$)

The synthesis, thermal, mechanical, and conduction properties of blends of a cationic polyelectrolyte, poly-(diallyldimethylammonium-dihydrogenphosphate), 'PAMA$^+$ $H_2PO_4^-$', and phosphoric acid are reported [115]. Blends of 'PAMA$^+$·$H_2PO_4^-$·$x$$H_3PO_4$' with $0.5 \leq x \leq 2.0$ can be cast into amorphous films, which are stable up to 150°C. DSC results show that the softening temperatures of the blends decrease from 126°C for $x = 0.5$ to −23°C for $x = 2.0$. Furthermore, the dc conductivity increases with x and reaches 10^{-4} S cm^{-1} at ambient temperature and 10^{-2} S cm^{-4} at 100°C for PAMA$^+$ H$_2$PO$_4^-$·2H$_3$PO$_4$. The ^{1}H- and ^{31}P-self-diffusion coefficients of PAMA$^+$·H$_2$PO$_4^-$·xH$_3$PO$_4$ for x = 1,2 were determined by PFG-NMR. The general result is that the phosphate moieties are considerably more immobilized in the blends as compared with H$_3$PO$_4$. This immobilization effect is more pronounced in blends with low phosphoric acid content and decreases with increasing temperature [115].

11.4.3.3 Polyacrylamide

Rodriguez and co-workers [116] showed that anhydrous mixtures of polyacrylamide (PAAM) with phosphoric acid H$_3$PO$_4$ have conductivity better than 10^{-3} S cm^{-1} at 27°C for acid concentrations of 1.5–2 moles per polymer repeat unit. Infrared spectroscopic studies of these blends indicate that the amide groups of PAAM are not protonated by H$_3$PO$_4$, but the inter- and intra-chain C = O...H-N interactions are replaced by hydrogen bonds with the acid, which has as result only a moderate perturbation to the hydrogen-bond network needed for proton migration. This is the reason why the high conductivity of molten H$_3$PO$_4$ is less lowered with PAAM than with other basic polymers.

Furthermore, protonic transport in polyacrylamide hydrogels doped with H$_3$PO$_4$ or H$_2$SO$_4$ has been studied [117]. These hydrogels exhibit room temperature conductivities greater than 10^{-2} S cm^{-1}, which is increased with temperature to 10^{-1} S cm^{-1} at 100°C. It is shown from conductivity and FT-IR studies that the concentration of acids and water influence the proton transport mechanisms. Long-time conductivity studies performed at temperatures between 70° and 100°C indicate "drying" of hydrogels, which results in a decrease in conductivity. Finally, Pyo and Bard [118] satisfactorily elaborated PAAM blend films.

11.4.3.4 Poly (Ethylenimine)

Hydrophilic polymers such as poly (vinyl alcohol) or polyoxyethylene can be impregnated with a mineral acid and behave as solid proton conductors [119]. Several researchers [120] showed that polyethylenimine–sulfuric acid and phosphoric acid systems were well-behaved solid proton conductors under anhydrous conditions. Further information regarding these membranes is available in the literature [120]. Finally, Cakmak and Bicaksi [121] satisfactorily elaborated PEEK/PEI blend films.

With the exception of the inorganic acid doped PBI system, no membrane (such as PEO [119], PVA [122], polyacrylamide (PAAM) [116,117,123], polyethylenimine (PEI) [120], and poly diallyl-dimethyl-ammonium-dihydrogen phosphate, PAMA$^+$-H$_2$PO$_4^-$ [115] has demonstrated conductivities superior to Nafion$^®$ at temperatures >100°C and low relative humidities (ambient pressure operating conditions). Most of these acid-polymers blend exhibit proton conductivity <10^{-3} S cm^{-1} at room temperature. High acid contents result in high conductivity but poor mechanical stability, especially at temperatures >100°C. To improve the mechanical strength, efforts in this field have been made by (1) cross-linking of polymers (e.g., PEI) [120]; (2) using high T_g polymers such as PBI and polyoxadiazole (POD) [87]; and (3) adding inorganic filler and/or plasticizer, as recently reviewed by Lassegues and co-workers [124].

The preceding review, although by no means comprehensive, amply demonstrates that there has been a tremendous amount of research in recent years targeted at preparing inexpensive PEMs with properties (especially proton conductivity, methanol permeability, and mechanical stability) superior to those of Nafion®.

11.5 Electrode Designs

The design of electrodes for PEFCs is a delicate balancing of transport phenomena. Conductance of gas, electrons, and protons must be optimized to provide efficient transport to and from the electrochemical reaction sites. This is accomplished through careful consideration of the volume of conducting media required by each phase and the distribution of the respective conducting network. This review is a survey of recent literature with the objective of identifying common components, designs, and assembly methods for PEFC electrodes.

There are two widely employed electrode designs: the PTFE-bound and thin-film electrodes. Emerging methods include those featuring catalyst layers formed with electrodeposition and vacuum deposition (sputtering) [125].

The most common electrode design currently employed is the thin-film design, characterized by the thin Nafion® film that binds carbon-supported catalyst particles. The thin Nafion® layer provides the necessary proton transport in the catalyst layer. However, this is a significant improvement over the PTFE-bound catalyst layer, which requires the less effective impregnation of Nafion®. Sputter deposited catalyst layers have been shown to provide some of the lowest catalyst loadings, as well as the thinnest layers. The short conduction distance of the thin sputtered layer dissipates the requirement of a proton-conducting medium, which can simplify production. The performance of the state of the art sputtered layer is only slightly lower than that of the present thin-film convention [125].

11.5.1 PTFE-bound Methods

In these catalyst layers, the catalyst particles are bound by a hydrophobic PTFE structure commonly cast to the diffusion layer. This method was able to reduce the platinum loading of prior PEM fuel cells by a factor of 10; from 4 to 0.4 mg cm^{-2} [126]. In order to provide ionic transport to the catalyst site, the PTFE-bound catalyst layers are typically impregnated with Nafion® by brushing or spraying. Even though the platinum utilization in PTFE-bound catalyst layers was approximately

20% [127,128], researchers continued working on developing new strategies for Nafion® impregnation [129].

Ticianelli and co-workers [126] fabricated the low-platinum loading PEM fuel cells featuring PTFE-bound catalyst layers. Cheng and co-workers [128] developed conventional PTFE-bound catalyst layer electrodes for direct comparison with the current thin-film method. The typical process employed for forming the PTFE-bound catalyst layer MEA in their study is detailed in the following [125].

1. Pt/C catalyst particles are mechanically mixed in a solvent.
2. PTFE emulsion is added.
3. The slurry is coated onto the wet-proofed carbon paper using a coating apparatus, such as screen printing.
4. The electrodes are dried at room temperature and then baked.
5. Nafion® solution is brushed onto the electrocatalyst layer.
6. The Nafion®-impregnated electrodes are dried in ambient air.
7. Once dry, the electrodes are bonded to the polymer electrolyte membrane through hot pressing to complete the membrane electrode assembly.

11.5.2   Thin-Film Methods

In his 1995 patent, Wilson and co-workers described the thin-film technique for fabricating catalyst layers for PEM fuel cells with catalyst loadings <0.35 mg cm^{-2}. In this method the hydrophobic PTFE employed to bind the catalyst layer is replaced with hydrophilic perfluorosulfonate ionomer (Nafion®). As a result, the binding material in the catalyst layer is composed of the same material as the membrane. Thin-film catalyst layers have been found to operate at almost twice the power density of PTFE-bound catalyst layers. This corresponds to an increase in active area utilization from 22% to 45.4% when a Nafion®-impregnated and PTFE-bound catalyst layer is replaced with a thin-film catalyst layer [128]. Moreover, thin-film MEA manufacturing techniques are more established and applicable to stack fabrication [130]. However, utilization of 45% suggests that there is still significant potential for improvement.

The typical procedure for forming a thin-film catalyst layer on a PTFE blank is as follows:

1. A solution of solubilized perfluorosulfonate ionomer (e.g., Nafion®) is combined with Pt/C support catalyst.
2. After the addition of water and glycerol, the solution is mixed and sonicated.
3. The viscosity and thickness of the ink thus prepared are adjusted by evaporating extra solvent.
4. The carbon–water–glycerol ink is printed onto the PTFE blank.
5. The blank is dried–the coating process is repeated until the desired loading is attained.
6. If desired, the thin film electrode can be applied directly onto the membrane.

11.5.3   Vacuum Deposition Methods

The vacuum deposition methods include chemical vapor deposition (CVD), physical (PVD) or thermal vapor deposition, and sputtering. PVD coatings involve atom-by-atom, molecule-by-molecule, or ion deposition of various materials on solid substrates in vacuum systems.

- **Thermal evaporation** uses the atomic cloud formed by the evaporation of the coating metal in a vacuum environment to coat all the surfaces in the line of sight between the substrate and the target (source). It is often used in producing thin 0.5 μm coatings or a very thick 1-mm layer of heat-resistant materials, such as MCrAlY—a metal, chromium, aluminum, and yttrium alloy.
- During **sputtering**, a thin catalyst layer is deposited onto either the membrane or gas diffusion layer and the product has high performance combined with a low Pt loading. The catalyst layer is in such intimate contact with the membrane that the need for ionic conductors in the catalyst layer is resolved [131]. The success of the sputtering method on reducing platinum loading depends on the reduction in the size of catalyst particles <10 nm. State of the art thin-film electrodes feature Pt loading of 0.1 mg cm^{-2} resolved [131]. However, the performance of a fuel cell with sputtered catalyst layer can vary depending on the thickness of the sputtered catalyst layer [132].
- **Chemical Vapor Decomposition (CVD)** is capable of producing thick, dense, ductile, and good adhesive coatings and in contrast to the PVD coating, the CVD can coat all surfaces of the substrate. Conventional CVD coating process requires a metal compound that will volatilize at a low temperature and decompose to a metal when in contact with the substrate at higher temperature. The most well-known example of CVD is the nickel carbonyl ($NiCO_4$) coating as thick as 2.5°mm.

11.5.4   Electrodeposition Methods

In their 1992 patent, Vilambi Reddy and co-workers [133] described the fabrication of electrodes with low platinum loading in which the platinum was electrodeposited into their uncatalyzed carbon substrate in a commercial plating bath. This carbon substrate consisted of a hydrophobic porous carbon paper that was impregnated with dispersed carbon particles and PTFE. Nafion® was also impregnated onto the side of the carbon substrate that was to be catalyzed. The Nafion®-coated carbon paper was placed in a commercial platinum acid-plating bath, along with a platinum counter electrode. The face of the substrate that was not coated with Nafion® was masked with some form of a non-conducting film, to ensure that platinum would only be deposited in regions impregnated with Nafion®. Hence, when an interrupted dc current was applied to the electrodes in the plating bath, catalyst ions would pass through the Nafion® to the carbon particles and be deposited only where protonic and electronic conduction coexists. This method is able to produce electrodes featuring platinum loadings of 0.05 mg cm^{-2}, which is a significant reduction in loading from the state of the art thin-film electrode.

11.6   MEA Development

The preparation of MEAs constitutes a vital part of fuel cell evaluation, with the performance of the fuel cell strongly dependent on the quality of the MEA prepared. The MEA consists of the PEM of a given thickness (usually between 25 and 200 μm), two electrodes made from Pt or Pt-Ru alloys (either as unsupported "blacks" or supported on carbon) combined with an ionomeric binder, and porous gas diffusion layers (GDLs) to facilitate reactant gas transport to the electrodes. The electrodes may be directly applied on to the surface of the PEM, or may be applied on to the porous carbon gas diffusion layer and subsequently attached by hot-pressing on to the PEM. In the latter case, the combined electrode and GDL is termed a "gas diffusion electrode." The presence of an ionomeric binder in the electrode is vital to ensure that proton transport from the reactive sites of the electrocatalyst to the membrane interface and vice versa proceeds with minimal resistive losses. In the interests of membrane electrode interfacial stability, it is advisable to use the same ionomeric material in the PEM and electrode.

The electrode is generally applied (either on the PEM or PTFE blank or on the GDL) by first preparing an ink comprising the catalyst particles dispersed in a solvent (typically methanol) and an appropriate quantity of the ionomeric binder. This ink may be applied directly on to either the PEM or GDL by spraying using an air brush. An alternate technique for MEA preparation involves first applying the ink to a Teflon® blank, evaporating the solvent to yield a film, and transferring the film on to the PEM by a hot-pressing process ("decal transfer"). The application of ink to the blank may be carried out by painting with a brush or a screen printing process. The screen printing process may also be employed for direct transfer of ink to the PEM or GDL. A good guide to MEA preparation methods is presented by Kocha [134].

The lifetime of the MEA is strongly dependent on the heat treatment imparted during preparation [135]. Preparing MEAs by using the membrane in the Na$^+$ form and catalyst containing tetra butyl ammonium (TBA$^+$) substituted Nafion® permits the MEA to be subject to high temperatures and pressures, following which the treated MEA can be protonated using dilute H$_2$SO$_4$ to enable the Nafion® in the membrane and electrode to regain its protonic form. Such pretreatment has been shown to improve endurance at 80°C. Using different ion exchanged forms of the ionomer in the membrane and the electrode improves the melt processability of the ionomer by reducing the PTFE-like crystallinity in the backbone and minimizing the strength of the "physical cross-links" formed due to ion aggregation in the hydrophilic phase of the ionomer [136]. The improved melt processability permits the application of higher temperatures and pressures without degrading the ionomer, and also causes the ionomer to flow at the interface, thereby leading to good intermixing and the formation of a stable interface upon cooling. The improved endurance is attributed to the improved interfacial stability realized by such treatment. By comparing the endurance of MEAs processed at different temperatures, it can be shown that the increased processing temperature was the predominant cause for improved endurance. Evidently, any new PEMs developed should consider and address the twin interlinked factors of compatibility of electrode ionomer and membrane/electrode interfacial stability to ensure that a MEA will be viable from an endurance standpoint.

To sum up, polymer electrolyte fuel cells (PEFCs) have a variety of applications in the stationary, mobile, and automotive power sectors. Existing membrane technology presently permits fuel cell operation at temperatures <100°C under fully saturated conditions. However, several advantages result by operating a PEFC at elevated temperatures (>100°C) and lower relative humidities, such as easier heat rejection rates, higher-quality waste heat, and improved CO tolerance by the anode electrocatalyst. In an attempt to extend the operating range of the polymer electrolyte membrane, modifications of the Nafion® membranes have been investigated extensively.

Sulfonated hydrocarbon membranes have been investigated with great interest in recent years as a potential substitute for Nafion®. It is interesting that some of the sulfonated hydrocarbons exhibit high conductivity for potential operation at temperatures >100°C, but it is unclear if the conductivity can be sustained at low relative humidities. This category of membranes includes organic-inorganic composite membranes based on these alternative polymers that demonstrate improved water retention, methanol permeation, and mechanical strength. Finally, another class of proton-conducting membranes with good performance at high temperature is the acid-base polymers. Phosphoric-acid–doped PBI and ionically cross-linked acid-base blends, among others, have received the most attention.

The advances made in the fabrication of electrodes for PEM fuel cells from the PTFE-bound catalyst layers of almost 20 years ago to the present technology are briefly discussed. The most common form of electrode today features a thin-film catalytic layer with a well-defined (although not 100% utilized) boundary between reactant gas pathways, ionomer, and active sites. These thin-film electrodes have significantly increased performance and reduced the level of platinum loading required. Thin sputtered layers have shown promise for low catalyst loading with adequate performance. Electrodeposition methods, whose main advantages lie in their ability to mass-produce electrodes in a commercial plating bath and deposit catalyst only where electronic and protonic conduction exists, are briefly discussed. Finally, the relationship between MEA processing and the durability of the membrane/electrode interface and hence the fuel cell as a whole is presented. Needless to say, much work in membrane, electrode, and MEA technology remains before fuel cells can be made sufficiently durable and inexpensive, but it is heartening to note that considerable research efforts are ongoing in academia and industry to help alleviate existing bottlenecks to permit fuel cell commercialization.

References

1. Blomen, L. J. M. J., and Mugerwa, M. N., 1993, *Fuel Cell Systems*, Plenum Press, New York, pp. 493–530.
2. Prater, K. B., 1994, Polymer electrolyte fuel cells: a review of recent developments, *J. Power Sources* **51**: 129–144.
3. Liebhafsky, H. A., and Cairns, E. J., 1969, *Fuel Cells and Fuel Batteries: A Guide to Their Research and Development*, John Wiley and Sons, New York, pp. 1–692.
4. Zawodzinski, T. A., Derouin, C., Radzinski, S., Sherman, R. J., Springer, T. E., Gottesfeld, S., and Smith, V. T., 1993a, Water uptake by and transport through Nafion® 117 membranes, *J. Electrochem. Soc.* **140**: 1041–1047.
5. Zawodzinski, T. A., Lopez, C., Jestel, R., Valerio, J., Gottesfeld, S., and Davey, J., 1993b, A Comparative study of water uptake by and transport through ionomeric fuel cell membranes, *J. Electrochem. Soc.* **140**: 1981–1985.
6. Randin, J. P., 1982, Ion-containing polymers ass semisolid electrolytes in WO_3-based electrochromic devices, *J. Electrochem Soc.* **129**: 1215–1220.
7. Anantaraman, A. V., and Gardner, C. L., 1996, Studies on ion-exchange membranes. Part 1. Effect of humidity on the conductivity of Nafion®, *J. Electroanal. Chem.* **414**: 115–120.
8. Zawodzinski, T. A., Neeman, M., Sellerud, T., and Gottesfeld, S., 1991, Determination of water diffusion coefficients in perfluorosulfonate ionomeric membranes, *J. Phys. Chem.* **95**: 6040–6044.
9. Sone, Y., Ekdunge, P., and Simonsson, D., 1996, Proton conductivity of Nafion® 117 as measured by a four-electrode AC impedance method, *J. Electrochem. Soc.* **143**: 1254–1259.
10. Fontanella, J. J., McLin, M. G., Wintersgill, M. C., Calame, J. P., 1993, Electrical impedance studies of acid form NAFION® membranes, *Solid State Ionics* **66**: 1–4.
11. Ludwigsson, M., Lindgren, J., and Tegenfeldt, T., 2000, FTIR study of water in cast Nafion films, *Electrochim. Acta* **45**: 2267–2271.
12. Haubold, H. G., Vad, Th., Jungbluth, H., and Hiller, P., 2001, Nano structure of NAFION: a SAXS study, *Electrochim. Acta* **46**: 1559–1563.
13. Deng, D., Wilkie, A., Moore, R. B., and Mauritz, K. A., 1998, TGA-FTIR investigation of the thermal degradation of Nafion® and Nafion®/[silicon oxide]-based nanocomposites, *Polymer* **39**: 5961–5972.
14. Zawodzinski, T. A., Springer, T. E., Uribe, F., and Gottesfeld, S., 1993c, Characterization of polymer electrolytes for fuel cell applications, *Solid State Ionics* **60**: 199–211.
15. Rikukawa, M., Inagaki, D., Kaneko, K., Takeoka, Y., Ito, I., Kanzaki, Y., and Sanui, K., 2005, Proton conductivity of smart membranes based on hydrocarbon polymers having phosphoric acid groups, *J. Molecular Structure* **739**: 153–161.
16. Eisenberg, A., 1970, Clustering of ions in organic polymers. A theoretical approach, *Macromolecules* **3**: 147–154.
17. Hsu, W. Y., and Gierke, T. D., 1982, Elastic theory for ionic clustering in perfluorinated ionomers, *Macromolecules* **15**: 101–105.
18. Hsu, W. Y., and Gierke, T. D., 1983, Ion transport and clustering in nafion perfluorinated membranes, *J. Membr. Sci.* **13**: 307–326.
19. Eisenberg, A., Hird, B., and Moore, R. B., 1990, A new multiplet-cluster model for the morphology of random ionomers, *Macromolecules* **23**: 4098–4107.
20. Mauritz, K. A., and Moore, R. B., 2004, State of understanding of nafion, *Chem. Rev.* **104**: 4535–4586.
21. Moore, R. B., and Martin, C. R., 1986, Procedure for preparing solution-cast perfluorosulfonate ionomer films and membranes, *Anal. Chem.* **58**: 2569–2570.
22. Moore, R. B., and Martin, C. R., 1988, Chemical and morphological properties of solution-cast perfluorosulfonate ionomers, *Macromolecules* **21**: 1334–1339.
23. Gebel, G., Aldebert, P., and Pineri, M., 1987, Structure and related properties of solution-cast perfluorosulfonated ionomer films, *Macromolecules* **20**: 1425–1428.

24. Zook, L. A., and Leddy, J., 1996, Density and solubility of Nafion: recast, annealed, and commercial films, *Anal. Chem.* **68**: 3793–3796.
25. Costamagna, P., and Srinivasan, S., 2001, Quantum jumps in the PEMFC science and technology from the 1960s to the year 2000: part I. Fundamental scientific aspects, *J. Power Sources* **102**: 242–252.
26. Li, L., Zhang, J., and Wang, Y., 2003, Sulfonated poly(ether ether ketone) membranes for direct methanol fuel cell, *J. Membr. Sci.* **226**: 159–167.
27. Alberti, G., and Casciola, M., 1997, Layered metal^{IV} phosphonates, a large class of inorgano-organic proton conductors, *Solid State Ionics* **97**: 177–186.
28. Alberti, G., and Casciola, M., 2001, Solid state protonic conductors, present main applications and future prospects, *Solid State Ionics* **145**: 3–16.
29. Clearfield, A., and Smith, J., 1969, Crystallography and structure of alpha-zirconium bis(monohydrogen orthophosphate) monohydra, *Inorg. Chem.* **8**: 431–436.
30. Abe, Y., Li, G., Nogami, M., Kasuga, T., and Hench, L. L., 1996, Superprotonic conductors of glassy zirconium phosphates, *J. Electrochem. Soc.* **143**: 144–147.
31. Glipa, X., El Haddad, M., Jones, D. J., and Rozière, J., 1997a, Synthesis and characterisation of sulfonated polybenzimidazole: a highly conducting proton exchange polymer, *Solid State Ionics* **97**: 323–331.
32. Glipa, X., Leloup, J. M., Jones, D. J., and Roziere, J., 1997b, Enhancement of the protonic conductivity of α-zirconium phosphate by composite formation with alumina or silica, *Solid state Ionics* **97**: 227–232.
33. Alberti, G., Casciola, M., and Palombari, R., 2000, Inorgano-organic proton conducting membranes for fuel cells and sensors at medium temperatures, *J. Membr. Sci.* **172**: 233–239.
34. Nakamura, O., Ogino, I., and Kodama, T., 1981, Temperature and humidity ranges of some hydrates of high-proton-conductive dodecamolybdophosphoric acid and dodecatungstophosphoric acid crystals under an atmosphere of hydrogen or either oxygen or air, *Solid State Ionics* **3–4**: 347–351.
35. Giordano, N., Staiti, P., Hocevar, S., and Arico, A. S., 1996, High performance fuel cell based on phosphotungstic acid as proton conducting electrolyte, *Electrochim. Acta* **41**: 397–403.
36. Staiti, P., Hocevar, S., and Giordano, N., 1997, Fuel cells with $H_3PW_{12}O_{40}$ $29H_2O$ as solid electrolyte, *Int. J. Hydrogen Energy* **22**: 809–814.
37. Staiti, P., Hocevar, S., and Passalacqua, E., 1997, P18 activity and stability tests in phosphotungstic acid electrolyte fuel cell, *J. Power Sources* **65**: 281–282.
38. Malhotra, S., and Datta, R., 1997, Membrane-supported nonvolatile acidic electrolytes allow higher temperature operation of proton-exchange membrane fuel cells, *J. Electrochem. Soc.* **144**: L23–L26.
39. Savadogo, O., and Tazi, B., 2000, Parameters of PEM fuel-cells based on new membranes fabricated from Nafion®, silicotungstic acid and thiophene, *Electrochim. Acta* **45**: 4329–4339.
40. Savadogo, O., and Tazi, B., 2001, Effect of various heteropolyacids (HPAs) on the characteristics of Nafion HPAS membranes and their H_2/O_2 polymer electrolyte fuel cell parameters, *J. New Mat. Electrochem. Syst.* **4**: 187–191.
41. Ramani, V., Kunz, H. R., and Fenton, J. M., 2005a, Effect of particle size reduction on the conductivity of Nafion®/phosphotungstic acid composite membranes, *J. Membr. Sci.* **266**: 110–114.
42. Ramani, V., Kunz, H. R., and Fenton, J. M., 2005b, Stabilized heteropolyacid/Nafion® composite membranes for elevated temperature/low relative humidity PEFC operation, *Electrochim. Acta* **50**: 1181
43. Baranov, A. I., Merinov, B. V., Tregubchenko, A. V., Khiznichenko, V. P., Shuvalov, L. A., and Schagina, N. M., 1989, Fast proton transport in crystals with a dynamically disordered hydrogen bond network, *Solid State Ionics* **36**: 279–282.
44. Ponomareva, V. G., Uvarov, N. F., Lavrova, G. V., and Hairetdinov, E. F., 1996, Composite protonic solid electrolytes in the $CsHSO_4$-SiO_2 system, *Solid State Ionics* **90**: 161–166.

45. Ponomareva, V. G., and Lavrova, G. V., 1998, Influence of dispersed TiO_2 on protonic conductivity of $CsHSO_4$, *Solid State Ionics* **106**: 137–141.

46. Crisholm, C. R. I., and Haile, S. M., 2000, Superprotonic behavior of $Cs_2 (HSO_4)(H_2PO_4)$ – a new solid acid in the $CsHSO_4$–CsH_2PO_4 system, *Solid State Ionics* **136–137**: 229–241.

47. Haile, S. M., Lentz, G., Kreuer, K. D., and Maier, J., 1995, Superprotonic conductivity in $Cs_3 (HSO_4)_2 (H_2PO_4)$, *Solid State Ionics* **77**: 128–134.

48. Boysen, D. A., Chrisholm, C. R. I., Haile, S. M., and Narayanan, R. S., 2000, Polymer solid acid composite membranes for fuel-cell applications, *J. Electrochem. Soc.* **147**: 3610–3613.

49. Bauer, F., and Willert-Porada, M., 2004, Microstructural characterization of Zr-phosphate–Nafion® membranes for direct methanol fuel cell (DMFC) applications, *J. Membr. Sci.* **233**: 141–149.

50. Datta, R., Jalani, N. H., and Dunn, K., 2005, Synthesis and characterization of Nafion-MO_2 (M = Zr, Si, Ti) nanocomposite membranes for higher temperature PEM fuel cells, *Electrochim. Acta* **51**: 553–560.

51. Watanabe, M., Uchida, H., Seki, Y., and Emori, M., 1998, Polymer electrolyte membranes incorporated with nanometer-size particles of pt and/or metal-oxides: experimental analysis of the self-humidification and suppression of gas-crossover in fuel cells, *J. Phys. Chem. B* **102**: 3129–3137.

52. Adjemian, K. T., Lee, S. J., Srinivasan, S., Benziger, J., and Bocarsly, A. B., 2002, Silicon oxide nafion composite membranes for proton-exchange membrane fuel cell operation at 80–140°C, *J. Electrochem. Soc.* **149**: A256–A261.

53. Mauritz, K. A., 1998, Organic-inorganic hybrid materials: perfluorinated ionomers as sol-gel polymerization templates for inorganic alkoxides, *Mater. Sci. Eng. C* **6**: 121–133.

54. Mauritz, K. A., and Warren, R. M., 1989, Microstructural evolution of a silicon oxide phase in a perfluorosulfonic acid ionomer by an in situ sol-gel reaction. 1. Infrared spectroscopic studies, *Macromolecules* **22**: 1730–1734.

55. Stefanithis, I. D., and Mauritz, K. A., 1990, Microstructural evolution of a silicon oxide phase in a perfluorosulfonic acid ionomer by an in situ sol-gel reaction. 3. Thermal analysis studies, *Macromolecules* **23**: 2397–2402.

56. Klein, L. C., Daiko, Y., Aparicio, M., and Damay, F., 2005, Methods for modifying proton exchange membranes using the sol-gel process, *Polymer* **46**: 4504–4509.

57. Deng, Q., Hu, Y., Moore, R. B., McCormick, C. L., and Mauritz, K. A., 1997, Nafion/ORMOSIL Hybrids via in situ sol-gel reactions. 3. Pyrene fluorescence probe investigations of nanoscale environment, *Chem. Mater.* **9**: 36–44.

58. Zoppi, R. A., Yoshida, I. V. P., and Nunes, S. P., 1997, Hybrids of perfluorosulfonic acid ionomer and silicon oxide by sol-gel reaction from solution: morphology and thermal analysis, *Polymer* **39**: 1309–1315.

59. Zoppi, R. A., and Nunes, S. P., 1998, Electrochemical impedance studies of hybrids of perfluorosulfonic acid ionomer and silicon oxide by sol-gel reaction from solution, *J. Electroanal. Chem.* **445**: 39–45.

60. Harmer, M. A., Farneth, W. E., and Sun, Q., 1996, High surface area Nafion® resin/silica nanocomposites: a new class of solid acid catalyst. *J. Am. Chem. Soc.* **118**: 7708–7715.

61. Sun, Q., Farneth, W. E., and Harmer, M. A., 1996, Dimerization of α-methylstyrene (AMS) catalyzed by sulfonic acid resins: a quantitative kinetic study, *J Catal.* **164**: 62–69.

62. Doyle, M., and Rajendran, G., Perfluorinated membranes, 2003, in: *Handbook of Fuel Cells, Volume 3*, W. Vielstichm, A. Lamm,and H. A. Gasteiger, eds., John Wiley & Sons Ltd., New York, pp. 351–395.

63. Frotts, S. D., Gervasio, D., Zeller, R. L., and Savinell, R. F., 1991, Investigation of H_2 gas transport in recast Nafion films coated on platinum in hydrogen saturated 85% phosphoric acid, *J. Electrochem. Soc.* **138**: 3345–3349.

64. Heinzel, A., and Barragan, V. M., 1999, A review of the state-of-the-art of the methanol crossover in direct methanol fuel cells, *J. Power Sources* **84**: 70–74.

65. Bibler, N. E., and Orebaugh, E. G., 1976, Iron-catalyzed dissolution of polystyrenesulfonate cation-exchange resin in hydrogen peroxide, *Ind. Eng. Chem. Prod. Res. Dev.* **15**: 136–138.

66. Lufrano, F., Gatto, I., Staiti, P., Antonucci, V., and Passalacqua, E., 2001, Sulfonated polysulfone ionomer membranes for fuel cells, *Solid State Ionics* **145**: 47–51.
67. Wang, F., Hickner, M., Kim, Y. S., Zawodzinski, T. A., and McGrath, J. E., 2002, Direct polymerization of sulfonated poly(arylene ether sulfone) random (statistical) copolymers: candidates for new proton exchange membranes, *J. Membr. Sci.* **197**: 231–242.
68. Mukerjee, S., Zhang, L., and Chengsong, M., 2004, Oxygen reduction and transport characteristics at platinum and alternative proton conducting membrane interface, *J. Electroanal. Chem.* **568**: 273–291.
69. McGrath, J. E., Zawodzinski, T. A., Kim, Y. S., Dong, L., and Pivovar, B. S., 2002, Processing induced morphological development in hydrated sulfonated poly(arylene ether sulfone) copolymer membranes, *Polymer* **44**: 5729–5736.
70. McGrath, J. E., Zawodzinski, T. A., Kim, Y. S., Wang, F., Hickner, M. A., McCartney, S., Hong, Y. T., and Harrison, W., 2003, Effect of acidification treatment and morphological stability of sulfonated poly(arylene ether sulfone) copolymer proton-exchange membranes for fuel-cell use above 100°C, *J. Polym. Sci., Part B: Polym. Phys.* **41**: 2816–2828.
71. McGrath, J. E., Kim, Y. S., Hickner, M. A., Dong, L., and Pivovar, B. S., 2004a, Sulfonated poly(arylene ether sulfone) copolymer proton exchange membranes: composition and morphology effects on the methanol permeability, *J. Membr. Sci.* **243**: 317–326.
72. McGrath, J. E., Sumner, M. J., Harrison, W. L., Weyers, R. M., Kim, Y. S., Riffle, J. S., Brink, A., and Brink, M. H., 2004b, Novel proton conducting sulfonated poly(arylene ether) copolymers containing aromatic nitriles, *J. Membr. Sc.* **239**: 199–211.
73. Kobayashi, T., Rikukawa, M., Sanui, K., and Ogata, N., 1998, Proton-conducting polymers derived from poly(ether-etherketone) and poly(4-phenoxybenzoyl-1,4-phenylene), *Solid State Ionics* **106**: 219–225.
74. Xing, P., Robertson, G. P., Guiver, M. D., Mikhailenko, S. D., Wang, K., and Kaliaguine, S., 2004, Synthesis and characterization of sulfonated poly(ether ether ketone) for proton exchange membranes, *J. Membr. Sci.* **229**: 95–106.
75. Swier, S., Ramani, V., Fenton, J. M., Kunz, H. R., Shaw, M. T., and Weiss, R. A., 2005, Polymer blends based on sulfonated poly(ether ketone ketone) and poly(ether sulfone) as proton exchange membranes for fuel cells, *J. Membr. Sc.* **256**: 122–133.
76. Yin, Y., Fang, J., Cui, Y., Tanaka, K., Kita, H., and Okamoto, K., 2003, Synthesis, proton conductivity and methanol permeability of a novel sulfonated polyimide from 3-(2′,4′-diaminophenoxy)propane sulfonic acid, *Polymer* **44**: 4509–4518.
77. Woo, Y., Oh, S., Kang, Y., and Jung, B., 2003, Synthesis and characterization of sulfonated polyimide membranes for direct methanol fuel cell, *J. Membr. Sci.* **220**: 31–45.
78. Genies, C., Mercier, R., Sillion, B., Cornet, N., Gebel, G., and Pineri, M., 2001, Soluble sulfonated naphthalenic polyimides as materials for proton exchange membranes, *Polymer* **42**: 359–373.
79. Gao, Y., Robertson, G. P., Guiver, M. D., Jian, X., Mikhailenko, K., and Wang, K., 2003, Sulfonation of poly(phthalazinones) with fuming sulfuric acid mixtures for proton exchange membrane materials, *J. Membr. Sci.* **227**: 39–50.
80. Jones, D. J., and Rozière, J., 2001, Recent advances in the functionalisation of polybenzimidazole and polyetherketone for fuel cell applications, *J. Membr. Sci.* **185**: 4158.
81. Staiti, P., Lufrano, F., Aricò, A. S., Passalacqua, E., and Antonucci, V., 2001a, Sulfonated polybenzimidazole membranes – preparation and physico-chemical characterization, *J. Membr. Sci.* **188**: 71–78.
82. Staiti, P., Aricò, A., Baglio, V., Lufrano, F., Passalacqua, E., and Antonucci, V., 2001b, Hybrid Nafion–silica membranes doped with heteropolyacids for application in direct methanol fuel cells, *Solid State Ionics* **145**: 101–107.
83. Guo, Q., Pintauro, P. N., Tang, H., and O'Connor, S., 1999, Sulfonated and crosslinked polyphosphazene-based proton-exchange membranes, *J. Membr. Sci.* **154**: 175–181.
84. Kreuer, K. D., 2001, On the development of proton conducting polymer membranes for hydrogen and methanol fuel cells, J. Membr. Sci. 185: 29–39.
85. Kerres, J. A., 2001, Development of ionomer membranes for fuel cells, *J. Membr. Sci.,* **185**: 3–27.

86. Alberti, G., Casciola, M., Massinelli, L., and Bauer, B., 2001, Polymeric proton conducting membranes for medium temperature fuel cells (110–160°C), *J. Membr. Sci.* **185**: 73–81.

87. Zaidi, S. M. J., Mikhailenko, S. D., Robertson, G. P., Guiver, M. D., and Kaliaguine, S., 2000, Proton conducting composite membranes from polyether ether ketone and heteropolyacids for fuel cell applications, *J. Membr. Sci.* **173**: 17–34.

88. Genova-Dimitrova, P., Baradie, B., Foscallo, D., Poinsignon, C., and Sanchez, J. Y., 2001, Ionomeric membranes for proton exchange membrane fuel cell (PEMFC): sulfonated polysulfone associated with phosphatoantimonic acid, *J. Membr. Sci.* **185**: 59–71.

89. Kim, Y., Wang, F., Hickner, M., Zawodzinski, T., and McGrath, J., 2003, Fabrication and characterization of heteropolyacid ($H_3PW_{12}O_{40}$)/directly polymerized sulfonated poly(arylene ether sulfone) copolymer composite membranes for higher temperature fuel cell applications, *J. Membr. Sci.* **212**: 263–282.

90. Ise, M., Kreuer, K. D., and Maier, F., 1999, Electroosmotic drag in polymer electrolyte membranes: an electrophoretic NMR study, *Solid State Ionics* **125**: 213–223.

91. Yeo, R. S., 1980, Dual cohesive energy densities of perfluorosulphonic acid (Nafion) membrane, *Polymer* **21**: 432–435.

92. Florjanczyk, Z., Wielgus-Barry, E., and Poltarzewski, Z., 2001, Radiation-modified Nafion membranes for methanol fuel cells, *Solid State Ionics* **145**: 119–126.

93. Kreuer, K. D., 1996, Proton conductivity: materials and applications, *Chem. Mater.* **8**: 610–641.

94. Wainright, J. S., Litt, M. H., and Savinell, R. F., 2001, High-temperature membranes, 2003, in: *Handbook of Fuel Cells, Volume 3*, W. Vielstichm, A. Lamm, and H. A. Gasteiger, eds., John Wiley & Sons Ltd., New York, pp. 436–446.

95. Choi, W. C., Kim, J. D., and Woo, S. I., 2001, Modification of proton conducting membrane for reducing methanol crossover in a direct-methanol fuel cell, *J. Power Sources* **96**: 411–414.

96. Wasmus, S., and Kuever, A., 1999, Methanol oxidation and direct methanol fuel cells: a selective review, *J. Electroanal. Chem.* **461**: 14–31.

97. Li, Q., He, R., Berg, R., Hjuler, H. A., and Bjerrum, N. J., 2004, Water uptake and acid doping of polybenzimidazoles as electrolyte membranes for fuel cells, *Solid State Ionics* **168**: 177–185.

98. Frank, G., 2003, *Proceedings of the 2nd European PEFC Forum*, Lucerne, Switzerland; pp. 749–752.

99. Bogdanovic, B., and Schwickardi, M., 1997, Ti-doped alkali metal aluminium hydrides as potential novel reversible hydrogen storage materials, *J. Alloys Compd.* **253–254**: 1–9.

100. Li, Q., He, R., Gao, J., Jensen, J. O., and Bjerrum, N. J., 2003, The CO Poisoning Effect in PEMFCs Operational at Temperatures up to 200°C, *J. Electrochem. Soc.* **150**: A1599–A1605.

101. Yang, C., Srinivasan, S., Arico, A. S., Creti, P., Baglio, V., and Antonucci, V., 2001, Composite Nafion/zirconium phosphate membranes for direct methanol fuel cell operation at high temperature, *Electrochem. Solid-State Lett.* **4**: A31–A34.

102. Kerres, J., Ullrich, A., Meier, F., and Häring, T., 1999, Synthesis and characterization of novel acid–base polymer blends for application in membrane fuel cells, *Solid State Ionics* **125**: 243–249.

103. Guo, Q., Fang, J., Watari, T., Tanaka, K., Kita, H., and Okamoto, K., 2002, Novel sulfonated polyimides as polyelectrolytes for fuel cell application. 2. Synthesis and proton conductivity of polyimides from 9,9-Bis(4-aminophenyl)fluorene-2,7-disulfonic acid, *Macromolecules* **35**: 6707–6713.

104. Fang, J., Guo, X., Harada, S., Watari, T., Tanaka, K., Kita, H., and Okamoto, K., 2002, Novel sulfonated polyimides as polyelectrolytes for fuel cell application. 1. Synthesis, proton conductivity, and water stability of polyimides from 4,4′-diaminodiphenyl ether-2,2′-disulfonic acid, *Macromolecules* **35**: 9022–9028.

105. Savinell, R., Yeager, E., Tryk, D., Landau, U., Wainright, J., Weng, D., Lux, K., Litt, M., and Rogers, C., 1994, A polymer electrolyte for operation at temperatures up to 200°C, *J. Electrochem. Soc.* **141**: L46–L48.

106. Wasmus, S., Valeriu, A., Mateescu, G., Tryk, D., and Savinell, R. F., 1995, Characterization of H_3PO_4-equilibrated Nafion® 117 membranes using ^1H and ^{31}P NMR spectroscopy, *Solid State Ionics* **80**: 87–92.

107. Wang, J., Savinell, R. F., Wainright, J., Litt, M., and Yu, H., 1996, A H_2/O_2 fuel cell using acid doped polybenzimidazole as polymer electrolyte, *Electrochim. Acta* **41**: 193–197.

108. He, R., Li, Q., Xiao, G., Bjerrum, N. J., 2003, Proton conductivity of phosphoric acid doped polybenzimidazole and its composites with inorganic proton conductors, *J. Membr. Sci.* **226**: 169–184.

109. Samms, S. R., Wasmus, S., and Savinell, R. F., 1996, Thermal stability of proton conducting acid doped polybenzimidazole in simulated fuel cell environments, *J. Electrochem. Soc.* **143**: 1225–1232.

110. Wainright, J. S., Wang, J. T., Weng, D., Savinell, R. F., and Litt, M., 1995, Acid-doped poly-benzimidazoles: a new polymer electrolyte, *J. Electrochem. Soc.* **142**: L121–L123.

111. Wainright, J. S., Savinell, R. F., and Litt, M., 1997, Acid doped polybenzimidazole as a polymer electrolyte for methanol fuel cells, in: *Proceedings of the Second International Symposium on New Materials for Fuel Cells and Modern Battery Systems*, O. Savadogo, and P. R. Roberge, eds., Montreal, Canada pp. 808–811.

112. Benicewicz, B. C., Duke, J. R., Hoisington, M. A., and Langlois, D. A., 1998, High temperature properties of poly(styreneco-alkylmaleimide) foams prepared by high internal phase emulsion polymerization, *Polymer* **39**: 4369–4378.

113. Lobato, J., Rodrigo, M. A., Linares, J. J., and Scott, K., 2006, Effect of the catalytic ink preparation method on the performance of high temperature polymer electrolyte membrane fuel cells, *J. Power Sources* **157**: 284–292.

114. Xing, B., and Savadogo, O., 2000, Hydrogen/oxygen polymer electrolyte membrane fuel cells (PEMFCs) based on alkaline-doped polybenzimidazole (PBI), *Electrochem. Commun.* **2**: 697–702.

115. Bozkurt, A., Ise, M., Kreuer, K. D., Meyer, W. H., and Wegner, G., 1999, Proton-conducting polymer electrolytes based on phosphoric acid, *Solid State Ionics* **125**: 225–233.

116. Rogriguez, D., Jegat, T., Trinquet, O., Grondin, J., and Lassegues, J. C., 1993, Proton conduction in poly (acrylamide)-acid blends, *Solid State Ionics* **61**: 195–202.

117. Stevens, J. R., and Raducha, D., 1997, Proton conducting gel/H3PO4 electrolytes, *Solid State Ionics* **97**: 347–358.

118. Pyo, M., and Bard, A. J., 1997, Scanning electrochemical microscopy. 35. Determination of diffusion coefficients and concentrations of $Ru(NH_3)_6^{3+}$ and methylene blue in polyacrylamide films by chronoamperometry at ultramicrodisk electrodes, *Electrochim. Acta* **42**: 3077–3083.

119. Donoso, P., Gorecki, W., Berthier, C., Defendini, F., Poinsignon, C., and Armand, M. B., 1988, NMR, conductivity and neutron scattering investigation of ionic dynamics in the anhydrous polymer protonic conductor $PEO(H_3PO_4)x$), *Solid State Ionics* **28–30**: 969–974.

120. Tanaka, R., Yamamoto, H., Shono, A., Kubo, K., and Sakurai, M., 2000, Proton conducting behavior in non-crosslinked and crosslinked polyethylenimine with excess phosphoric acid, *Electrochim. Acta* **45**: 1385–1389.

121. Cakmak, M., and Bicakci, S., 2002, Kinetics of rapid structural changes during eat setting of preoriented PEEK/PEI blend films as followed by spectral birefringence technique, *Polymer* **43**: 2737–2746.

122. Petty-Weeks, S., Zupancic, J. J., and Swedo, J. R., 1988, Proton conducting interpenetrating polymer networks, *Solid State Ionics* **31**: 117–125.

123. Wieczorek, W., and Stevens, J. R., 1996, Proton transport in polyacrylamide based hydrogels doped with H_3PO_4 or H_2SO_4, *Polymer* **38**: 2057–2065.

124. Lassegues, J. C., Grondin, J., Hernandez, M., and Maree, B., 2001, Proton conducting polymer blends and hybrid organic inorganic materials, *Solid State Ionics* **145**: 37–45.

125. Litster, S., and McLean, G., 2004, PEM fuel cell electrodes, *J. Power Sources* **130**: 61–76.

126. Ticianelli, E. A., Derouin, C. R., Redondo, A., and Srinivasan, S., 1988, Methods to advance technology of proton exchange membrane fuel cells, *J. Electrochem. Soc.* **135**: 2209–2214.

127. Murphy, O. J., Hitchens, G. D., and Manko, D. J., 1994, High power density proton-exchange membrane fuel cells, *J. Power Sources* **47**: 353–368.
128. Cheng, X., Yi, B., Han, M., Zhang, J., Qiao, Y., and Yu, J., 1999, Investigation of platinum utilization and morphology in catalyst layer of polymer electrolyte fuel cells, *J. Power Sources* **79**: 75–81.
129. Lee, S. J., Mukerjee, S., McBreen, J., Rho, Y. W., Kho, Y. T., and Lee, T. H., 1998, Effects of Nafion impregnation on performances of PEMFC electrodes, *Electrochim. Acta* **43**: 3693–3701.
130. Chun, Y. G., Kim, C. S., Peck, D. H., and Shin, D. R., 1998, Performance of a polymer electrolyte membrane fuel cell with thin film catalyst electrodes, *J. Power Sources* **71**: 174–178.
131. Cha, S. Y., and Lee, W. M., 1999, Performance of proton exchange membrane fuel cell electrodes prepared by direct deposition of ultrathin platinum on the membrane surface, *J. Electrochem. Soc.* **146**: 4055–4060.
132. O'Hayre, R., Lee, S. J., Cha, S. W., and Prinz, F. B., 2002, A sharp peak in the performance of sputtered platinum fuel cells at ultra-low platinum loading, *J. Power Sources* **109**: 483–493.
133. Vilambi-Reddy, N. R. K., Anderson, E. B., and Taylor, E. J., 1992, High utilization supported catalytic metal-containing gas-diffusion electrode, process for making it, and cells utilizing it, US Patent No. 5,084,144.
134. Kocha, S. S., Principles of MEA preparation, 2003, in: *Handbook of Fuel Cells, Volume 3*, W. Vielstichm, A. Lamm, and H. A., Gasteiger, eds., John Wiley & Sons Ltd., New York, pp. 538–565.
135. Wilson, M. S., Valerio, J. A., and Gottesfeld, S., 1995, Low platinum loading electrodes for polymer electrolyte fuel cells fabricated using thermoplastic ionomers, *Electrochim. Acta* **40**: 355–363.
136. Moore, R. B., Cable, K. M., and Croley, T. L., 1992, Barriers to flow in semicrystalline ionomers. A procedure for preparing melt-processed perfluorosulfonate ionomer films and membranes, *J. Membr. Sci.*, **75**: 7–14.

Chapter 12
Carbon-Filled Polymer Blends for PEM Fuel Cell Bipolar Plates

Leon L. Shaw

Abstract Carbon-filled polymer blends with a triple-continuous structure, consisting of a binary (or ternary) polymer blend and carbon particles, have great potential to provide injection moldable PEM fuel cell bipolar plates with superior electrical conductivity and sufficient mechanical properties. Four carbon nanotube (CNT)-filled polymer blends, i.e., CNT-filled polyethylene terephthalate (PET)/ polyvinylidene fluoride, PET/polypropylene, PET/nylon 6,6, and PET/high-density polyethylene blends, have been injection molded and characterized in terms of their microstructures, electrical conductivities, and mechanical properties. Effects of the thermodynamic driving force, rheology of the polymer blend, and injection molding conditions on the distribution of CNTs in the blends have been examined. The simultaneous improvements in the electrical conductivity and mechanical properties of carbon-filled polymer blends over carbon-filled polymers have been investigated based on the CNT distribution in the polymer blends. The results unambiguously indicate that the preferential location of CNTs in one of the continuous polymer phases in the polymer blend is highly desirable for both mechanical and electrical properties. Future directions in this emerging area are discussed.

12.1 Introduction

Polymer electrolyte membrane or proton exchange membrane (PEM) fuel cells are expected to provide power to automobiles, personal computers, wireless phones, and outdoor electric tools. This is thanks to their low temperature of operation, perfect CO_2 tolerance by the electrolyte, and a combination of high power density and high-energy conversion efficiency [1]. However, significant barriers are present before this fuel cell technology can be fully embraced for commercial applications. One of the major barriers is the high cost of PEM fuel cells, estimated at about $200 kW^{-1} [2]. Reductions in the cost to about $25 kW^{-1} are needed for PEM fuel cells to become economically advantageous over the internal combustion engine [2,3]. Recent technical cost analyses [2,4] indicate that the cost of either the platinum electrode or bipolar plates ranks first in the PEM fuel cell stack, depending on the

S.M.J. Zaidi, T. Matsuura (eds.) *Polymer Membranes for Fuel Cells*,
doi: 10.1007/978-0-387-73532-0, © Springer Science+Business Media, LLC 2009

model used to estimate the cost. Thus, widespread applications of PEM fuel cells rely heavily on the breakthrough in the cost reduction of both the electrode and bipolar plates.

Bipolar plates, the subject in this study, are significant parts of PEM fuel cell stacks. The overall efficiency of fuel cells is greatly dependent on the performance of the bipolar plates in the stack [3]. Accounting for about 80% of the total weight [4], bipolar plates are multifunctional components in a PEM fuel cell stack. One of the primary functions of bipolar plates is to provide the electrical connection between the cells, carrying the electrical current from cell to cell while preventing leakage of reactants and coolant. Because of this requirement, good electrical conductivity, especially in the direction perpendicular to the plate plane, is essential for bipolar plates. Bipolar plates with a series of flow channels also distribute fuel gases uniformly to a gas-diffusion layer and remove the heat from the active areas to achieve a homogeneous temperature distribution over the cells. Furthermore, some bipolar plates incorporate cooling channels to control the stack temperature. Thus, the bipolar plate material is required to be thermally conductive, impermeable to the fuel gases and air, resistant to the operation temperature, and chemically inert in the fuel cell environment. To reduce the weight and volume of the fuel cell stack, the bipolar plate material also needs to have reasonable mechanical strength. Finally, to make the PEM fuel cell a viable technology, bipolar plates should be fabricated at low cost. In short, the following criteria/properties for the desired bipolar plates and their materials have been proposed [5–7]:

- Electrical conductivity: plate resistance $<0.01 \ \Omega \ cm^2$.
- Thermal conductivity: as high as possible.
- Hydrogen/gas permeability: $<10^{-4} \ cm^3 \ s^{-1} \ cm^{-2}$.
- Corrosion resistance: corrosion rate $<0.016 \ A \ cm^{-2}$.
- Compressive strength: $>22 \ lb \ in^{-2}$.
- Density: $<5 \ g \ cm^{-2}$.
- Stack volume: $<1 \ l \ kW^{-1}$.
- Cost: material + fabrication $<US\$0.0045 \ cm^{-2}$.

12.2   Materials Investigated for Bipolar Plates

Extensive efforts have been made to investigate different materials that may satisfy all of the requirements of bipolar plates. Traditionally, PEM fuel cells have been constructed from Poco graphite bipolar plates, which have a high bulk electrical conductivity ($680 \ S \ cm^{-1}$) and are resistant to corrosion in the fuel cell environment. However, graphite suffers from being brittle, expensive, bulky, and difficult to machine [3,8]. Because of the brittleness of graphite, the bipolar plates have to be made several millimeters thick, which makes a fuel cell stack heavy and voluminous; the brittleness of graphite also drives the machining cost of the flow channels on bipolar plates to a prohibitive level-about $10/plate if both the material and machining costs are included [3]. In light of these issues, extensive research to replace graphite bipolar plates has been conducted. These different materials are briefly summarized in the following. For detailed accounts of these materials, readers are referred to original publications and recent review articles [3,5–26].

Flexible graphite foil is a thin, low-density, inexpensive material made from expanded natural graphite. Foils with a thickness between 0.2 and 2 mm have been produced for bipolar plate applications. The foil can be cut and embossed into the desired form at room temperature. It is electrically conducting, corrosion resistant, and self-sealing, which obviates the requirement for separate gaskets [8]. However, it is not mechanically strong, so that a stack built from this material would have to be protected from impact [8]. Other disadvantages of flexible graphite foils include the very limited formability, poor dimensional stability, and the slight permeability to gases [8,9].

Carbon-carbon composites have been investigated by Oak Ridge National Laboratory. The composite is fabricated by slurry molding a chopped-fiber perform, followed by sealing with chemically vapor-infiltrated carbon at $1,500°C$ for 4 h. The bulk conductivity achieved is $200–300$ S cm^{-1} and the surface resistivity is 12.2 ± 4.2 Ω cm^{-1} [3]. The PEM fuel cell testing indicates equivalent performance to that provided by the Poco graphite plates. Although low-weight, good conductivity, and high biaxial flexure strength, carbon-carbon composite is not expected to achieve ambitious cost targets for PEM fuel cell applications because of the slow and costly processing step of chemical vapor infiltration [9].

Metals such as titanium, aluminum, nickel, and stainless steels have been pursued for bipolar plate applications [5,8,10–12]. However, these research efforts met limited success because of the chemical instability of the metals in the fuel cell environment, especially when in contact with the acidic electrolytic membrane. Corrosion of the metal bipolar plate leads to a release of cations, which can both lead to an increase in membrane resistance and poisoning of the electrode catalysts [12]. The oxide film formed on the surface of the self-passivating metals also results in high voltage losses across the plate/macro-diffuser interface [8,11].

Coated metals have the potential to solve the corrosion problem described in the preceding. The coatings investigated include graphite, diamond-like carbon, conductive polymer, noble metals, metal nitrides, and metal carbides [7,8,11–13]. The key to the success of the coating approach is the formation of conductive defect-free coatings. Pinhole-free NbN/TiN and CrN/Cr_2N coatings on model Nb-Ti and Ni-Cr alloys have been demonstrated using a preferential thermal nitridation process [14,15]. The result shows promise with excellent corrosion resistance and negligible contact resistance. However, the high cost of the Nb-Ti and Ni-Cr alloys limits their applications for PEM fuel cell bipolar plates [14,15]. The formation of this type of coating on stainless or any other steel is yet to be demonstrated.

Carbon/graphite-filled polymers as alternatives for graphite bipolar plates have seen increasing interest because of their potential to offer the advantages of low cost, low weight, and easy manufacturing [16–23]. With proper selection of the polymer matrix the composite can provide chemical inertness and gas tightness. To obtain adequate electrical conductivities (>20 S cm^{-1}), high carbon and graphite loadings in the composite (typically > 50 vol%) are required [18–23]. As a result of such high carbon and graphite concentrations, compression molding or extrusion often becomes the choice of processing method [18–23], which adds costs to the composite. Injection molding has also been pursued but with lower carbon and graphite concentrations and thus at the expense of electrical conductivities [9].

Carbon-filled polymer blends have been explored recently as a novel approach to address issues of properties and processability simultaneously, thereby producing low-cost and high-performance bipolar plates [24–26]. The approach is aimed at the production of injection moldable carbon-filled polymer blends through a hierarchical microstructural design that leads to the formation of a triple-continuous microstructure. The results obtained so far have demonstrated the efficacy of the approach in improving the electrical conductivity and mechanical strength simultaneously, while maintaining injection moldability [24–26]. The principle and major results for carbon-filled polymer blends are described in the remainder of this chapter.

12.3   Concept of Carbon-Filled Polymer Blends

It is well known that the electrical conductivity of polymers loaded with conductive fillers, such as carbon black (CB), graphite particles (GPs), carbon nanotubes (CNTs), and metallic particles, exhibits a discontinuous increase with the filler loading. The phenomenon is explained in terms of the percolation theory [27]. When the concentration of the conductive filler reaches a critical value, termed the percolation threshold, a conductive path is formed in the composite along with a sudden jump in the electrical conductivity by several orders of magnitude [27]. However, even with this jump the conductivity obtained is still too low for bipolar plate applications. As such, filler concentrations much higher than the percolation threshold are required to raise the conductivity to the level suitable for bipolar plate applications [18–23]. Unfortunately, the conductivity derived in this way is obtained at the expense of processability and thus increases the manufacturing cost of bipolar plates.

An additional problem associated with high filler concentrations in polymers is the substantial reduction in the strength and ductility of the polymer composite. It is reported that the tensile strength of polymer composites increases initially with the addition of a small amount of fillers (~5–20 vol%), but decreases with higher filler loading [28–30]. Such phenomena have normally been attributed to the weak filler-matrix interface [28]. Thus, the high conductive polymer composites are obtained at the expense of the manufacturing cost as well as the desirable mechanical properties.

To address the issues of the manufacturing cost and concurrent reduction in mechanical properties when a high filler concentration is used, carbon-filled polymer blends containing a triple-continuous structure in 3D space have been pursued recently [24–26]. Shown in Fig. 12.1 is the schematic of the carbon-filled polymer blend with a triple-continuous structure, consisting of a binary polymer blend (i.e., Phases A and B) and CB or CNT particles. Both polymer phases are continuous in 3D space. The conductive carbon is preferentially located in Phase A and its concentration is

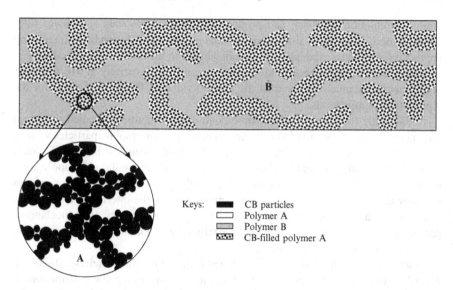

Fig. 12.1 Schematic of the microstructure of carbon-filled polymer blends with a triple-continuous structure. (From M. Wu and L. Shaw, A novel concept of carbon-filled polymer blends for applications of PEM fuel cell bipolar plates, *Int. J. Hydrogen Energ.* 2005; **30**(4):373–380, with permission.)

high enough to form a continuous structure (i.e., at least higher than the percolation threshold in Phase A). As such, a continuous electrical conductive path is present in the carbon-filled polymer blend. Such a triple-continuous structure has the advantage of achieving conductive polymer composites at lower carbon concentrations since only the percolation threshold in one phase, rather than the entire polymer blend, needs to be exceeded. Alternatively, heterogeneous distribution of carbon at the interface of the two polymer phases can also result in conductive polymer blends with minimal carbon concentrations. Two additional advantages offered by the triple-continuous structure are: (a) the improved processability because of the low carbon concentration used, and (b) minimal degradation in tensile properties because of the presence of a continuous neat polymer phase (Phase B in Fig. 12.1). The former allows the use of injection molding and thus reduces the manufacturing cost of bipolar plates, whereas the latter permits the utilization of thin bipolar plates to increase the power density of a fuel cell stack.

The concept of co-continuous polymer blends with carbon black preferentially located in one of the continuous polymer phases or at the polymer-blend interface has been studied for more than a decade with an aim to reduce the percolation threshold. Examples of this kind are the work by Geuskens et al. in as early as 1987 [31], which shows that for the same carbon loading, the resistivity of the co-continuous polymer/rubber blends is several orders of magnitude smaller than that of the single polymer/carbon black composites. Recent works on polymer/elastomer combinations [32,33] and on polymer/polymer systems [34–41] have also shown that the double percolation approach can produce conducting materials at a lower filler concentration. All of these studies suggest that polymer blends can be an interesting approach for making conductive polymers. However, all of the work cited above is limited to low carbon concentration systems with the resistivity at 10^2 Ω cm or higher, which is much higher than the desired values ($10^{-1} - 10^{-3}$ Ω cm) for the bipolar plate applications.

In this study the concept of the triple-continuous structure is applied to several polymer/polymer blends loaded with carbon nanotubes or graphite particles with an aim to obtain injection moldable carbon-filled polymer blends with high electrical conductivities and sufficient mechanical properties for the bipolar plate application of PEM fuel cells. The distribution of CNTs in the polymer blends is examined in terms of their wetting coefficients and minimization of the interfacial energy. The relationships among the microstructure, electrical conductivity, and mechanical properties are studied with an emphasis on achieving simultaneous improvement in both conductivity and tensile strength.

In what follows, the major results obtained from the previous studies [24–26] are described first. Then the underlying principle for controlling the distribution of CNTs in the polymer blends is presented, which is followed by discussion of the mechanisms responsible for the simultaneous improvements in the electrical conductivity and mechanical properties observed in the CNT-filled PET/PVDF blends. Finally, future directions in this emerging area are presented.

12.4 Experimental

Four polymer blends were used as the matrix for conductive carbon nanotubes. Each polymer blend system was composed of two kinds of immiscible polymers. They were: (a) polyethylene terephthalate (PET)/polyvinylidene fluoride (PVDF), (b) PET/polypropylene (PP), (c) PET/high-density polyethylene (HDPE), and (d) PET/nylon 6,6.

Two types of PET were used in this study; one was neat PET and the other the CNT-filled PET. The latter was obtained from Hyperion Catalysis International, Inc. and prefilled with 15 wt% (i.e., 12 vol%) carbon nanotubes through a twin-screw extruder. The CNT-filled PET came in a cylindrical pellet form with sizes of 3 mm in diameter and 2.5 mm in height. The CNTs used to prepare master batches of the CNT-filled PET pellets were hollow, multi-walled tubes with 8–15 walls and a graphitic microstructure. The outside diameter of the tube was approximately 10–15 nm, whereas the inside diameter was about 5 nm. The tube had a very large aspect ratio with the tube length in the range of 10–15 μm.

The PVDF used in this study was Kynar 720 pellets in a biconvex-lens shape with 5 mm in diameter and about 2 mm in thickness at the center of the lens, obtained from Atofina Chemicals, Inc. The nylon 6,6 was Celanese Nylon 6/6 1,000–1 pellets in a cylindrical shape with sizes of 2 mm in diameter and 2.5 mm in height, whereas the PP and HDPE pellets came in the similar shape and size to that of PVDF.

Each CNT-filled polymer blend was prepared in the same way. First, the CNT-filled PET was dried at 150°C for 5 h, and the second polymer was dried under 100°C for 1 h. The dried CNT-filled PET and second polymer pellets were then mixed in a 1 to 1 volume ratio using a rotating bottle for 5 min. As a result of this ratio, the CNT concentration in each composite system was 6.0 vol.%. Final composite samples were prepared using an injection-molding machine (Arburg 221–75–350). Two types of injection-molded samples were fabricated, as shown in Fig. 12.2, with one for electrical conductivity measurements and the other for mechanical testing. The processing conditions were determined based on the melting points, T_m, and thermal decomposition temperatures, T_D, of the polymers in each polymer blend. Differential scanning calorimetry (DSC) and thermogravimetric analysis (TGA) were utilized to establish T_m and T_D of all the polymers used in this study. In all of these simultaneous DSC/TGA analyses, a heating rate of 10°C min^{-1} was employed using a TA instrument (SDT 2960 Simultaneous DTA/TGA) under a flowing argon atmosphere from ambient temperature to 400°C. Table 12.1 lists the melting and thermal decomposition temperatures of all the polymers determined in this study. The thermal decomposition temperature in Table 12.1 was defined as the temperature at which weight loss reached 0.5%. Based on these T_m and T_D, the processing temperatures for injection molding of each polymer blend were selected and are summarized in Table 12.2. Note that whenever possible, the nozzle temperature was selected to be between T_m and T_D of the two polymer constituents. This way both polymers in the blend were in a molten state during injection molding.

The microstructure of and the CNT distribution in polymer blends were observed with an environmental scanning electron microscope (Phillips ESEM 2020). The SEM samples were prepared in four different approaches, depending on the purpose of the observation. The first approach entailed fracturing specimens in liquid nitrogen

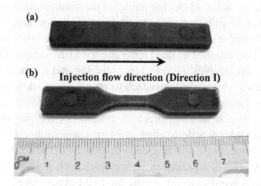

Fig. 12.2 Samples for (**a**) conductivity measurements and (**b**) tensile tests. The direction parallel to the injection flow direction is termed Direction I, whereas the direction perpendicular to the injection flow direction is called Direction II in the text. (From M. Wu and L. Shaw, A novel concept of carbon-filled polymer blends for applications of PEM fuel cell bipolar plates, *Int. J. Hydrogen Energ.* 2005; **30** (4):373–380, with permission.)

Table 12.1 Thermal analysis data of neat polymers

Materials	PET	PVDF	HDPE	PP	Nylon 6,6
$T_m{}^a$ (°C)	247	176	133	148	263
$T_D{}^a$ (°C)	372	386	255	262	350

^aT_m and T_D represent the melting temperature and thermal decomposition temperature, respectively. See the text for details.

Table 12.2 Processing parameters in injection molding

	Nozzle temperature (°C)	Barrel temperature (°C)			Mold temperature (°C)	Screw speed (rpm)	Injection speed (mm s⁻¹)
		Pumping section	Melting section	Feeding section			
CNT-filled PET/PVDF	285	275	270	265	Ambient	200	80
CNT-filled PET/PP	270	265	260	255	Ambient	200	80
CNT-filled PET/HDPE	260	255	245	240	Ambient	200	80
CNT-filled PET/nylon 6,6	275	268	260	250	Ambient	200	80
CT-filled PET/PET	280	270	265	260	Ambient	200	80

to reveal the fracture surface for microstructural observations. The second sample preparation approach consisted of cutting the injection-molded samples with a diamond blade, followed by polishing with Al_2O_3 suspensions down to 0.05 μm, and then ion-etching using an Argon Ion Sputter Gun (Physical Electronic Industry, Inc.) with a 3 kV voltage and a 45-degree angle of the sputter gun with respect to the fracture surface for 45 min to reveal the position of carbon nanotubes. This set of SEM samples allowed examination of the microstructure with minimum loading before the SEM observation. The third approach for SEM sample preparation was to cut the tension-tested samples, followed by polishing and ion etching with the same process parameters as the second approach to reveal the crack initiation and propagation patterns on the cross-section parallel to the tensile loading axis. The last sample preparation approach was the direct observation of the fracture surface of the samples fractured under tensile loading at room temperature. This set of samples offered another perspective regarding deformation and fracture mechanisms under tensile loading. All the SEM samples were coated with gold-palladium before the SEM observation to avoid charging during the SEM observation.

Tensile specimens were in a dog-bone shape and had a gauge length of 10 mm (Fig. 12.2). The tensile test was conducted at a constant crosshead speed of 6 mm min⁻¹ using a servo-hydraulic loading frame. An extensometer was attached to the gauge length of the sample to provide the strain value as a function of loading.

To measure the electrical conductivity, a QuadTech 1880 Milliohmmeter was utilized to get the resistance of samples with a certain cross-section area and thickness. Based on the current, I, and the voltage, V, recorded, the electrical conductivity, σ, was calculated with the aid of:

$$\sigma = \frac{dI}{AV} \qquad (12.1)$$

where d is the specimen thickness between the two electrodes and A is the cross-sectional area perpendicular to the current direction in the sample. Silver paste was utilized in all the measurements to ensure good contact of the sample surface with the electrodes. Furthermore, the electrical conductivity was measured in two directions for the injection-molded rectangular plates (Fig. 12.2); one was parallel to the major flow direction of injection molding (called Direction I hereafter), and the other was perpendicular to Direction I (called Direction II).

12.5 Results and Discussion

12.5.1 Microstructures of Carbon-Filled Polymer Blends

Figure 12.3 shows the fracture surfaces of CNT-filled PET/PVDF, PET/PP, and PET/HDPE blends fractured at the liquid nitrogen temperature. Note that there are two distinct regions in each polymer blend: one contains carbon nanotubes (Region A) and the other does not (Region B). To estimate the area fractions of Regions A and B in each polymer blend, 100 SEM images randomly selected at a magnification of 10,000× have been examined for each polymer blend. The average area fraction of Region A counted from these 100 images is found to be 53, 57, and 57% for CNT-filled PET/PVDF, PET/PP, and PET/HDPE, respectively. Since these CNT-filled polymer blends are fabricated via mixing 50 vol% of the CNT-filled PET with 50 vol% of the second polymer (i.e., PVDF, PP, or HDPE), these data indicate that a small amount of CNTs has transferred to the second polymer phase during the injection molding process. Furthermore, there is more CNT transfer in the CNT-filled PET/PP and PET/HDPE systems than that in the CNT-filled PET/PVDF system. Such a CNT transfer phenomenon is believed to be related to two mechanisms; one is the CNT transfer forced mechanically due to the shearing action derived from the screw rotation during the mixing stage of the injection molding process, and the other is the CNT transfer driven by the thermodynamic driving force to minimize the interfacial energy of the CNT-filled polymer blend. It is argued that the former mechanism plays a key role in the CNT transfer for the CNT-filled PET/PVDF system, whereas both mechanisms are operational in the CNT transfer for CNT-filled PET/PP and PET/HDPE systems. This viewpoint is supported by the thermodynamic analysis discussed in the following.

For a polymer blend, the distribution of carbon particles (or nanotubes) can be predicted by the state of the minimum interfacial energy if the equilibrium state is reached. Such a minimum interfacial energy state can be determined by Young's equation [42]:

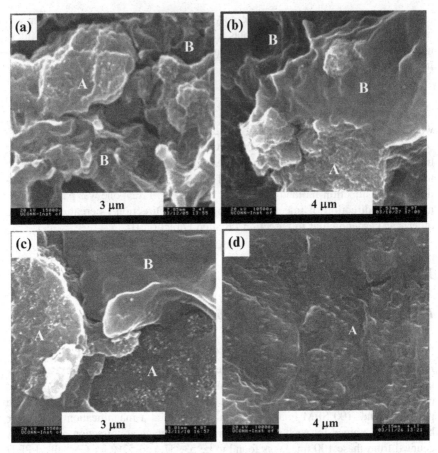

Fig. 12.3 ESEM images of the fracture surface of (**a**) CNT-filled PET/HDPE, (**b**) CNT-filled PET/PP, and (**c**) CNT-filled PET/PVDF polymer blends fractured at the liquid nitrogen temperature. Region A contains CNTs, whereas Region B is free from CNTs. (**d**) Region A in the CNT-filled PET/PVDF after argon ion etching for 45 min. The sticks protruded from the surface are carbon nanotubes. (From M. Wu and L. Shaw, A novel concept of carbon-filled polymer blends for applications of PEM fuel cell bipolar plates, *Int. J. Hydrogen Energ.* 2006; **99**:477–488, with permission.)

$$\omega_a = \frac{\gamma_{C-B} - \gamma_{C-A}}{\gamma_{A-B}} \qquad (12.2)$$

where ω_a is the wetting coefficient, and γ_{C-A}, γ_{C-B}, and γ_{A-B} are the interfacial energy between carbon and polymer A, carbon and polymer B, and polymers A and B, respectively. When $\omega_a > 1$, carbon particles distribute within polymer A. When $-1 < \omega_a < 1$, carbon particles distribute at the interface of the polymer blend. Finally, when $\omega_a < -1$, carbon particles distribute within polymer B. The interfacial energy between two phases, γ_{12} (for phases 1 and 2), in (12.2) can be estimated using the harmonic-mean equation [43]:

$$\gamma_{12} = \gamma_1 + \gamma_2 - 4\left[\frac{\gamma_1^d \gamma_2^d}{\gamma_1^d + \gamma_2^d} + \frac{\gamma_1^p \gamma_2^p}{\gamma_1^p + \gamma_2^p}\right] \qquad (12.3)$$

where γ stands for the surface tension and subscripts 1 and 2 refer to phases 1 and 2, respectively. Further, $\gamma = \gamma^d + \gamma^p$, γ^d is the dispersion component of surface tension, and γ^p is the polar component. The harmonic-mean equation has been shown experimentally to be suitable for estimating the interfacial energy between low-energy materials such as polymers, organic liquids, water, etc. [43]. Thus, (12.3) is utilized here to evaluate the interfacial energy between carbon and polymers.

Based on the surface tension data of carbon, PET, PVDF, PP, HDPE, and nylon 6,6, as well as their dispersion and polar components at 180°C [43], the ω_a values for all the CNT-filled polymer blends investigated in this study are calculated and listed in Table 12.3. As shown in the table, there are two different situations for CNT distribution in these immiscible polymer blends. For CNT-filled PET/PVDF and PET/nylon 6,6 blends, the consideration of the interfacial energy alone predicts that CNTs should stay in the PET phase. In contrast, the predicted location for CNTs in the CNT-filled PET/PP and PET/HDPE blends is at the interface between the PET phase and the second polymer phase. Thus, there is a thermodynamic driving force for CNTs to transfer from the PET phase to the interface for the CNT-filled PET/PP and PET/HDPE systems, while this is not the case for the CNT-filled PET/PVDF and PET/nylon 6,6 systems. It is this difference that has resulted in more CNT transfer into the second polymer phase in the CNT-filled PET/PP and PET/HDPE than that in the CNT-filled PET/PVDF. Furthermore, the thermodynamic analysis performed in the preceding also suggests that a small amount of the CNT transfer into the second polymer phase in the CNT-filled PET/PVDF blend, as evidenced by the increase of the CNT-filled region from 50 to 53 vol%, is not driven by the thermodynamic driving force, but due to the shearing action derived from the screw rotation during the mixing stage of the injection molding process.

12.5.2 Effects of Polymer Blend Rheology on Microstructure

It is well known that the microstructure and morphology of polymer blends, both miscible and immiscible, are strongly affected by material properties and process

Table 12.3 The ω_a value and the predicted CNT location for four CNT-filled polymer blends

Materials	CNT-filled PET/PVDF	CNT-filled PET/PP	CNT-filled PET/HDPE	CNT-filled PET/nylon 6,6
Wetting coefficient (ω_a)	5.10	0.98	0.05	3.33
Predicted location of CNTs in the composite	PET	Interface	Interface	PET

conditions. It has been established that the morphology of immiscible polymer blends produced via melt blending is dependent on many parameters, including: (a) viscosities of the constituent phases, (b) the ratio of the viscosities of the constituent phases, (c) the volume fraction of each constituent phase, (d) the interfacial energies among the constituent phases, (e) shear stresses during melt blending, and (f) the duration of blending [44–49]. All of these parameters except (f) change with the temperature and mixing velocity during melt blending. What kinds of roles do these parameters play in the distribution of CNTs in the polymer blends studied here? This topic is addressed in this section.

The formation of polymer blends via melt blending typically starts with pellets or powders that can be turned into particulate, fibrillar, lamellar, or co-continuous morphologies, depending on the materials and processing conditions [44–54]. When one of the constituent phases is in the molten state, the second constituent phase can be elastic solid, deformable solid, or viscoelastic fluid. In the case of the elastic solid, pellets (or powders) of the second constituent phase come out of the extruder (or mixer) without melting and deformation [47]. When the second constituent phase is a deformable solid, the viscosity ratio of the deformable solid to the melt can range from 500 to well above 1,000, and the morphology of the second constituent phase can be films, ribbons, fibrils, or particles, depending on the viscosity ratio, shear rate, and blending time [47,55]. Increasing the blending time and shear rate and reducing the viscosity ratio favor the formation of particles, whereas film and ribbon morphologies are enhanced with a high viscosity ratio and short blending time. When both constituent phases are in the molten state, the morphology change and phase size reduction normally occur in the first several minutes (i.e., <6 min) of mixing in conjunction with melting of the polymers [44,47,56]. After this transition stage, the morphology and phase dimensions of the immiscible polymer blend are mainly determined by the volume fraction of each constituent phase, the viscosities and viscosity ratio of the constituent phases, the shear rate, and the interfacial energy between the constituent phases [44,46,47,49,51,56].

Once all of the polymer constituents are in a fully molten state, the morphology of the polymer blend can be divided into three general groups, namely, the droplet/matrix morphology, co-continuous morphology, and mixture of the two [46]. The last morphology occurs at and near the blend composition at which the percolation threshold appears. Polymer blends with such a morphology contain droplets as well as the percolating structure [46,57]. As the volume fraction of the second constituent phase increases further above the percolation threshold, all of the second phase will fully contribute to the 3D structure, leading to the formation of a co-continuous morphology. The minimum volume fraction of the second constituent phase required for the formation of the co-continuous structure is determined by the viscosity of the major constituent phase, interfacial tension between the constituents, and shear rate during blending [57]. Specifically, the following semiempirical relation has been established to give the lower limit of the volume fraction for a co-continuous structure to form, $\phi_{d,cc}$ [57]:

$$\frac{1}{\phi_{d,cc}} = 1.38 + 0.0213 \left[\frac{\eta_m \dot{\gamma}}{\sigma_i} R_0 \right]^{4.2} \tag{12.4}$$

where η_m is the viscosity of the major constituent of the polymer blend, $\dot{\gamma}$ the shear rate, σ_i the interfacial tension, and R_0 the equivalent sphere diameter for a cylindrical rod of the minor (second) constituent with a L/B aspect ratio. L and B are the average length and diameter of the cylindrical rods of the minor constituent of the polymer blend, respectively. A full co-continuous structure forms when all the cylindrical rods of the minor constituent phase touch one another [57]. Equation (12.4) shows that the minimum volume fraction of the minor phase for the formation of a co-continuous structure decreases with an increase in the viscosity of the major constituent and the shear rate and a decrease in the interfacial tension. Thus, with proper blending conditions and interfacial tension, the minimum volume fraction for forming co-continuous structures can be reduced to as low as ~25 vol% [51,58]. For example, for a blend with $\sigma = 5$ mN m^{-1}, $R_0 = 1$ μm, $\eta_m = 1,000$ Pa. s, and $\dot{\gamma} = 20$ s^{-1}, the minimum volume fraction for forming co-continuous structures, $\phi_{d,cc}$, would be 11.7% according to (12.4). The upper limit for disappearing of the co-continuous structure can also be found via (12.4), but with the two components of the blend changing the role.

For most of the immiscible polymer blends, a co-continuous structure can be formed with relative ease through melt blending if both constituents of the blend are 50 vol% [46,51,57,58]. This is the case in the present study because all of the CNT-filled polymer blends investigated are composed of two polymers each with about 50 vol%. The co-continuous structure has indeed been formed in the CNT-filled PET/PVDF, PET/PP, and PET/HDPE blends, as discussed in elsewhere in this chapter. Furthermore, the PET pellets and other polymer (PVDF, PP, and HDPE) pellets have all been reduced from millimeter to micrometer scales during injection molding. These results indicate that the morphology and size of the CNT-filled polymer blends investigated here are governed by the rheological behavior of polymer blends. Therefore, the CNT-filled PET phase during injection molding can be treated as a viscoelastic fluid with a high viscosity owing to the addition of CNTs. Based on this trend, it can be argued that the distribution of CNTs in PET or the other polymer phase (i.e., PVDF, PP, or HDPE, designated as Polymer B hereafter) is determined by the nature of the physical process taking place at the interface between the PET and Polymer B viscoelastic fluids, as schematically shown in Fig. 12.4. It is clear that at the interface between two viscoelastic fluids the only driving force that can systematically force CNTs to transfer out of the PET viscoelastic fluid is the difference in the interfacial energy between CNTs and Polymer B and between CNTs and PET. Thus, it can be concluded that the viscosities of PET and Polymer B, the interfacial energy between the CNT-filled PET and Polymer B, and the shear stress during melt blending mainly affect the morphology and sizes of the CNT-filled PET and Polymer B, whereas the distribution of CNTs within the polymer blend is predominately dictated by the thermodynamic driving force, i.e., the minimization of the interfacial energy.

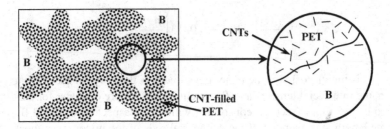

Fig. 12.4 Schematic of the physical process for the distribution of CNTs at the interface between PET and Polymer B viscoelastic fluids and within these viscoelastic fluids

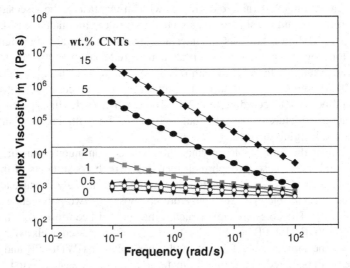

Fig. 12.5 Complex viscosity of polycarbonate with and without CNTs as a function of angular frequency. (Courtesy of Hyperion Catalysis International, Inc.)

Finally, it should be pointed out that the addition of CNTs to PET is not expected to increase the viscosity of the CNT-filled PET by several orders of magnitude. Otherwise, the CNT-filled PET phase cannot be treated as a viscoelastic fluid during injection molding. Shown in Fig. 12.5 is the complex viscosity of polycarbonate (PC) as a function of the CNT concentration and the angular frequency, provided by Hyperion Catalysis International, Inc. Note that at angular frequency of 1 rad s^{-1}, the viscosity of PC with 15 wt% CNTs (which is the same weight percent of CNTs in PET) is about two orders of magnitude higher than that of neat PC. However, at the high shear rate range (i.e., 100 rad s^{-1}), the increase in the viscosity by addition of 15 wt% CNTs is only one order of magnitude. Thus, the viscosity ratio of the CNT-filled PET to Polymer B at the fully molten state during injection molding is expected to be within the typical range (i.e., <1,000) investigated previously for many polymer blend systems [44,46,47,49,51,55,56]. As such, the rheology of polymer blends during injection molding of the CNT-filled polymer blends at the fully molten state affects the morphology and sizes of the CNT-filled PET and Polymer B, but not the distribution of CNTs in the blend. However, if one of the polymer phases is not in the fully molten state during injection molding, the morphology and sizes of the polymer phases cannot be predicted based on the behavior of viscoelastic fluids, and the distribution of CNTs may also be affected, as discussed in the following using the CNT-filled PET/nylon 6,6 as an example.

Shown in Fig. 12.6 are SEM images of the cross-sections of the CNT-filled PET/nylon 6,6 blend. It is obvious that the microstructure of the CNT-filled PET/ nylon 6,6 is quite different from that of the CNT-filled PET/PVDF, PET/PP, and PET/HDPE. First, large nylon regions free from CNTs at millimeter scales (Region C marked in Fig. 12.6) are present. Second, there are microcracks at the interface between the large nylon region and the CNT/PET/nylon mixed region (Region A + B in Fig. 12.6). Third, the CNT/PET/nylon mixed region is mainly composed of the CNT-filled PET (Region A) and CNT-free nylon phase (Region B). Fourth, the area fraction of Region A is found to be about 53%, while the area fraction of the CNT-free regions (i.e., Region B plus Region C) is about 47%.

The markedly different microstructure obtained in the CNT-filled PET/nylon 6,6 blend is attributed to the insufficient breakdown and mixing during the injection-molding process. Recall that for CNT-filled PET/PVDF, PET/PP, and PET/HDPE blends, the size of the neat polymer region (i.e., Region A) is typically in the range of $1–40 \mu m^2$ (Fig. 12.3), which is substantially smaller than the original sizes of the PVDF, PP, and HDPE pellets (several millimeters). In contrast, the large neat nylon regions found in the CNT-filled PET/nylon 6,6 blend have sizes similar to the

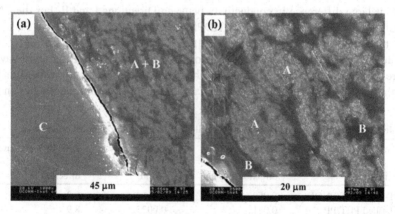

Fig. 12.6 SEM images of the cross sections of the CNT-filled PET/nylon 6,6 blend prepared by cutting with a diamond blade, polishing with Al_2O_3 suspensions, ion etching with an argon ion sputter gun, and finally coating with gold-palladium. Region A contains CNTs, whereas Regions B and C are free from CNTs. (**a**) A low magnification image, showing the presence of a large nylon region (marked as C) and a microcrack at the interface between Region C and Region (A + B), and (**b**) a high magnification image of the (A + B) region in (**a**). (From M. Wu and L. Shaw, A novel concept of carbon-filled polymer blends for applications of PEM fuel cell bipolar plates, *Int. J. Hydrogen Energ.* 2006;**99**:477–488, with permission.)

original size of nylon 6,6 pellets. This insufficient breakdown of nylon 6,6 regions and lack of uniform mixing is related to the relatively low processing temperatures used in the injection molding process. As shown in Tables 12.1 and 12.2, the difference between the processing temperatures (275°C at the nozzle and 268°C at the pumping section) and the melting temperature of nylon 6,6 (263°C) is <15°C. The small overheating and the presence of the temperature gradient within the barrel of the injection molding machine result in partial melting of nylon 6,6 pellets. As a result, some nylon 6,6 pellets are not broken up and mixed with the CNT-filled PET phase well, while some are. In the fully molten region, the distribution of CNTs is determined by the minimization of the interfacial energy (Fig. 12.6b), whereas the re-distribution of CNTs is impossible in the non-molten region (Fig. 12.6a). The presence of microcracks at the interface between the large neat nylon region and the CNT/PET/nylon mixed region has been identified to be due to the mismatch in the coefficient of thermal expansion (CTE) between PET and nylon and the presence of large nylon regions [26].

12.5.3 Electrical Conductivity

The electrical conductivities of injection-molded, CNT-filled polymer blends are summarized in Table 12.4. For comparison, Table 12.4 also includes the literature value of conductivities of neat polymers used in this study. For each polymer blend, electrical conductivities are measured in two directions (i.e., Directions I and II in Fig. 12.2) to determine whether the specimen is isotropic or not. It is found that there is large difference in conductivity between Directions I and II. For the CNT-filled PET/PVDF, PET/PP, and PET/HDPE, the conductivity in Direction I is about 4–8 times higher than that in Direction II. For the CNT-filled PET/nylon 6,6, the conductivity difference in the two directions is even larger, with Direction I having more than 22 times higher conductivity than Direction II. The anisotropy found in all the specimens is related to the partial alignment of carbon nanotubes in the

Table 12.4 Electrical conductivities of CNT-filled polymer blends and neat polymers

Materials	Conductivity in direction I (S cm^{-1})	Conductivity in direction II (S cm^{-1})	Ratio of conductivity in direction I to direction II
CNT-filled PET/PVDF	0.059	0.0114	5.2
CNT-filled PET/PP	0.021	0.0023	8.9
CNT-filled PET/HDPE	0.011	0.0014	7.7
CNT-filled PET/nylon 6,6	0.011	0.0005	23.4
PET [59]	>1.0 × 10^{-14}	N/A	N/A
PVDF [60]	1.0 × 10^{-13}	N/A	N/A
HDPE [27]	1.0 × 10^{-19}	N/A	N/A
PP [27]	1.0 × 10^{-19}	N/A	N/A
Nylon 6,6 [27]	1.0 × 10^{-15}	N/A	N/A

polymer blend caused by the shear stress induced by the drag force of the die surface during injection flow.

It is also noticeable that even though all the polymer blends have the same CNT concentration of 6.0 vol%, different polymer blends display different conductivities. The highest conductivity obtained from the CNT-filled PET/PVDF in Direction I is 2.8–5.4 times of the highest conductivities obtained from other CNT-filled polymer blends in Direction I. The better conductivity obtained from the CNT-filled PET/PDVF blend in comparison with the CNT-filled PET/PP and PET/HDPE blends is attributed to its less CNT transfer to the second polymer phase. Such reasoning is supported by the following analysis.

Given that all of the CNT-filled polymer blends in this study are prepared with 50 vol% of the CNT-filled PET phase (with 12 vol% CNTs) plus 50 vol% of the second immiscible polymer phase (with no CNTs), it is reasonable to assume that both the CNT-filled PET phase and the second immiscible neat polymer phase have formed self-continuous 3D networks in the polymer blends. This expectation is confirmed by the microstructure examination (see Sect. 12.1), which reveals that the area fractions of the CNT-filled region and the CNT-free region are both near 50%. Furthermore, the electrical conductivity data suggest that the carbon nanotubes within the PET phase have also formed a 3D conductive path because the electrical resistivity has been reduced from the neat polymer blends to the CNT-filled polymer blends by about 12 orders of magnitude. With such a triple-continuous structure, the conductive CNT-filled PET network and the non-conductive second polymer phase can be treated as parallel conductors, and the resulting resistivity, ρ, of the CNT-filled polymer blend can be estimated using the statistical percolation model proposed by Bueche [61]:

$$\rho = \frac{\rho_c \rho_n}{V_n \rho_c + \omega(1 - V_n)\rho_n} \tag{12.5}$$

where ρ_c and ρ_n are the resistivities of the conductive and non-conductive phases, respectively, V_n is the volume fraction of the non-conductive phase, and ω is the fraction of the conductive phase being incorporated in the conducting network. The largest possible value for ω is 1, which corresponds to the case in which all the CNT-filled PET regions are incorporated into the conductive network. For the present CNT-filled polymer blends, Equation (12.5) can be reduced to:

$$\rho \approx \frac{\rho_c}{\omega(1 - V_n)} \tag{12.6}$$

because $\rho_n \gg \rho_c$. For example, ρ_n is 10^{13} Ω cm for PVDF [60] and ρ_c is only 4 Ω cm [25] for the PET phase with 12 vol% CNTs. Equation (12.6) can be utilized to qualitatively explain the electrical conductivity data obtained in this study. The CNT transfer from the PET phase to the second polymer phase will increase the resistivity of the CNT-filled PET, ρ_c, but at the same time will decrease the volume fraction of the non-conductive phase, V_n. However, the change in V_n is very small (e.g., 50–43 vol% for the CNT-filled PET/PP and PET/HDPE blends). In contrast, the change in ρ_c can be potentially very large with changes of several orders of magnitude if the original concentration of CNTs in the PET phase is near the percolation threshold. The CNT transfer to the second polymer phase could also lead to a reduction in ω because the newly CNT-filled polymer regions may not be incorporated into the conductive network. Even when they are incorporated, their resistivities are unlikely to be as low as that of the CNT-filled PET regions because these newly CNT-filled polymer regions are most likely to have lower carbon nanotube concentrations than the CNT-filled PET regions. Therefore, based on the possible range of change for the parameters in (12.6), it can be stated that the CNT transfer to the second polymer phase during the injection molding process, in general, will increase the resistivity of the resulting composite, and thus is undesirable for improving electrical conductivity. The present set of experiments supports this theoretical analysis, showing that the CNT-filled PET/PVDF has the highest electrical conductivity because it has the least CNT transfer to the second polymer phase. It is noted that the CNT-filled PET/nylon 6,6 blend exhibits the lowest electrical conductivity among all the systems investigated. Although the CNT transfer to the second polymer phase for the CNT-filled PET/nylon 6,6 is small (i.e., similar to the CNT-filled PET/PVDF blend), the presence of microcracks in this composite is believed to be responsible for the lowest electrical conductivity observed.

Finally, it is important to compare the electrical conductivities of the CNT-filled polymer blends with that of the CNT-filled polymers. Figure 12.7 shows the resistivities of the CNT-filled PET/PVDF and CNT-filled PET as a function of the carbon

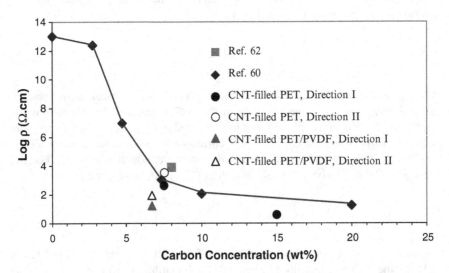

Fig. 12.7 The resistivities of the CNT-filled PET and CNT-filled PET/PVDF as a function of the carbon concentration. The resistivities of several carbon black-filled PVDF composites from the literature are also included for comparison. The line is added as a visual guide only. (From M. Wu and L. Shaw, On the improved properties of injection-molded, carbon nanotube-filled PET/PVDF blends, *J. Power Sources* 2004;**136**:37–44, with permission.)

concentration. Note that the CNT-filled PET is prepared by injection molding of 50 vol% of PET pellets containing 12 vol% CNTs with 50 vol% neat PET. Thus, the injection molded CNT-filled PET has the same nominal CNT concentration as the CNT-filled PET/PVDF blend. Several electrical conductivities of carbon black (CB)-filled PVDF polymers in the open literature have also been included for comparison. It can be seen that PET with 12 vol% CNT (i.e., 15 wt%) has a lower electrical resistivity than the carbon black-filled PVDF with the same carbon loading. The resistivity of PET with 6 vol% CNT (i.e., 7.5 wt%), however, is similar to that of the carbon black)-filled PVDF. In contrast, the PET/PVDF blend with 6 vol% CNT exhibits about two orders of magnitude reduction in the resistivity over the CNT-filled PET and the carbon black-filled PVDF with the same carbon loading (in volume percent), showing the efficacy of the segregation of carbon in one of the phases in a binary polymer blend.

12.5.4 Mechanical Properties

Figure 12.8 compares the stress—strain curves of neat PET, neat PVDF, PET with 6 vol% CNT and PET/PVDF with 6 vol% CNT. Several features are noted from Fig. 12.8. First, the addition of CNT to the polymers has resulted in reductions in both tensile strength and elongation at break. For instance, the addition of 6 vol% CNT to PET has led to a decrease in the tensile stress at break from 34 to 25 MPa and in the elongation at break from 2.2% to 1.2%. Second, as expected, the addition

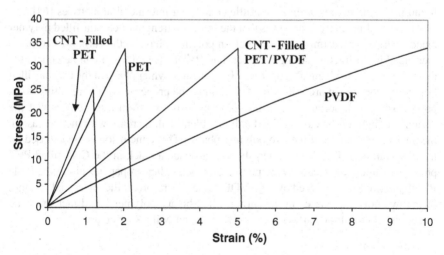

Fig. 12.8 Stress-strain curves of neat PET, PET with 6 vol% CNT, and PET/PVDF with 6 vol% CNT. For comparison, part of the stress-strain curve of neat PVDF (having an elongation at break = 1,400%) is also included. (From M. Wu and L. Shaw, On the improved properties of injection-molded, carbon nanotube-filled PET/PVDF blends, *J. Power Sources* 2004;**136**:37–44, with permission.)

of CNT has increased the elastic modulus of PET. Furthermore, the modulus of the CNT-filled PET/PVDF blend falls between that of the CNT-filled PET and neat PVDF, as would be expected from the rule of mixtures for composites [63]. Third, the CNT-filled PET/PVDF exhibits a 325% improvement in elongation and a 36% improvement in fracture strength over the CNT-filled PET with the same CNT loading. These improvements are due to the presence of the clean PVDF phase free from CNT in the CNT-filled PET/PVDF blend, as discussed in the following.

Shown in Fig. 12.9 are SEM images of crack paths in the CNT-filled PET/PVDF blend right before the fracture of the specimen. Note that cracks preferentially initiate and propagate within the CNT-filled PET phase. This is consistent with the stress-strain behaviors of the CNT-filled PET and neat PVDF shown in Fig. 12.8; that is, at strains slightly higher than 1.2%, the CNT-filled PET phase will have cracks, whereas the clean PVDF phase can still carry the load because its elongation at break is ~1,400%. Thus, the presence of the clean PVDF phase free from CNT has provided strengthening mechanisms for the CNT-filled PET/PVDF blend. Such strengthening mechanisms are manifested as crack bridging and crack deflection at the PET/PVDF interface, as revealed in Fig. 12.9.

Assuming that the load after cracking of the CNT-filled PET phase is carried only by the clean PVDF phase, the fracture strength of the CNT-filled PET/PVDF would be ~27 MPa because of the presence of 50 vol% of the PVDF phase and its fracture strength of ~54 MPa. The measured fracture strength of the CNT-filled PET/PVDF is 34 MPa (the average of three specimens), slightly higher than the prediction of the simple rule of mixtures. This discrepancy is likely due to the presence of the CNT-filled PET phase that imposes deformation constraints to the clean PVDF phase. Under constraints the fracture strength of a ductile phase can increase, as previously found in ductile metals within the brittle ceramic or intermetallic matrices [64].

The preceding analysis suggests that the fracture strength of carbon-filled polymer blends can be further improved if the clean polymer phase in the carbon-filled polymer blend has a fracture strength higher than PVDF. Therefore, it can be concluded that the intrinsic mechanical properties of the clean polymer phase in the carbon-filled binary polymer blend are critical in determining the properties of the resulting composite. Another important conclusion that can be drawn from this study is that the fracture strength of the carbon-filled polymer blend will increase with the increase in the interfacial strength of the two polymer phases. This conclusion is made based on the observation of Fig. 12.9, which shows that when a crack in the CNT-filled PET phase encounters the clean PVDF phase, the crack either propagates along the PET/PVDF interface or is bridged by the PVDF phase. Therefore, if the interfacial strength of the two polymer phases in the carbon-filled binary polymer blend increases, the fracture strength of the carbon-filled polymer blend will also increase.

12.6   Concluding Remarks

The efficacy of carbon-filled polymer blends in improving electrical conductivity and mechanical strength simultaneously has been demonstrated in this study. The CNT-filled PET/PVDF blend exhibits 2,500% improvement in electrical

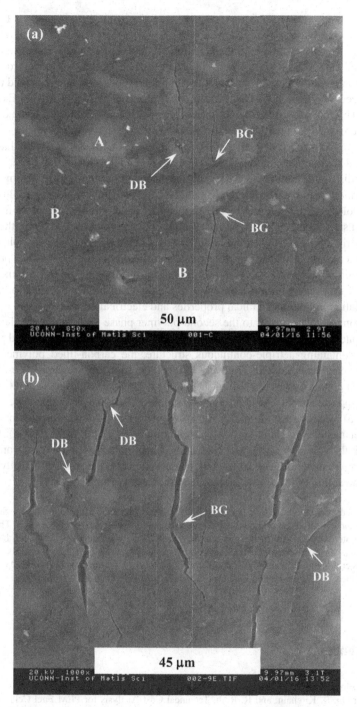

Fig. 12.9 SEM secondary electron image of crack paths in the CNT-filled PET/PVDF blend with (**a**) from the region near the shoulder of the tensile specimen, and (**b**) from the central region of the tensile specimen. Region A is the PVDF phase, whereas Region B is the PET with CNT. BG stands for bridging, while DB represents debonding. The loading axis is horizontal. (From M. Wu and L. Shaw, On the improved properties of injection-molded, carbon nanotube-filled PET/PVDF blends, *J. Power Sources* 2004;**136**:37–44, with permission.)

conductivity, 36% increase in tensile strength, and 320% improvement in elongation over the CNT-filled PET with the same carbon loading. Such improvements have been related to the formation of a triple-continuous structure achieved through the forced segregation of CNT in the PET phase of the CNT-filled PET/PVDF blend. This CNT-filled PET phase offers an electrical short circuit for the composite, whereas the clean PVDF phase provides the strength and elongation for the composite. As a result of such a combination, the CNT-filled PET/PVDF has better electrical conductivity, strength, and elongation than the CNT-filled PET.

The distribution of CNTs in the polymer blend is predominantly determined by the thermodynamic driving force when the polymer blend is in a fully molten state. The rheology of the polymer blend, on the other hand, mainly affects the morphology and sizes of the polymer phases, but not the distribution of CNTs in the blend. The distribution of CNTs in the polymer blend plays a very important role in both mechanical and electrical properties of carbon-filled polymer blends. The preferential location of CNTs in one of the continuous polymer phases in the polymer blend is highly desirable from the viewpoint of the mechanical and electrical properties. Degradation in both mechanical properties and electrical conductivity are observed when some CNTs transfer to the second polymer phase in the polymer blend.

In spite of the substantial progress made with the concept of carbon-filled polymer blends containing a triple-continuous structure, the carbon-filled polymer blends studied so far only contain relatively low carbon concentrations. As a result, their electrical conductivities are still far below the desired values (such as >100 S cm^{-1}) for the application of PEM fuel cell bipolar plates. Therefore, it is imperative to investigate: (a) whether such a triple-continuous structure can still be injection molded for polymer blends with high carbon concentrations (e.g., >30 vol% carbon); and (b) whether the polymer blends with high carbon concentrations and a triple-continuous structure, if injection moldable, still possess superior electrical conductivity and mechanical properties. Both issues will be the topics of future studies.

Acknowledgments The author is indebted to Professors Frano Barbir, Montgomery Shaw, and Lei Zhu for fruitful discussion over a wide range of the topics related to this research. The author is also grateful to many of his former and current students, especially Ms. Man Wu, Dr. Daniel Goberman, Mr. Hong Luo, and Dr. Juan Villegas, for carrying out various experiments related to this project. Finally, the financial support from the US Army (contract #: DAAB07–03–3-K415) through the Connecticut Global Fuel Cell Center is greatly appreciated.

References

1. L. J. Blomen and M. N. Mugerwa, *Fuel Cell Systems* (Plenum, New York, NY, 1993).
2. I. Bar-On, R. Kirchain, and R. Roth, Technical Cost Analysis for PEM Fuel Cells, *J. Power Sources* **109**, 71–75 (2002).
3. T. M. Besmann, J. W. Klett, J. J. Henry, and E. Lara-Curzio, Carbon/Carbon Composite Bipolar Plate for Proton Exchange Membrane Fuel Cells, *J. Electrochem. Soc.* **147**(11), 4083–4086 (2000).

4. H. Tsuchiya and O. Kobayashi, Mass Production Cost of PEM Fuel Cell by Learning Curve, *Int. J. Hydrogen Energ.* **29**(10), 985–90 (2004).
5. A. Kumar and R. G. Reddy, Materials and Design Development for Bipolar/end Plates in Fuel Cells, *J. Power Sources* **129**, 62–67 (2004).
6. V. Mehta and J. S. Cooper, Review and Analysis of PEM Fuel Cell Design and Manufacturing, *J. Power Sources* **114**, 32–53 (2003).
7. A. Hermann, T. Chaudhuri, and P. Spagnol, Bipolar Plates for PEM Fuel Cells: A Review, *Int. J. Hydrogen Energ.* **30**, 1297–1302 (2005).
8. P. L. Hentall, J. B. Lakeman, G. O. Mepsted, P. L. Adcock, and J. M. Moore, New Materials for Polymer Electrolyte Membrane Fuel Cell Current Collectors, *J. Power Sources* **80**, 235–241 (1999).
9. W. Middelman, W. Kout, B. Vogelaar, J. Lenssen, and E. de Waal, Bipolar Plates for PEM Fuel Cells, *J. Power Sources* **118**, 44–46 (2003).
10. D. P. Davies, P. L. Adcock, M. Turpin, and S. J. Rowen, Bipolar Plate Materials for Solid Polymer Fuel Cells, *J. Appl. Electrochem.* **30**, 101–105 (2000).
11. R. Hornung and G. Kappelt, Bipolar Plate Materials Development using Fe-Based Alloys for Solid Polymer Fuel Cells, *J. Power Sources* **72**, 20–21 (1998).
12. R. C. Makkus, A. H. H. Janssen, F. A. de Bruijn, and R. K. A. M. Mallant, Use of Stainless Steel for Cost Competitive Bipolar Plates in the SPFC, *J. Power Sources* **86**, 274–282 (2000).
13. H. J. Davis, Metal-Cored Bipolar Separator and End Plates for Polymer Electrolyte Membrane Electrochemical and Fuel Cells, U.S. Patent # 2003/0027028 A1.
14. M. P. Brady, K. Weisbrod, I. Paulauskas, R. A. Buchannan, K. L. More, H. Wang, M. Wilson, F. Garzon, and L. R. Walker, Preferential Thermal Nitridation to Form Pin-Hole Free Cr-Nitrides to Protect Proton Exchange Membrane Fuel Cell Metallic Bipolar Plates, *Scripta Mater.* **50**, 1017–1022 (2004).
15. M. P. Brady, K. Weisbrod, C. Zawodzinski, I. Paulauskas, R. A. Buchannan, and L. R. Walker, Assessment of Thermal Nitridation to Protect Metal Bipolar Plates in Polymer Electrolyte Membrane Fuel Cells, *Electrochem. Solid-State Lett.* **5**(11), A245–A247 (2002).
16. C. Del Rio, M. C. Ojeda, J. L. Acosta, M. J. Escudero, E. Hontanon, and L. Daza, New Polymer Bipolar Plates for Polymer Electrolyte Membrane Fuel Cells: Synthesis and Characterization, *J. Appl. Polym. Sci.* **83**(13), 2817–2822 (2002).
17. G. Marsh, Fuel Cell Materials, *Mater. Today* **4**(2), 20–24 (2001).
18. F. Barbir, J. Braun, and J. Neutzler, Properties of Molded Graphite Bipolar Plates for PEM Fuel Cell Stacks, *J. New Mater. Electrochem. Syst.* **2**, 197–200 (1999).
19. J. Braun, J. E. Zabriskie, Jr., J. K. Neutzler, M. Fuchs, and R. C. Gustafson, Fuel Cell Collector Plate and Method of Fabrication, US Patent # 6,180,275.
20. A. Bonnet and J.-F. Salas, Microcomposite Powder Based on Flat Graphite Particles and on a Fluoropolymer and Objects made from Same, U.S. Patent # 2004/0262584 A1.
21. C.-C. M. Ma, K. H. Chen, H. C. Kuan, S. M. Chen, M. H. Tsai, Y. Y. Yan, and F. Tsau, Preparation of Fuel Cell Composite Bipolar Plate, U.S. Patent # 2005/0001352 A1.
22. C. W. Extrand, Lyophilic Fuel Cell Component, U.S. Patent # 2005/0008919 A1.
23. K. I. Bulter, Highly Conductive Molding Compounds for Use as Fuel Cell Plates and the Resulting Products, U.S. Patent # 2003/0042468 A1.
24. M. Wu and L. Shaw, A Novel Concept of Carbon-Filled Polymer Blends for Applications of PEM Fuel Cell Bipolar Plates, *Int. J. Hydrogen Energ.* **30**(4), 373–380 (2005).
25. M. Wu and L. Shaw, On the Improved Properties of Injection-Molded Carbon Nanotube-Filled PET/PVDF Blends, *J. Power Sources* **136**, 37–44 (2004).
26. M. Wu and L. Shaw, Electrical and Mechanical Behaviors of Carbon Nanotube-Filled Polymer Blends, *J. Appl. Polym. Sci.* **99**, 477–488 (2006).
27. K. Miyasaka, K. Watanabe, E. Jojima, H. Aida, M. Sumita, and K. Ishikawa, Electrical Conductivity of Carbon-Polymer Composites as a Function of Carbon Content, *J. Mater. Sci.* **17**, 1610–1616 (1982).
28. J. C. Grunlan, W. W. Gerberich, and L. F. Francis, Electrical and Mechanical Behavior of Carbon Black-Filled Poly(Vinyl Acetate) Latex-Based Composites, *Polym. Eng. Sci.* **41**(11), 1947–1962 (2001).

29. J.-C. Huang, Review Carbon Black Filled Conducting Polymers and Polymer Blends, *Adv. Polym. Tech.* **21**(4), 299–313 (2002).
30. C. Xu, Y. Agari, and M. Matsuo, Mechanical and Electric Properties of Ultra-High- Molecular Weight Polyethylene and Carbon Black Particle Blends, *Polym. J.* **30** (5), 372–380 (1998).
31. G. Geuskens, J. L. Gielens, D. Geshef, and R. Deltour, The Electrical Conductivity of Polymer Blends Filled with Carbon-Black, *Eur. Polym.* **23**, 993–995 (1987).
32. B. G. Soares, F. Gubbels, R. Jérôme, and Ph. Teyssié, Electrical Conductivity in Carbon Black-Loaded Polystyrene-Polyisoprene Blends. Selective Localization of Carbon Black at the Interface, *Polym. Bull.* **35**, 223–228 (1995).
33. B. G. Soares, F. Gubbels, R. Jérôme, E. Vanlanthem, and R. Deltour, Electrical Conductivity of Polystyrene-Rubber Blends Loaded with Carbon Black, *Rubb. Chem. Technol.* **70**, 60–70 (1997).
34. F. Gubbels, S. Blacher, E. Vanlanthem, R. Jérôme, R. Deltour, F. Brouers, and P. Teyssié, Design of Electrical Conductive Composites: Key Role of the Morphology on the Electrical Properties of Carbon Black Filled Polymer Blends, *Macromolecules* **28**, 1559–1566 (1995).
35. R. Tchoudakov, O. Breuer, M. Narkis, and A. Siegmann, Conductive Polymer Blends with Low Carbon Black Loading: Polypropylene/Polyamide, *Polym. Eng. Sci.* **36**, 1336–1346 (1996).
36. Ye. P. Mamunya, Morphology and Percolation Conductivity of Polymer Blends Containing Carbon Black, *J. Macromol. Sci. Phys.* **B38**, 615–622 (1999).
37. K. Cheah, G. P. Simon, and M. Forsyth, Effects of Polymer Matrix and Processing on the Conductivity of Polymer Blends, *Polym. Int.* **50**, 27–36 (2001).
38. J. G. Mallette, A. Márquez, O. Manero, and R. Castro-Rodríguez, Carbon Black Filled PET/ PMMA Blends: Electrical and Morphological Studies, *Polym. Eng. Sci.* **40**, 2272–2278 (2000).
39. S. H. Foulger, Electrical Properties of Composites in the Vicinity of the Percolation Threshold, *J. Appl. Polym. Sci.* **72**, 1573–1582 (1999).
40. J. Feng and C. M. Chan, Carbon Black-filled Immiscible Blends of Poly(vinylidene fluoride) and High Density Polyethylene: Electrical Properties and Morphology, *Polym. Eng. Sci.* **38**, 1649–1657 (1998).
41. G. J. Lee, K. D. Suh, and S. S. Im, Effect of Incorporating Ethylene-Ethylacrylate Copolymer on the Positive Temperature Coefficient Characteristics of Carbon Black Filled HDPE Systems, *Polym. Eng. Sci.* **40**, 247–255 (2000).
42. M. Sumita, K. Sakata, S. Asai, K. Miyasaka, and H. Nakagawa, Dispersion of Fillers and the Electrical Conductivity of Polymer Blends Filled with Carbon Black, *Polym. Bull.* **25**, 265–271 (1991).
43. S. Wu, *Polymer Interface and Adhesion* (Dekker, New York, NY, 1982).
44. C. E. Scott and C. W. Macosko, Morphology Development during the Initial Stages of Polymer-Polymer Blending, *Polymer* **36**(3), 461–470 (1995).
45. M. Evstatiev, J. M. Schultz, S. Petrovich, G. Georgiev, S. Fakirov, and K. Friedrich, In Situ Polymer/Polymer Composites from Poly(ethylene Terephthalate), Polyamide-6, and Polyamide-66 Blends, *J. Appl. Polym. Sci.* **67**, 723–737 (1998).
46. M. Castro, C. Carrot, and F. Prochazka, Experimental and Theoretical Description of Low Frequency Viscoelastic Behavior in Immiscible Polymer Blends, *Polymer* **45**, 4095–4104 (2004).
47. J. S. Hong, J. L. Kim, K. H. Ahn, and S. J. Lee, Morphology Development of PBT/PE Blends during Extrusion and Its Reflection on the Rheological Properties, *J. Appl. Polym. Sci.* **97**, 1702–1709 (2005).
48. W. Yu, C. Zhou, and Y. Xu, Rheology of Concentrated Blends with Immiscible Components, *J. Polym. Sci.* **43**, 2534–2540 (2005).
49. H. A. Khonakdar, S. H. Jafari, A. Yavari, A. Asadinezhad, and U. Wagenknecht, Rheology, Morphology and Estimation of Interfacial Tension of LDPE/EVA and HDPE/EVA Blends, *Polym. Bull.* **54**, 75–84 (2005).
50. R. Ratnagiri and C. E. Scott, Effect of Viscosity Variation with Temperature on the Compounding Behavior of Immiscible Blends, *Polym. Eng. Sci.* **39**(9), 1823–1835 (1999).

51. R. C. Willemse, A. Posthuma de Boer, J. van Dam, and A. D. Gotsis, Co-continuous Morphologies in Polymer Blends: the Influence of the Interfacial Tension, *Polymer* **40**, 827–834 (1999).
52. N. Marin and B. D. Favis, Co-continuous Morphology Development in Partially Miscible PMMA/PC Blends, *Polymer* **43**, 4723–4731 (2002).
53. J. Li and B. D. Favis, Characterizing Co-continuous High Density Polyethylene/Polystyrene Blends, *Polymer* **42**, 5047–5053 (2001).
54. J. K. Lee and C. D. Han, Evolution of Polymer Blend Morphology during Compounding in an Internal Mixer, *Polymer* **40**, 6277–6296 (1999).
55. C.-K. Shih, D. G. Tynan, and D. A. Denelsbeck, Rhrological Properties of Multicomponent Polymer Systems Undergoing Melting or Softening during Compounding, *Polym. Eng. Sci.* **31**(23), 1670–1673 (1991).
56. B. D. Favis, The Effect of Processing Parameters on the Morphology of an Immiscible Binary Blend, *J. Appl. Polym. Sci.* **39**, 285–300 (1990).
57. R. C. Willemse, A. Posthuma de Boer, J. van Dam, and A. D. Gotsis, Co-continuous Morphologies in Polymer Blends: a New Model, *Polymer* **39**(24), 5879–5887 (1998).
58. R. C. Willemse, Co-continuous Morphologies in Polymer Blends: Stability, *Polymer* **40**, 2175–2178 (1999).
59. Technical Data from Goodfellow Corporation Home Page, 2004. Available from World Wide Web: http://www.goodfellow.com/csp/active/gfHome.csp.
60. G. Wu, C. Zhang, T. Miura, S. Asai, and M. Sumita, Electrical Characteristics of Fluorinated Carbon Black-Filled Poly(vinylidene Fluoride) Composites, *J. Appl. Polym. Sci.* **80**(7), 1063–1070 (2001).
61. F. Bueche, Electrical Resistivity of Conducting Particles in an Insulating Matrix, *J. Appl. Phys.* **43**(11), 4837–4838 (1972).
62. Z. Zhao, W. Yu, X. He, and X. Chen, The Conduction Mechanism of Carbon Black-Filled Poly(vinylidene Fluoride) Composite, *Mater. Lett.* **57**, 3082–3088 (2003).
63. B. D. Agarwal and L. J. Broutman, *Analysis and Performance of Fiber Composites, 2nd Edition* (Wiley, New York, NY, 1990).
64. L. Shaw and R. Abbaschian, On the Flow Behavior of Constrained Ductile Phases, *Metall. Trans.* **24A**, 403–415 (1993).

Chapter 13
Critical Issues in the Commercialization of DMFC and Role of Membranes

Hyuk Chang, Haekyoung Kim, Yeong Suk Choi, and Wonmok Lee

Abstract Mobile telecommunication devices in the next generation require a new concept of quick charging and a long-lasting mobile energy source. The direct methanol fuel cell (DMFC) is becoming attractive, but there are critical issues involved in its commercialization with regard to the core technologies of catalyst, membrane, membrane electrode assembly (MEA), stack, and system. More importantly, the main role of the proton-conducting membrane is enhancing the energy and power density and affecting the other components in DMFC systems. Functions, current status, and technical approaches are discussed in terms of protonic conductivity, methanol permeability, water permeability, life cycle, and processing cost as well as interaction with other compartments. Materials such as perfluorinated and partially fluorinated membranes, hydrocarbon membranes, composite membranes, and other modified ionomers have been studied in connection with technology roadmap of membrane and mobile DMFC systems. These would explain the critical issues of DMFC and the role of membranes for commercialization.

13.1 Direct Methanol Fuel Cell

13.1.1 Needs and Status

Technical advances and user-friendly functions of mobile telecommunication devices beyond the third-generation (B3G) multimedia era call upon a mobile energy source greater than the limit of conventional technology. Since those B3G mobile devices perform more functions of talking and exchanging information via the Internet as well as digital mobile broadcasting (DMB) systems, more power and energy are demanded. Table 13.1 indicates the expected load and power demand of one example of a B3G phone. This shows that the device will require nearly 3790 mAh per day, which will need three 5-h charges for a 1-day use [1]. In order to avoid these problems in future devices, direct methanol fuel cell (DMFC) technology is becoming attractive. Also, there is a demand for lower weight and longer operation power sources for mobile Note PCs (personal computers). Figure 13.1 explains these expectations schematically in the mobile phone as well as in the Note PC [2].

S.M.J. Zaidi and T. Matsuura (eds.) *Polymer Membranes for Fuel Cells*,
doi: 10.1007/978-0-387-73532-0, © Springer Science+Business Media, LLC 2009

Table 13.1 Load and power demand of B3G phone

	Call	DMB	MP3	Game	VOD	DC	Bell	Etc.	Total
Load (mA)	400	500	225	490	390	250	350	120	—
Power (W)	1.45	1.8	0.8	1.8	1.4	0.9	1.3	—	—
Operation (h)	2	1.5	1	1	0.5	0.5	0.05	17.45	24
Capacity (mAh)	800	750	225	490	180	125	20	1,200	3,790

(a)

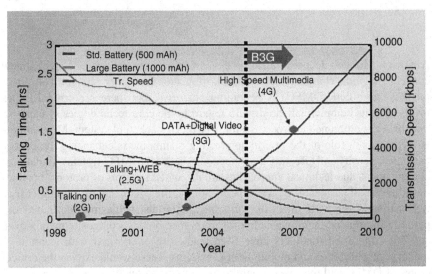

(b)

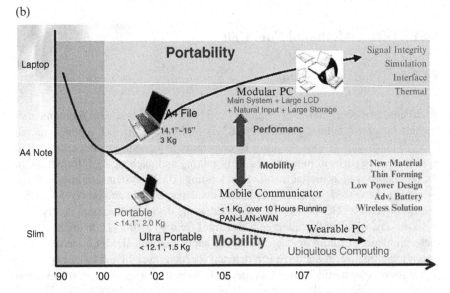

Fig. 13.1 Power demands of next generation (**a**) mobile phone and (**b**) note PC

Of all the system characteristics of DMFC with regard to commercialization as a mobile device energy source, energy efficiency, energy density, power density, and cost have to be competitive with rechargeable batteries. These requirements are somewhat different from other types of fuel cells for residential or vehicle application, which can utilize auxiliary compartments more freely. These compartments include a balance of plant (BOP), such as a compressor or pump, for fuel and oxidant supply, humidifier, heat exchanger, and so on. These compartments have to be miniaturized effectively or eliminated. This means that the materials in the stack should themselves have the functions of BOP.

Many technical advances have been achieved, and the application to a small portion or niche market of mobile devices with DMFC as their energy source will be on the market within the next couple of years. This commercial device will affect life styles through electrochemical engineering. However, DMFC is not a simple electrochemical device anymore because it can not come up to the commercial level without the aid of materials science, chemical engineering, mechanical engineering, and electrical engineering. This chapter describes how DMFC integrates all those technologies, including catalyst and membrane materials science, system energy balance chemical engineering, fluid dynamics mechanical engineering, and hybrid circuit design electrical engineering.

The preceding text discussed the required functions and critical issues on the road to commercialization with regard to catalyst, membrane, MEA, stack, and system technologies. However, much of this chapter focuses on details of the proton-conducting membrane, which has a major role in enhancing energy and power density and also affects the catalyst, mass balance, and energy balance of DMFC systems for mobile applications.

13.1.2   Functions and Critical Issues of Core Technologies

13.1.2.1   Catalyst

Platinum (Pt) has been used as the best catalyst in most cases. Hydrogen oxidation or oxygen reduction is not an exception and this is explained by the activation energy for these reactions, as shown in the following:

H_2 dissociation: $\Delta G_{298} = 412$ kJ mol^{-1}

O_2 dissociation: $\Delta G_{298} = 468$ kJ mol^{-1}

$H + O_2$ or $H_2 + O$:activation 40 kJ mol^{-1}

$2Pt + H_2 \rightarrow 2Pt\text{-}H$: zero activation

$2Pt + O_2 \rightarrow 2Pt\text{-}O$: zero activation

$1/2\ O_2 + H2 \rightarrow H_2O:\Delta G_{298} = -232$ kJ mol^{-1}

However, the intermediate product during the oxidation of methanol makes the catalysis complicated, and the reaction rate for making CO_2 out of methanol solution is slow. Moreover, a second metal such as Ruthenium (Ru) is required, which is explained by the bifunctional mechanism. In other words, activation of water or surface oxides at lower potentials makes the CO absorption bond weaker on the PtRu alloy catalyst. In the meantime, oxidative methanol dehydrogenation occurs on Pt by oxygen-like species on Ru, so that OH_{ads} species on Pt-Ru pair sites enables the continuous oxidation of CO to CO_2.

Although this PtRu is a currently available catalyst for finishing methanol oxidation, its reaction rate is still low, causing most polarization to occur within the activation polarization region, as shown in Fig. 13.2, and an alternative catalyst is demanded. In that manner, an electrochemical catalyst capable of efficient methanol oxidation, CO oxidation for anodic reactions, and oxygen reduction with a methanol-tolerant catalyst are critical. Of course, the high use of these materials is another issue; they are support materials, which require large surface area, high electrical conductivity, and chemical stability such as various micromorphologies and carbon chemistries. Those supported catalysts have their own critical issues in terms of the catalyst particle size of several nanometers having uniform distribution on the surface of support materials with less agglomeration at the high loading condition of more than 60 wt%.

All of these issues are directly related to electrode performance and DMFC system final cost.

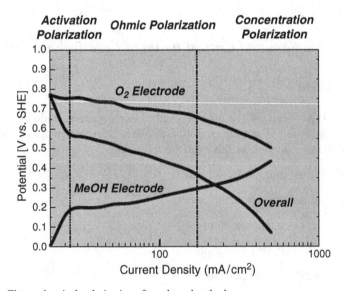

Fig. 13.2 Electrochemical polarization of anode and cathode

13.1.2.2 Membrane

With regard to the DMFC membrane, the technical goal has been well defined, i.e., maintaining the ionic conductivity at the current technical level and reducing the methanol crossover as much as possible, preferably down to 0%. There are several reasons why the perfluorinated sulfonic acid ionomer membrane is still the best, even though it permeates significant amounts of methanol during operation. First, its chemical and mechanical stability with high ionic conductivity is far better than the other types of membranes that are still struggling to reach 0.1 S cm^{-1} at ambient temperatures with rigidity. Second, and fortunately for PFSI membranes, most of the membrane and electrode assembly (MEA) processes have been designed perfectly for utilizing PFSI-based membranes.

Those critical functions of membrane for DMFC are simple but most important. Required functions are ionic conductivity, electrical insulation, gas and liquid (especially methanol) tightness, and chemical and mechanical stability. As indicated in Fig. 13.2, ohmic polarization is mainly due to the ionic resistance of membranes, but the low open circuit potential of cathode is also mainly coming from the voltage drop by mixed potential made of fuel crossover through the membrane. The low cost of material and process is also another factor in terms of commercialization. Especially for mobile applications, membranes have the additional function for mass balance of liquid fuel and water products circulated out of or through the membrane. In this manner, alternative membranes are under development and researchers are focused on four types: perfluorinated and partially fluorinated membranes; hydrocarbon; and composite and other ionomer modifications; inorganic materials. The current state of the art and technical approaches to these materials are discussed in detail elsewhere in this volume.

13.1.2.3 Membrane Electrode Assembly

Needless to say MEA is the core compartment of DMFC with its electrochemical reaction function. Figure 13.3 shows the typical microstructure of MEA, demonstrating the interface of catalyst and membrane. Its function is to deliver materials, such as catalyst and membrane, and physical functions for fuel delivery and recovery. Mobile application MEAs' minor functions, such as fuel delivery and recovery, have become more important. MEA can be defined as three compartments of membrane, a catalyst layer and diffusion electrode with a microporous layer. The catalyst layer consists of catalyst and interface materials with membrane. This layer has to be designed for effective utilization of the catalyst in order to minimize the use of precious metals while maintaining the produced proton path to the membrane. For this reason, this layer has to be electron- and ion-conductive with low fuel flow resistance. The membrane is located at the center of the MEA, with the catalyst layer coated (catalyst-coated membrane, CCM) in some cases. Its ion conduction would be made a lot easier by reducing the impedance at the interface with the catalyst.

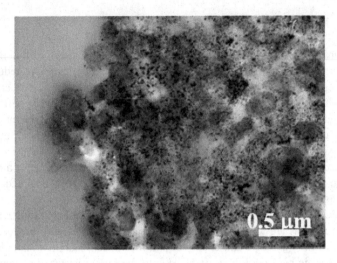

Fig. 13.3 Microstructure of MEA interface showing membrane, catalyst and micro porous layer

Generally, diffusion electrode functions as a path for liquid methanol, water, and CO_2 gas, and is composed of a microporous layer coated on a porous substrate. The materials and design of this compartment affect the slope of concentration polarization in Fig. 13.2. Importantly, in terms of commercialization, the diffusion electrode becomes more important since this will take all the bulky, heavy, and noisy compartments of BOP and the critical issue is somewhat serious in that all the BOP has to be removed by the functional materials and design of MEA.

13.1.2.4 Stack

The size of the stack is the first issue regarding commercialization. There are two types of stacking: one is stacking with bipolar plates, and the other is flat design with a monopolar structure. If the system allows some volume, it can go with a bipolar structure since this is more efficient in terms of reducing cell resistance and tightness. However, if the system does not allow much volume and prefers a thinner design, the stack has to be in the shape of a sheet with the cells arranged in one plane. Figure 13.4 shows the bipolar stack design for Note PC application and monopolar design for cellular phone application, respectively. The state of the art is somewhat prototype, but these will be the basic structures of DMFC stacks for mobile applications.

13.1.2.5 System

With all the descriptions of the functions and critical issues of materials and components, those are to be assembled in a system for typical application. Figure 13.5

(a)   (b)

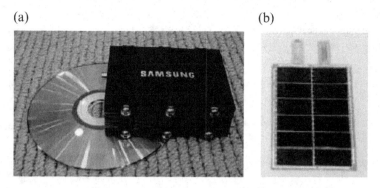

Fig. 13.4 Example of stacks having (**a**) bipolar and (**b**) monopolar design

(a)

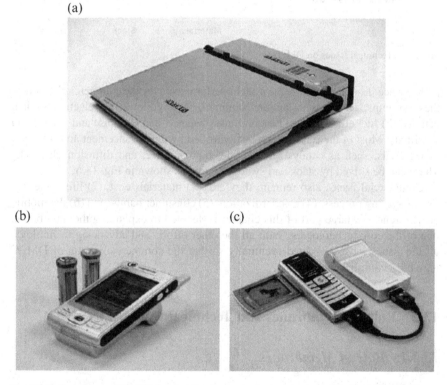

(b)   (c)

Fig. 13.5 Prototypes of DMFC for mobile electronic devices (**a**) note PC, (**b**) PDA phone, and (**c**) mobile charger for DMB phone

shows the state of the art DMFC for Note PC and cellular phones as a form of mobile charger and hybrid power for PDA phones. Depending upon the system requirements, the materials and components have to be designed, but the current status of technology stays in the level of system design decided by materials and component

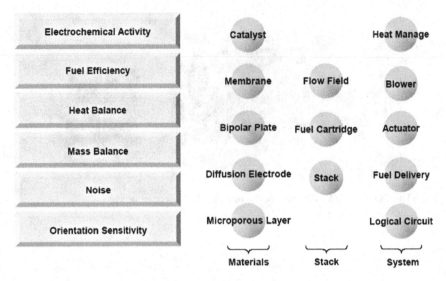

Fig. 13.6 Technical issues on DMFC

performance. Looking through the system performance, for the commercialization of these systems, technical issues are summarized as follows: (a) electrochemical activity, (b) fuel efficiency, (c) heat and mass balance, and (d) noise and orientation sensitivity. Most of these issues can be controlled by the advancement and selection of materials, such as catalyst, membrane, bipolar plate, and diffusion electrode. These can be solved by stack and system design, as shown in Fig. 13.6.

Commercial issues also remain; they are: (1) materials cost, (2) life cycle, (3) manufacturing process, (4) standardization, (5) consumer habits, and (6) the mobile device trends. A large part of this chapter is devoted to explaining the membrane, one of the core technologies. With all the other efforts and advancing technology, all the issues must be solved eventually so that the commercialization of DMFC may be accomplished.

13.2   Role of Membranes and Technical Approaches

13.2.1   Role of Membranes

Proton exchange membranes (PEMs) in DMFC should transport protons as an electrolyte and prevent fuel and oxidant mixing as a separator. Proton transport capacity affects the resistance and performance of fuel cells. The ability to separate influences the long-term stability and fuel efficiency. The insufficiency of function in separation, which is called methanol crossover, leads to deterioration of cathode catalysts, and thus generates mixed potential and decreases the performance and fuel efficiency of DMFC.

Proton transports in PEM have been explained by the vehicular and Grotthus mechanisms [3–5]. Proton transports take place with water molecules, and water membrane sustainability increases proton transport. Ionic conductivity is a function of mobility (stiffness) and capability of absorbing ions. Highly concentrated ionic sites and lower activation energy for proton exchange in PEM give high proton conductivity. The design of ionic sites concentration and stiffness of polymer matrix are issues for PEM development. Based on the preceding mechanisms, development of PEM for DMFC has the contradiction in ionic conductivity and methanol crossover. For high ionic conductivity PEM should be swollen enough with water while it should not for low methanol crossover. In DMFC, strategies for reducing methanol crossover must be considered in designing PEM. Methanol crossover occurs by the absorption of fuel (methanol) by membranes, diffusion through membranes, and desorption from membranes. To diminish methanol crossover, three stages of methanol crossover are considered. Diffusion through membrane related to electro-osmotic water drag and methanol with proton transport [6–9]. Electro-osmotic drag should be reduced by new electrolyte design [10]. Rigid polymer matrix and small free volume in membrane can reduce the electro-osmosis methanol diffusion. Besides ionic conductivity and methanol crossover, PEM criteria are comprised of low electronic conductivity, good mechanical properties in dry and hydrated status, capability for improvement with MEA fabrications, chemical and hydrolysis stability, and low cost. However, PEM, which satisfies both high ionic conductivity and low methanol crossover for DMFC, is not easily developed.

With current technology in PEM, the priority of criteria is chosen, followed by the operation methods of DMFC system. Ionic conductivity and fuel crossover of PEM are dependent on the operation temperature and fuel supply methods [11]. Ionic conductivity and methanol crossover increase with temperature. Concentration and rate of supplied fuel/oxidant affect methanol crossover [12]. As mentioned, DMFC system can be applied to automobile, stationary, and mobile energy sources [13]. Depending on applications, differences in operation temperature and fuel supply method leads to the strategic development of materials and designs of components. Higher flow rate of fuel/oxidant promotes methanol crossover because of high concentration gradient. Passive condition of DMFC application, which has a lower flow of fuel/oxidant, tends to make PEM developments complicated. And the method and amount of supplied water will decide fuel cell efficiency and system volume. Development of PEM will be directed by the application and water supplement methods. There have been many approaches for solving problems that are induced by methanol crossover. Methanol tolerant catalysts and design of electrodes are examples of these approaches [14–17]. The combination of membrane, electrode, and diffusion layer achieve the best fuel cell performance. For high fuel cell efficiency, DMFC systems have to have small volume and high performance. The water management approach is another way to resolve the issue among methanol concentration, power density, and fuel efficiency in a DMFC system. Water at the cathode side is used to dilute the anode fuel to achieve high-energy density [18] and long-term stability [19–21]. The methods of water reuse from the cathode are one of the important factors in DMFC efficiency.

PEM for DMFC is mainly characterized with ionic conductivity and methanol crossover. Ionic conductivity is tested with the four-point probe method. Through plane and in plane ionic conductivity is characterized with temperature and humidity [22]. Methanol crossover can be measured with diffusion cell method, pervaporation of methanol solution, dictating of CO_2 amount from cathode out stream [23]. An oxidant-impermeable property is also one of the important factors in PEM for DMFC. The properties of PEM in DMFC are also characterized with an interfacial resistance, an electro-osmotic drag, methanol crossover through MEA, and so on. In this chapter, development of PEM for DMFC are discussed. Approaches for PEM in DMFC are classified as fluorinated polymer, hydrocarbon polymer, modification of polymer materials, and the technical approaches are described as well.

13.2.2   Technical Approaches

13.2.2.1   Perfluorinated and Partially Fluorinated Membranes

By virtue of exceptional chemical inertness of fluoroalkyl group, fluorine containing ionomers can survive during a harsh fuel cell operating condition, and therefore mostly preferred in DMFC as well as PEMFC. On the other hand, environmental and safety issues remain serious problems in the manufacturing process. As summarized in Fig. 13.7, they can be classified into two major categories based on the chemical structure of their backbone. Conceptually, perfluorinated membranes consist of tetrafluoroethylene (PTFE)-like backbone where perfluorinated vinyl ethers with acid terminal group are co-polymerized. Other types of fluorine containing ionomers which include hydrocarbon moiety in their structure fall into partially fluorinated membranes.

Nafion and Other Perfluorinated Membranes

As often represented by Nafion, which was commercialized by Dupont in the late 1960s, perfluorinated membrane has been the most commonly used ionomer material

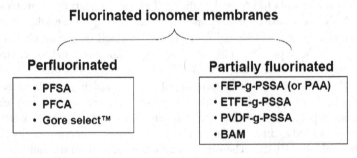

Fig. 13.7 Classification of fluorinated ionomers

due to its exceptional stability. It has been reported that the membrane is durable up to 60,000 h [24]. In fact, its first application was not toward fuel cell systems but for a permselective membrane in chlor-alkali cells. Its usefulness as a fuel cell membrane was verified in 1966 for NASA's space mission project [25].

In perfluorinated ionomers, a PTFE-based polymeric backbone offers chemical stability from the radical species or acid-base, which causes hydrolytic degradation of the polymer chain. Ionic conductivity is provided by pendant acidic moiety in carboxylate or sulfonate form. There are some reports on perfluorinated carboxylic acid (PFCA) materials, most of which are derived from Nafion [26–29]. However, PFCA is not suitable for fuel cell application due to its low proton conductivity. Perfluorosulfonic acid (PFSA) is the most favored choice among not only perfluorinated membranes but all other ionomers in fuel cell applications. Sulfonic acid form of Nafion is a representative PFSA and thus has been intensively studied since 1960s. Reported chemical structure of Nafion membrane is given in Fig. 13.8.

Such a non-ionic form of Nafion is further treated with a base followed by an acid to give a highly ion-conducting sulfonic acid ($-SO_3H$) form (>0.1S cm^{-1} at ambient temperature under humid conditions). By changing the co-monomer ratio, ion content and thus membrane ion exchange capacity (IEC = mequiv./g of polymer) can be varied over a wide range. However, too much ion content will cause excessive swelling of the membrane, which deteriorates its mechanical strength. Generally, Nafion with an equivalent weight of 1,100 is popularly used in most applications, although equivalent weights (EWs) of 900 and 1,200 are commercially available. Nafion has also stimulated the mass production of other PFSAs, such as Flemion (Asahi Glass), Aciplex-S (Asahi Kasei Corp., formerly Asahi Chemicals), and Dow membranes (Dow Chemical). These membranes are currently available in the market except Dow membrane, which is no longer manufactured. They are similar in structure since the PTFE backbone and PFSA side chains are main constituents for all of them [30]. The difference come from the details of side chains, as shown by co-monomer structures, as shown in Table 13.2.

The neat PFSA ionomer has the tendency to swell undesirably when wet, which results in poor mechanical properties. The swelling problem is particularly serious in fuel cells operating at elevated temperatures. In 1995, W.L. Gore & Associates developed Gore-Select membrane, which is a micro-reinforced composite structure of PTFE fabrics and PFSA ionomers. A similar reinforced PFSA was also reported by Asahi Glass, and these membranes are as thin as 5 mm with good mechanical strength. Since Nafion has received the greatest interest recently as a proton-conducting

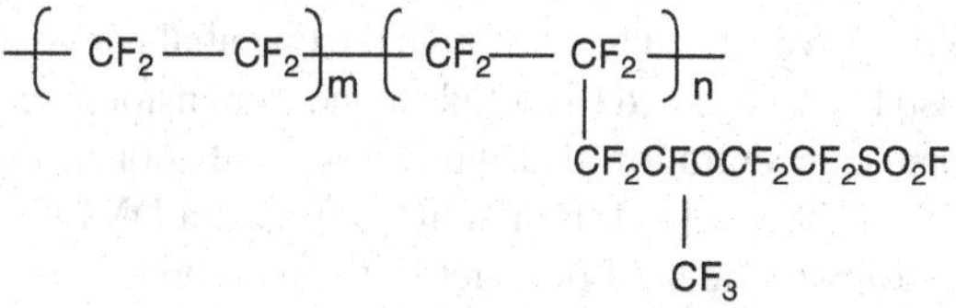

$$\left(\!CF_2\!-\!CF_2\!\right)_m\!\left(\!CF_2\!-\!CF_2\!\right)_n$$
$$CF_2CFOCF_2CF_2SO_2F$$
$$|$$
$$CF_3$$

Fig. 13.8 Chemical structure of Nafion

Table 13.2 Typical functional monomers of commercial PFSA

Membrane Co-monomer structure
Nafion, Flemion $CF_2=CFOCF_2CF(CF_3)OCF_2CF_2SO_2F$
Aciplex $CF^2=CFOCF_2CF(CF_3)OCF_2CF_2CF_2SO_2F$
Dow $CF_2=CFOCF_2CF_2SO_2F$

membrane for fuel cells, researchers are trying to understand water transport phenomena within the membrane from the microstructural point of view. There has been astonishing progress in theoretical modeling of transport mechanisms related to Nafion's nanoscale morphology. In early 1980s, Gierke and Hsu proposed a cluster network model that is thought to be the basis of other theories [31]. They described the polymeric membrane in terms of an inverted micellar structure in which the sulfonyl ion sites are spatially separated from the fluorocarbon backbone to form spherical clusters that are interconnected. According to this model, a dry Nafion membrane consists of clusters having a diameter of 1.8 nm containing 26 sulfonyl groups distributed on the inner pore surface. Upon swelling with water, the pore size becomes 4 nm and each pore contains 70 sulfonyl groups surrounding 1,000 water molecules in the pore. Gierke and Hsu also adopted a percolation theory to explain a nonlinear relationship between ionic conductivity and swelling ratio of the Nafion membrane. Recently, Eikerling et al. developed a random network model that assumes that the hydrated regions are randomly distributed in the polymer matrix to facilitate quicker transport of protons upon the movement of pendant sulfonyl groups [32]. Basically, their work was a modified version of the cluster network model. Some reports verified random network models by experimental techniques such as small angle X-ray scattering (SAXS) or atomic force microscopy (AFM) using Nafion membranes [33,34].

Meanwhile, transport theories from the microstructural point of view are being developed to understand the general transport mechanism of water molecules and protons in fuel cells. There have been efforts to explain methanol transport in DMFC by semi-empirical methods [23,35–42]. Verbrugge [43] developed a simple diffusion model for methanol through an ionomer membrane, assuming the dilute solution. He validated his model by experimentally showing that the methanol diffusion through the membrane occurred nearly as readily as through water. Cruickshank and Scott [44] presented a simple model to describe the methanol crossover in a vapor feed DMFC and its effect on the cathodic overpotential. The measured permeation rates of water and methanol through a Nafion117 membrane under varied pressure differentials across the PEM were used to determine the essential parameters in the model. This model has also been extended to include a one-dimensional model of the potential distribution and concentration distribution of methanol in the anode catalyst layer for a vapor feed DMFC [44]. Ren et al. [45] used a DMFC to experimentally determine the electro-osmotic drag of water in the membrane.

Although Nafion-based PFSA membranes still dominate most PEM fuel cell systems, there are critical drawbacks, such as high cost (US$700 m^{-2}), limited temperature range, and high methanol crossover. Particularly, the methanol crossover problem hampers its practical application as a DMFC membrane for long-term operation. The methanol permeability of the Nafion membrane is about 2×10^{-6} cm^2 s^{-1} at room temperature, which gets more serious when the temperature increases upon DMFC operation. This irreversibly contaminates the Pt catalyst at the cathode side, consequently aggravating cell performance. Methanol crossover problems limit the actual fuel concentration, which is as low as 1 M in DMFC experiments. Ren et al. [46] have reported that the methanol crossover rate can be significantly diminished by designing the cell anode to adjust the concentration of methanol between the flow field and membrane surface. They performed a 3,000 h lifetime test on a DMFC single cell at 75°C with 0.3 M methanol fuel where 40% loss of initial current was shown.

There are efforts to modify Nafion-based membranes to improve their properties. Various composite materials to improve disadvantageous properties of the PFSA ionomer are introduced later in another section.

Partially Fluorinated Membranes

Research has been actively conducted for the past 20 years in search of alternatives to perfluorinated membranes for PEMFC and DMFC applications. Partially fluorinated ionomers are of great interest among other various candidates in such efforts. Like perfluorinated membranes, partially fluorinated ionomers also have a PTFE-like polymer backbone as a main part to resist chemical attacks. However, ionic groups are attached to styrene moieties instead of perfluorinated side chains. Due to its availability and easy sulfonation, styrene is a reasonable choice for ionomeric materials. In the late 1990s, Ballard Power Systems introduced a partially fluorinated low-cost membrane for fuel cell applications [24,47,48]. Ballard Advanced Materials (BAM) membrane is a family of sulfonated styrenic co-polymers of α,β,β-trifluorostyrene and substituted α,β,β-trifluorostyrene co-monomers by emulsion polymerization. The chemical structure of BAM membrane is shown in Fig. 13.9.

Due to the excellent chemical stability of the fluorinated backbone structure of the BAM membrane, its performance is reported to be comparable to perfluorinated membranes, and more than 100,000 h of PEMFC operation is possible [48]. Similarly, by co-polymerizing trifluorostyrene and substituted vinyl monomers,

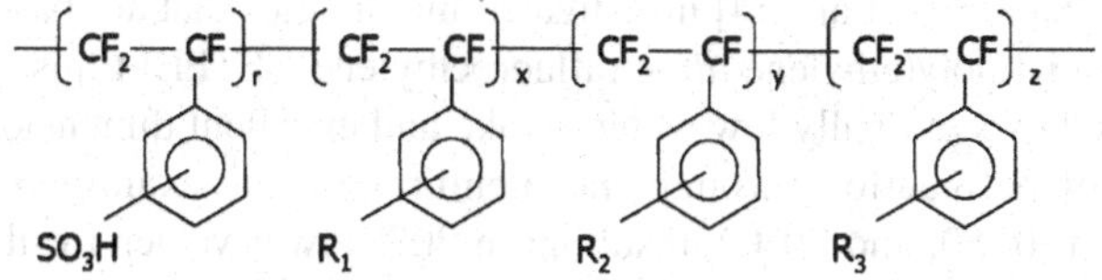

Fig. 13.9 Chemical structure of BAM membrane. R_1, R_2, R_3 are alkyls, halogens, OR, CF=CF$_2$, CN, NO$_2$, OH

Kim has improved the mechanical properties of partially fluorinated membranes and confirmed their performance by PEMFC and DMFC operation [49].

Another approach toward partially fluorinated membranes is the use of the radiation-grafting technique to cross-link styrenic polymers on the fluorinated base polymer film followed by sulfonation of the phenyl group. This approach has the advantage of: (1) easy modification of both surface property and bulk property, and (2) flexibility of introducing a variety of monomers with suitable microstuctural properties that would be generally difficult to obtain by classical synthesis routes. Recently, Gubler et al. developed radiation-grafted poly(tetrafluoro-ethylene-co-hexa-fluoropropylene)-g-polystyrene (FEP-g-PS) membranes and conducted PEMFC performance using those membranes [50]. The preparation of the membrane starts with irradiation of base polymer film of FEP (DuPont) with an electron beam followed by grafting PS in the presence of a divinylbenzene (DVB) cross-linker. Subsequently, the grafted films were sulfonated with chlorosulfonic acid. Using sulfonated FEP-sPS (sulfonated polystyrene) membranes, they performed a PEMFC single-cell durability test over 7,900 h at a cell temperature of 80–85°C, which resulted in 500 mA cm^{-2} of current density. Under this condition, the cell showed negligible degradation during the first 4,000 h. Other research groups developed a FEP-g-sAA (sulfonated acrylic acid) and tested on PEMFC [51]. Since acrylic acid itself has low ionic conductivity, further sulfonation at the α-carbon of the carboxylic group was performed to give comparable ion conductivity to the membrane. They reported a successful PEMFC MEA test, with the membrane having maximum 32% sulfonation degree. Radiation-grafted membrane has been also applied for DMFC system. While maintaining conductivity, a substantial decrease in methanol crossover was achieved by using low-cost ethylene-tetrafluoroethylene (ETFE)–based polymer membranes [52]. However, poor membrane-electrode interface and delamination were the problems to be solved.

Shen et al. [53] reported the performance of DMFC with radiation grafted polymer electrolyte membranes. The membranes used poly ethylene—tetrafluoroethylene (ETFE), polyvinylidene fluoride (PVDF), and low-density polyethylene (LDPE) as base polymer films. The base polymer films were grafted with polystyrene sulfonic acid (PSSA) as proton-conducting groups. They reported that varying the intrinsic properties of the co-polymer by increasing the degree of grafting (DOG) could increase the conductivity, methanol diffusion coefficient, and surface expansion rate of the membrane. As membrane thickness is increased, the resistance of the membrane also increased, which reduced DMFC performance. However, at the same time it limited methanol crossover, which counteracted this problem. They reported the effects of the substrate and DOG for composite membrane in DMFC performance. Saarinen et al. [54] investigated membranes that are based on 35-μm thick commercial poly(ethylene-*alt*-tetrafluoroethylene) (ETFE) films. These membranes showed exceptionally low water uptake and excellent dimensional stability. The methanol permeation through the membranes was examined at different temperatures (30, 50, and 70°C), resulting in 90% lower values for the ETFE-SA than for the Nafion 115, which was more than three times thicker than ETFE-SA.

Recently, Arico et al. investigated durability of ETFE-based membranes during DMFC operation. They performed DMFC operation at temperatures of 90–130°C [55]. Their MEA showed an initial performance of 92 mW cm^{-2} at 110°C, with 1 M methanol and 2 atm air as anode and cathode fuel, respectively. After 20 days the performance still remained the same as the initial value, and the authors confirmed a good adhesion of electrodes to the membrane by observation of MEA cross-section using optical microscopy.

Although perfluorinated or partially fluorinated ionomers are mostly favored as fuel cell membrane materials due to their unique resistance to chemical attack from their surroundings, fundamental solution to overcome inherent shortcomings of Nafion and other fluorinated membranes would be the development of chemically different types of ionomers. Alternative membranes for PFSA can reduce methanol crossover as low as two orders of magnitude. Such membranes are discussed in the following sections.

13.2.2.2 Hydrocarbon Membrane

As oxidation-reduction cycles of fuels produce electricity in fuel cell systems, fuel cells demand tough materials against harsh electrochemical and mechanical environments. Proton exchange membranes for direct methanol fuel cells strongly relate with the operation lifetimes. As discussed before, PFSA is widely used membranes due to its stability against electrochemical reactions and high acidity to confer high proton conductivity. Alternatives of PFSA, however, are required for the DMFC application because the fluorinated polymers have a few obstacles for communization: high cost of the membranes, safety of fluorinated monomers, methanol crossover through the membranes, low mechanical properties including high swelling ratio in organic solvents. The high cost and safety of fluorinated monomers might be the main stimulus for producing non-fluorinated membranes, because the methanol crossover and low mechanical properties would not be major concerns if anodic or cathodic catalyst efficiencies would be increased.

Hydrocarbon membranes are defined as polymeric materials consisting of non-fluorinated main chains and side chains containing functional groups. The functional groups are mainly sulfonic acids relating to proton conductivity. Such hydrocarbon polymers have merits on cost and structural diversity to fluorinated polymer membranes, but the hydrocarbon polymers should meet with other requirements for real applications, because sulfonated polystyrene-divinylbenzene co-polymer membranes adopted for the fuel cells for the Gemini space program in the early 1960s were extremely expensive and had short lifetimes due to oxidative degradation. Widely accepted criteria for high performance of proton exchange polymer membranes include: (1) high ionic conductivity, (2) low electronic conductivity, (3) low permeability to fuel and oxidant, (4) low water transport through diffusion and electro-osmosis, (5) oxidative and hydrolytic stability, (6) good mechanical properties in both dry and hydrated states, (7) cost, and (8) capability for fabrication into MEAs [56].

Membrane formation is usually done via casting polymer solutions on glasses or coating polymer melts or polymer solutions on substrates using coating machines. In casting polymer solution methods, membranes are prepared by dissolving co-polymers or homopolymers in effective solvents (e.g., dimethylacetamide), filtering the polymer solutions to remove undissolved polymers or impurities, and casting the solutions onto clean glass substrates. The polymer membranes were subjected to acid treatment to convert to the required acid form if the cast polymers were in the salt form. During film formation, the dissolved polymer chains adhere to the substrates—glass in this case—experiencing shear and tensile stresses [57]. The residual stresses of cast membranes cause cracks or deformation on membranes in long-term performance. Therefore, the stresses in cast films need to be removed through annealing or thermal treatments. The factors involved in film castings are solubility parameter of solvents, solvent concentration, rate of evaporation, and annealing times, whereas coating methods have rheological behaviors, such as viscosity of polymer solutions, gap size of extruding dies, pressure of extruders.

Sulfonation Methods for Hydrocarbon Membrane [22]

Polymer membranes have low proton conductivity until they contain enough proton conductive functional groups, e.g., sulfonic acid, in polymer backbones or side chains. Sulfonic acid in polymer chains shows a high dissociation constant, resulting in high proton conductivity in aqueous environments compared with other acids; therefore, most DMFC membranes use the sulfonic moiety as the proton transfer medium. Sulfonation methods of polymer membranes are generally classified as post-sulfonation in the presence of polymers and in situ sulfonation through co-polymerization of sulfonated monomers and nonsulfonated monomers.

The degree of sulfonation via post-sulfonation relates with the electrophilic characteristics of pristine polymers and sulfonating agents. For example, aromatic polymers, the most widely researched alternatives to fluorinated polymer membranes, are sulfonated using concentrated sulfuric acid, fuming sulfuric acid, chlorosulfonic acid, or sulfur trioxide (or complexes thereof), a trimethylsilylchlorosulfonate sulfonating agent, and a sulfur trioxide-triethyl phosphate complex. Since the sulfonation reactions are based on an electrophilic substitution reaction, the reaction depends on the substituents present on the aromatic rings. The electrophilic substitution reactions are carried out favorably with electron-donating substituents, but not with electron-withdrawing substituents. The sulfonation degree is also affected by the activated position on the aromatic rings. For example, in bisphenol A—based polymers, more than one sulfonic acid group can not be attached in a repeat unit due to the stability of the aromatic ring, the charge of existing sulfonic acid, and steric hindrances. Other methods for post-sulfonation are methylation (Li), sulfination by SO_2 gas, and oxidation. For lithium-based sulfonation, the choice of oxidant to convert the lithium sulfinate to sulfonic acid is a critical step. Disadvantages of post-sulfonation reactions are lack of precise control over the degree and location of functionalization, the possibility of side reactions, and degradation of the polymer backbone.

Sulfonated organic molecules have been a subject of research for their flame-retarding properties, because the sulfonated organic molecules show enhanced thermal characteristics. The acidities and thermal stabilities of sulfonic acid groups covalently bound to electron-deficient aromatic rings might be higher than those connected to electron-donating aromatic rings. Therefore, monomer sulfonations are generally carried out with aromatic monomers containing electron-deficient groups (dihalide: chloride, fluoride) using fumaric sulfonic acids and bases (NaOH, KOH) for neutralization. In situ sulfonations were preceded by co-polymerization of sulfonated monomers and neat monomers under the similar conditions of condensation polymerizations and nucleophilic substitution reactions. The resultant co-polymers afforded random co-polymers that had hydrophilic/hydrophobic portions. The hydrophilic portions were originated from the sulfonated monomers and the hydrophobic portions were from the non-sulfonated monomers. Degree of the sulfonation could be controlled by sulfonated monomer amounts.

Types of Hydrocarbon Membranes

Types of hydrocarbon membranes are classified as new hydrocarbons, cross-linked hydrocarbons (covalent cross-linking, ionic cross-linking through acid-base interaction), polymer blends, and grafted polymers. More rigorously, the methods for synthesizing membranes are categorized as nucleophilic substitution reactions (condensation polymerizations), and radical polymerization using vinyl monomers and radical initiators (grafted polymerizations). The nucleophilic substitution or condensation polymerization rates depend on intermediate nucleophilic substitution rates that involve the acidity/basicity of monomers, stoichiometric balance of monomers, and removal of impurities. The radical polymerization rates are affected by the initiator concentrations, monomer concentration, polarity of monomers (structures), and radical transfer constant of solvents. The grafted polymerizations depend on the radical formation rates of polymers, grafting initiator types, and catalyst types.

Sulfonated phenol-formaldehyde polymer is the first hydrocarbon-based polymer membrane in the literature. The phenolic polymer membrane was prepared by condensation polymerization of sulfonated phenol with formaldehyde, but the sulfonated phenolic polymer had low chemical and mechanical stability for fuel cell applications.

The first type of hydrocarbon membrane for fuel cell applications was the sulfonated polystyrene-divinylbenzene co-polymer membranes equipped for the power source in NASA's Gemini space flights, but the sulfonated polystyrene had low chemical stability for long-term applications, because the proton on the tertiary carbons and benzylic bonds are easily dissociated in an oxygen environment forming hydroperoxide radicals. Since a styrene monomer is easily co-polymerized with other vinyl monomers via radical polymerization methods, various styrenic polymers were researched intensively. Two commercial polystyrene-based/related membranes are available: BAM (Ballard), and Dais Analytic's sulfonated styrene-ethylene-butylene-styrene (SEBS) membrane. Dais membranes are produced using well-known commercial block co-polymers containing SEBS blocks. Due to the low chemical stability of styrene block, Dais membranes are designed for portable fuel cell power sources of 1 kW or less and <60°C operation temperatures.

Polyphosphazene-based polymers have high chemical and thermal stabilities and structural diversity through attaching various side chains and cross-linking onto the polymer backbones, so they could be potential materials for fuel cell applications. A typical synthesis of polyphosphazene is shown in Fig. 13.10. The strong point of polyphosphazenes is related to the synthetic and technological concerns because the structures of side groups can be diversified to unlimited molecular structures. Polyphosphazene films show low mechanical properties under hydrated conditions, so the mechanical property should be enhanced through modifications or polymer blends [58].

Arylene-based polymers have been intensively developed, because they have high mechanical and chemical stabilities next to fluorinated polymers [59]. Typical polymerizations of arylene-based polymers are carried out via nucleophilic substitution reactions in the presence of sulfonated monomers (Fig. 13.11) and none-sulfonated monomers (Fig. 13.12). Degree of sulfonation of the polymers can be controlled by varying amounts of sulfonated monomers. The arylene-based polymers are categorized

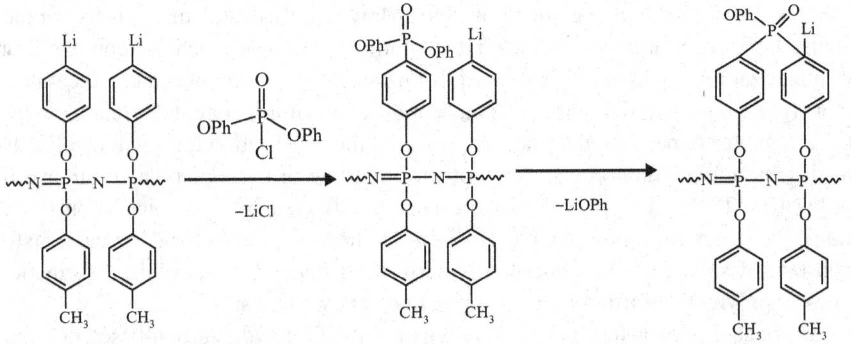

Fig. 13.10 Intramolecular coupling of polyphosphazene with diphenyl chlorophosphate

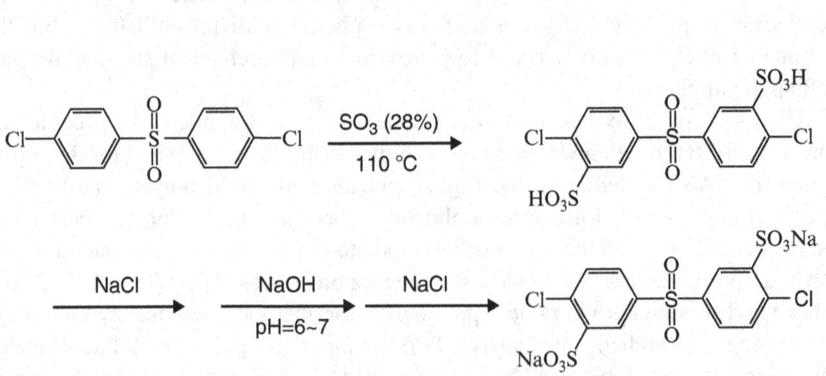

Fig. 13.11 Synthesis of 3,3-disulfonated 4,4-dichlorodiphenyl sulfone and its sodium salt

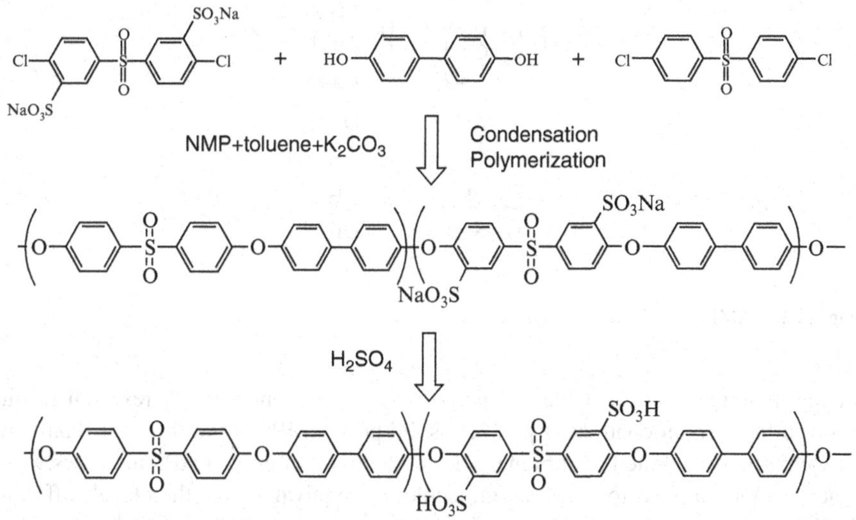

Fig. 13.12 Synthesis of directly co-polymerized wholly aromatic sulfonated poly(arylene ether sulfone), BPSH-*xx*, where *xx* is the ratio of sulfonated/unsulfonated activated halide

as following: poly(phenylene ether), poly(2,6-dimethyl-1,4-phenylenether), poly (2,6- diphenyl-1,4-poly(phenyl-quinoxaline)), poly(ether sulfone), poly(etherketone), poly(phenylenesulfide), and poly(phenyl-quinoxaline). Among the arylene-based polymers, poly(arylene ether) materials such as poly(arylene ether ether ketone), poly(arylene ether sulfone), and their derivatives attract huge attention because they have good chemical stability under harsh conditions.

The synthetic procedure for a most studied arylene-based co-polymer is carried out via step polymerization. For example, the first step in the synthesis involves preparation of short sequences of 4,4-diamino-2,2-biphenyl disulfonic acid condensed with 1,4,5,8-tetracarboxylic dianhydride. An adjusted ratio of these two monomers allows one to create different block lengths of the sulfonated sequence [60]. In the second polymerization step, the degree of sulfonation can be precisely controlled by regulating the molar ratio of BDA and the unsulfonated diamine, which is 4,4'-oxydianiline (ODA) in SPI. Degree of sulfonation of polyamides should be carefully controlled, because highly sulfonated polyamides became excessively swelling or dissolved in aqueous solutions. The stabilities of sulfonated polyamides also involve the structures. Five-membered ring polyamides, having high performances in non-sulfonated forms, showed expected quick degradation in sulfonated forms, because phthalic imides tend to hydrolyze, leading to chain scissions, but naphthalenic polyamides showed much higher stability in fuel cell applications.

Due to sulfonic acid group in 2-acrylamido-2-methylpropanesulfonic acid (AMPS) molecules, homo-poly(AMPS), or (AMPS)-based co-polymers (Fig. 13.13) would be applicable to proton-conductive membranes for fuel cell applications [61]. The amphiphilic monomer can be easily polymerized with radical initiators and the price

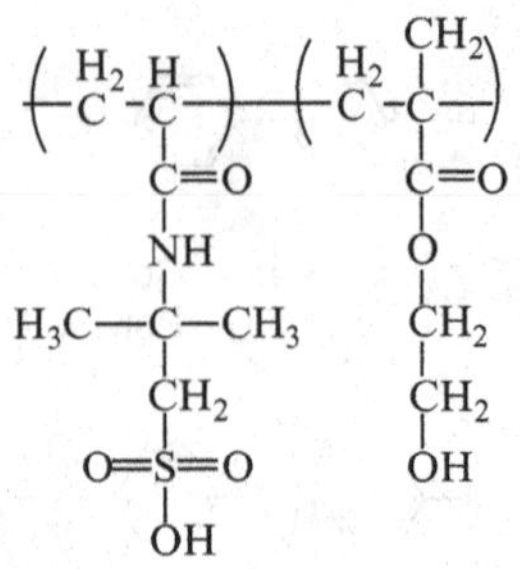

Fig. 13.13 AMPS—HEMA co-polymer

of the monomer is reasonable compared with other monomers. A research result showed that proton conductivity of semisolid poly-AMPS showed higher than that of the partially hydrated membrane. The conductivity of poly-AMPS increases with water content only up to about six molecules per equivalent and then levels off, but Nafion was hydrated to 15 water molecules per sulfonic acid group. The conductivity illustrated that poly-AMPS may have high proton conductivity under low water conditions. Since poly-AMPSs are highly swollen in aqueous media and become gels that are too fragile for DMFC membranes, the hydrophilicity of poly-AMPSs need to be controlled by introducing hydrophobic portions into the polymer backbones via co-polymerization of the hydrophobic monomer.

The basic concept to use block co-polymer for the application to the DMFC is that ordered hydrophilic/hydrophobic phase separations offer a route for the selective transport of proton ions with reduced methanol crossover in the hydrophilic domains, because block co-polymers can be selectively sulfonated using post-sulfonation methods, and the block co-polymers can be verified over a wide range of structures during anionic polymerization. For example, methanol transport behaviors of a tri-block co-polymer ionomer, sulfonated poly(styrene-isobutylene-styrene) (S-SIBS), were compared with Nafion to determine whether the sulfonated block co-polymer could serve as a viable alternative membrane for application to the DMFC [62]. The S-SIBS membranes showed approximately 5–10 times more methanol selectivity than that of Nafion117, although the S-SIBS membranes exhibited low conductivity compared with Nafion 117.

Chitosan is an inexpensive and abundant natural material with high molecular weight, but it is water soluble or highly swelling in aqueous solutions. Therefore, chitosan needs to be modified in solubility by blending or cross-linking using ionic characters. Recently, chitosan-based membranes (Fig. 13.14) were prepared by the spontaneous formation of a polyion complex due to the occurrence of ionic cross-linking [63].

Covalently cross-linked ionomer membranes or blend membranes are expected to have dimensional and chemical stability to reduce methanol crossover. Cross-linking agents such as divinylbenzene, sulfonyl n-imidazolide, 4,4′-diaminodiphenylsulfone can imbibe in the polymer main chain during polymerizations. Cross-linked polymer membranes showed reduced methanol crossover; however, it is questionable whether the cross-linking bridges are stable in the strongly acidic environment of the fuel cell.

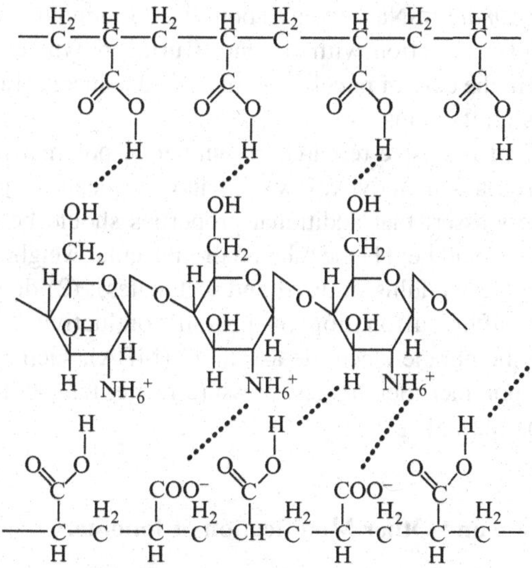

Fig. 13.14 Formation of polyelectrolyte complex with ionic interaction and hydrogen bonding

Another problem is that the cross-linked polymers become very brittle upon drying out, which is a severe problem when these membranes are used as fuel cell membranes. The brittleness is possibly caused by the inflexibility of covalent networks.

Flexible ionomer networks can be built up via mixing polymeric acids and polymeric bases, obtaining networks that contain ionic cross-links formed by proton-transfer from the polymeric acid onto the polymeric base. It was observed that upon drying, acid-base blend membranes show reduced brittleness compared with un—cross-linked or covalently cross-linked ionomer membranes, which is possibly caused by the flexibility of ionic network. The developed acid-base blend membranes show outstanding thermal stabilities determined by DSC and TGA.

Graft co-polymers, in which ion-containing polymer grafts are attached to a hydrophobic backbone, could be suitable structures for studying structure—property relationships in ion-conducting membranes if the length of the graft and number density of graft chains can be controlled. In principle, the grafting length would determine the size of ionic domains, whereas the number density of graft chains would determine the number of ionic domains per unit volume. Collectively, the size and number density of ionic aggregates/clusters are expected to control the degree of connectivity between ionic domains. Recently, several researchers have shown that it is possible to synthesize graft co-polymers possessing ionic grafts bound to hydrophobic backbones using macromonomers formed by stable free radical polymerization (SFRP) techniques [64]. The detailed synthesis and characterization of this class of co-polymer that comprises a styrenic main chain and sodium styrenesulfonate graft chains was reported by Holdcroft et al. PS-*g*-PSSNa was prepared by: (1) pseudo-living, tempo-mediated free radical polymerization of sodium styrenesulfonate (SSNa), and (2) termination with divinylbenzene (DVB).

The macromonomer, *mac*PSSNa, serves as both the co-monomer and emulsifier in the emulsion co-polymerization with styrene. During polymerization, the DVB terminus is located in the core of micellar particles and is incorporated into growing polystyrene (PS) as graft chains.

From the results of intensive research, a number of polymer membranes were reported on high proton conductivity, low methanol permeation, and water uptake; but many researchers assert that additional properties should be considered. The additional properties would be: (1) the relation of molecular weights and mechanical properties (e.g., tensile modulus in dried and wet states); (2) dependence of acid treatment methods after sulfonation on proton conductivity; (3) introducing optimized way for membrane electrode assembly (MEA), such as proper binders for new developed polymer membranes, pressure range, temperature, and catalyst layer configurations [22,65].

13.2.2.3 Composite and Other Modification of Ionomers

In order to diminish the methanol crossover through the membrane, main directions can be followed: a development of new polymer electrolytes, a modification of the structure of the conventional membranes, or a modification of new polymer electrolytes. Much research has been carried out to improve ionic conductivity and methanol crossover of PEM. Hydrocarbon-based new electrolytes have been described and sulfonated polyarylsulfone [22] shows better results as an electrolyte for DMFC. However, many kinds of membranes have contradictions in ionic conductivity and methanol crossover up to now. Ionic conductivity is decreased with reduction of fuel crossover. For application of the DMFC system, this contradiction has to be solved. To overcome this drawback, modification of conventional membrane or new membrane has been approached. Approach for modification has been achieved by various methods such as physical or chemical modifications and arrangements. Plasma etching [66] or palladium film [67] were tried to reduce methanol crossover by physical modification of perfluorinated polymer membrane itself. Yoon et al. [68] have carried out the modification of Nafion membrane with Pd film by sputtering. Pd film, which is thinner than 300+, acts as a barrier to methanol crossover and at the same time it also reduces proton conductivity. Nano silica layer has also been deposited by the Plasma enhanced chemical vapor deposition (PECVD) technique by Kim et al. [69]. They deposited nanoscale layer of silica (10, 32, and 68 nm) on Nafion membranes. The ionic conductivity of the Nafion/silica composite membranes was declined by about 7–22% to the unmodified Nafion membrane, but its methanol permeability was reduced by about 40–70%. Layered double hydroxides (LDH) were incorporated into polyelectrolyte membranes in order to investigate the electrochemical reaction processes affected by transport rates of methanol and proton in DMFC applications by Lee et al. [70]. Depending on different ion exchange capacities and the composition of LDH, the polyelectrolyte membranes gave different diffusion coefficients of methanol and proton conductivities, which were correlated with the maximum power density of a single-cell DMFC. They observed an optimal property condition when the proton conductivity and diffusion coefficient were balanced in a desirable manner providing a 24% increase of the maximum power density compared with the pristine Nafion membrane at 5 M of methanol feed concentration. The developed nanocomposite technique identified the effects of the diffusion coefficient and proton conductivity of polyelectrolyte membranes in the electrochemical reactions of DMFC. The TiO_2 layer has been coated for the methanol barrier [71]. Ren et al. [72] have modified hydrocarbon polymer, SPEEK (sulfonated poly ether ether ketone) with Nafion ionomer for decreasing methanol crossover with SPEEK and low interfacial resistance with electrode with Nafion ionomer. Also, a porous substrate that is not permeable to methanol has been impregnated with ionomeric material, usually a Nafion ionomer. In this approach, by reducing resistance with a thinner membrane, performance in DMFC can be improved with lowering fuel crossover. Inorganic materials have been adapted for reducing fuel crossover and increasing water sustainability. Morphology, size, and properties of inorganic materials can affect PEM performance in a DMFC system. Here, modified composite membranes will be classified as: (1) multilayer composite membranes with methanol impermeable substrate, (2) composite membranes with various kinds of organic materials and blends, and (3) composite membranes with inorganic materials.

Multilayer Composite Membrane

Composite membranes using a methanol impermeable substrate have been achieved to reduce methanol crossover for DMFC. With thin porous substrate, a composite membrane can reduce an internal resistance of membrane and reduce the material cost. Methanol impermeable porous membrane such as Teflon, PVDF, PTFE, and polyethylene-terephthalate has been used for multilayer composite membrane. Multilayer composite membranes have also been used in PEMFC for self-humidified membranes for water management. Water management of thin membranes can be used in DMFC systems for particular applications. Impregnation of ionomer to substrate was affected by the kinds of ionomer and substrate. A composite membrane with a thin porous substrate was initiated by Nafion ionomer with a porous PTFE [18] or micro PTFE fibril [73] for PEMFC applications. This approach was applied in DMFC. Lin found that Nafion/PTFE composite membrane caused reduction of not only in methanol diffusion crossover but also in the electro-osmosis of methanol crossover in the membrane. To improve membrane performance, impregnated ionomer and substrate should be well-connected for good interface formation. In DMFC, a methanol permeable ionomer and methanol impermeable substrate can be delaminated in highly concentrated fuel. The instability delaminating problem was solved by using chemical bonding between ionomer and substrate. Grafted ionomers to fluoropolymer backbone can also fall into a partially fluorinated polymer. This was reviewed in the previous section.

Yamaguchi et al. [74,75] reported the pore-filling membranes that reduce methanol crossover in a wide range of methanol concentrations due to the suppression effect of the substrate matrix. They used porous cross-linked high-density polyethylene (CLPE), on which proton conductivity was given by acrylamide tert-butyl sulfonate sodium. The MEA with pore-filling membrane successfully generated electricity, and showed excellent fuel cell performance with high concentration methanol fuel. To reduce the methanol crossover using a new type of membrane, high concentration methanol of 32 wt% (10 M) fuels can be used for the operation.

Multilayer composite membrane can have low resistance by thinner thickness and low methanol crossover by impermeable substrate. The correlation between impregnated ionomer and substrate are the factors that control performance of PEM in DMFC. Process is also an important design factor to obtain good fuel cell performance.

Composite Membrane with Organic Materials

Other approaches to develop PEM in DMFC are composite membranes, which are composed of organic-organic or organic-inorganic systems. Each system has been achieved by various processes or materials. Organic-organic systems consist of an ionic conducting phase and non-conducting phase or methanol-permeable and - impermeable phase. An organic-organic system can control the morphology, properties, and size of each phase. The process of membrane fabrication is one of the important factors to control performance of PEM for DMFC. The ratio and structure of phases control ionic conductivity and methanol crossover. These properties are driving the development of block co-polymer for DMFC. A block co-polymer can be applied for good control of morphology in polymer membranes. However, until now detailed study of blend and block co-polymer has not been investigated. More studies should be carried out to develop morphology of PEM for controlling methanol—water diffusion and electro-osmosis effects. In recent years, several polymer blends have been developed as effective proton exchange membrane alternatives for DMFCs [76–79]. Ren et al. [80] have investigated polymer blends of sulfonated poly(ether ether ketone) (SPEEK) with polyvinylidene fluoride (PVDF). The liquid uptake of the 9/1 SPEEK/PVDF blend membrane is about 20%, and the value of proton conductivity of this membrane is 1.75×10^{-3} S cm^{-1}, which is about 9% of Nafion 115. The methanol permeability of 9/1 SPEEK/PVDF blend membrane is about 1/20 that of Nafion 115 on the same testing conditions. Composite membrane with SPEEKK/polyaniline (PANI) was also investigated by Li et al [81].

Qiao et al. [82] have investigated a new type of chemically cross-linked polymer blend membranes consisting of poly(vinyl alcohol) (PVA), 2-acrylamido-2-methyl-1-propanesulfonic acid (PAMPS), and poly(vinyl pyrrolidone) (PVP) as proton-conducting polymer electrolytes. Through chemical cross-linking reactions between the hydroxyl groups (−OH) of the polyhydroxy polymer PVA and the aldehyde groups (−CHO) of glutaraldehyde cross-linkers, the swelling property of PVA was effectively controlled because of the increased cross-linking density. The resultant PVA—PAMPS—PVP blend membranes are thus capable of possessing all the required properties of a proton exchange membrane; namely, reasonable swelling and a good PVA—PAMPS—PVP blend. The proton conductivity of the membranes was investigated as a function of cross-linking time, blending composition, water content, and ion exchange capacity (IEC). The membranes attained 0.088 S cm^{-1} of the proton conductivity and 1.63 mequiv/g of IEC at 25°C for a polymer composition PVA—PAMPS—PVP of 1 1 0.5 in mass, and a methanol permeability of 6.1×10^{-7} cm^2 s^{-1}, which showed a comparable proton conductivity to Nafion 117, but only one third of Nafion 117 methanol permeability under the same measuring conditions. Polymer blending with PVDF has been carried out by many researchers because it is impermeable to PVDF methanol [83–85].

Nafion membranes were modified by the in situ electrodeposition of polypyrrole inside the membrane pores on the anode side only, in order to prevent the crossover of methanol in the direct methanol fuel cell (DMFC) [86]. The modified membranes were studied in terms of morphology, electrochemical characteristics, and methanol permeability. FTIR and SEM confirmed the presence of the polypyrrole on the anode side of the Nafion membrane. SEM showed the polymer to be present both on the membrane surface and inside the membrane pores. It was found to be deposited as small grains, with two distinct sizes; the smallest particles had a diameter of around 100 nm, whereas the larger particles had diameters of around 700 nm. Methanol permeability was determined electrochemically and was shown to be effectively reduced. Controlling the phase through blend or block co-polymer will be a good approach in PEM development for satisfying requirement in various DMFC systems.

Composite Membrane with Inorganic Materials

To use inorganic materials is another effective way for reducing methanol crossover. PEM in fuel cells is very permeable or prone to swelling in methanol fuel. High swelling of membrane in fuel solution reinforces or accelerates fuel crossover. Transport or crossover of fuel through the membrane is progressed by adsorption of fuel to membrane, diffusion through membrane, and desorption from membrane. Efforts to reduce methanol crossover can be achieved to suppress the phenomena of these three stages. Organic-inorganic systems have been tried to reduce methanol crossover through suppression of fuel crossover in three stages. Inorganic material hinders the swelling of PEM. Lower swelling of membrane reduces the diffusion rate of methanol. Inorganic materials play a role as a barrier or obstacle for diffusion of fuel through membrane in diffusion of methanol and electro-osmotic diffusion in DMFC. Electro-osmotic drag of methanol in DMFC can not be easily reduced by just inorganic materials. To reduce electro-osmotic drag effectively, ionic cluster or electrostatic forces inside ionic channel and the interaction of ionic and mobile phase should be controlled. Nano-sized metal oxides, heteropolyacid, layered inorganic materials, and modified inorganic materials have been used and well dispersed for composite membranes in polymer matrix. Morphology and size are important membrane properties.

An organic-inorganic system was first reported by Watanabe [87] in 1994. Watanabe reported that hygroscopic metal oxide enhanced the water-absorbing ability of PEMFC. Thereafter, many articles were written and patents granted. Antonucci et al. [88] tested oxide-containing membranes in a DMFC at 145°C, and obtained 350 mA cm^{-2} at 0.5 V. Mauritz [89] developed a sol-gel process to introduce SiO2 into the fine hydrophilic channels (50+ diameter). Detailed investigations on microstructure and fundamental properties of the composite membranes by a sol-gel process have been carried out [90–93]. A modification of the method is proposed by using a Nafion solution instead of Nafion membrane, mixed with tetraethyl orthosilicate (TEOS) or TMDES [94,95]. Modified inorganic materials have been also introduced into Nafion membranes by a sol-gel process [96]. Nunes et al. [97] reported the results of composite membranes with various inorganic materials. They developed inorganic modification by in situ hydrolysis of different alkoxides of Si, Ti, and Zr, and organically modified silanes with basic groups (I-silane). A remarkable reduction of methanol and water permeability was achieved by inorganic modification of SPEK and SPEEK. With zirconium phosphate and $Zr(OPr)_4/ACAC$ as inorganic compounds, water and methanol crossover was reduced drastically without diminishing conductivity to the same extent. Dimitrova et al. [98] also used inorganic materials for DMFC. They studied composite membranes that have different kinds of inorganic materials. They found that ionic conductivity of composite membranes had no correlation with change of crystallinity. Increase of crystallinity of composite membrane improves the mechanical properties. They concluded that addition of inorganic materials to the polymer matrix increased the water-absorbing ability and lowered methanol crossover by increasing crystallinity.

Nanocomposite membranes have been tested as a barrier of methanol fuel. Jung et al. [99] studied nanocomposite membranes with layered structure of inorganic materials. In this work, they made a nanocomposite membrane with various contents of montmorillonite (MMT), and modified montmorillonite (m-MMT) by the melt intercalation method using an internal mixer. Nanocomposite membranes were characterized by XRD and intercalation or exfoliation of layered structures was checked. Ionic conductivity of composite membranes was measured to be $8.9–6.7 \times 10^{-2}$ Scm^{-1} at 110°C. They confirmed that nanocomposite membranes have lower methanol crossover than Nafion membranes in MEA performance at high temperature.

Inorganic materials in composite membrane decrease not only methanol crossover, but also ionic conductivity. Without decreasing ionic conductivity, inorganic materials that have ionic sites have been investigated. Heteropolyacids and solid acids have been used as composite membranes. Also, inorganic materials modifications have been carried out for not decreasing ionic conductivity. Ponce et al. [100] have used an organic matrix of sulfonated polyetherketone (SPEK), different heteropolyacids, and an inorganic network of ZrO_2 or $RSiO_{3/2}$. The inorganic oxide network decreased methanol and water permeability across the membrane, as well as decreasing the bleeding out of the heteropolyacid. Hybrid membranes with SPEK and ZrO^2-TPA had the highest conductivity values (0.110–0.086 S cm^{-1}). Rhee et al. [101] reported nanocomposite membranes using modified montmorillonite. A modification method is given in Fig. 13.15.

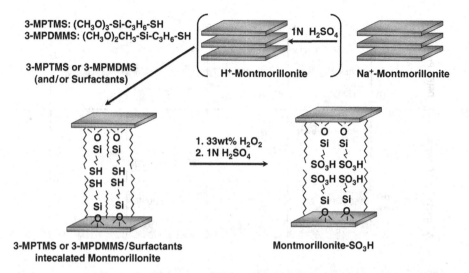

Fig. 13.15 Idealized representation of the processes of preparing the Nafion/sulfonated Montmorillonite composite

In their work, nanocomposite membranes decreased methanol crossover without reducing ionic conductivity. They developed the method of modification of inorganic materials to have ionic capability. Modified inorganic materials need to be more developed and ionic capability in polymer and inorganic materials will be maximized. And ionic capability in inorganic materials should be freer for the transport of proton in composite membranes. Development in PEM seems to be very difficult for getting high ionic conductivity and zero methanol crossover. Development of enhancing Grotthus mechanism at high temperature should be considered for selection of ionomer and inorganic materials in designing composite membrane. Vehicular mechanism of ionic sites in inorganic materials should be reinforced for low temperature application.

Composite membranes have been investigated for optimizing ionic conductivity and methanol crossover. Both properties can be represented as the ratio, and the ratio can be expressed as selectivity for membrane. Compositions of polymer and inorganic materials are the design factors in composite membranes. More subtle research needs to be advanced to examine the correlation between composite membranes and DMFC performance.

13.3   Technology Roadmap

13.3.1   Direct Methanol Fuel Cell for Mobile Application

Figure 13.16 shows the technology roadmap of DMFC for mobile applications. As shown in the roadmap, the energy density of the system reaches 500 Wh L^{-1}, which overcomes the rechargeable battery limit in terms of energy capacity by single charging.

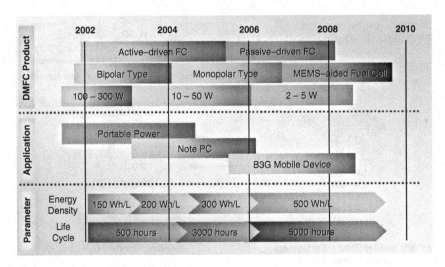

Fig. 13.16 Technology roadmap of DMFC for mobile application

13.3.2   Membranes for Direct Methanol Fuel Cell

PEM for DMFC has been investigated by developing fluorinated polymer, hydro-carbon polymer, and modification of ionomeric polymer. Increasing ionic conductivity and lowering methanol crossover are the main developing points for PEM with various DMFC applications. The membrane selectivity (ionic conductivity/methanol crossover) is a decisive point in active or semipassive DMFC systems. With proton transport mechanisms, methanol is easily diffused with proton and water inside ionic clusters, which is formed with ionic sites in DMFC. As mentioned, PEM should be altered depending on DMFC applications. For high-energy density and fuel cell efficiency, water produced by cathodes should be reutilized as anode fuel. For water management, PEM for active and passive DMFC system has high ionic conductivity and low water/methanol crossover. However, for water management in passive systems, PEM has high ionic conductivity, low methanol crossover, and high water crossover [21,102]. Even though strategies of PEM development should be decided with system application, no methanol crossover is certainly an important factor that should be solved.

To diminish methanol crossover, PEMs have been researched for decreasing three stages of methanol crossover (adsorption, diffusion, and desorption). Rigidity and lower swelling of hydrocarbon-based polymers can induce minimization of adsorption and diffusion of methanol. Hydrocarbon-based PEM is needed for optimization in mobility and stiffness of ionic groups. Control of ionic and non-ionic phases also can be developed for satisfactory ionic conductivity and methanol crossover at the same time. A smaller number of ionic groups reduce methanol adsorption to PEM. Distribution of ionic sites in PEM affects the morphology and size of ionic cluster. Block co-polymer is a good approach to control ionic cluster formation due to its unique microphase separation. The size and morphology of ionic clusters can be designed for reducing the diffusion rate of methanol through membrane. Some block or segmented co-polymer systems have been synthesized, but the subtle difference or effects are not yet well established in DMFC systems [103]. Properties such as polymer electrolytes can be investigated by altering polymer designs. Especially for micro-fuel cell of passive system, supply of highly concentrated fuel will be fulfilled with re-utilization of water produced by electrochemical reduction at cathode. Water re-utilization can be approached by self-humidified MEA or water collection from the cathode side. Self-humidified MEA will achieve water back diffusion through membrane from cathode to anode. Here, membranes for passive systems need to be methanol impermeable and water permeable. Active systems need membranes that are impermeable to either methanol or water, and MEA developments for new electrolytes should be pursued.

We would like to make one last but important comment. With all the new membranes, the current MEA technology that uses a Nafion ionomer as an electrode binder should be optimized for new electrolyte and DMFC systems. Novel MEA processes for new electrolytes will reduce the interfacial resistance of membranes and electrodes. Although conventional membranes, such as Nafion, have an elastomeric property and low glass transition temperature, methanol-impermeable membranes have some degree of rigidity. Thus, the adhesion process has to be optimized so that those efforts of membrane development can maximize MEA performance.

Figure 13.17 shows a roadmap for the development of membranes for DMFC and explains the advance of materials and performance. Maintaining ionic conductivity with methanol permeability down to zero would be the direction via those technologies available through various approaches discussed in this chapter.

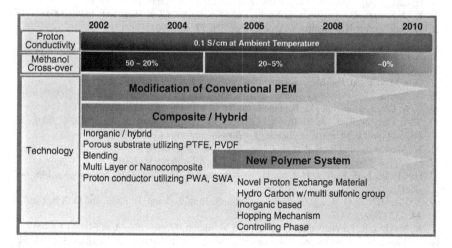

Fig. 13.17 Technology roadmap of membrane for DMFC

13.4 Conclusion

Microfuel cells, mostly recognized as a form of DMFC, are defined at the International Electrotechnology Committee (IEC) as follows: Small-voltage, low-power fuel cell systems that include the fuel container/cartridge and are connected to handheld or wearable electronic devices by flexible cords and plug arrangements or termination connectors integrated into the casing of electric DC unit power electric devices such as laptops, cell phones, and PDAs and fuel cartridges are removable articles or devices that contain and supply fuel cell power units or internal reservoirs, not to be refilled by the user. This kind of standardized activity explains that this technology is outside clean energy issues. Commercialization of DMFC for mobile electronics applications has to be considered in a different way compared with those fuel cell systems for residential or distributed power and vehicle applications, since there will not be any support from international regulation or any national policy related to clean or alternative energy that could accelerate its development.

Customers will need new energy sources for their convenience, and it is thought that this will exert action on even more powerful resources for the realization of DMFC. Of all the core technologies, membrane is in the core of MEA, Stack, and System. Achievement of the proper level of ionic conductivity, methanol permeability, water permeability, life cycle, and processing cost will reduce the burden of the other interfacing components of the DMFC system and it will move this system up in customers' hands.

References

1. H. Chang, paper presented at the Small Fuel Cells for Portable Applications, 2005.
2. H. Chang, paper presented at the International Fuel Cell Expo, Tokyo 2004.
3. K. D. Kreuer, S. J. Paddison, E. Spohr, and M. Schuster, *Chem. Rev.* **104**, 4637 (2004).
4. H. Tang, and P. N. Pintauro, *J. Appl. Polym. Sci.* **79**, 49 (2001).
5. R. Wycisk, and P. N. Pintauro, *J. Membr. Sci.* **119**, 155 (1996).
6. X. Ren, T. E. Springer, T. A. Zawodzinski Jr., and S. Gottesfeld, *J. Electrochem. Soc.* **147**, 466 (2000).
7. G. J. M. Janssen, and M. L. J. Overvelde, *J. Power Sources* **101**, 117 (2001).
8. B. Smitha, D. Suhanva, S. Sridhar, and M. Ramakrishna, *J. Membr. Sci.* **241**, 1 (2002).
9. C. A. Edmondson, and J. J. Fontanella, *Solid State Ionics* **152–153**, 355 (2002).
10. S. Y. Cha, N. Tran, A. T. Duong, G. Hou, M. Lefebvre, and A. Attia, *PBFC 2003, 1st International Conference of Polymer for Battery and Fuel Cell* (2003).
11. K. A. Mauritz, and R. B. Moore, *Chem. Rev.* **104**, 4535 (2004).
12. H. Kim, J. Cho, J. Yoon, and H. Chang, *2001 ECS Conference* (2001).
13. H. Chang, J. R. Kim, J. H. Cho, H. K. Kim, and K. H. Choi, *Solid State Ionics* **148**, 601 (2002).
14. J. Guo, G. Sun, O. Wang, G. Wang, Z. Zhou, S. Tang, L. Jiang, B. Zhou, and O. Xin, *Carbon* **44**, 152 (2006).
15. K. Makino, K. Furukawa, K. Okajima, and M. Sudoh, *Electrochim. Acta* **51**, 961 (2005).
16. S. Baranton, C. Coutanceau, J. M. Leger, C. Roux, and P. Capron, *Electrochim. Acta* **51**, 517 (2005).

17. R. Dillon, S. Srinivasan, A. S. Arico, and V. Antonucci, *J. Power Sources* **127**, 112 (2004).
18. W. L. Gore, B. Bahar, A. R. Hobson, J. A. Kodle, and A. Zuckerbrod, (W. L. Gore & Associates, Inc., US patent 5547551, 1996).
19. C. Xie, J. Bostaph, and J. Pavio, *J. Power Sources* **136**, 55 (2004).
20. K. H. Kim, S. J. Choi, and H. Chang, *203rd ECS Paris Meeting* (2003).
21. G. Q. Lu, F. Q. Liu, and C. Y. Wang, *Electrochem. Solid-State Lett.* **8**, 1099 (2005).
22. M. A. Hickner, H. Ghassemi, Y. S. Kim, B. R. Einsla, and J. E. McGrath, *Chem. Rev.* **104**, 4587 (2004).
23. J. A. Drake, W. Wilson, and K. Killeen, *J. Electrochem. Soc.* **151**, A413 (03 2004, 2004).
24. A. E. Steck, Proceedings of 1st International Symposium on New Material Fuel Cell Systems, 74, (1995, 1995).
25. W. G. Grot, *Chem. Ind.* **82**, 161 (1994).
26. H. L. Yeager, B. Kipling, and R. L. Dotson, *J. Electrochem. Soc.* **127**, 303 (1980).
27. M. A. F. Robertson, and H. L. Yeager, *Macromolecules* **29**, 5166 (1996).
28. M. Doyle, M. E. Lewittes, M. G. Roelofs, S. A. Perusich, and S. A. Lowrey, *J. Membr. Sci.* **184**, 257 (2001).
29. S. A. Perusich, P. Avakian, and M. Y. Keating, *Macromolecules* **26**, 4756 (1993).
30. R. G. Rajendran, *MRS Bullet.* **30**, 587 (2005).
31. T. D. Gierke, and W. Y. Hsu, *Perfluorinated Ionomer Membranes*. A. Eisenberg, H. L. Yeager, Eds., ACS Symposium Series No. 180. (American Chemical Society, Washington, DC, 1982), p 283.
32. M. Eikerling, A. A. Kornyshev, and U. Stimming, *J. Phys. Chem. B* **101**, 10807 (1997).
33. H. G. Haubold, T. Vad, H. Jungbluth, and P. Hiller, *Electrochim. Acta* **46**, 1559 (2001).
34. P. J. James, T. J. McMaster, N. J. M, and M. J. Miles, *Polymer* **41**, 4223 (2000).
35. H. A. Every, M. A. Hickner, J. E. McGrath, and T. A. Zawodzinski Jr., *J. Membr. Sci.* **250**, 183 (2005).
36. P. Choi, N. H. Jalani, and R. Datta, *J. Electrochem. Soc.* **152**, A1548 (2005).
37. P. Argyropoulos, K. Scott, A. K. Shukla, and C. Jackson, *J. Power Sources* **123**, 190 (2003).
38. V. M. Barragan, and A. Heinzel, *J. Power Sources* **104**, 66 (2002).
39. J. Divisek, J. Fuhrmann, K. Gartner, and R. Jung, *J. Electrochem. Soc.* **150**, A811 (2003).
40. G. Murgia, L. Pisani, A. K. Shukla, and K. Scott, *J. Electrochem. Soc.* **150**, A1231 (2003).
41. Z. H. Wang, and C. Y. Wang, *Proceedings of International Symposium on Direct Methanol Fuel Cells.* **286** (2001).
42. Z. H. Wang, and C. Y. Wang, *J. Electrochem. Soc.* **150**, A508 (2003).
43. M. W. Verbrugge, *J. Electrochem. Soc.* **136**, 417 (1989).
44. J. Cruickshank, and K. Scott, *J. Power Sources* **70**, 40 (1998).
45. X. Ren, and S. Gottesfeld, *J. Electrochem. Soc.* **144**, L267 (2001).
46. X. Ren, M. S. Wilson, and S. Gottesfeld, *J. Electrochem. Soc.* **143**, L12 (1996).
47. J. Wei, C. Stone, and A. E. Steck, (Power Systems, Inc., US Patent 5422411, 1995).
48. A. E. Steck, and C. Stone, Proceedings of 2nd International Symposium On New Material Fuel-Cell and Modern Battery Systems II, 792 (1997).
49. H. Kim, (Samsung Electronics, US Patent, 6774150B2, 2004).
50. L. Gubler, H. Kuhn, T. J. Schmidt, G. G. Scherer, H. P. Brack, and K. Simbeck, *Fuel Cells* **4**, 196 (2004).
51. M. Patri, V. R. Hande, S. Phadnis, B. Somaiah, S. Roychoudhury, and P. C. Deb, *Polym Adv. Technol.* **15**, 270 (2004).
52. K. Scott, W. M. Taama, and P. Argyropoulos, *J. Membr. Sci.* **171**, 119 (2000).
53. M. Shen, S. Roy, J. W. Kuhlmann, K. Scott, K. Lovell, and J. A. Horsfall, *J. Membr. Sci.* **251**, 121 (2005).
54. V. Saarinen, T. Kallio, M. Paronen, P. Tikkanen, E. Rauhala, and K. Kontturi, *Electrochim. Acta* **50**, 3453 (2005).
55. A. S. Arico, V. Baglio, P. Creti, A. Di Blasi, V. Antonucci, J. Brunea, A. Chapotot, A. Bozzi, and J. Schoemans, *J. Power Sources* **123**, 107 (2003).
56. J. A. Kerres, *J. Membr. Sci.* **185**, 3 (2001).

57. M. S. Tirumkudulu, and W. B. Russel, *Langmuir* **21**, (2005).

58. H. R. Allcock, M. A. Hofmann, C. M. Ambler, and R. V. Morford, *Macromolecules* **35**, 3484 (2002).

59. F. Wang, M. Hickner, Y. S. Kim, T. A. Zawodzinski, and J. E. McGrath, *J. Membr. Sci.* **197**, 231 (2002).

60. C. Genies, R. Mercier, B. Sillion, N. Cornet, G. Gebel, and M. Pineri, *Polymer* **42**, 359 (2001).

61. C. W. Walker Jr., *J. Power Sources* **110**, 144 (2002).

62. Y. A. Elabd, E. Napadensky, J. A. Sloan, D. M. Crawford, and C. W. Walker, *J. Membr. Sci.* **217**, 227 (2003).

63. B. Smitha, S. Sridhar, and A. A. Khan, *Macromolecules* **37**, 2233 (2004).

64. J. Ding, C. Chuy, and S. Holdcroft, *Macromolecules* **35**, 1348 (2002).

65. Y. S. Kim, F. Wang, M. Hickner, S. McCartney, Y. T. Hong, W. Harrison, T. A. Zawodzinski, and J. E. McGrath, *J. Polym. Sci. B* **41**, 2816 (2003).

66. W. C. Choi, J. D. Kim, and S. I. Woo, *J. Power Sources* **96**, 411 (2001).

67. E. Smotkin, T. Mallouk, M. Wardchael, and K. Ley, (US Patent 5846669, 1998).

68. S. R. Yoon, G. H. Hwang, W. I. Cho, I. H. Oh, S. A. Hong, and H. Y. Ha, *J. Power Sources* **106**, 215 (2002).

69. D. Kim, M. A. Scibioh, S. Kwak, I. H. Oh, and H. Y. Ha, *Electrochem. Commun.* **6**, 1069 (2004).

70. K. Lee, J. H. Nam, J. H. Lee, Y. Lee, S. M. Cho, C. H. Jung, H. G. Choi, Y. Y. Chang, J. U. Kwon, and J. D. Nam, *Electrochem. Commun.* **7**, 113 (2005).

71. D. Kim, J. Lee, T.-H. Lim, I.-H. Oh, and H. Y. Ha, *J. Power Sources* **155**, 203 (2006).

72. S. Ren, C. Li, X. Zhao, Z. Wu, S. Wang, G. Sun, Q. Xin, and X. Yang, *J. Membr. Sci.* **247**, 59 (2005).

73. Y. Higuchi, N. Terada, H. Shimoda, and S. Hommura, (Ashai Glass Co., EP1139472, 2001).

74. T. Yamaguchi, F. Miyata, and S. Nakao, *Adv. Mater.* **15**, 1198 (2003).

75. T. Yamaguchi, H. Kuroki, and F. Miyata, *Electrochem. Commun.* **7**, 730 (2005).

76. J. A. Kerres, A. Ullrich, F. Meier, and T. Haering, *J. Membr. Sci.* **206**, 443 (2002).

77. J. A. Kerres, and A. Ullrich, *Sep. Purif. Technol.* **22–23**, 1 (2001).

78. J. Lin, M. Ouyang, J. M. Fenton, H. R. Kunz, J. T. Koberstein, and M. B. Cutlip, *J. Appl. Polym. Sci.* **70**, 121 (1998).

79. C. Hasiotis, *Electrochim. Acta* **46**, 2401 (2001).

80. S. Ren, G. Sun, C. Li, Z. Wu, W. Jin, and W. Chen, *Mater. Lett.* **60**, 44 (2006).

81. X. Li, D. Chen, D. Xu, C. Zhao, Z. Wang, H. Lu, and H. Na, *J. Membr. Sci.* **275**, 134 (2006).

82. J. Qiao, T. Hamaya, and T. Okada, *Polymer* **46**, 10809 (2005).

83. K.-Y. Cho, J.-Y. Eom, H.-Y. ung, N.-S. Choi, Y. M. Lee, J.-K. Park, J.-H. Choi, K.-W. Park, and Y.-E. Sung, *Electrochim. Acta* **50**, 583 (2004).

84. H. J. Kim, H. J. Kim, Y. G. Shul, and H. S. Han, *J. Power Sources* **135**, 66 (2004).

85. G. K. S. Prakash, M. C. Smart, Q.-J. Wang, A. Atti, V. Pleynet, B. Yang, K. McGrath, G. A. Olah, S. R. Narayanan, and W. Chun, *J. Fluorine Chem.* **125**, 1217 (2004).

86. M. A. Smit, A. L. Ocampo, M. A. Espinosamedina, and P. J. Sebastian, *J. Power Sources* **124**, 59 (2003).

87. M. Watanabe, paper presented at the The Electrochemical Society Meeting PV 94–2, Pennington NJ 1994.

88. P. L. Antonucci, A. S. Arico, P. Creti, E. Ramunni, and V. Antonucci, *Solid State Ionics* **125**, 431 (1999).

89. K. A. Mauritz, *Mater. Sci. Eng. C* **6**, 121 (1998).

90. Q. Deng, R. B. Moore, and K. A. Mauritz, *Chem. Mater.* **7**, 2259 (1995).

91. Q. Deng, Y. Hu, R. B. Moore, C. L. McCormick, and K. A. Mauritz, *Chem. Mater.* **9**, 36 (1997).

92. Q. Deng, C. A. Wilkie, R. B. Moore, and K. A. Mauritz, *Polymer* **39**, 5961 (1998).

93. D. H. Jung, S. Y. Cho, D. H. Peck, D. R. Shin, and J. S. Kim, *J. Power Sources* **106**, 173 (2002).

94. R. A. Zoppi, I. V. P. Yoshida, and S. P. Nunes, *Polymer* **39**, 1309 (1997).

95. R. A. Zoppi, and S. P. Nunes, *Electroanal. Chem.* **445**, 39 (1997).

96. H. Kim, J. Cho, J. Yoon, and H. Chang, (Korea Patent KR-P0413801, 2004).
97. S. P. Nunes, B. Ruffmann, E. Rikowski, S. Vetter, and K. Richau, *J. Membr. Sci.* **203**, 215 (2002).
98. P. Dimitrova, K. A. Friedrich, U. Stimming, and B. Vogt, *Solid State Ionics* **150**, 115 (2002).
99. D. H. Jung, S. Y. Cho, D. H. Peck, D. R. Shin, and J. S. Kim, *J. Power Sources* **118**, 205 (2003).
100. M. L. Ponce, L. Prado, B. Ruffmann, K. Richau, R. Mohr, and S. P. Nunes, *J. Membr. Sci.* **217**, 5 (2003).
101. C. Rhee, H. Kim, J. S. Lee, and H. Chang, *Chem. Mater.* **17**, 1691 (2005).
102. S. Gottesfeld, Proceedings of Small Fuel Cells for Portable Applications, 6th Edition (2005).
103. C. K. Shin, G. Maier, B. Andreaus, and G. G. Scherer, *J. Membr. Sci.* **245**, 147 (2004).

Chapter 14
Modified Nafion as the Membrane Material for Direct Methanol Fuel Cells

Jun Lin, Ryszard Wycisk, and Peter N. Pintauro

Abstract There have been numerous studies on modifying DuPont's Nafion (a perfluorosulfonic acid polymer) in order to improve the performance of this membrane material in a direct methanol fuel cell. Modifications focused on making Nafion a better methanol barrier, without sacrificing proton conductivity, so that methanol crossover during fuel cell operation is minimized. In this chapter, a brief literature survey of such modifications is presented, along with recent experimental results (membrane properties and fuel cell performance curves) for: (1) thick Nafion films, (2) Nafion blended with Teflon-FEP or Teflon-PFA, and (3) Nafion doped with polybenzimidazole.

14.1 Introduction

DuPont's Nafion (perfluorosulfonic acid) is the membrane material of choice for moderate temperature ($T \leq 80°C$) hydrogen/air fuel cells, but it does not perform well in a direct methanol fuel cell (DMFC) due to high methanol crossover [4,23,24]. Membrane permeation of methanol during DMFC operation results in low power output due to chemical oxidation of methanol at the cathode, causing: (1) electrode depolarization, (2) consumption of O_2, (3) cathode catalyst poisoning by CO (an intermediate of methanol oxidation), and (4) excessive water build-up at the cathode (water being produced by methanol oxidation), which limits O_2 access to cathode catalyst sites. Additionally, the overall fuel utilization efficiency of the fuel cell is lowered when there is excessive methanol crossover. The effect of methanol crossover on fuel cell performance plots of voltage (V) vs. current density (i) and power density vs. current density is shown in Figure 14.1 for Nafion 117 at two different methanol feed concentrations (1 and 5 M) and in Fig. 14.2 for Nafion membranes of different thickness (Nafion 117 with a wet thickness of 215 μm and Nafion 112 with a wet thickness of 60 μm) at a single methanol feed concentration (1.0 M). In Fig. 14.1, the lowering of the V-i curve at 5.0 M and the consequential reduction in power output, as compared with 1.0 M methanol data, is due to excessive methanol crossover. Similarly, in Fig. 14.2, there is a drop in power when a Nafion 117 membrane is replaced by the thinner Nafion 112.

S.M.J. Zaidi, T. Matsuura (eds.) *Polymer Membranes for Fuel Cells*,
doi: 10.1007/978-0-387-73532-0, © Springer Science+Business Media, LLC 2009

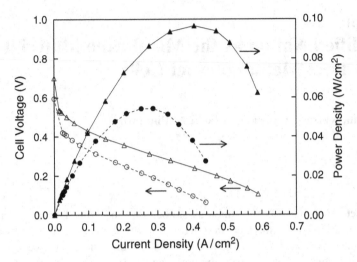

Fig. 14.1 Effect of methanol feed concentration on direct methanol fuel cell performance with a Nafion 117 membrane. $T = 60°C$, ambient air at 500 sccm. (*triangle*) and (*filled rectangle*) 1.0 M methanol; (*circle*) and (*filled circle*) 5.0 M methanol

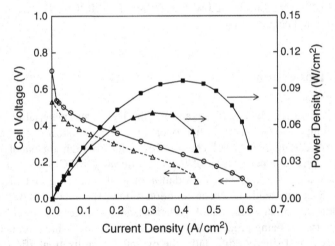

Fig. 14.2 The effect of membrane thickness on direct methanol fuel cell performance. $T = 60°C$, 1.0 M methanol feed, ambient pressure air at 500 sccm. (*circle*) and (*filled rectangle*) Nafion 117 (215 μm wet thickness); (*triangle*) and (*filled triangle*) Nafion 112 (60 μm wet thickness)

There has been considerable research on modifying Nafion, so as to improve its properties for use in a direct methanol fuel cell. In this chapter, a review of Nafion-based DMFC membranes is presented, including a literature survey followed by recent results by the present authors on improving Nafion by: (1) using thick stacked Nafion membranes, (2) blending Nafion with Teflon-FEP or Teflon-PFA, and (3) doping Nafion with polybenzimidazole.

14.2   Literature Survey

14.2.1   Bulk Modification of Nafion

Perhaps the simplest modification of Nafion for DMFCs was carried out by Tricoli [28], who doped Nafion 117 with Cs^+ cations. The decrease in size of hydrated micellar domains, upon substituting a portion of the membranes sulfonate sites with cesium, caused the methanol permeability to fall. The ratio of proton conductivity to methanol permeability, which reflects membrane selectivity, increased by a factor of 2.5 when Nafion 117 in the H^+ form was fully exchanged with cesium. To prevent cesium from leaching out of the membrane during fuel cell operation, Tricoli proposed adding a suitable amount of Cs^+ salt to the methanol/water feed mixture.

In another bulk modification [1], Nafion 112 membrane was impregnated with 1-vinylimidazole and a photoinitiator followed by in situ UV polymerization. It was found that complexation of the poly(1-vinylimidazole) (PVI) base with the sulfonic groups of Nafion led to a reduction of water sorption and an improvement in the membrane's methanol barrier property. Although incorporation of PVI produced a drop in proton conductivity (0.03 S cm^{-1} at 36% PVI vs. 0.08 S cm^{-1} for neat Nafion at room temperature under fully humidified conditions), the overall membrane selectivity (the ratio of proton conductivity to methanol permeability) was greater than that of commercial Nafion. Fuel cell membrane electrode assemblies (MEAs) with a Nafion membrane containing 3% PVI generated 190 mW cm^{-2} (at 60°C with 2 M methanol and oxygen), as compared with 170 mW cm^{-2} with a Nafion 112 MEA. In a similar fashion, Nafion/polypyrrole composite membranes were prepared by absorption of pyrrole into Nafion 115 membranes and subsequent in situ polymerization upon immersion in a H_2O_2 solution [21]. Nano-sized polypyrrole particles were incorporated mainly in the membrane's ionic domains with some evidence of membrane surface enrichment. Membrane water uptake, proton conductivity, and methanol permeability decreased with increasing polypyrrole content. DMFC power output (at room temperature with 2.0 or 5.0 M methanol and oxygen) with the modified membrane, however, was not as high as that with a pristine Nafion 115 MEA.

High-energy radiation (electron beam or gamma radiation) has been used to modify Nafion either in the bulk [29] or in a thin subsurface region [6]. Cross-linking and/or a reduction in pore size, along with the replacement of sulfonic groups with carboxylic acid species, led to an improvement in methanol selectivity. When Nafion 117 was exposed to a low dose of electron beam radiation (9.2 µC cm^{-2}) and then fabricated into an MEA, it performed well in a DMFC (60°C, 10% methanol and air), with a maximum power output >51% greater than with commercial Nafion 117 [7].

Sauk et al. [26] grafted styrene onto Nafion 115 membranes using supercritical CO_2 as a swelling agent. The embedded polystyrene was then sulfonated using concentrated sulfuric acid. The resultant membranes showed greater proton conductivity and lower methanol permeability as compared with unmodified Nafion 115. At 14% grafting, the membrane ion exchange capacity (IEC) was 1.02 mmol g^{-1} (vs. 0.909 mmol g^{-1} for commercial Nafion), the proton conductivity increased to

0.046 S cm^{-1} (from 0.039 S cm^{-1}), and the methanol permeability was lowered to 2.15×10^{-6} cm^2 s^{-1} (from 3.29×10^{-6} cm^{2}s^{-1} for commercial Nafion). In a methanol fuel cell, there was a modest improvement in power output (a current density of 140 mA cm^{-2} at 0.35 V vs. 120 mA cm^{-2} for an MEA with an unmodified Nafion 115 membrane).

Nafion/PTFE composite membranes were fabricated [17] by impregnation of a porous PTFE film with a solution of Nafion in a 2-propanol/water mixture. After solvent evaporation, the impregnated film was annealed at 120°C for 1 h. The resultant membrane was 20 μm in thickness, with a proton conductivity of 0.033 S cm^{-1} at 25°C (30% that of Nafion 117), while the methanol flux was 4.43×10^{-4} mol cm^{-2} s^{-1} (as compared with 1.62×10^{-4} mol cm^{-2} s^{-1} for Nafion 117 and 6.20×10^{-4} mol cm^{-2} s^{-1} for Nafion 112). DMFC performance of the composite membrane, measured at 70°C with 2.0 M methanol feed and pure oxygen, was superior to that of both Nafion 112 and 117.

As a substitute to Nafion impregnation into a fluoropolymer matrix, solution cast films have been prepared from blends of Nafion and vinylidene fluoride-hexafluoropropylene copolymer [16]. Upon addition of 20-60% vinylidene copolymer, a steep decrease in proton conductivity was observed, from 10^{-2} S cm^{-1} to 10^{-3}–10^{-4} S cm^{-1}, which was accompanied by a significant decrease in methanol permeability. In a more recent paper [27], Si et al. solution cast membranes composed of Nafion and poly(vinylidene fluoride), henceforth denoted as PVDF, where characterized where the PVDF content was varied from 0 to 65 wt%. As expected, the conductivity and methanol permeability decreased with increasing PVDF content, with a much more dramatic drop in transport parameters when the PVDF content was >50%. From proton conductivity and methanol permeability data, the relative selectivity can be determined:

$$\text{Relative selectivity} = \frac{\left[\dfrac{\kappa}{P}\right]}{\left[\dfrac{\kappa}{P}\right]_{\text{Nafion}}} \tag{14.1}$$

where κ is the membrane proton conductivity (S cm^{-1}) and P is the methanol permeability (cm^2 s^{-1}). A plot of relative selectivity vs. PVDF membrane content is shown in Fig. 14.3. As the wt% of PVDF increases, the membrane selectivity rises non-linearly, indicating that the decrease in methanol permeability with increasing PVDF is greater than the drop in proton conductivity. According to the results in Fig. 14.3, the best membrane would have ≤65 wt% PVDF. Unfortunately, the decrease in membrane conductivity with PVDF content (see Fig. 14.3 for κ values) means that the thickness of a blended membrane must decrease with increasing PVDF so that its sheet/areal resistance (the membrane thickness divided by conductivity) is sufficiently low for DMFC applications (the areal resistance should be comparable to that of a Nafion 117 membrane, about 0.2 Ω cm^2). Thus, the highly selective 65 wt% PVDF membrane must be very thin, ≈5 μm if reasonably high power outputs are to be realized in a DMFC. Such thin membranes are difficult to convert into MEAs.

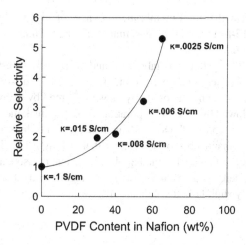

Fig. 14.3 The effect of PVDF content on the relative selectivity of Nafion/PVDF blended membranes (data from [27])

14.2.2 Addition of Inorganic Particles

There exists extensive literature data on Nafion modification through hybridization with inorganic particles, including the use of silica [8], sulfonated silica [15], phosphotungstic acid [32], titania [2], zirconium and titanium phosphates [3], montmorillonite [10], sulfonated montmorillonite [25], hydroxyapatite [20], and calcium hydroxyphosphate [22]. Various effects of particle addition were reported, such as a change in polymer crystallinity and micellar structure, an increase in tortuosity, a reduction in equilibrium water uptake, and a reduction in freezable water content. In general, DMFC performance with these membranes did not improve significantly over that with a pristine Nafion membrane; at best the improvement in maximum power density was 20–25%.

14.2.3 Layered Membranes

Kim et al. [14] modified Nafion 117 membranes by means of layer-by-layer deposition of clay-nanocomposite thin films. Membranes containing 20 nanocomposite bilayers exhibited a reduction in methanol permeability by ≈50%, while the proton conductivity remained unchanged. Unfortunately, no fuel cell performance was reported for this membrane material. A three layer membrane system consisting of a sulfonated poly(ether ether ketone) film sandwiched between two recast Nafion films was studied by Yang and Manthiram [33], where the middle layer was either

25- or 45-μm thick with a total membrane thickness of 115 or 135 μm. DMFC performance of the 115-μm membrane was marginally better than that with a commercial Nafion membrane.

Another three-layer system was fabricated and extensively investigated by Si et al. [27]. The middle, barrier layer was solution cast from a mixture of Nafion and poly(vinylidene fluoride) and the outer layers were formed by spraying a commercial Nafion solution followed by solvent evaporation. A 50-μm thick membrane comprising a 10-μm barrier layer (55% PVDF loading) had the same thickness as Nafion 112 membrane, but its methanol crossover was 36% that of Nafion 112. The membrane was tested in a DMFC at 60°C with 1M methanol and ambient pressure air. Its performance (maximum power density of 80 mW cm^{-2}) was significantly better than that with Nafion 112, but only slightly better than an MEA with a Nafion 117 membrane.

14.3   The Use of Thick Nafion Films

The simplest and most straightforward way to decrease the methanol flux across a Nafion membrane is to increase its thickness. The effect of Nafion thickness on DMFC performance was studied by fabricating membrane electrode assemblies (MEAs) with multiple (stacked) Nafion 112 films (where each Nafion 112 layer had a wet thickness of 60 μm). Using this approach, one can effectively create Nafion fuel cell membranes that are thicker than that which is available commercially. Figure 14.4 shows the effect of Nafion membrane thickness (stacking

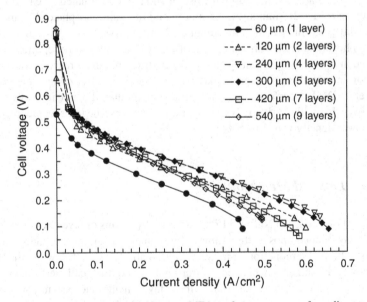

Fig. 14.4 The effect of membrane thickness on MEA performance curves for a direct methanol fuel cell operating at 60°C with a 1.0 M methanol feed

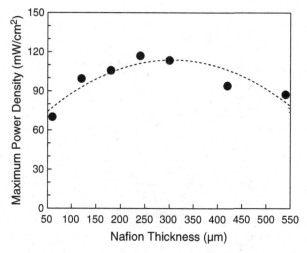

Fig. 14.5 The effect of Nafion thickness on the maximum power density of a direct methanol fuel cell, operating at 60°C with a 1.0 M methanol anode feed

between 1 and 9 Nafion 112 films) on the voltage-current density plots of fuel cell performance for a 1.0 M methanol anode feed. The methanol crossover flux decreased with membrane thickness, and MEAs with an intermediate number of Nafion 112 films (four or five layers) worked best. This is further illustrated in Fig. 14.5, where the maximum power density (product of voltage and current density) is plotted against total membrane thickness. As can be seen, there appears to be an optimum membrane thickness for a 1.0 M methanol feed at about 240 μm. For thinner membranes fuel cell power is lost due to excessive methanol crossover, while for membrane of greater thickness, there is a drop in power due to membrane resistance losses.

The effect of membrane thickness on DMFC performance for three different methanol feed concentrations (1.0, 5.0, and 10.0 M) is shown in Fig. 14.6. A 215–μm membrane is much more sensitive to increases in methanol feed concentration (a more significant downshift in the V-i curves with methanol concentration) than the thicker membranes (420 and 540 μm). The V-i curves for the 540 μm Nafion were nearly independent of methanol feed concentration (indicating that methanol crossover was small), but were suppressed as compared with those with a thinner Nafion film due to a high IR drop.

14.4   Blended Membranes of Nafion and Teflon-FEP or Teflon-PFA

The underlying premise of this study was that the Teflon phase in a proton conducting thin film will act as a methanol blocker and restrict the swelling of the Nafion component of the blend in methanol solutions. Both Teflon-FEP and Teflon-PFA

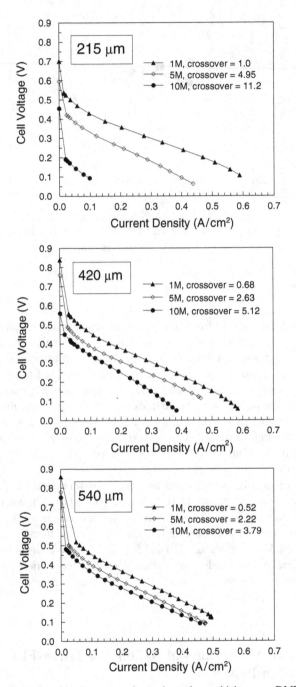

Fig. 14.6 The effect of methanol concentration and membrane thickness on DMFC performance curves. Methanol crossover fluxes are given relative to Nafion 117 (215 μm) at 1.0 M and 60°C. Crossover is expressed as a relative fraction of the methanol flux observed for a Nafion 117 membrane at the fuel cell operation conditions with Nafion 117 and 1.0 M methanol

(see Fig. 14.7) are melt processible, with moderately high melting points (275°C for FEP and 305°C for PFA). They have excellent mechanical, thermal, and chemical stability properties, but are not soluble in common membrane casting solvents. Nafion is not melt processible due to the presence of ionic aggregates that prevent flow [5], but Nafion's sulfonyl fluoride precursor (see Fig. 14.7) is a conventional thermoplastic that can be readily melt-processed and then hydrolyzed to convert SO_2F groups to sulfonic acid ion-exchange sites.

A series of membranes, 50- to 100-μm in thickness, were prepared from blends of Nafion and Teflon FEP/PFA by melt processing. Blended membranes were prepared by melt-mixing and extruding pellets of Nafion precursor in the sulfonyl fluoride form (R1100, from Ion-power, Inc.). The extruded blends were hot pressed into thin films at 300°C and 5,000 psi, and then immersed in an aqueous 15% NaOH solution at 80°C for 24–72 h to hydrolyze the sulfonyl fluoride moieties of the precursor into sulfonate ion-exchange groups. After thorough washing with deionized water, the membranes were boiled in 1.0 M H_2SO_4 (to condition the membrane and exchange Na^+ counterions for H^+) and rinsed thoroughly with deionized water.

A freeze-fractured SEM cross-section of one blended membrane, with 50 wt% FEP, is shown in Fig. 14.8. For this and all compositions, the Nafion and FEP phases were immiscible. At lower FEP content (10–20%), well-isolated spherical FEP domains of diameter ranging from 0.5 to 3 μm were seen, whereas elongated cylindrical domains were present at 30 wt% FEP. The origin of the oriented and elongated morphology was associated with the significantly higher melt viscosity of FEP, as compared with Nafion precursor, and the strong in-plane stresses during membrane formation by hot pressing.

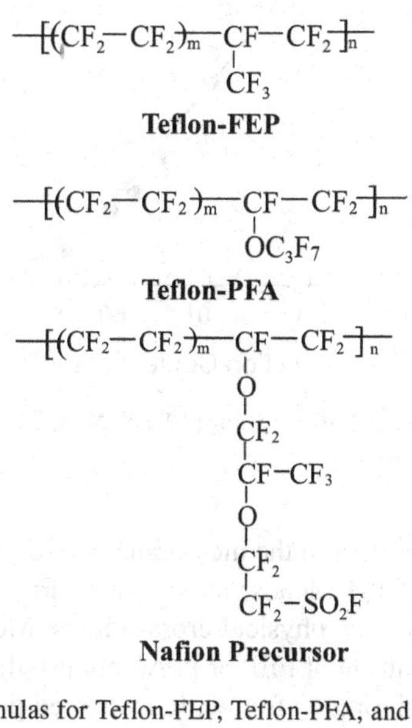

$$-\!\!\{(CF_2\!-\!CF_2)_m\!-\!\underset{\underset{CF_3}{|}}{CF}\!-\!CF_2\}_n\!\!-$$

Teflon-FEP

$$-\!\!\{(CF_2\!-\!CF_2)_m\!-\!\underset{\underset{OC_3F_7}{|}}{CF}\!-\!CF_2\}_n\!\!-$$

Teflon-PFA

$$-\!\!\{(CF_2\!-\!CF_2)_m\!-\!\underset{\underset{\underset{\underset{\underset{\underset{\underset{CF_2\!-\!SO_2F}{|}}{CF_2}}{|}}{O}}{|}}{CF\!-\!CF_3}}{CF_2}}{O}}\!-\!CF_2\}_n\!\!-$$

Nafion Precursor

Fig. 14.7 The chemical formulas for Teflon-FEP, Teflon-PFA, and Nafion precursor

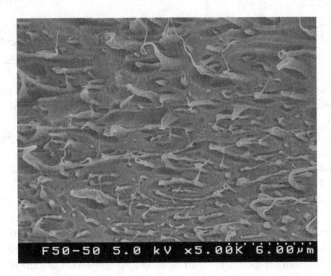

Fig. 14.8 Freeze-fractured cross section of a Nafion-FEP membrane (50 wt% Teflon-FEP). The elongated domains parallel to the upper/lower membrane surfaces is FEP. (From [18], reproduced by permission of The Electrochemical Society, Inc.)

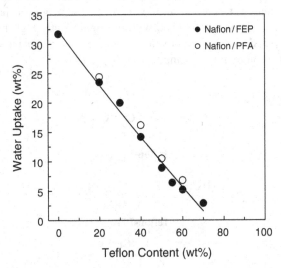

Fig. 14.9 Equilibrium water swelling of Nafion-FEP and Nafion-PFA membranes at $T = 25°C$ as a function of Teflon (FEP or PFA) content

The presence of FEP or PFA in the membrane strongly restricted the swelling of the Nafion component of the blends, as shown in Fig. 14.9, indicating that the Teflon domains were acting as physical cross-linkers. Methanol permeability was greatly lowered by the addition of FEP or PFA, more so than the decrease in proton conductivity caused by diluting Nafion with uncharged polymer. Thus, the relative selectivity of the blended films, as defined by (14.1), was always >1.0. As can be seen in Figs. 14.10 and 14.11, the dependence of methanol permeability and through-plane proton conductivity on membrane Teflon content is qualitatively similar, with an abrupt drop in these two transport parameters when the membrane FEP or PFA content exceeded 42 vol% (50 wt%). The experimental data points in Figs. 14.9 and 14.10 were fitted to a Maxwell equation [19] for either a dispersion of impermeable polymer spheres in a conducting matrix (for FEP contents < 50 wt%) or for semi-infinite and impermeable slabs (flakes) oriented parallel to the membrane surface (for FEP > 50 wt%) [18].

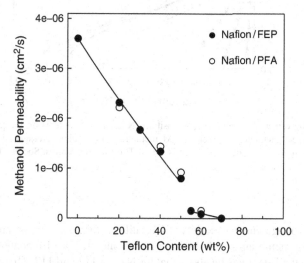

Fig. 14.10 Methanol permeability vs. wt% Teflon content in Nafion-FEP and Nafion-PFA membranes (for 1.0 M methanol at 60°C)

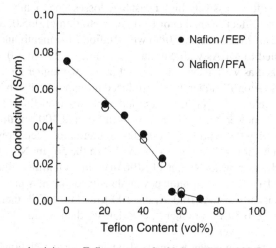

Fig. 14.11 Proton conductivity vs. Teflon content for Nafion-FEP and Nafion- PFA membranes (conductivity measured in water at 25°C)

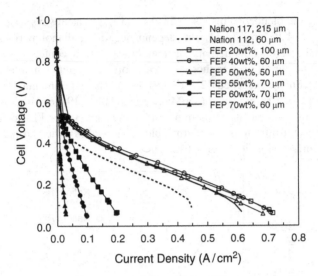

Fig. 14.12 The effect of membrane composition (Nafion/FEP content) on the performance of MEAs in a direct methanol fuel cell. 1.0 M methanol and ambient pressure air at 500 sccm. $T = 60°C$. (From [18], reproduced by permission of The Electrochemical Society, Inc.)

Typical voltage-current density fuel cell performance plots are shown in Fig. 14.12 for membranes with different FEP contents. For FEP contents of 20, 40, and 50 wt%, the V-i curves lie above that for Nafion 112 and 117. These membranes had an acceptably low sheet/areal resistance and a low methanol permeability. When the membrane contained >50% FEP, fuel cell performance fell below that of Nafion 112 and 117, due to excessively high resistance losses (the membrane conductivity was too low). The effect of methanol feed concentration (1.0, 5.0, and 10.0 M) on the performance of MEAs with a 50/50 wt% Nafion/PFA membrane (70 μm thickness) is contrasted to MEAs with Nafion 117 in Fig. 14.13. At 1.0 M, the blended membrane worked as well as Nafion 117, but at the two higher methanol feed concentrations, the Nafion/PFA membrane produced more power due to lower methanol crossover (see Table 14.1 of open circuit methanol crossover fluxes).

The maximum power density of MEAs with 50- and 100-μm thick membranes containing 50/50 wt% Nafion/PFA at three methanol feed concentrations is contrasted with Nafion 117 in Fig. 14.14. At 1.0 M the 50-μm Nafion/FEP blended membrane works as well as Nafion 117 (96 mW cm^{-2}) but this film outperformed Nafion at 5.0 M (with a maximum power density of 69 mW cm^{-2}, one and half times that of Nafion 117). At 10.0 M, the 100-μm Nafion/FEP membrane worked best (59 mW cm^{-2}, 6.5-times higher power density than Nafion 117) due to a significantly lower crossover.

Significant cost savings can be realized when using the Nafion/Teflon blended films rather than Nafion 117 in a DMFC. For example, a 50-μm thick membrane with

50 wt% FEP uses seven times less Nafion polymer, as compared with commercial Nafion 117 (due to the reduction in membrane thickness and dilution of Nafion with FEP). For a 100-μm blended membrane, cost savings are realized by the higher power density at 10.0 M methanol and the decrease in Nafion polymer content.

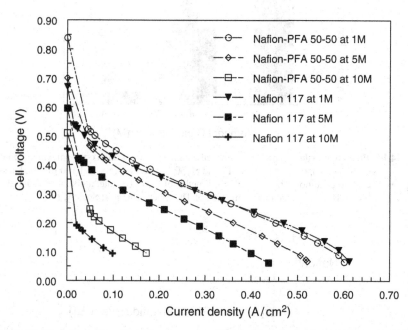

Fig. 14.13 The effect of methanol feed concentration on the performance of MEAs with Nafion 117 and Nafion/PFA (50/50 wt%) membranes in a direct methanol fuel cell. Temperature = 60°C and ambient pressure air at 500 sccm

Table 14.1 Methanol crossover flux through Nafion 117 and a Nafion/PFA-blended membranes (50/50 wt%, 70 μm thickness) in a fuel cell MEA at open circuit, as measured by CO_2 in the cathode air exhaust

Methanol concentration (M)	Methanol crossover flux (mol cm^{-2} min^{-1})	
	Nafion 117	Nafion/PFA
1.0	1.0×10^{-5}	7.8×10^{-6}
5.0	5.0×10^{-5}	2.9×10^{-5}
10.0	1.1×10^{-4}	6.0×10^{-5}

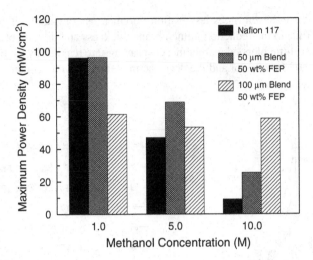

Fig. 14.14 Effect of membrane thickness and methanol feed concentration on the maximum power density from MEAs composed of Nafion 117 and 50/50 Nafion/FEP (50- and 100-μm thickness). Fuel cell operating conditions: 60°C, 2 ml min^{-1} methanol feed flow rate, ambient pressure air at 500 sccm. (From [18], reproduced by permission of The Electrochemical Society, Inc.)

Fig. 14.15 Acid-base complex formation mechanism between the sulfonic acid group of Nafion and the imidazole nitrogen of PBI. (reprinted from [31], with permission from Elsevier).

14.5   Nafion Doped with Polybenzimidazole

In contrast to the Nafion-FEP system, in which the immiscibility of the polymer components is clearly shown in Fig. 14.8, greater compatibility of Nafion and polybenzimidazole (PBI) was expected due to acid-base interactions between imidazole nitrogens and sulfonic acid protons. PBI should function in such a system as a crosslinker (see Fig. 14.15), producing a reduction in membrane swelling and permeability by methanol. Although the idea of examining acid-base membrane blends in fuel cells is not new [9,11–13,30], little work has been carried out with Nafion.

Membranes were prepared using dry Nafion powder [31] that was obtained by evaporating the solvent from a commercial Nafion solution. The powder was equilibrated in an aqueous NaCl/HCl solution with a preset Na$^+$/H$^+$ concentration ratio to adjust the protonation degree of the Nafion polymer. The substituted Nafion was dried again and then mixed with an appropriate amount of PBI in DMAc solvent. Membranes were solution cast into a glass dish. After solvent evaporation at 80°C, the films were annealed at 150°C for 3 h. The resultant membranes (50–100 μm in dry thickness) were conditioned by boiling in 1M H$_2$SO$_4$ followed by extensive washings with de-ionized water. Membranes were stored at room temperature in water for later use.

As can be seen in Figs. 14.16 and 14.17, membrane proton conductivity and methanol permeability were dependent on the Nafion protonation degree during blending and the PBI content. At a given initial protonation degree of the Nafion powder, there was a decrease in conductivity and permeability with increasing PBI content. Membranes were less conductive and better methanol barriers for a given PBI content when the initial proton degree was high (i.e., when there were more SO$_3$H groups in Nafion, there was more complexation with PBI).

DMFC performance of the PBI-doped membranes was superior to that of Nafion 117 at 1.0 M and 5.0 M methanol feed solutions. Representative DMFC voltage-current density curves for three Nafion-PBI membranes of different PBI content are shown in Fig. 14.18, along with reference curves of Nafion 112 and 117. The doped membranes were fabricated from fully protonated Nafion and their

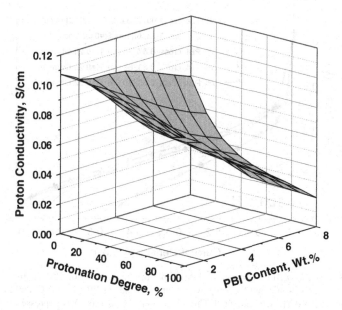

Fig. 14.16 Membrane proton conductivity of PBI-doped Nafion membranes (in water at 25°C) as a function of the Nafion powder protonation degree and PBI content. (Reprinted from [31], with permission from Elsevier.)

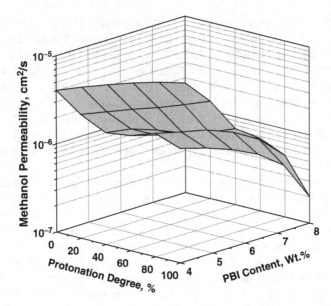

Fig. 14.17 Dependence of the membrane methanol permeability (at 60°C) on the Nafion powder protonation degree and PBI content of PBI-doped Nafion membranes. (Reprinted from [31], with permission from Elsevier.)

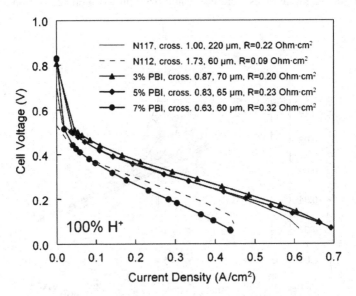

Fig. 14.18 Effect of PBI content on the DMFC performance of Nafion-PBI membrane samples prepared from fully protonated Nafion powder, with 3, 5 and 7% PBI. $T = 60°C$, ambient pressure air at 500 sccm, 1.0 M methanol feed. Crossover (denoted as cross.) is expressed as a relative fraction of the methanol flux observed for a Nafion 117 membrane at the fuel cell operation conditions. R denotes the membrane areal resistance. (Reprinted from [31], with permission from Elsevier.)

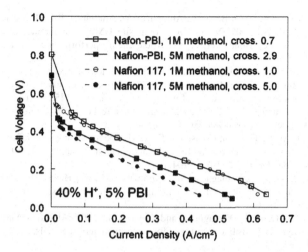

Fig. 14.19 DMFC performance comparison of Nafion117 and Nafion-PBI membranes at two methanol feed concentrations (1.0 and 5.0 M at a flow rate of 2 ml min⁻¹) at 60°C and ambient pressure air. The Nafion-PBI membrane contained 5% PBI and was prepared from 40% protonated Nafion powder, with a wet thickness of 80 μm. Crossover is expressed as a relative fraction of the methanol flux observed for a Nafion 117 membrane at the fuel cell operation conditions with a 1.0 M methanol feed. (Reprinted from [31], with permission from Elsevier.)

thickness was between 60 and 70 m. The MEA with 3 wt% PBI membrane delivered the highest power density, which was greater than that with either of the two Nafion MEAs. As the membrane PBI content increased, its ohmic resistance increased and performance in a fuel cell decreased. The methanol crossover flux in all of the Nafion-PBI MEAs was lower than that in Nafion 117, even though the doped films were 2–4 times thinner than Nafion. The high open-circuit voltage for the PBI-doped membranes (>0.8 V) is another indication of lower methanol crossover.

Due to the unacceptably high methanol permeability of commercial Nafion, most DMFC tests are performed with relatively dilute aqueous methanol feed solutions (typically, 0.5 or 1 M). A Nafion-PBI membrane (5 wt% PBI with Nafion polymer that was 40% in the protonated form) performed well at low (1 M) and high (5 M) methanol feed concentrations, as shown in Fig. 14.19. As was the case for the Nafion-FEP and Nafion-PFA blends discussed previously, PBI-doped Nafion films contain much less fluoropolymer (2–4 times less) as compared with commercial 117 membranes from DuPont, yet their performance is equal or superior to any Nafion material in a DMFC.

References

1. Bae, B., Ha, H. Y., and Kim, D., 2005, Preparation and characterization of Nafion/poly(1-vinylimidazole) composite membrane for direct methanol fuel cell application, *J. Electrochem. Soc.* **152**(7):A1366–A1372.

2. Baglio, V., Arico, A. S., Di Blasi, A., Antonucci, V., Antonucci, P. L., Licoccia, S., Traversa, E., and Fiory, F. S., 2005, Nafion-TiO₂ composite DMFC membranes: physico-chemical properties of the filler versus electrochemical performance, *Electrochim. Acta* **50**:1241–1246.
3. Bauer, F., and Willert-Porada, M., 2005, Characterization of zirconium and titanium phosphates and direct methanol fuel cell (DMFC) performance of functionally graded Nafion composite membranes prepared out of them, *J. Power Sources* **145**:101–107.
4. Cruickshank, J., and Scott, K., 1998, The degree and effect of methanol crossover in the direct methanol fuel cell, *J. Power Sources* **70**(1):40–47.
5. Gierke, T. D., Munn, G. E., and Wilson, F. C., 1981, The morphology in Nafion perfluorinated membrane products, as determined by wide-angle and small-angle X-ray studies, *J. Polym. Sci. Polym. Phys. Ed.* **19**(11):1687–1704.
6. Hobson, L. J., Ozu, H., Yamaguchi, M., Muramatsu, M., and Hayase, S., 2002, Nafion 117 modified by low dose EB irradiation: surface structure and physical properties, *J. Mater. Chem.* **12**:1650–1656.
7. Hobson, L. J., Ozu, H., Yamaguchi, and Hayase, S., 2001, Modified Nafion 117 as an improved polymer electrolyte membrane for direct methanol fuel cells, *J. Electrochem. Soc.* **148**(10):A1185–A1190.
8. Jiang, R., Kunz, H. R., and Fenton, J. M., 2006, Composite silica/Nafion membranes prepared by tetraethylorthosilicate sol-gel reaction and solution casting for direct methanol fuel cells, *J. Membr. Sci.* **272**:116–124.
9. Jorissen, L., Gogel, V., Kerres, J., and Garche, J., 2002, New membranes for direct methanol fuel cells, *J. Power Sources* **105**:267–273.
10. Jung, D. H., Cho, S. Y., Peck, D. H., Shin, D. R., and Kim, J. S., 2003, Preparation and performance of Nafion/montmorillonite nanocomposite membrane for direct methanol fuel cell, *J. Power Sources* **118**:205–211.
11. Kerres, J., Ullrich, A., Meier, F., and Haring, T., 1999, Synthesis and characterization of novel acid-base polymer blends for application in membrane fuel cells, *Solid State Ionics* **125**:243–249.
12. Kerres, J. A., 2001, Development of ionomer membranes for fuel cells, *J. Membr. Sci.* **185**:3–27.
13. Kerres, J. A., 2005, Blended and cross-linked ionomer membranes for application in membrane fuel cells, *Fuel Cells* **5**(2):230–247.
14. Kim, D. W., Choi, H.-S., Lee, C., Blumstein, A., and Kang, Y., 2004, Investigation on methanol permeability of Nafion modified by self-assembled clay-nanocomposite multilayers, *Electrochim. Acta* **50**:659–662.
15. Li, C., Sun, G., Ren, S., Liu, J., Wang, Q., Wu, Z., Sun, H., and Jin, W., 2006, Casting Nafion-sulfonated organosilica nano-composite membranes used in direct methanol fuel cells, *J. Membr. Sci.* **272**:50–57.
16. Lin, J.-C., Ouyang, M., Fenton, J. M., Kunz, H. R., Koberstein, J. T., and Cutlip, M. B., 1998, Study of blend membranes consisting of Nafion and vinylidene fluoride-hexafluoropropylene copolymer, *J. Appl. Polym. Sci.* **70**:121–127.
17. Lin, H.-L., Yu, T. L., Huang, L.-N., Chen, L.-C., Shen, K.-S., and Jung, G.-B., 2005, Nafion/PTFE composite membranes for direct methanol fuel cell applications, *J. Power Sources* **150**:11–19.
18. Lin, J., Lee, J. K., Kellner, M., Wycisk, R., and Pintauro, P. N., 2006, Nafion-fluorinated ethylene-propylene resin membrane blends for direct methanol fuel cells, *J. Electrochem. Soc.* **153**:A1325–A1331.
19. Maxwell, J. C., 1881, *A Treatise on Electricity and Magnetism*, Vol. 1, 2nd Ed., Claredon Press, London, p. 440
20. Park, Y.-S., and Yamazaki, Y., 2005, Novel Nafion/hydroxyapatite composite membrane with high crystallinity and low methanol crossover for DMFC, *Polym. Bull.* **53**:181–192.

21. Park, H. S., Kim, Y. J., Hong, W. H., and Lee, H. K., 2006, Physical and electrochemical properties of Nafion/polypyrrole composite membrane for DMFC, *J. Membr. Sci.* **272**:28–36.
22. Park, Y.-S., and Yamazaki, Y., 2006, Low water/methanol permeable Nafion/CHP organic-inorganic composite membrane with high crystallinity, *Eur. Polym. J.* **42**:375–387.
23. Ren, X., Wilson, M. S., and Gottesfeld, S., 1996, High performance direct methanol polymer electrolyte fuel cells, *J. Electrochem. Soc.* **143**(1):L12–L15.
24. Ren, X., Zelenay, P., Thomas, S., Davey, J., and Gottesfeld, S., 2000, Recent advances in direct methanol fuel cells at Los Alamos National Laboratory, *J. Power Sources* **86**(1–2):111–116.
25. Rhee, C. H., Kim, H. K., Chang, H., and Lee, J. S., 2005, Nafion/sulfonated montmorillonite composite: a new concept electrolyte membrane for direct methanol fuel cells, *Chem. Mater.* **17**:1691–1697.
26. Sauk, J., Byun, J., and Kim, H., 2004, Grafting of styrene on to Nafion membranes using supercritical CO_2 impregnation for direct methanol fuel cells, *J. Power Sources* **132**:59–63.
27. Si, Y., Lin, J.-C., Kunz, H. R., and Fenton, J. M., 2004, Trilayer membranes with a methanol-barrier layer for DMFCs, *J. Electrochem. Soc.* **151**(3):A463–A469.
28. Tricoli, V., 1998, Proton and methanol transport in poly(perfluorosulfonate) membranes containing Cs^+ and H^+ cations, *J.Electrochem. Soc.* **145**(11):3798–3801.
29. Tsao, C.-S., Chang, H.-L., Jeng, U.-S., Lin, J.-M., and Lin, T.-L., 2005, SAXS characterization of the membrane nanostructure modified by radiation cross-linkage, *Polymer* **46**:8430–8437.
30. Walker, M., Baumgartner, K.-M., Kaiser, M., Kerres, J., Ullrich, A., and Rauchle, E., 1999, Proton-conducting polymers with reduced methanol permeation, *J. Appl. Polym. Sci.* **74**:67–73.
31. Wycisk, R., Chisholm, J., Lee, J., Lin, J., and Pintauro, P. N., 2006, Direct methanol fuel cell membranes from Nafion-polybenzimidazole blends, *J. Power Sources*, **163**:9–17.
32. Xu, W., Lu, T., Liu, C., Xing, W., 2005, Low methanol permeable composite Nafion/silica/PWA membranes for low temperature direct methanol fuel cells, *Electrochim. Acta* **50**:3280–3285.
33. Yang, B., and Manthiram, A., 2004, Multilayered membranes with suppressed fuel crossover for direct methanol fuel cells, *Electrochem. Commun.* **6**:231–236.

Chapter 15
Methanol Permeation Through Proton Exchange Membranes of DMFCs

M. Bello, S.M. Javaid Zaidi, and S.U. Rahman

Abstract This chapter presents efforts and progress being made by researchers worldwide to develop membranes with low methanol permeability, without compromising on other qualities, such as high ionic conductivity, good chemical and thermal stability, and cost. Three approaches have been pursued —Nafion membranes modification, development of alternative membranes, and development of high activity anode catalysts or methanol-tolerant cathode catalysts. Success has been made in developing membranes with permeability values of 10–70 times lower than the pure Nafion membranes. Various techniques, both electrochemical and non-electrochemical, for measuring methanol permeation through the membranes are also discussed. It has been found that electrochemical techniques are more accurate. Potentiometric technique in particular has ease of reproducibility of results, and getting more data points.

15.1 Introduction

Fuel cells are promising energy conversion devices with numerous possible applications. Their most appealing properties include high energy efficiency, reduced polluting emissions, versatility, flexibility in power supply, and promise of low cost. These properties make them attractive when compared with the existing conventional energy conversion technologies, especially in this era of strict environmental regulations, deregulation of power sectors, and increased energy insecurity.

Hydrogen as an energy carrier is the common fuel for fuel cells but is not readily available. In addition, there are no infrastructural networks to support the easy distribution of hydrogen. Research efforts are geared toward designing a fuel cell that would directly oxidize a liquid fuel at the anode and have improved overall cell performance. Direct methanol fuel cells (DMFCs), in which methanol fuel is supplied directly to the anode of the cell, is one of the promising candidates. It has the advantages of not requiring a fuel reformer, allowing simple and compact designs.

S.M.J. Zaidi, T. Matsuura (eds.) *Polymer Membranes for Fuel Cells*,
doi: 10.1007/978-0-387-73532-0, © Springer Science+Business Media, LLC 2009

An infrastructure built for gasoline can be used without significant alteration, and changes in power demand can be accommodated simply by an alteration in the supply of methanol feed. In addition, due to the low operating temperature of the cell there is no production of NO_x [1]. DMFC technology is especially attractive as an energy source for portable power applications in addition to its use for the automobile industry. In the field of portable power applications, DMFCs could provide longer usage time before refueling and could be refueled in minutes compared with hours of recharging for batteries [12].

In spite of these advantages, the low activity of methanol electro-oxidation catalysts, methanol permeation through polymer electrolyte membranes, and cost of components inhibit large-scale commercialization of DMFCs. The current perfluorosulfonate polymer electrolyte membranes used in DMFCs such as Dupont 's Nafion, permits methanol crossover from the anode to the cathode [32]. The methanol transported through the membrane causes loss of fuel and reduces the effective area of the cathode by competing for the available cathode Pt sites, thus affecting oxygen reduction reaction. This leads to secondary reactions, mixed potentials, decreasing energy and power density, generation of additional water that must be managed, increasing oxygen stoichiometric ratio, and hence reduced overall performance [16]. Thus, DMFCs technology would gain much in terms of both performance and efficiency by solving the methanol permeation problem. The issue of methanol permeation could be addressed by modifying the Nafion membranes, developing alternative methanol-impermeable membranes, and developing very active anode methanol electro-oxidation catalysts or methanol-tolerant oxygen reduction catalysts.

Nafion membranes could be modified by blending with either organic or inorganic materials. Alternative membranes, such as polymer/inorganic mineral acid composite membranes, partially fluorinated polymers, nonfluorinated polymers and their combinations, etc., with low methanol permeability have been developed for DMFC application. In the development of membranes for DMFCs, it is important to have accurate, reliable, and convenient methods of measuring the membrane 's methanol permeability. A number of techniques have been developed to determine methanol permeation in DMFCs or its simulated systems. This chapter discusses approaches adopted in order to provide membranes with low methanol permeability. It also highlights different techniques developed for measuring methanol permeation through the membranes.

15.2 Approaches to Reduce Methanol Permeation Through the Membranes

Early studies and modeling of methanol permeation through commercial perfluorinated membranes revealed that methanol readily diffuses through them [45]. Since then researchers worldwide have been making serious efforts to develop membranes with low methanol permeability without compromising on other qualities, such as high ionic conductivity, good chemical and thermal stability, and cost. Some of these attempts are discussed here.

15.2.1 Membrane Modification

The current perfluorosulfonate polymer electrolyte membranes used in DMFCs, such as Dupont 's Nafion, permit methanol crossover from the anode to the cathode. Research efforts are underway to modify the Nafion membranes to make them less permeable to methanol and improve their water retention capacity at higher temperatures. Nafion has a microphase-separated structure with a hydrophobic region interspersed with ion-rich hydrophilic domains. Because of its structure, phase separation occurs in the hydrated state between the hydrophilic and hydrophobic regions, and ionic clusters are formed. This morphology with discrete hydrophobic and hydrophilic regions gives the material good properties. The well-connected hydrophilic domain is responsible for its excellent ability to allow transport of protons and water easily, whereas the hydrophobic domain provides the polymer with the morphologic stability and prevents the polymer from dissolving in water. This explains the exceptional transport efficiency in hydrated Nafion membranes. Figure 15.1 illustrates the chemical structure of Nafion.

Water is necessary for the transport of protons, but it also has high affinity for methanol transport. One way of reducing methanol permeation is to modify the Nafion membranes by blending with either organic or inorganic materials.

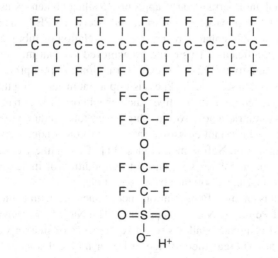

Fig. 15.1 Chemical structure of Nafion [42]. (This figure was published in *Journal of Membrane Science*, Vol. 259, B. Smitha, S. Sridhar, A. A. Khan, Solid Polymer Electrolyte Membranes for Fuel Cell Applications—a review, page 15, Copyright Elsevier, 2005)

15.2.1.1 Nafion/Organic Composite Membranes

Nafion/polypyrrole [47], Nafion/poly(1-vinylimidazole) [7], Nafion/polyfurfuryl alcohol (PFA) [20], Nafion/polyvinylidene fluoride (PVdF) [17], and Nafion/polyvinyl alcohol (PVA) [48], composite membranes have low methanol permeability compared with bare Nafion membranes, and their proton conductivity is comparable to that of Nafion membranes. However, choice of the blending ionomer material and thickness of the composite membranes is important in achieving the desired methanol crossover reduction. Multilayered membranes containing a thin inner layer of SPEEK as a barrier and two outer layers of recast Nafion fabricated by hot pressing could significantly reduce methanol crossover in DMFCs [19]. The Nafion-SPEEK-Nafion composite membranes can also be prepared by immersing the SPEEK in Nafion-containing casting solution. In order to have Nafion-SPEEK-Nafion composite membranes with improved qualities for DMFCs application, appropriate sulfonation degree and thickness of the inner SPEEK layer are particularly important.

15.2.1.2 Nafion/Inorganic Composite Membranes

Incorporating inorganic materials such as silica and titanium oxide into Nafion membrane enhances the water retention capacity of the membranes at high temperatures. This enables the operation of the direct methanol fuel cell at high temperatures. Although the membranes could achieve a significant improvement in water retention and proton conductivity, it may not retard methanol permeation [46]. However, elsewhere it has been indicated that Nafion/silica hybrid membranes can decrease the methanol permeation if appropriate silica content is used [30]. Also, multilayers of clay nanoparticles and ionic polyacetylene PEPy-C18 deposited on Nafion membrane could enhance resistance of the Nafion membrane against methanol crossover significantly. This could be achieved without much negative effect on its proton conductivity [22]. However, formation of appropriate bilayer nanocomposite films with a suitable thickness is important in achieving the desired goal. Nafion membrane 'containing additives such as silicon dioxide particles (Aerosil) and molybdophosphoric acid also have higher proton conductivity, but the combined parameter of methanol crossover rate and proton conductivity could be less than that of commercial Nafion membranes [11]. This is likely due to a structural modification of the membranes because of the addition of inorganic components and having new interfacial polymer-particle particles.

Another means of modifying Nafion membranes attracting interest is that of impregnation of Pd on the Nafion membranes. The Nafion membrane modified by impregnating Pd-nanophases allows selective transport of smaller water molecules or hydrogen ions, whereas the passage of larger molecules would be restricted. A well-dispersed Pd nanophase in the Nafion is effective in preventing or reducing methanol crossover through the membrane even at high methanol concentration while at the same time maintaining good proton conductivity [23]. Several other studies using different deposition or coating techniques show that Nafion/Pd composite membranes can significantly reduce methanol crossover compared with bare Nafion membranes and do not change the membranes conductivity. This is expected because during the deposition of Pd, the 'SO$_3$H group is not affected but the presence of the Pd reduces the methanol permeation. Thus, Nafion/Pd composites membranes would provide better cell performance.

Many other inorganic materials such as Ca$_3$ (PO$_4$)$_2$, BPO$_4$, ZrPO$_4$, etc., when blended with Nafion membranes will show improvement on the membranes properties such as thermal stability, proton conductivity, and lower methanol permeation. Nafion doped with cesium cations also show evidence of good performance in the operation of DMFCs. The presence of cesium ions in the membrane, specifically in the water-rich domains, causes remarkable reduction of methanol permeation [43]. However, the proton conductivity could be depressed to a lesser extent by the presence of the cesium ions in the membrane. Nevertheless, at ambient conditions, the combined parameter of both proton conductivity and methanol permeability shows better performance of Cs$^+$-doped membranes than the Nafion 117 membranes in the operation of DMFC. Table 15.1, shows permeability values for Nafion 117 membrane and four different Nafion 117 modified membranes. A reduction in the permeability values when the Nafion 117 membrane is modified with appropriate methanol inhibiting material can be seen. A very sharp reduction is particularly noticed in the case of Cs$^+$-doped membrane. This indicates that exchange of H$^+$ by Cs$^+$ cations is very effective in reducing methanol content in the hydrophilic domains of Nafion.

Table 15.1 Methanol permeability values for Nafion 117 and four Nafion 117-modified membranes

Membrane type	Thickness (μm)	Methanol concentration (M)	Temperature (°C)	Permeability (cm^{2}s^{-1})	Reference
Nafion 117	177	2	22	1.15×10^{-6}	[44]
Nafion 117/Clay nanocomposite multilayer	178	12.5	25	7.58×10^{-7}	[22]
Pd impregnated nano composite in Nafion 117	—	2 & 10	30	4.3×10^{-7}	[23]
Plasma-etched and Pd-Sputte- red Nafion 117	—	2	Room temperature	1.598×10^{-6}	[10]
Nafion 117 doped with cesium ion	199	2	21	3.34×10^{-8}	[43]

15.2.2 Development of Alternative Membranes

15.2.2.1 SPEEK Membranes

Membranes based on aromatic poly(ether ether ketone) (PEEK) seem very promising. In addition to their low methanol permeability, they have good mechanical properties, and high thermal stability, and the proton conductivity can be controlled through the degree of sulfonation. The main difference between Nafion and SPEEK membranes, which makes the latter less permeable to methanol, can be attributed to the difference in their microstructure. In SPEEK membranes, there is less pronounced hydrophilic/hydrophobic separation as compared with Nafion membrane, and the flexibility of the polymer backbone of SPEEK produce narrow proton channels and a highly branched structure that baffles the transfer of methanol [18]. Thus, SPEEK membranes have lower electro-osmotic drag and methanol permeability values.

However, proton conductivity of many SPEEK membranes is lower than that of Nafion membranes, but SPEEK membranes, with 39% and 47% degrees of sulfonation, have been found to have proton conductivity values close to that of Nafion 115 membranes [26]. Their methanol diffusion coefficient value is an order lower than that of Nafion 115 membranes under the same conditions. At 80°C, the overall DMFC performance of the SPEEK membranes is better than that of Nafion 115 membrane. Furthermore, SPEEK membranes with methanol diffusion coefficient values of 5×10^{-8}–3×10^{-7} cm^{2}s^{-1} at 25°C—which is lower compared with that of Nafion membranes (10^{-5}–10^{-6} cm^{2}s^{-1})—have been developed [28].

15.2.2.2 SPEEK Composite Membranes

In order to reduce methanol permeability, SPEEK membranes can be blended with an appropriate amount of inorganic materials such as BPO_4, SiO_2, ZrO_2, etc., which act both as a barrier to methanol permeation and enhance conductivity of the membranes [29]. The inorganic network can remarkably reduce methanol permeation through the SPEEK membranes, even if heteropolyacids are incorporated into the SPEEK polymer matrix, which usually increases water and methanol permeation. Incorporating different proportions of solid proton conductors, TPA- and MPA-loaded MCM41 and TPA and MPA-loaded Y-zeolite into the SPEEK polymer matrix, provide composite membranes with low methanol permeability [5,6]. The reduction in the methanol permeability is better achieved with lower loadings of solid inorganic materials (TPA- or MPA-loaded MCM41/Y-zeolite) than with higher loadings. These composite membranes offer less expensive alternatives, and have low methanol permeability, high proton conductivity, and high thermal stability, which make them suitable for DMFC application.

Furthermore, composite membranes prepared by incorporating laponite and montmorillonite (MMT) into a partially sulfonated PEEK polymer also help to

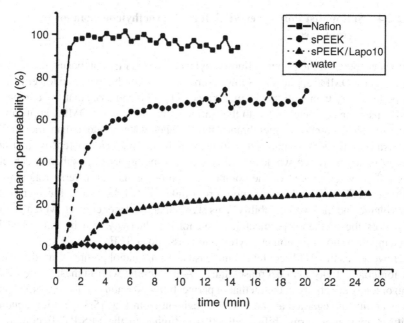

Fig. 15.2 Methanol permeability in Nafion 115 and composite membranes [9]. (This figure was published in *Journal of Power Sources*, Vol. 124, J. H. Chang, J. H. Park, C-S. Kim, O-Ok Park, Proton-Conducting Composite Membranes Derived from Sulfonated Hydrocarbon and Inorganic Materials, page 22, Copyright Elsevier, 2003)

reduce swelling in hot water significantly and reduce methanol permeability without a serious reduction in the proton conductivity [9]. As shown in Fig. 15.2, the methanol flux across SPEEK/Lapo10 composite membrane is low compared with that of Nafion 115 membrane. The SPEEK membrane also displays lower methanol permeability compared with the Nafion 115 membrane due to the difference in their microstructure.

15.2.2.3 Sulfonated Poly-Styrene Membranes

Commercial non—cross-linked poly (styrene) can be partially sulfonated to various degrees, obtaining a homogeneous distribution of the sulfonic acid groups in the polymer. The membranes prepared from these polymers showed proton conductivity similar to that of Nafion membrane [24]. In the same way, the permeability to methanol also increases with the density of the sulfonate groups in the polymer. However, even at the highest degree of sulfonation, the permeability of SPS is comparatively small compared with that of Nafion membranes, which makes SPS membranes attractive for the DMFC application.

15.2.2.4   Sulfonated Polystyrene/Polytetrafluoroethylene Composite Membranes

Sulfonated polystyrene/polytetrafluoroethylene (SPS/PTFE) composite membranes pre-pared for DMFC application show comparable ion conductivity and lower methanol permeability than Nafion 117 membrane [41]. This indicates that the composite membranes can compete well with the Nafion membranes in the DMFC application. The sulfonated composite membranes have higher water content than the Nafion membrane. This is presumably due to their high sulfonic acid content with strong affinity to water; which would be expected from the higher ion-exchange capacity (IEC). The water content of the composite membranes has an inverse relationship with the PTFE content because more cross-linked networks reduce the membrane free volume and the swelling ability. This shows that the water content, which greatly influences the methanol permeability and ion conductivity, can be controlled by changing the ratio of monomer (styrene) to cross-linker TFE.

However, as the PTFE content is increased, the methanol permeability decreases but ion conductivity also decreases fairly. A membrane with highest value of the ratio of ion conductivity to the methanol permeability (Φ = ion conductivity/methanol permeability) is regarded as the best. It is evident from Fig. 15.3, that ion conductivity to methanol permeability ratio (Φ) is higher in the SPS/PTFE composite membranes than in Nafion membranes. It increases with decreasing Styrene/PTFE ratio. This indicates that desired SPS/PTFE composite membranes with high potential for DMFC application can be prepared.

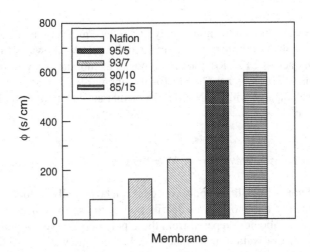

Fig. 15.3 Comparison of the Φ (S cm^{-1}) parameter of Nafion and the SPS/PTFE composite membranes [41]. (This figure was published in *Journal of Membrane Science*, Vol. 251, J-P. Shin, B-J. Chang, J-H. Kim, S-B. Lee, D. H. Suh, Sulfonated Polystyrene/Composite Membranes, page 253, Copyright Elsevier, 2005.)

15.2.2.5   Poly(vinylidene) Fluoride Composite Membranes

Composite poly(vinylidene) fluoride (PVdF) — based proton conducting membranes have low methanol crossover, high proton conductivity (which shows less dependence on the water content), and low cost [33]. In addition, polymer blends of appropriate ratios of PVdF/SPEEK produces composite membranes with low methanol permeability and low proton conductivity compared with Nafion 115 membranes under the same testing condition. Polystyrenesulfonic acid (PSSA), which is a good blending material for proton exchange membranes, can be used to prepare PVdF/Polystyrenesulfonic acid (PSSA) composite membranes. This provides membranes with low methanol permeability, good proton conductivity, high stability, and excellent water management. The PVdF/PSSA composite membranes are expected to improve the DMFC performance.

15.2.2.6   Poly-Ethylene Tetrafluoroethylene-Based Membranes

Methanol permeation through poly(ethylene tetrafluoroethylene) (ETFE) membrane investigated at temperatures of 30, 50, and, 70°C showed 90 % reduction in the permeability values compared with the Nafion 115 membrane [38]. In addition, the ETFE-based membranes are cheaper, have good chemical and mechanical stability. However, the efficiency of the DMFC with this membrane is about 40–60 % lower than its efficiency with the Nafion 115 membrane due to lower proton conductivity and electrodes polarization. But radiation grafted polymer electrolyte membranes prepared using polyethylene tetrafluoroethylene, polyvinylidene fluoride, and low-density polyethylene as base polymer films and polystyrene sulfonic acid as proton-conducting groups have good performance in the DMFCs [40]. With an appropriate degree of grafting and membrane thickness, membranes with suitable conductivity and low methanol permeability superior to Nafion membranes can be produced.

15.2.2.7   Polyphosphazene-Based Membranes

Studies of methanol permeation through proton-conducting polyphosphazene membranes show that polyphosphazene-based proton-exchange membranes have low methanol permeability and good proton conductivity [15]. These membranes have been tested in the DMFCs and have shown good performance.

15.2.2.8   Polyvinyl Alcohol/Mordenite

Polyvinyl alcohol/mordenite composite membrane prepared using appropriate quantity of mordenite and heat treatment could significantly reduce the permeation of methanol molecules [27]. However, to achieve this improvement, there should be a suitable tailoring of the transport properties between the polymer and zeolite

Table 15.2 Methanol permeability values for Nafion 117 membrane and different alternative membranes

Membrane type	Thickness (µm)	Methanol concentration (M)	Temperature (°C)	Permeability (cm^{2}s^{-1})	Reference
Nafion 117	177	2	22	1.15×10^{-6}	[44]
SPEEK (39% DS)	78	1	80	1.321×10^{-7}	[26]
30wt % (60wt % MCM41 + 40wt % TPA) and 70wt % SPEEK	160	2.5	22	1.63×10^{-8}	[4]
30wt % (60wt % Y-zeolite + 40wt % TPA) and 70wt % SPEEK	160	2.5	22	3.34×10^{-8}	[5]
15 mol % Sulfonated polystyrene/ poly(ethylene- ran-butylene) copolymer	313	1.5	Room temperature	2.1×10^{-8}	[24]
Sulfonated poly- styrene/poly- tetrafluoroethylene (85/15)	50	5	25	1.3×10^{-7}	[41]
Sulfonated polyphosphazene	175	12.5	22	3.09×10^{-7a}	[15]
Phosphonated polyphosphazene	114	12.5	22	2.49×10^{-8a}	[15]
BPSH 40 (40% disul-fonated wholly aroma-tic polyarylene ether sulfone)	132	1 & 5	30	6.4×10^{-7}	[14]

^{a}Diffusivity.

phases. The zeolite phases allow the transport of electrons but inhibit that of methanol molecules, thus making the membrane less permeable to methanol. Table 15.2, shows methanol permeability values for Nafion 117 membrane and some alternative membranes developed for use in DMFCs. It is clear that many alternative membranes with low methanol permeability have been developed. SPEEK/TPA/MCM41 or Y-zeolite composite membranes prepared in our laboratory are among the membranes with the lowest permeability values. These membranes can withstand high temperature (~160°C) in an acidic medium, as observed during pretreatment before permeability measurements. Thus, in addition to low methanol permeability they also have good thermal stability.

15.2.3   Improvement on Methanol Anode Catalysts and Methanol Tolerant Cathode Catalysts

Some attempts have been made to improve the activity of the electro-oxidation catalyst, which is also seen as helping to reduce methanol permeation. When the catalyst activity is very high, more of the methanol is expected to be oxidized instantly, thereby reducing the amount permeating to the cathode side. Pt-based electro-oxidation catalysts display the necessary stability in the acidic environment of DMFC. Although much progress is still needed, there has been significant activity. A reaction mechanism for methanol oxidation in aqueous electrolyte has been suggested as follows [21]:

$$Pt\text{-}CH_3OH \rightarrow Pt\text{-}CH_2OH + H^+ + e^- \rightarrow Pt\text{-}CHOH + H^+ + e^-$$
$$\rightarrow Pt\text{-}COH + H^+ + e^- \rightarrow Pt\text{-}CO + H^+ + e^- \qquad (15.1)$$

$$H_2O + Pt \rightarrow Pt\text{-}OH + H^+ + e^- \qquad (15.2)$$

$$Pt\text{-}CO + Pt\text{-}OH \rightarrow 2Pt\text{-}CO_2 + H^+ + e^- \qquad (15.3)$$

One problem associated with methanol dehydrogenation on the Pt-electrode surface is poisoning of the electrode by CO formed as an oxidation intermediate. The CO can adsorb very strongly on the Pt surface, blocking the active sites and causing a large decrease in electrode performance. Pt-Ru alloy catalysts have been used to overcome the poisoning problem. Ru can dissociate water at lower potential to create the oxygen-containing surface groups needed to convert CO to CO_2. It forms Ru-OH$_{ads}$ species at lower potentials, which helps to oxidize the adsorbed CO through the bifunctional mechanism. Weakening of the Pt-CO bond takes place in the alloy, resulting in lower CO coverage and increased mobility similar to what happens with other CO-tolerant Pt alloys, via the modification of the Pt electronic structure by alloying.

So far, the Pt-Ru alloy has shown the most promising performance for the oxidation of methanol and hydrogen oxidation reaction in the presence of CO. Carbon black has been used as support for the metal nanoparticles, particularly Vulcan XC-72 (Cabot), which has a surface area of $240 \text{ m}^2 \text{ g}^{-1}$. Methanol oxidation starts at lower potential values for all the Pt-Ru/C catalysts than for Pt/C anodes. Comparison of electrodes prepared with Pt and Pt-Ru as the electrocatalyst supported on nanotubes and those prepared with the most usual support, Vulcan XC-72 showed that multi-wall carbon nanotubes produce catalysts (Pt-Ru/MWNT) with better performance than on other supports, particularly with respect to those prepared with the traditional Vulcan XC-72 carbon powder [18].

On the other hand, efforts are also being made to provide methanol-tolerant cathode catalysts. Electrochemically pretreated $Bi_2Ir_2O_7$ is reported to be inactive toward methanol oxidation [3]. In addition, electrochemically pretreated $Bi_2Pt_{0.6}Ir_{1.4}O_7$ exhibits negligible methanol oxidation activity and shows activity for oxygen reduction reaction. The methanol tolerance increases as the Pt content decreases. Methanol-tolerant cathode catalysts will reduce the negative effects being caused by the diffused methanol.

15.3   Methanol Permeation Measurement Techniques

It is important to have accurate, reliable, and convenient techniques of measuring the methanol permeation in the development of membranes for DMFC. A number of techniques have been developed to measure methanol permeation through the membranes. This section describes some of the techniques (electrochemical and non-electrochemical) giving their merits and limitations.

15.3.1   Monitoring CO_2 from Cathode of DMFC

Early studies on the methanol permeation in the DMFC revealed that the methanol permeating to the cathode side oxidizes to CO_2 [45]. Since then, determination of CO_2 content in the cathode exhaust using an optical IR CO_2 sensor, gas chromatography, and mass spectrometry is widely used to measure methanol permeation in a DMFC [12]. However, during the operation of DMFC part of the CO_2 produced in the anodic catalyst layer permeates through the perfluorosulfonate membrane to the cathode side as shown in Fig. 15.4.   In addition, there is incomplete oxidation of the methanol at the cathode side. Studies have confirmed the presence of anodic CO_2 at the cathode and incomplete oxidation of methanol at the cathode side [13]. For low methanol concentration, methanol permeation decreases with increasing current density, which implies low production of CO_2 at the cathode. Thus, at high current densities and low methanol concentration the amount of CO_2 crossing from the anode to the cathode can even be higher than the amount of CO_2 produced at the cathode by methanol oxidation. Therefore, the presence of anodic CO_2 and incomplete oxidation of methanol at the cathode side have to be considered in order to avoid reporting wrong values for the methanol permeability. The exact amount of anodic CO_2 permeating through the membrane to the cathode side can be obtained using a methanol-tolerant cathode layer, which does not oxidize the permeated methanol. In addition, using gravimetric determination of $Ba\,CO_3$, the CO_2 produced both at the anode and the cathode can be analyzed:

$$\text{Anode: } CH_3OH + H_2O \Rightarrow CO_2 + 6H^+ + 6e^- \tag{15.4}$$

$$\text{Cathode: } 3/2(O_2) + 6H^+ + 6e^- \Rightarrow 3H_2O \tag{15.5}$$

$$\text{Overall: } CH_3OH + 3/2(O_2) \Rightarrow CO_2 + 2H_2O \tag{15.6}$$

15.3.2   Open Circuit Voltage Measurements

In an attempt to provide an easy way of determining methanol permeability of DMFCs membranes, a model equation has been developed [2]. The model equation relates open circuit voltage (OCV) values measured in a DMFC with cathode

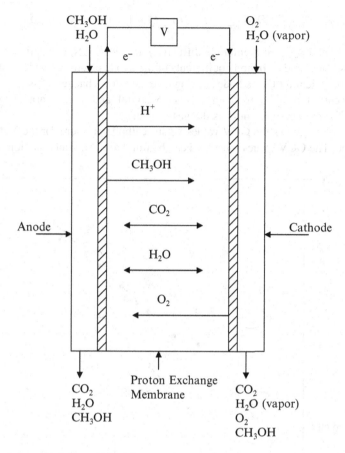

Fig. 15.4 Schematic diagram of DMFC showing CO_2 crossover

pressure using membrane diffusivity as a parameter. The cathode pressure was varied between 1 and 5 atm in order to study the effect on the OCV of a DMFC while maintaining an atmospheric pressure at the anode. The model equation is given as:

$$OCV = E^0_{cell} - D^0 \ln(P_2/P_1) - \gamma/\{M + G(P_2 - P_1)\}, \qquad (15.7)$$

where P_1 and P_2 are the anode and cathode pressures, respectively, E^0_{cell} and D^0 are cell voltage at pressure P_1 and temperature and reaction potential, respectively.

$$\gamma = \chi D_{mem} C^*/\delta_{mem},$$

$$D^0 = \Delta NRT / \eta_{cat} F,$$

$$M = D_{mem} \delta_{el} / D_{el} \delta_{mem} + (D_{mem} / \delta_{mem} k) + 1 \text{ and } G = K_{mem} / k \delta_{mem},$$

where D_{mem} and δ_{mem} are membrane diffusivity and thickness, respectively, δ_{el} and D_{el} are thickness and diffusion coefficients of the anode electrode, and k is the mass transfer coefficient for the cathode backing layer and flow channel. C^* is the methanol feed concentration, η_{cat} is the cathode over potential, and is the Faraday constant, whereas T is temperature and R is the gas constant.

Figure 15.5 shows the OCV values versus cathode pressure for the Nafion 115 membrane. The OCV values measured can be fitted to the model equation by using

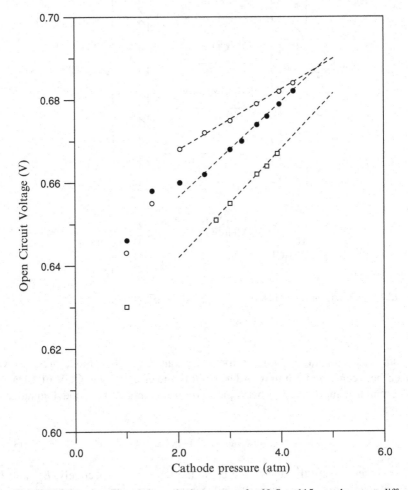

Fig. 15.5 OCV as a function of the cathode pressure for Nafion 115 membrane at different DMFC operation times. The dotted lines correspond to the model equation; (*circle*) first measurement; (*filled circle*) second measurement; and (*rectangle*) third measurement [2]. (This figure was published in *Journal of Power Sources*, Vol. 104, V. M. Barragan, Estimation of the Membrane Methanol Diffusion Coefficient from Open Circuit Voltage Measurements in a Direct Methanol Fuel Cell, page 71, Copyright Elsevier, 2002.)

a three-parameter (γ, M, and G) nonlinear regression method for the membranes. The parameters are a function of membrane diffusivity. A diffusivity value of Nafion membranes on the order of 10^{-5} cm^2 s^{-1} at 120°C using 5 M methanol solution was obtained.

Although OCV values are useful in determining membrane diffusivity, care should be taken to make sure that correct values are recorded. When a cell is on load before it is put to the open circuit condition, the OCV jumps and reaches a peak in seconds. The voltage at this peak is just a transient value. After some minutes, the voltage will drop until it stabilizes to a constant value, which should be taken as the real OCV value [35].

15.3.3 Limiting Current Density

Methanol crossover flux and diffusivity can be determined by measuring the limiting current in a DMFC. Assuming the methanol crossover to be due to diffusion only, then the diffusivity is given as [37]:

$$D_m = J_{lim} \delta_m / 6FC_m, \tag{15.8}$$

where
D_m = membrane diffusivity (cm^2 s^{-1}),
J_{lim} = Steady-state limiting current (A cm^{-2})
δ_m = membrane thickness (cm)
F = Farady constant (96,484 C mol^{-1})
C_m = CH$_3$OH concentration at the feed edge (M)

However, corrections for electro-osmotic drag effects are necessary even for low methanol concentrations for accurate methanol crossover flux measurement in a DMFC at open circuit voltage. With correction, the diffusivity becomes:

$$D_m = J_{lim} \delta_m / 6FC_m k_{dl}, \tag{15.9}$$

where
$$k_{dl} = \text{drag correction factor for } J_{lim}.$$

15.3.4 Measurements Using Diffusion Cell Set-up

It is a common practice to use a two-compartment diffusion cell set-up to determine the permeability of membranes in methanol solutions. Normally, a methanol solution of known concentration is put in one of the compartments with the membrane separating the two compartments, and the methanol concentration in the other

Fig. 15.6 Schematic of methanol permeation through a membrane from reservoir compartment (A) to receiving compartment (B)

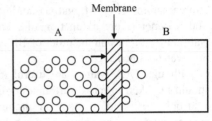

compartment is then determined with time. Theoretically, the concentration, flux, and permeability relationship is derived using a simple sketch of a two-compartment diffusion cell, Fig. 15.6, and some mass transfer equations.

Unsteady state material balance on the receiving compartment (B) yields:

$$j_A = \frac{V_B}{A} - \left(\frac{dC_B}{dt} \right),$$ (15.10)

where

J_A = Methanol flux from the reservoir compartment (A) to receiving compartment (B)
V_B = Volume of compartment B
A = Membrane area
C_B = Methanol concentration in the receiving compartment
t = Time

Simplified expression for Fick's law is given as:

$$j_A = DK / \delta_m (C_A - C_B),$$ (15.11)

where

D = Methanol diffusivity (assumed to be constant inside the membrane)
k = Partition coefficient (constant).
DK = Membrane permeability.
δm = Membrane thickness
C_A = Methanol concentration in the reservoir compartment

Combining (15.10) and (15.11) gives:

$$V_B (dC_B / dt) = A(DK / \delta_m)(C_A - C_B).$$ (15.12)

Taking methanol molar balance for the two compartments gives:

$$C_{A_0} \times V_A = C_A \times V_A + C_B \times V_B$$ (15.13)

Assuming $V_A \approx V_B$ always, then:

$$V_B (dC_B / dt) = A(DK / \delta_m)(C_{A_0} - 2C_B).$$ (15.14)

Thus,

$$\int_{C_{B_0}}^{C_{B(t)}} \left(\frac{1}{\left(C_{A_0} - 2C_B(t) \right)} \right) dC_B = A(DK / V_B \delta_m) \int_{t_0}^{t} dt. \tag{15.15}$$

This simplifies to:

$$-\ln\{1 - (2C_B(t)/C_{A_0})\} = 2A(DK / V_B \delta_m)(t - t_0), \tag{15.16}$$

where

$$C_{A0} / C_{B(t)} > 2.$$

If $C_B(t)$ is known, then from the slope of linear plot of (15.16), the permeability (DK) of the membrane could be obtained. Average methanol crossover flux can be determined using dC_B / d_t values and equation (15.10). $C_B(t)$ is expressed as:

$$C_B(t) = \frac{C_{A_0}}{2} \left[1 - e^{-\frac{2ADK}{V_B \delta_m}(t - t_0)} \right]. \tag{15.17}$$

If $C_A(t) \gg C_B(t)$, then:

$$C_B(t) = A(DK / V_B \delta_m) C_{iA_0} (t - t_0) \tag{15.18}$$

C_{iA_0} = Methanol initial concentration in the reservoir compartment,

t_0 = Time lag related to diffusivity as $t_0 = L^2 / 6D$.

There are different ways for determining $C_B(t)$ experimentally, some of which are discussed in the following.

15.3.4.1 Gas Chromatography

Gas chromatography (GC) is a non-electrochemical technique that is being used to evaluate methanol permeability through polymer electrolyte membranes [44]. Samples of the solution from the receiving compartment are taken at various periods and analyzed to quantify the concentration of methanol. The technique apparently works fine; however, the samples taken are prone to contamination before the GC analysis if an online GC is not available. If the supporting electrolyte is aggressive, such as H_2SO_4, then a HP-Innowax capillary column inserted in a GC using a FID should be used [39]. This technique is useful for preliminary screening of membranes in a diffusion cell set-up. However, in a DMFC, the permeated methanol oxidizes to CO_2. Thus, relying on measuring the methanol concentration using GC is unsuitable in a real DMFC.

Similar to the gas chromatographic method where the methanol concentration of the receiving compartment $(C_B(t))$ is monitored with time, in a variation of this method the receiving compartment can be connected to a vacuum system that allows pervaporation of the crossed methanol [34]. The vaporized methanol is collected in cryogenic traps. This method could be cumbersome, and like gas chromatography is useful only for membrane screening since it does not simulate fuel cell conditions well.

15.3.4.2   Mass Spectrometry

Mass spectrometry of liquid samples of the cathode outlet stream is another way of determining the methanol crossover flux. For mass spectrometric measurements of methanol crossover, a clear description of the respective system could be achieved by measuring the background methanol signal of a cell filled with distilled water and equipped with the membrane sample, and subsequently adding well-adjusted portions of aqueous or pure methanol to this liquid [25]. The slopes of mass signal vs. time curves are typical for diffusion-controlled processes and with the help of the calibration lines, the diffusion coefficient of methanol through the membrane can be calculated. Online analysis of the cathode exhaust gas with multipurpose electrochemical mass spectrometry can also be employed to determine methanol permeability. However, as mentioned, the assumptions that the entire permeated methanol is converted to CO_2 and that there is no anodic CO_2 contribution are contentious.

15.3.4.3   Cyclic Voltammetry and Chronoamperometry

Voltammetric and chronoamperometric techniques are widely used recently to determine the methanol permeability through proton exchange membranes intended for use in direct methanol fuel cell [36]. These techniques can be used with either a real fuel cell or a simulated one (diffusion cell). It is expected that the only electron transfer reaction that can occur at the working electrode surface, within the potential range or potential value under consideration, is the methanol oxidation.

$$CH_3OH \rightarrow CO_2 + 6e^- + 6H^+ \qquad (15.19)$$

$$\text{Rate of oxidation} = k(C_{CH_3OH})_{\text{at the electrode surface}} \qquad (15.20)$$

In the diffusion cell set-up, the methanol concentrations in the receiving compartment $(C_B(t))$ are obtained by recording the cyclic voltammograms (CVs) and chronoamperometric curves using AUTOLAB or EG and G PARC potentiostat/galvanostat with a programmable power supply. These techniques allow both qualitative and quantitative study of the methanol permeation through the membrane. Figure 15.7 shows the permeability curves at different methanol concentration

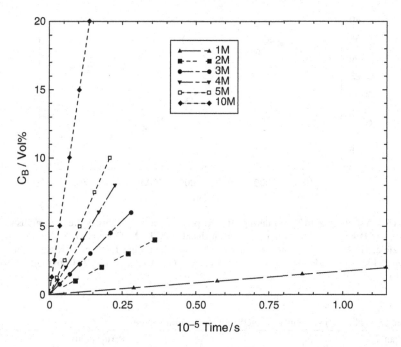

Fig. 15.7 Permeability curves at various concentrations. (This figure was published in *Journal of Electroanalytical Chemistry*, Vol. 542, K. Ramya, K. S. Dhathathreyan, Direct Methanol Fuel Cells: Determination of Fuel Crossover in a Polymer Electrolyte Membrane, page 114, Copyright Elsevier, 2003.)

obtained by Ramya et al. using these methods. The methanol permeability is dependent on the initial concentration of the methanol in the methanol-added compartment. An optimum concentration of 1 –2 M solution for operation of fuel cells is suggested.

15.3.4.4   Potentiometric Technique

A potentiometric technique is being developed for measuring methanol permeability through polymer electrolyte membranes of DMFC [31]. In this technique, the variation of potential (E) of the working electrode in the methanol receiving compartment is measured at OCV during the CH_3OH permeation. The slope (dE / dt) of E vs. T (time) curve has been found to be proportional to the permeability. Figure 15.8 shows a plot of (dE / dt) vs. T (time), indicating higher permeability initially, which reduces with time.

The potential (E) of a working electrode measured in the methanol-receiving compartment has an inverse relationship with the methanol concentration. As such, the potential (E) values obtained with time can be used to determine the methanol concentrations ($C_B(t)$) as a function of time. Plot of ($C_B(t)$) vs. time will give

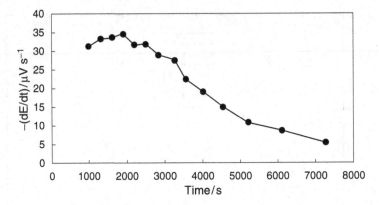

Fig. 15.8 Variation of (dE / dt) during CH$_3$OH permeation [31]. (This figure was published in *Journal of Power Sources*, Vol. 117, N. Munichandraiah, K. McGrawth, G. K. S. Prakash, R. Aniszfeld, G. A. Olah, A Potentiometric Method of Monitoring Methanol Crossover through Polymer Electrolyte Membranes of Direct Methanol Fuel Cells, page 101, Copyright Elsevier, 2003.)

Table 15.3 Nafion 117 membrane methanol permeability values

Methanol concentration (M)	Temperature (°C)	Technique	Permeability (cm^2 s^{-1})	Reference
–	25	Radioactive tracer	1.15×10^{-5}	[45]
12.5	25	Gas chromatography (diffusion cell)	1.91×10^{-6}	[22]
2 & 10	Room temperature	DMFC	3.2×10^{-6}	[23]
1.5	Room temprature	Differential refractive index detector (diffusion cell)	2.6×10^{-6}	[24]
1	30	Limiting current density (DMFC)	4.27×10^{-6}	[37]
5	120	OCV model equation (DMFC)	1.56×10^{-5}	[2]
2	22	Gas chromatography (diffusion cell)	1.15×10^{-6}	[44]
2.5	22	Potentiometry (diffusion cell)	1.13×10^{-6}	[6]

values of dC_B/dt. This slope will initially be high as the permeation starts, but decreases as the flux reduces until it becomes zero at equilibrium concentration. Average methanol crossover flux can be obtained using a material balance—flux relationship (15.10). Thus, the measurement of the electrode potential is useful in determining the methanol crossover flux through proton exchange membranes of DMFC. Table 15.3 shows Nafion 117 membrane permeability values determined by different authors using different techniques. There are some variations in the values because of differences in the experimental conditions and probably because some techniques are better than others. Generally, Nafion 117 membrane has methanol permeability values in the order of 10^{-5} cm^2/s to 10^{-6} cm^2/s.

In the methanol permeation study using a diffusion cell set-up, the positioning of the working electrode is very important. Attaching the working electrode on the membrane enhances methanol oxidation current detection. Even a small distance away from the membrane surface can affect the results because the methanol oxidation current may not be detected. The best way is to attach the working electrode on the membrane or to prepare it on the membrane by a suitable method, such as electroless deposition.

15.4 Conclusions

Despite advancement in the development of direct methanol fuel cells (DMFCs), some restrictions still inhibit their large-scale commercialization. This chapter has discussed one of the primary constraints, that is, identification of appropriate membrane materials. Nafion membranes that dominate the market of polymer electrolyte membranes allow methanol permeation from the anode to the cathode side of a DMFC. This results in serious negative consequences. Three approaches have been pursued in order to resolve the methanol permeation problem. These include Nafion membranes modification, development of alternative membranes and provision of high activity anode catalysts or methanol tolerant cathode catalysts. All the three options have achieved certain degree of success in solving the problem. Of particular interest are the Nafion membranes modification and development of alternative membranes in which membranes with permeability values of 10 to 70 times lower than the pure Nafion membranes have been developed. In general, based on the tremendous research efforts being made to develop DMFCs membranes with the best qualities, we are optimistic that very soon the issue of methanol permeation shall become a history.

The chapter also talked on various techniques developed for determining the membranes methanol permeability. Assessment of these techniques has shown that electrochemical techniques are more accurate. Among the electrochemical techniques, potentiometry has additional advantages of easier reproducibility of results, obtaining more data points, and convenience.

References

1. Baldauf, M., and Preidel, W., 1999, "Status of the development of a direct methanol fuel cell ", *J. Power Sources*, **84** (2):161–166.
2. Barragan, V. M., and Heinzel, A., 2002, "Estimation of the membrane methanol diffusion coefficient from open circuit voltage measurements in a direct methanol fuel cell ", *J. Power Sources* **104** (1):66–72.

3. Beck, N. K., Steiger, B., Scherer, G. G., and Wokaun, A., 2006, "Methanol tolerant oxygen reduction catalysts derived from electrochemically pretreated $Bi_2Pt_{2-y}Ir_yO_7$ pyrochlores ", *Fuel Cells*, No.1:26 –30.
4. Bello, M., Zaidi, S. M. J., and Rahman, S. U., 2008, "Proton and methanol transport behavior of SPEEK/TPA/MCM-41 composite membranes for fuel cell application", J. Membr. Sci. **322**: 218–224..
5. Bello, M., Zaidi, S. M. J., and Rahman, S. U., 2006, "Evaluation of methanol crossover through SPEEK/TPA/Y- zeolite composite membranes by an electrochemical method ", Fuel cells seminar a677, Hawaii, USA.
6. Bello, M., Rahman, S. U., and Zaidi, S. M. J., 2007, "Comparative study on measurement techniques for methanol crossover through polymer electrolyte membranes of DMFCs ", 7th International conference and exhibition on chemistry in industry, Manama, Kingdom of Bahrain.
7. Byungchan, B., Yong, H. H., and Dukjoon, K., 2005 "Preparation and characterization of Nafion/poly (1- vinylimidazole) composite membrane for direct methanol fuel cell application ", *J. Electrochem. Soc.* **152** (7):A1366–A1372.
8. Carmo, M., Paganin, V. A., Rosolen, J. M., and Gonzacez, E. R., 2005, "Alternative supports for the preparation of catalyst for low temperature fuel cells: The use of carbon nanotubes ", *J. Power Sources* **142** (1–2):169–176.
9. Chang, J. H., Park, J. H., Park, G-G., Kim, C-S., and Park, O-OK., 2003, "Proton- conducting composite membranes derived from sulfonated hydrocarbon and inorganic materials ", *J. Power Sources* **124** (1):18–25.
10. Choi, W. C., Kim, J. D., and Woo, S. I., 2001, "Modification of proton conducting membrane for reducing methanol crossover in a direct methanol fuel cell ", *J. Power Sources* **96** (2):411–414.
11. Dimitrova, P., Friedrich, K. A., Stimming, U., and Vogt, B., 2002, "Modified Nafion — based membranes for use in direct methanol fuel cells ", *Solid State Ionics* **150** (1–2):115–122
12. Dohle, H., Divisek, J., Mergel, J., Oetjen, H. F., Zingler, C., and Stolten, D., 2002, "Recent developments of the measurement of the methanol permeation in a direct methanol fuel cell ", *J. Power Sources*, **105**(2):274–282.
13. Drake, J. A., Wilson, W., and Killen, K., 2004, "Evaluation of the experimental model for methanol crossover in DMFCs ", *J. Electrochem. Soc.* **151** (3):A413–A417
14. Every, H. A., Hickner, M. A., McGrath, J. E., and Zawodzinski Jr, T. A., 2005, "An NMR study of methanol diffusion in polymer electrolyte fuel cell membranes ", *J. Membr. Sci.* **250** (1–2):183 –188.
15. Fedkin, M. V., Zhou, X., Hofmann, M. A., Chalkova, E., Weston, J. A., Allcock, H. R., Lvov, S. N., and Serguie, N., 2002, "Evaluation of methanol crossover in proton-conducting polyphosphazene membranes ", *Mater. Lett.* **52** (3):192–196.
16. Hamnett, H., 2003, "Direct methanol fuel cell " in: Handbook of fuel cells; Fundamental technology and applications ", edited by: Wolf, V., Arnold, L., and Hubert, A. G., Wiley, Vol. **1**: 305–322.
17. Hong, W., Yuxin, W., and Shichang, W., 2002, "Study on the preparation and properties of PVdF — based — blend Nafion membranes ", *Gaofenzi Xuebao* 4:540–543.
18. Ise, M., 2000, "Polymer elecktrolyt membranen: untersuchungen zur mikrostruktur und zu den transporteingenschaften fur protonen und wasser", *PhD Thesis*, University of Stuttgart.
19. Jiang, R., Kunz, H. R., and Fenton, J. M., 2005, "Improvement of DMFC performance and stability using multilayer structure membranes", *230th ACS National Meeting*, Washington DC, USA.
20. Jin, L., Huanting, W., Shaoan, C., and Kwong-Yu, C., 2005, "Nafion — Polyfurfuryl alcohol nanocomposite membranes for direct methanol fuel cells", *J Membr. Sci.* **246** (1):95–101.
21. Jung, I., Kim, D., Yun, Y., Chung, S., Lee, J., and Tak, Y., 2004, "Electro-oxidation of methanol diffused through proton exchange membrane on Pt surface: crossover rate of methanol ", *Electrochem. Acta* **50** (2 –3):607–610.

22. Kim, D. W., Choi, H-S., Lee, C., Blumstein, A., and Kang, Y., 2005, "Investigation on methanol permeability of Nafion modified by self — assembled clay — nanocomposite membrane for DMFC", *Solid State Ionics*, **176** (11–12):1079–1089.

23. Kim, Y-M., Park, K-W., Choi, J-H., Park, I. S., and Sung, Y-E., 2003, "A Pd- impregnated nanocomposite Nafion membrane for use in high—concentration methanol fuel in DMFC", *Electrochem. Com.* **5** (7):571–574.

24. Kim, J., Kim, B., and Jung, B., 2002, "Proton conductivity and methanol permeability of membranes made from partially sulfonated polystyrene-block-poly (ethylene-ran-butylene)-block-polystyrene copolymers", *J. Membr. Sci.* **207** (1):129–137.

25. Kuver, A., and Potje-Kamloth, K., 1998, "Comparative study of methanol crossover across electro-polymerized and commercial proton exchange membrane electrolytes for the acid direct methanol fuel cell", *Electrochem. Acta* **43** (16–17):2527–2535.

26. Li, L., Zhang, J., and Wang, Y., 2003, "Sulfonated poly (ether ether ketone) membranes for direct methanol fuel cell", *J. Membr. Sci.* **226** (1–2):159–167.

27. Libby, B., Smyrl, W. H., and Cussler, E. L., 2003, "Polymer-zeolite composite membranes for direct methanol fuel cells", *AIChE J.* **49** (4):991–1001.

28. Maria, G., Xiangling, J., Xianfeng, L., Hui, N., Hampsey, J. E., and Yunfeng, L., 2004, "Direct synthesis of sulfonated aromatic poly (ether ether ketone) proton exchange membranes for fuel cell applications", *J. Membr. Sci.* **234** (1–2):75–81.

29. Mikhailenko, S. D., Zaidi, S. M. J., and Kaliaguine, S., 2001, "Sulfonated polyether ether ketone based composite polymer electrolyte membranes", *Cat. Today* **67** (1–3):225–236.

30. Miyake, N., Wainright, J. S., and Savinell, R. F., 2001, "Evaluation of a sol-gel derived Nafion/silica hybrid membrane for polymer electrolyte membrane fuel cell applications 11: Methanol uptake and methanol permeability", *J. Electrochem. Soc.* **148** (8):A905–A909.

31. Munichandraiah, N., McGrath, K., Prakash, G. K. S., Aniszfeld, R., and Olah, G. A., 2003, "A potentiometric method of monitoring methanol crossover through polymer electrolyte membranes of direct methanol fuel cells", *J. Power Sources*, **117** (1–2):98–101.

32. Narayanan, S. R., Valdez, T., Rohatgi, N., Chun, W., and Hoover, G., 1999, "Recent advances in direct methanol fuel cells", *Proceedings of 11th Annual Battery Conference on Applications and Advances*, California, USA (12–15):73–77.

33. Navarra, M. A., Materazzi, S., Panero, S., and Scorsese, B., 2003, "PVdF-based membranes for DMFC applications", *J. Electrochem. Soc.* **150** (11):A1528–A1532.

34. Ponce, M. L., Prado, L., Ruffmann, B., Richau, K., Mohr, R., and Nunes, S. P., 2003, "Reduction of methanol permeability in polyetherketone-heteropoly acid membranes", *J. Membr. Sci.* **217** (1–2):5–15.

35. Qi, Z., and Kaufman, A., 2002, "Open circuit voltage and methanol crossover in DMFCs", *J. Power Sources* **110** (1):177–185.

36. Ramya, K., and Dhathathreyan, K. S., 2003, "Direct methanol fuel cells: Determination of fuel crossover in a polymer electrolyte membrane", *J. Electroanal. Chem.* **542**:109–115.

37. Ren, X., Springer, T. E., Zawodzinski, T. A., and Gottesfeld, S., 2000, "Methanol transport through Nafion membranes electro-osmotic drag effects on potential step measurements", *J. Electrochem. Soc.* **147** (2):466–474.

38. Saarinen, V., Kallio, T., Paronan, M., Tikkanen, P., Rauhala, E., and Kontturi, K., 2005, "New ETFE-based membrane for direct methanol fuel cell", *Electrochem. Acta* **50** (1):3453–3460.

39. Schaffer, T., Hacker, V., Tschinder, T., Besenhand, J. O., and Prenninger, P., 2005, "Introduction of an improved gas chromatographic analysis and comparison of methods to determine methanol crossover in DMFCs", *J. Power Sources* **145** (2):188–198.

40. Shen, M., Roy, S., Kuhlmann, J. W., Scott, K., Lovell, K., and Horsfall, J. A., 2005, "Grafted polymer electrolyte membrane for direct methanol fuel cells", *J. Membr. Sci.* **251** (1–2):121–130.

41. Shin, J-P., Chang, B-J., Kim, J-H., Lee, S-B., and Suh, D. H., 2005, "Sulfonated polystyrene/PTFE composite membranes", *J. Membr. Sci.*, **251**(1–2):247–254.

42. Smitha, B., Sridhar, S., and Khan, A. A., 2005, "Solid polymer electrolyte membranes for fuel cell applications — a review, *J. Membr. Sci.* **259** (1–2):10–26.

43. Tricoli, V., 1998, "Proton and methanol transport in poly (perfluorosulfonate) membranes containing Cs⁺ and H,⁺ cations", *J. Electrochem. Soc.* **145** (11):3798–3801.

44. Tricoli, V., Carretta, N., and Bartolozzi, M., 2000, "A comparative investigation of proton and methanol transport in fluorinated ionomeric membranes", *J. Electrochem. Soc.* **147** (4):1286–1290.

45. Verbrugge, M. W., 1989, "Methanol diffusion in per-fluorinated ion-exchange membranes", *J. Electrochem. Soc.*, **136** (2):417–423.

46. Weilin, X., Tianhong, L., Changpeng, L., and Wei, X., 2005, "Low methanol permeable composite Nafion/Silica/PWA membranes for low temperature direct methanol fuel cells", *Electrochem. Acta*, **50** (16–17):3280–3285.

47. Xu, F., Innocent, C., Bonnet, B., Jones, D. J., and Roziere, J., 2005, "Chemical modification of per fluorosulfonated membranes with pyrrole for fuel cell application: Preparation, characterization and methanol transport", *Fuel Cells* **5** (3):398–405.

48. Zhi-Gang, S., Xin, W., and I-Ming, H., 2002, "Composite Nafion/polyvinyl alcohol membranes for the direct methanol fuel cells", *J Membr. Sci.* **210** (1):147–153.

Chapter 16
Systematic Design of Polymer Electrolyte Membranes for Fuel Cells Using a Pore-Filling Membrane Concept

Takeo Yamaguchi

Abstract In this chapter, systematic membrane design and development using our pore-filling membrane concept is described. Pore-filling electrolyte membranes for use as electrolyte membranes for polymer electrolyte membrane fuel cells (PEMFCs) or direct methanol fuel cells (DMFCs) are described. The pores of a porous substrate are filled with a polymer electrolyte and the membrane swelling is suppressed by the substrate matrix. Proton conductivity is achieved through the impregnated electrolyte polymer. Fuel crossover is reduced by suppression of the electrolyte polymer swelling and mechanical strength is maintained by the substrate. Using this concept, high proton conductivity has been shown to exist with reduced membrane fuel crossover and good dimensional stability. In this chapter, high performance pore-filling electrolyte membranes and their DMFC or PEMFC performances are shown. To achieve a high energy density DMFC device, we must use a high concentration of methanol aqueous solution as the fuel and crossover must be reduced. The membrane also showed almost no dimensional change with variation in humidity. The DMFC and PEMFC performances are also described with several varieties of pore-filling membranes for each application.

16.1 Introduction

Fuel cells are of interest as an energy source, because they can directly convert chemical energy into electrical energy and therefore have high electrical energy conversion efficiency. In particular, polymer electrolyte membrane fuel cells (PEMFCs) can be operated at low temperatures with low exhaust emissions and they can be miniaturized. Therefore, PEMFCs are expected to come into practical use as power sources in automobiles, as decentralized power sources for facilities such as domestic houses, hospitals and offices, and as mobile power sources.

Each application requires a different fuel supply system and operating temperature for the required power density. In addition, their target costs will be different. For example, automobile fuel cells should aim to reduce the cost of the current fuel cell system by hundreds of times to compete with mature internal combustion engines, and portable fuel cells can compete economically with current lithium ion batteries.

S.M.J. Zaidi, T. Matsuura (eds.) *Polymer Membranes for Fuel Cells*,
doi: 10.1007/978-0-387-73532-0, © Springer Science+Business Media, LLC 2009

The portable device can be used at a low temperature, less than 60°C, although auto-mobile or stationary fuel cells operate at above 70 or 80°C. The durability require-ments are also different. Therefore, we must systematically design a different polymer electrolyte membrane for each application. In most cases, a new polymer electrolyte is made and then the fuel cell performance is tested. However, we are proposing a systematic design methodology for the membrane to suit each application.

In this chapter, we focus on mobile applications to explain the systematic design thinking, because the mobile application will probably become the first widely used fuel cell device from the many possible applications. In our laboratory, we are also developing new polymer electrolyte membranes for automobiles, stationary power sources, and other applications.

16.2 Mobile Fuel Cell Technologies

For mobile applications, both the energy and power density of the device are very important. Methanol will be used as the hydrogen carrier because methanol is liq-uid at normal temperatures and the energy density is much higher than that of pres-surized hydrogen. Methanol aqueous solution can be directly converted to protons and electrons, as shown in Fig. 16.1, and the potential energy density is more than 1,000 Wh Kg^{-1}, even though a voltage of only 0.5 V was used. This value is much higher than those of current battery devices, such as approximately 200 Wh Kg^{-1} for Li ion batteries. Current and future mobile phones, digital cameras, videos, or laptop computers require a greater energy density because many functions are being installed in the one small device. Therefore, direct methanol fuel cells (DMFCs) have great potential for use in portable devices, because of their low weight and simple system features [1]. Figure 16.1a, b show a PEMFC using hydrogen as its fuel and a DMFC. However, current DMFCs do not achieve high

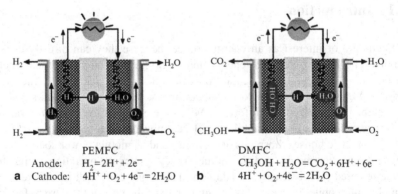

PEMFC
Anode: $H_2 = 2H^+ + 2e^-$
a   Cathode: $4H^+ + O_2 + 4e^- = 2H_2O$

DMFC
$CH_3OH + H_2O = CO_2 + 6H^+ + 6e^-$
b $4H^+ + O_2 + 4e^- = 2H_2O$

Fig. 16.1 Schematic illustrations of PEMFC using hydrogen fuel and direct methanol fuel cell (DMFC)

energy densities compared with the Li ion batteries that are the current portable power sources [2]. The problem is methanol fuel crossover through the membrane. The crossover leads to poor fuel cell performance and serious energy density reduction because most of the methanol fuel will not become electrical energy and just pass through the membrane. To create portable DMFCs with high energy density, the methanol crossover through the membrane must be reduced. In addition, a high concentration of methanol in aqueous solution must be used as the fuel even when a recycling system is used.

Usually, methanol concentrations of 2 M or less are used in DMFCs, due to the serious problem of crossover of methanol through the electrolyte membranes. When this occurs, the transported methanol reacts directly on the cathode and seriously reduces the DMFC voltage. As a result of catalyst poisoning and mixed potential loss at the cathode the energy density using low concentration, methanol fuel cannot match that of current batteries. The anode reaction is:

$$CH_3OH + H_2O \rightarrow 6H^+ + 6e^- + CO_2 \qquad (16.1)$$

The stoichiometric methanol concentration required for complete reaction of all molecules to produce protons is around 64 wt%. However, the system requires more water for efficient proton conductivity. To achieve a power density several times larger than that of a Li ion battery, DMFCs should be operated with approximately 10 M methanol in aqueous solution (32 wt%).

The desired membrane must also have the following characteristics: (1) the proton conductivity of the membrane material should be high and the membrane should be thin to minimize membrane resistance; (2) the membrane must be mechanically strong and the change in membrane area between the dry and the swollen states should be negligible, to minimize any membrane/electrode interface resistance; (3) the membrane material must be chemically stable during DMFC operation; and (4) the membrane must be economical to manufacture.

To date, many membranes have been developed [3–12] and research has been carried out to develop a high performance membrane to replace perfluorinated polymer electrolyte membranes, such as Nafion membranes. However, no membranes developed to date satisfy all of the above criteria required for DMFC membranes [2–12].

16.3 Pore-Filled Membrane Concept and Membrane Performance

We are proposing the concept of a pore-filling membrane to the design and fabrication of an electrolyte membrane suitable for a PEMFC or DMFC. Most polymer electrolyte membranes require water for proton migration, but membrane swelling leads to a high solvent permeation [13]. To control membrane swelling and solvent permeation, we proposed the pore-filling concept [13,14]. The first application was as a separation membrane for liquid separation applications [13].

This concept can be applied to polymer electrolyte membranes for fuel cell applications. In this case, the pore-filling electrolyte membrane is composed of two materials: a porous substrate and a graft- or gel-type polymer electrolyte that fills the pores of the substrate, as shown in Fig. 16.2. The porous substrate is completely inert to liquid fuels or gas, has mechanically strong matrices, and the filling polymer electrolyte can contain water for proton migration. The filling polymer provides proton conductivity and the porous substrate matrix prevents excess swelling of the filling polymer mechanically, which can lead to high methanol crossover. In addition, the substrate matrix restricts changes in the membrane area from the dry to the swollen state. In this concept, the suppression of swelling is important, unlike in the previously reported perfluorinated sulfonic acid polymers impregnated on porous support membranes [15,16], and the substrate matrix must possess strong mechanical properties to restrict the very swollen polymer gel.

In this study, poly(acrylic acid) (poly(AA)), poly(acrylic acid-co-vinylsulfonic acid) (poly(AAVS)) or poly(acrylamide tert-butyl sulfonic acid) (poly(ATBS)) was used as the filling polymer, and porous polytetrafluoroethylene (PTFE) [17, 18], porous cross-linked high-density polyethylene substrate (CLPE) or a porous polyimide substrate (PI) was used as the substrate. Polyacrylic acid is a weak acid and the poly(AAVS) copolymer has a sulfonic acid group of 0.7 mmol g^{-1}-polymer, because the vinylsulfonic acid content in the copolymer is 5 mol%. Poly(ATBS) has a 4.5 mmol g^{-1}-polymer sulfonic acid content.

The pore-filling ratio ϕ_f [%] was estimated using the following equation:

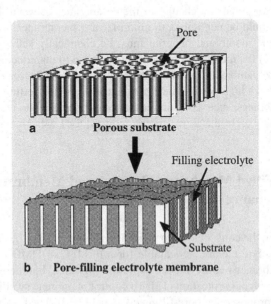

Fig. 16.2 The concept of a pore-filling electrolyte membrane. The real substrate matrix has an isotropic structure, and the pores are not cylindrical and they are interconnected

$$\phi_f = \frac{\left(w_{mem}^d - w_{sub}^d\right)}{w_{sub}^d} \times 100 \qquad (16.2)$$

where w_{mem}^d = dry membrane weight and w_{Sub}^d = dry substrate weight. The membrane area was measured in water and the change in the ratio of membrane areas between the swollen state and the dry state, ϕ_{al} [%], was evaluated at 25°C,

$$\phi_{al} = \frac{S_{mem}^s - S_{mem}^d}{S_{mem}^d} \times 100 \qquad (16.3)$$

where S_{men}^s and S_{men}^d are the membrane areas in the swollen and dry states, respectively. The entire membrane area was measured in the experiments, including that of the substrate.

The dependence of the pore-filling ratio on the methanol permeability through the pore-filling membranes is shown in Fig. 16.3a. Below a 60% filling ratio, the methanol permeability decreased with an increasing pore-filling ratio for the samples with both substrates. The filling polymer density was low and some leakage flux occurred through the membranes. Above an 80% filling ratio, the methanol permeation was almost independent of the filling ratio because filling electrolyte almost filled the pores. The reason methanol permeation through membranes increases somewhat with an increase in the filling ratio is due to an increase of the methanol permeation region. This is because methanol can permeate only through the filling polymer region. Figure 16.3b shows the dependence of the filling ratio on the proton conductivity at 25°C. The membrane was in the swollen state when in the water. The proton conductivity increased as the filling ratio increased for both

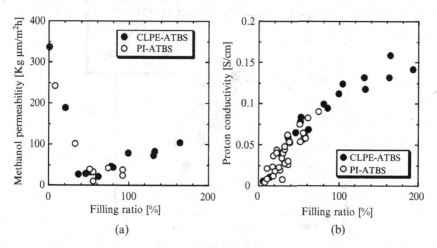

Fig. 16.3 The relationship between the pore-filling ratio and the membrane performances: (a) methanol permeability through the membrane at 50°C, (b) the proton conductivity through the membrane at 25°C in the humid state

substrates. The filling ratio changed the filling polymer content in the membrane and resulted in an increase in the sulfonic acid content and proton conduction region. This effect led to increasing proton conductivity with an increasing pore-filling ratio. The proton conductivity of Nafion117 was 0.08 S cm^{-1}, whereas that of the pore-filling membranes was 0.15 S cm^{-1}.

Figure 16.4 shows a comparison of the performance of the pore-filling electrolyte membranes with Nafion 112 and Nafion 117. Pore-filling membranes with a PTFE substrate and polyacrylic acid (AA) or poly(acrylic acid-co-vinylsulfonic acid) (AAVS) are also included [14]. Results using the low filling ratio membranes were omitted because of the leakage methanol flux. The ordinate axis shows the inverse of the methanol permeability values that were obtained by pervaporation at 50°C and the abscissa shows the proton conductivity at 25°C. A point located towards the upper right-hand side of Fig. 16.4 implies a higher performance membrane. Poly(acrylic acid) pore-filling membranes showed low proton conductivities because of the weak acidity of the carboxyl group. For the pore-filling membranes with AAVS copolymers, the vinyl sulfonic acid content, which is strongly acidic, was approximately 5 mol% and the sulfonic group content was low. The pore-filling membranes with poly(ATBS), which has a higher sulfonic acid group content, showed higher conductivity than the poly(AAVS) pore-filling membranes because of the higher strong acid group content, but the methanol permeability values did not differ greatly. This is because CLPE or PI substrates have a mechanically strong matrix and the swelling of the poly(ATBS) was effectively suppressed. The methanol

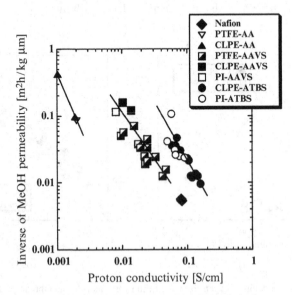

Fig. 16.4 The relationship between the proton conductivity and the inverse of the methanol permeability

permeability through the pore-filling membranes showed lower values compared with the Nafion 117 membrane. The substrate matrix effectively suppressed the membrane swelling and resulted in a lower methanol permeation value.

If we use a single polyelectrolyte as the filling polymer with different substrate and pore-filling ratios, the relationship between the proton conductivity and the inverse of the methanol permeability would be a single line. Assuming that this line is straight, then the order of performance would be: poly(AA) < poly(AAVS) < poly (ATBS), as shown in Fig. 16.4, and this order corresponds to the sulfonic acid content. Therefore, we can obtain better performance membranes by combining a filling polymer having a high strong-acid-group content with a mechanically strong matrix to suppress the swelling. At the same time, it is possible to achieve a balance in the methanol permeability and proton conductivity that is suitable for realizing a DMFC by controlling the substrate strength and filling ratio [19].

In summary, the CLPE and PI substrates can suppress membrane swelling and the change in membrane area between the dry and swollen states is negligible for pore-filling membranes containing those substrates. An ATBS filling polymer having a high sulfonic acid content shows a high proton conductivity of 0.15 S cm^{-1} at 25°C. The relationship between proton conductivity and methanol permeability of a single pore-filling polymer can be controlled by changing the substrate strength and pore-filling ratio. This enables us to control the membrane performance for a given fuel cell application.

16.4 Membrane Performance and DMFC Applications

To create portable DMFCs with high energy density, a high concentration of methanol in aqueous solution must be used as the fuel, as described. We developed a new type of membrane and its membrane electrode assembly (MEA) gives high performance with a high concentration methanol fuel (approximately 10 M). The results were compared with a commercial Nafion membrane MEA.

The electrodes consisted of a backing layer, a gas diffusion layer, and a catalyst layer. Toray's graphite fiber paper was used as the backing layer. The catalyst layer was formed on the diffusion layer by painting on a mixture of metal supported on a carbon black catalyst. The catalysts used on the anode and cathode were 49.4 wt% PtRu/Ketjenblack (TEC60E50E, Tanaka Kikinzoku Kogyo KK, Japan) and 55.3 wt% Pt/Ketjenblack (TEC10E60E, Tanaka Kikinzoku Kogyo KK, Japan), respectively. The anode, membrane, and cathode were stacked in that order and then treated at a pressure of 2 MPa and a temperature of 130°C.

The change in the ratio of the membrane areas is important in the practical operation of a fuel cell, especially for cycling. A carbon electrode can become detached from a membrane that exhibits a large change in membrane area, because carbon electrodes do not swell. Figure 16.5 shows the effect of methanol concentration on the membrane area change ratio for a pore-filling membrane and a commercial Nafion 117 membrane. Clearly, the Nafion membrane changed in area

Fig. 16.5 Comparison between the pore-filling membrane (CLPE-ATBS) and the Nafion 117 membrane: effect of methanol concentration on change in membrane area ratio at 25°C

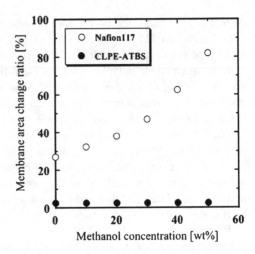

between the swollen and dry states and the degree of change increased exponentially with methanol concentration. However, in the case of the pore-filling membrane, the area change was negligible over the entire methanol concentration range, because the substrate effectively suppresses membrane swelling, as expected with the membrane concept.

The performances of MEAs with CLPE-ATBS or Nafion 117 membranes were examined using a single cell with the same electrodes at 50°C. Aqueous methanol solution and oxygen gas at ambient pressure were used as the fuel and oxidizing reagent, respectively. The flow rates of the methanol solution and oxygen were set at 10 and 500–1,000 mL min^{-1}, respectively, to avoid the flow rate effect. Fuel cell performance was measured and the current-potential (I–V) curve was recorded under galvanostatic control. The current density was increased in steps from zero to a given current density over time and then decreased to zero again using the same rate. This cycle was repeated until no difference between the forward and the backward curve was observed.

Figures 16.6a, b show the DMFC performance using MEAs with the Nafion 117 membrane and the pore-filling membrane, respectively. Initially we tried to measure DMFC performance of a MEA with a Nafion 117 membrane, but DMFC performances were extremely low at methanol fuel concentrations above 15 wt%, as shown in Fig. 16.6a. In contrast, a MEA using the pore-filling membrane gave high DMFC performance with 16 wt% methanol in aqueous solution and the results were not significantly different from the results obtained using an 8 wt% methanol fuel system. Even when 32 wt% (10 M) aqueous methanol fuel was employed, approximately 50 mW cm^{-2} was obtained at 50°C using the MEA with the pore-filling membrane. Although the pore-filling membrane used has a thickness of only 23 μm, the membrane showed high resistance to methanol permeation, compared with the 200 μm, thick Nafion 117 membrane [20].

MeOH conc.	Voltage	Power density
8 wt%	□	■
16 wt%	△	▲
32 wt%	○	●

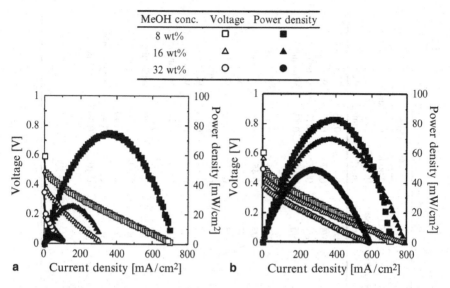

Fig. 16.6 DMFC performance of MEAs using the pore-filling membrane or Nafion 117 membrane at 50°C under atmospheric pressure. Methanol concentration in the fuel was varied. (PtRu loading at the anode and Pt loading at the cathode were 1.9–2.0 mg cm^{-2} and 0.9–1.1 mg cm^{-2}, respectively. Dry oxygen was used as the cathode gas.)

16.5 PEMFC Application

We fabricated an MEA using pore-filling electrolyte membranes and measured their fuel cell performance in PEMFC operation using pure hydrogen gas as a fuel. The membrane was prepared using a cross-linked polyethylene (CLPE) film as the porous substrate and poly(acrylamide *tert*-butyl sulfonic acid) (ATBS) as the filling electrolyte.

The fuel cell performance of the MEA containing the CLPE-ATBS membrane was examined as a single cell at ambient pressure and 60°C. The relevant I–V curves are shown in Fig. 16.7. The current density (I) and the power density (P) were calculated by dividing the observed current and power values by the geometric area of the electrode (5 cm^2). The open circuit voltage (OCV) was approximately 1.0 V and the cell voltage and power density at a current density of 2.0 A cm^{-2} reached 0.5 V and 1.0 W cm^{-2}, respectively. It was found that the MEA containing the CLPE-PATBS membrane achieved a relatively high fuel cell performance during H$_2$-O$_2$ PEFC operation [21].

We summarized the results of fuel cell performance with different types of electrolyte membranes under H$_2$-O$_2$ PEFC operation that have been reported in the literature; these are shown in Table 16.1. The operating conditions, such as cell temperature and pressure as well as materials used as the electrolyte membrane, are also included in Table 16.1. The values of the cell voltage at current densities given in Table 16.1 are either transcribed from the referenced articles, or deduced from

Table 16.1

Electrolyte membrane	Cell voltage [V] Current density				Conditions			Pt loadings		References
	0 [A cm⁻²]	0.5 [A cm⁻²]	1.0 [A cm⁻²]	2.0 [A cm⁻²]	Membrane thickness [μm]	Cell temperature [°C]	H_2/O_2 pressure [atm/atm]	Anode [mg min⁻²]	Cathode [mg min⁻²]	
CLPE-PATBS	1.00	0.76	0.68	0.50	20*	60	1/1	0.12	0.38	This work
Membrane"C"	0.90	0.76	0.70	0.50	175*	80	3/5	0.17	0.17	Wilson[19]
Membrane"C"	0.90	0.68	0.60	0.40	175*	80	1/1	0.17	0.17	Wilson[19]
DOWXUS-13204.10	0.90	0.83	0.75	0.60	125*	80	3/5	0.13	0.13	Wilson[19]
Membrane"C"	1.02	0.77	0.68	0.47	175*	80	3/5	0.45	0.45	Wilson[20]
Membrane"C"	1.02	0.72	0.64	0.38	175*	80	3/5	0.45	0.15	Wilson[20]
Nafion®112	1.01	0.80	0.73	0.56	58*	80	3/3	0.25	0.70	Slade[21]
Nafion®115	1.03	0.78	0.67	–	161*	80	3/3	0.25	0.70	Slade[21]
Nafion®/PTFE	0.975	0.70	0.55	–	45	80	2/2	0.3	0.3	Liu[2]
S-PPBP	0.90	0.54	0.30	–	–	80	2/2	0.40	0.40	Bae[8]
recast Nafion®	1.0	0.80	0.73	0.54	50	80	1/1	0.70	0.70	Watanabe[6]
Pt-TiO₂(TIP)-recast Nafion®	0.99	0.76	0.67	0.53	50	80	1/1	0.70	0.70	Watanabe[6]
10% silicon oxide/ Recast Nafion®	0.979	0.60	0.46	–	125–150	80	3/3	0.40	0.40	Adjemian[3]
10% silicon oxide/ Recast Nafion®	0.932	0.57	0.38	–	125–150	130	3/3	0.40	0.40	Adjemian[3]
Nafion® 112/6% silicon oxide	0.918	0.63	0.50	–	72	130	3/3	0.40	0.40	Adjemian[4]
Aciplex®1004/6% silicon oxide	0.975	0.68	0.57	–	112	130	3/3	0.40	0.40	Adjemian[4]
Recast Nafion® Zircoriumphosphate (36%)	1.007	0.75	0.65	0.45	90	80	3/3	0.40	0.40	Costmagna[5]
Recast Nafion®/Zircoriumphosphate (36%)	0.951	0.71	0.60	0.38	90	120	3/3	0.40	0.40	Costmagna[5]

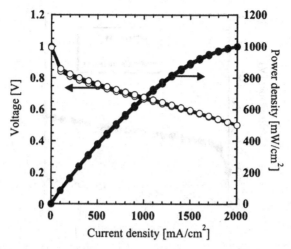

Fig. 16.7 The fuel cell performance during H₂-O₂ PEFC operation using an MEA containing CLPE-PATBS membranes at an ambient pressure and a temperature of 60°C

figures given in the articles. The membranes, "Membrane C" and "DOW XUS-13204.10" from Wilson et al. [22,23], "Nafion112" from Slade et al. [24], "recast Nafion" and "Pt-TiO₂ (TIP) recast Nafion" from Watanabe et al. [25] and "zirconium phosphonate/recast Nafion" from Costamagna et al. [26] all show excellent fuel cell performances. The MEAs used in those studies show fuel cell performances that are equivalent to our MEA that contains the CLPE-PATBS membrane, although there are also many differences, such as the catalyst metal loading value and the electrode structure. Typically, fuel cell performance can be improved by running it at elevated cell temperatures [27,28] and with higher pressure cathode feed gases [22,29]. Furthermore, fuel cell performance is increased to some degree with the mass of catalyst contained in the electrodes [22,23,30]. Therefore, we need to interpret a fuel cell's performance with these factors in mind. Both the operational temperature of T = 80–130°C and the operational pressure of 1–5 atm in the results reported in the preceding are higher than those we used. Additionally, it is interesting that the CLPE-ATBS membrane shows a relatively high OCV of approximately 1.0 V. In general, H₂ fuel crossover decreases the cell voltage. However, our membrane shows a relatively high OCV value, despite its thinness (approximately 20 μm). This indicates that the CLPE-ATBS membrane suppresses fuel crossover. The fact that the MEA that contains the CLPE-ATBS membrane shows excellent fuel cell performance at relatively low temperatures and pressures suggests its suitability for future mobile fuel cell applications.

To evaluate the fuel cell performance in detail, the ohmic drop (η_{IR}); and the cathode overpotential (η_c) were determined and are shown in Fig. 16.8. The ohmic drop was measured using an in situ current interruption method during the H₂-O₂ PEFC operation and η_c was calculated using the equation:

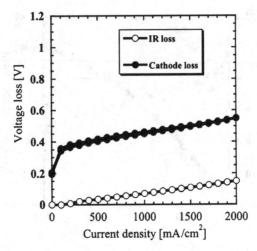

Fig. 16.8 The cathode loss and the ohmic drop measured during H_2-O_2 PEFC operation of an MEA containing CLPE-ATBS membranes at an ambient pressure and a temperature of 60°C

$$E = E_0 - (\eta_a + \eta_c + \eta_{IR}),\qquad(16.4)$$

where E and η_{IR} are obtained experimentally and E_0 is the open circuit voltage calculated from the Nernst equation. In this reaction system, the anode overpotential (η_a) is negligible and therefore η_c is obtained from the preceding equation.

The ohmic loss is regarded as a reference index of a membrane's performance. The value of the proton conductance at 60°C obtained from the ohmic drop data was calculated to be 15 S cm^{-2}. This value is approximately 66% of the proton conductance at 25°C of the soaked CLPE-ATBS membrane of 22 S cm^{-2}, indicating that the water content of the CLPE-ATBS membrane during PEMFC operation is lower than that of the soaked membrane. In this study, the fuel cell was operated under conditions that prevented the electrodes from flooding by the water formed by the oxygen reduction reaction at the cathode catalyst layer, i.e., the water activity in the gases supplied to the electrodes was adjusted to be less than unity. Therefore, it is considered that the decrease in proton conductance of the CLPE-ATBS membrane containing MEA is reasonable.

16.6 Conclusions

To develop polymer electrolyte membrane for fuel cells, our conceptual approach is a pore-filling membrane.

The pore-filling membranes can reduce methanol crossover in wide range of methanol concentrations due to the suppression effect of the substrate matrix. A membrane-electrode assembly using pore-filling electrolyte membranes

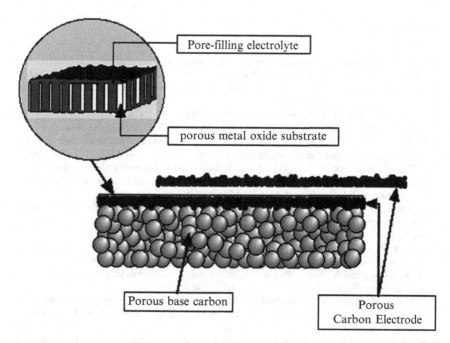

Fig. 16.9 The concept of an electrode-electrolyte membrane integrated system using pore-filling membrane with inorganic substrate

successfully generated electricity and showed excellent fuel cell performance with a high concentration methanol fuel. To reduce the methanol crossover using the new type of membrane a high concentration methanol of 32 wt% (10 M) fuel can be directly used. This results in high energy density DMFCs that can be compared with current battery devices. Furthermore, the pore-filling membrane concept can be applied to higher-temperature PEMFCs using different substrates and filling polymer materials having thermal and electrochemical durability. In addition, using an inorganic substrate, such as silica, makes it possible to further enhance the thermal stability and a thin fragile ceramic substrate can be used for an integrated membrane-electrode assembly system as shown in Fig. 16.9 [31]. As a filling polymer, nano-dispersed zirconia hydrogen phosphate with poly(sulfonated arylene ether) can be used to produce high proton conductivity with low humidity [32].

This concept can be used to design a suitable membrane for different applications by choosing a substrate and filling polymer electrolyte. We are also developing a design methodology based on the material's physical properties [33].

Acknowledgments We wish to thank the Ube Industry Co. Ltd., for supplying the porous polyimide substrate; Nitto Denko Co. Ltd., for supplying the porous PTFE and porous CLPE substrate; and the Toa Gosei Co. Ltd., for supplying the ATBS-Na monomer.

References

1. X. M. Ren, P. Zelenay, S. Thomas, J. Davey and S. Gottesfeld, Recent advances in direct methanol fuel cells at Los Alamos National Laboratory, *J. Power Sources*, **86**(1–2), 111–116 (2000).
2. K. M. McGrath, G. K. S. Prakash and G. A. Olah, Direct methanol fuel cells, *J. Ind. Eng. Chem.*, **10**(7), 1063–1080 (2004).
3. T. A. Zawodzinski, T. E. Springer, F. Uribe and S. Gottesfeld, Characterization of polymer electrolytes for fuel-cell applications,*Solid State Ionics*, **60**(1–3), 199–211 (1993).
4. B. Baradie, J. P. Dodelet and D. Guay, Hybrid Nafion (R)-inorganic membrane with potential applications for polymer electrolyte fuel cells, *J. Electroanal. Chem.*, 489(1–2), 101–105 (2000).
5. C. Yang, S. Srinivasan, A. S. Arico, P. Creti, V. Baglio and V. Antonucci, Composition Nafion/zirconium phosphate membranes for direct methanol fuel cell operation at high temperature, *Electrochem. Solid State Lett.*, **4**(4), A31–A34 (2001).
6. N. Miyake, J. S. Wainright and R. F. Savinell, Evaluation of a sol-gel derived Nafion/silica hybrid membrane for polymer electrolyte membrane fuel cell applications — II. Methanol uptake and methanol permeability, *J. Electrochem. Soc.*, **148**(8), A905–A909 (2001).
7. K. Scott, W. M. Taama and P. Argyropoulos, Performance of the direct methanol fuel cell with radiation-grafted polymer membranes, *J. Membr. Sci.*, **171**(1), 119–130 (2000).
8. M. Rikukawa and K. Sanui, Proton-conducting polymer electrolyte membranes based on hydrocarbon polymers, *Prog. Polym. Sci.*,**25**(10), 1463–1502 (2000).
9. L. Depre, M. Ingram, C. Poinsignon and M. Popall, Proton conducting sulfon/sulfonamide functionalized materials based on inorganic-organic matrices, *Electrochim. Acta*, **45**(8–9), 1377–1383 (2000).
10. K. Okamoto, Y. Yin, O. Yamada, M. N. Islam, T. Honda, T. Mishima, Y. Suto, K. Tanaka and H. Kita, Methanol permeability and proton conductivity of sulfonated co-polyimide membranes, *J. Membr. Sci.*, **258**(1–2), 115–122 (2005).
11. K. Miyatake, H. Zhou, T. Matsuo, H. Uchida and M. Watanabe, Proton conductive polyimide electrolytes containing trifluoromethyl groups: synthesis, properties, and DMFC performance, *Macromolecules*, **37**(13), 4961–4966 (2004).
12. Y. S. Kim, M. J. Sumner, W. L. Harrison, J. S. Riffle, J. E. McGrath and B. S. Pivovar, Direct methanol fuel cell performance of disulfonated poly-arylene ether benzonitrile) copolymers, *J. Electrochem. Soc.*, **151**(12), A2150–A2156 (2004).
13. T. Yamaguchi, S. Nakao, S. Kimura, Plasma-graft filling polymerization: preparation of a new type of pervaporation membrane for organic liquid mixtures, *Macromolecules*, **24**(20), 5522–5527 (1991).
14. T. Yamaguchi, S. Nakao and S. Kimura, Swelling behavior of the filling type membrane, *J. Polym. Sci., Polym. Phys. Ed.*, **35**(3), 469–477 (1997).
15. K. D. Kreuer, On the development of proton conducting polymer membranes for hydrogen and methanol fuel cells, *J. Membr. Sci.*, **185**(1), 29–39 (2001).
16. R. M. Penner and C. R. Martin, Ion transporting composite membranes.1. nafion-impregnated gore-tex, *J. Electrochem. Soc*, **132**(2), 514–515 (1985).
17. T. Yamaguchi, H. Hayashi, S. Kasahara and S. Nakao, Plasma-graft pore-filling electrolyte membranes using a porous poly(tetrafluoroethylene) substrate, *Electrochemistry*, **70**(12), 950–952 (2002).
18. T. Yamaguchi, F. Miyata and S. Nakao, Pore-filling type polymer electrolyte membranes for a direct methanol fuel cell, *J. Membr. Sci.*, **214**(2), 283–292 (2003).
19. T. Yamaguchi, F. Miyata and S. Nakao, Polymer electrolyte membranes with pore-filling structure for a direct methanol fuel cell,*Adv. Mater.*, **15**(14), 1198–1201 (2003).
20. T. Yamaguchi, H. Kuroki and F. Miyata, DMFC performances using a pore-filling polymer electrolyte membrane for portable usages, *Electrochem. Commun.*, **7**(7), 730–734 (2005).
21. H. Nishimura and T. Yamaguchi, Performance of a pore-filling electrolyte membrane in hydrogen-oxygen polymer electrolyte fuel cell, *Electrochem. Solid-State Lett.*, **7**(11), A385–A388 (2004).

22. M. S. Wilson and S. Gottesfeld, High-performance catalyzed membranes of ultra-low pt load-ings for polymer electrolyte fuel cells, *J. Electrochem. Soc.*, **139**(2), L28–L30 (1992).
23. M. S. Wilson and S. Gottesfeld, Thin-film catalyst layers for polymer electrolyte fuel-cell electrodes, *J. Appl. Electrochem.*, **22**(1), 1–7 (1992).
24. S. Slade, S. A. Campbell, T. R. Ralph and F. C. Walsh, Ionic conductivity of an extruded Nafion 1100 EW series of membranes, *J. Electrochem. Soc.*, **149**(12), A1556–A1564 (2002).
25. M. Watanabe, H. Uchida, Y. Seki, M. Emori and P. Stonehart, Self-humidifying polymer electrolyte membranes for fuel cells, *J. Electrochem. Soc.*, **143**(12), 3847–3852 (1996).
26. P. Costamagna, C. Yang, A. B. Bocarsly and S. Srinivasan, Nafion (R) 115/zirconium phos-phate composite membranes for operation of PEMFCs above 100 degrees C, *Electrochim. Acta*, **47**(7), 1023–1033 (2002).
27. I. Honma, S. Nomura and H. Nakajima, Protonic conducting organic/inorganic nanocompos-ites for polymer electrolyte membrane, *J. Membr. Sci.*, **185**(1), 83–94 (2001).
28. Y. G. Chun, C. S. Kim, D. H. Peck and D. R. Shin, Performance of a polymer electrolyte membrane fuel cell with thin film catalyst electrodes, *J. Power Sources*, **71**(1–2), 174–178 (1998).
29. O. J. Murphy, G. D. Hitchens and D. J. Manko, High-power density proton-exchange mem-brane fuel cells, *J. Power Sources*, **47**(3), 353–368 (1994).
30. Z. Qi and A. Kaufman, Quick and effective activation of proton-exchange membrane fuel cells, *J. Power Sources*, **114**(1), 21–31 (2003).
31. T. Yamaguchi, M. Ibe, B. N. Nair and S. Nakao, Pore-filling electrolyte membrane-electrode integrated system for direct methanol fuel cell application, *J. Electrochem. Soc.*, **149**(11), A1448–A1453 (2002).
32. G. M. Anilkumar, S. Nakazawa, T. Okubo and T. Yamaguchi, Proton conducting phosphated zirconia-sulfonated polyether sulfone nanohybrid electrolyte for low humidity, wide tempera-ture PEMFC operation, *Electrochem. Commun.*, **8**(1), 133–136 (2006).
33. T. Yamaguchi, Y. Miyazaki, S. Nakao, T. Tsuru and S. Kimura, Membrane design for pervapo-ration or vapor permeation separation using a filling-type membrane concept, *Ind. Eng. Chem. Res.*, **37**(1), 177–184 (1998).

Chapter 17
Research and Development on Polymeric Membranes for Fuel Cells: An Overview

Dipak Rana, Takeshi Matsuura, and S.M. Javaid Zaidi

Abstract This review is intended to provide the recent status in the development of polymeric-electrolyte (proton-exchange) membranes for the improvement of fuel cell performance based primarily on the preceding chapters of this book. Special attention is paid to the modification of present membranes, recent novel strategies for preparation of membranes, conceptual design of new membrane materials, and also promising approaches to overcome issues that severely restrict commercialization. The critical role of the materials and membranes and also relevant infrastructure of electrode is addressed. The new possibilities to improve technologies for implementation, and future trends are briefly examined.

17.1 Introduction

Fuel cell is the electrochemical design, which converts chemical energy to electrical energy. Sir William Robert Grove [1] first expounded the fuel cell principles in February 1839 in *Philosophical Magazine* consisting of hydrogen and oxygen in the presence of two platinized platinum electrodes. Fuel cells are mainly classified into five types according to the electrolyte material used: (1) polymer electrolyte (also known as proton exchange) membrane (PEM), (2) alkaline, (3) phosphoric acid, (4) molten carbonate, and (5) solid oxide. If methanol is used as a fuel instead of hydrogen or hydrogen-rich gas, then the special type of PEM fuel cell (PEMFC) is designated as direct methanol fuel cell (DMFC). There are many recent papers [2–16] that elaborately demonstrate the present activities of fuel cells. The membrane, heart of the PEMFC and DMFC, acts as a proton conductive medium, and also as barrier to avoid the direct contact between the fuel and oxidant. The fuel cell core, the efficiency and durability (i.e., long-term stability), depends upon the membrane. Figures 17.1 and 17.2 depict a cross-sectional view of the PEMFC [17] and the working principle of DMFC [18], respectively. The gas diffusion media, GDM, is basically porous electrical conductive materials, in general, carbon paper and/or carbon cloth. The ionomer membrane is sandwiched between catalyst and GDM layers, as shown in enlarged view for PEMFC in Fig. 17.1. The proton (or $(H_2O)_n H^+$ ion where n is typically 1 to 2), produced by the release of electron

S.M.J. Zaidi, T. Matsuura (eds.) *Polymer Membranes for Fuel Cells*,
doi: 10.1007/978-0-387-73532-0, © Springer Science+Business Media, LLC 2009

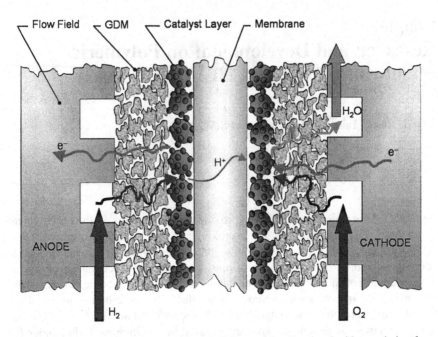

Fig. 17.1 Enlarged cross-section view of PEMFC materials. (Reprinted with permission from [17]. Copyright 2005 Taylor & Francis, Inc)

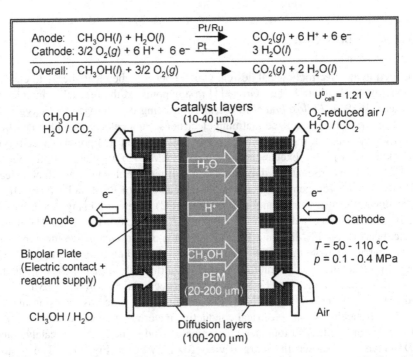

Fig. 17.2 Detailed working principle of DMFC. (Reprinted with permission from [18]. Copyright 2005 Elsevier)

from hydrogen at the anode, pass through the channels of membrane to the catalytic site of the cathode. Proton reacts with oxygen to produce water by receiving electron at the cathode. Thus, electric energy is generated by PEMFC, while water is produced as a byproduct. Hydrogen and water are replaced by methanol and carbon dioxide in DMFC, as shown schematically in Fig. 17.2.

Fuel cell technology will be essential to the community as significant contributions for renewable and environmentally friendly energy sources. The development of new polymeric materials for PEMFC is one of the most state-of-the-art research areas, aiming at the new energy sources, of the scientific community, although it started more than four and half decades ago. The material mainly used for fuel cell research is perfluorinated ionomer membranes (Nafion of du Pont de Nemours & Co., Inc. [19]; Aciplex-S of Asahi Chemicals Industry Co. Ltd. [20]; Flemion of Asahi Glass Co. Ltd. [21]; Hyflon of Dow Chemicals [22]; GoreSelect of Gore and Associates [23]; etc.), partially fluorinated ionomer membranes (Ballard Advanced Materials of 3rd Generation Membranes, BAM3G of Ballard Power Systems, Inc. [24]; etc), and non-fluorinated membranes (Dais of Dais Corp. [25]; BAM of Ballard Power Systems Inc. [24]; etc). The most popular one is Nafion [19,26], which is basically copolymer of tetrafluoroethylene and perfluorovinyl ether terminated by a sulfonic acid group. Protons become mobile when dissociated from the sulfonic acid groups in aqueous environment and, as a result, the membrane becomes a proton-conducting electrolyte membrane. DuPont invented the Nafion membrane for chloralkali processes in early 1960s; however, General Electric [27] used it for PEM fuel cells in 1960s, although the company initiated development of the power supply program as early as the 1950s.

17.2 Current Topic in the Fuel Cells Research on Polymer Membranes

17.2.1 Research Objective

To date, many membranes have been developed and research has been carried out to develop a high performance membrane for replacing perfluorinated or partially fluorinated PEMs, such as Nafion membranes. However, no membranes developed to date satisfy all of the following criteria required for the membranes. The primary requirements of PEMFC should have the following characteristics: (1) the proton conductivity of the membrane material should be high and the membrane should be thin to minimize membrane resistance, (2) the membrane must be mechanically and thermally stable, and the change in membrane area between the dry and the swollen states should be negligible, to minimize any membrane/electrode interface resistance, (3) the membrane material must be chemically stable with low methanol permeability for DMFC operation, and (4) the membrane must be moderately economical to manufacture.

There has been considerable research on modifying Nafion and novel development of membranes, so as to improve its properties for use in a DMFC. Despite advancement in the development of DMFCs, severe restrictions still inhibit their large-scale commercialization due to the deficient of appropriate membrane materials. Nafion membranes that still dominate the market of PEMs allow methanol permeation from the anode to the cathode of a DMFC, resulting in serious negative consequences. Three approaches have been pursued in order to resolve the methanol permeation problem: (1) modification of Nafion membranes, (2) development of alternative membranes, and (3) provision of high activity anode catalysts or methanol tolerant cathode catalysts.

There are various techniques to modify the exiting Nafion membranes or to synthesis the novel polymeric materials for fuel cell applications. The sulfonated poly(ether ether ketone ketone) (sulfonated PEEKK) is one of the alternative materials for fuel cell. The artist-view [28] of the nanoscopic hydrated structure of Nafion, and sulfonated PEEKK resulting from small angle X-ray scattering is presented in Fig. 17.3. Comparing the models for the two materials, Nafion has wider water channels, whereas sulfonated PEEKK has narrower channels with some channels blocked. The hydration pockets are more separated and well connected with less branching for the former membrane. On the other hand, the latter membrane is with more branching and with less connectivity of the hydrated group due to the rigidity of the backbone chain and less acidic nature of sulfonic group. The preparation of hybrid-Nafion membranes with in-situ growth of colloidal silica was demonstrated first by Mauritz's [29] and Nunes's [30] group. The various heteropolyacids (HPA) additives [31–33], such as phosphotungstic acid (PTA) (also known as tungstophosphoric acid, TPA), phosphomolybdic acid (PMA) (also known as molybdophosphoric acid, MPA), silicotungstic acid (STA), silicomolybdic acid (SMA), were used to modify the Nafion membrane. The basic structural unit of the HPA is $[PM_{12}O_{40}]^{3+}$ cluster, which is called Keggin unit. For example, the formula of PTA (same as TPA), and PMA (same as MPA) are $H_3PW_{12}O_{40}$ $29H_2O$, and $H_3PMo_{12}O_{40}$ $29H_2O$, respectively. The HPA additive, for example, phosphatoantimonic acid [34], was also used to associate with conventional polymer, such as, sulfonated polysulfone. Most recently, various synthetic routes have been developed for novel polymeric material preparation and also various additives [35–63], for example, carbon nano-tube (CNT), single-walled carbon nano-tube (SWNT), multi-walled carbon nano-tube (MWNT), conducting polymer (CP), liquid crystalline polymer (LCP), etc., have been added during the preparation of fuel cell membranes. Presently, the nano-composite [64–69] membrane is also an important research topic for betterment of performance.

Radiation-induced graft copolymerization is an alternative modification approach for preparation of PEMs, commonly prepared by grafting of styrene or its derivatives onto fluoro-polymer films and subsequent sulfonation. These membranes have been found to possess excellent combinations of physicochemical and thermal properties. However, chemical stability remains the main challenge precluding the implementation of such membranes in commercial fuel cell systems despite many successful laboratory tests in the short and medium terms. Various approaches have

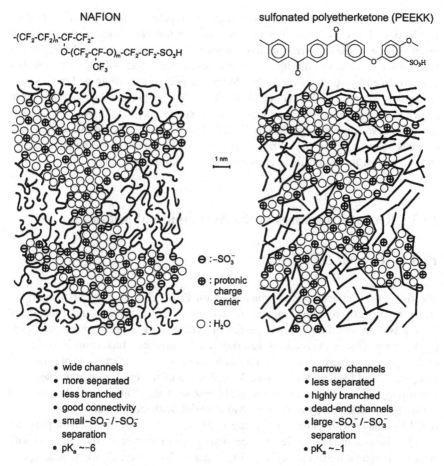

NAFION

$-(CF_2-CF_2)_n-CF-CF_2-$
$\quad O-(CF_2-CF-O)_m-CF_2-CF_2-SO_3H$
$\qquad CF_3$

sulfonated polyetherketone (PEEKK)

1 nm

$\ominus : -SO_3^-$

$\oplus$: protonic charge carrier

$O : H_2O$

- wide channels
- more separated
- less branched
- good connectivity
- small $-SO_3^-/-SO_3^-$ separation
- $pK_a \sim -6$

- narrow channels
- less separated
- highly branched
- dead-end channels
- large $-SO_3^-/-SO_3^-$ separation
- $pK_a \sim -1$

Fig. 17.3 Stimulated view of the microstructures of Nafion, and alternative polymeric material, sulfonated PEEKK. (Reprinted with permission from [28]. Copyright 2001 Elsevier)

been proposed to boost the stability, including cross-linking of the grafted moiety, grafting substituted styrene monomers, grafting onto cross-linked fluoro-polymer films, and direct introduction of the functional groups to irradiated polymer backbone without a monomer host. However, more substantial efforts have to be undertaken to further improve the quality of the membranes obtained from different starting materials and their interfacial adhesion properties with the electrodes in membrane electrode assembly (MEA).

Most PEMs require water for proton migration but membrane swelling leads to high solvent permeation. To control membrane swelling and solvent permeation, the pore-filling concept is proposed. A systematic membrane design and development based on the pore-filling concept was made to prepare electrolyte membranes for PEMFCs or DMFCs. The pores of a porous substrate are filled with a polymer electrolyte, and the substrate matrix suppresses the membrane swelling. Proton conductivity is achieved through the impregnated electrolyte polymer. Fuel crossover is reduced by suppression of the electrolyte polymer swelling and mechanical strength is maintained by the substrate. The high proton conductivity has been shown to exist with reduced membrane fuel crossover and good dimensional stability. To achieve a high energy density DMFC device, a high concentration of ethanol aqueous solution as the fuel is used and crossover is reduced. The membrane also showed almost no dimensional change with variation in humidity.

17.3 Recent Progresses in the Academic Laboratories

17.3.1 Overall Outline

The following characteristics is considered for PEMFC in the present context: the structure–property relationship, current state-of-the-art research of modification of exiting Nafion membrane, novel design of synthesis of materials and membrane modification, cost of the relevant materials and techniques which contribute significantly to the commercialization. The synthesis of novel membranes is provided by Dang et al. and by Kim and Guiver. Modification of the existing membranes with heteropolyacids is introduced by Lin et al., by Zhu et al., Trogadas and Ramani, and is briefly summarized by Zaidi et al. Modification with carbon nano-tubes is elucidated by Shaw, with radiation grafting is discussed by Nasef, and with polymer blending is presented by Kerres. Conceptual design of membrane modification with organic-inorganic hybrid materials is implemented by Nunes. Conceptual design of membrane modification with pore-filling is emphasized by Yamaguchi. Methanol permeability determination for DMFC by various instrumental techniques is elaborately demonstrated by Zaidi et al. and crucial issues for commercialization are examined by Chang et al. They are further elaborated in the following section.

17.3.2 Recent Achievements

A series of high molecular weight, highly sulfonated poly(arylene thio ether sulfone) (SPTES) polymers were synthesized by a polycondensation of sulfonated aromatic difluorosulfone, aromatic difluorosulfone, and 4,4′-thiobisbenzenthiol in sulfolane solvent at the temperature up to temperature 180°C, which allowed up to 100 mol% sulfonation as mentioned by Dang et al. The end-capping groups were introduced in the SPTES polymers to control the molecular weight distribution and reduce the water solubility of the polymers. Tough and ductile films were formed via solvent casting method, and water absorption was increased with increasing degrees of sulfonation. The polymerization conducted with the introduction of end-capping groups resulted in a wide variation in the polymer proton conductivity, which spanned a range of 100–300 mS cm^{-1}, measured at 65°C and at 85% relative humidity. The measured proton conductivities are up to three times higher than Nafion PEM under nearly comparable conditions, indicating that these polymers are promising candidates for PEMs in fuel cells. The thermal and mechanical properties of SPTES polymers, as investigated by thermo-gravimetric analysis, dynamic mechanical analysis, and tensile measurements indicated that these membranes were useful for PEM operation. The SPTES polymers show high glass transition temperature, T_g at ~220°C, depending on the degree of copolymerization. Wide-angle X-ray scattering of the polymers shows two broad scattering features centered at 4.5 and 3.3 Å, the latter peak being attributed to the presence of water molecules. MEAs fabricated using SPTES-50 polyelectrolyte membrane incorporating conventional electrode application techniques, exhibit high proton mobility. The electrochemical performance of SPTES-50 membrane in the MEA was superior to that of Nafion. Novel SPTES copolymers were synthesized by co-polymerization of the disulfonated dihalidesulfone, unsulfonated dihalidesulfone, and 4,4′-thiobisbenzenethiol monomers. All the sulfonated random copolymers were soluble in polar, aprotic solvents at room temperature. The free acid forms of the polymers could be fabricated into tough, flexible films from dimethylacetamide solvent. The proton conductivities of cast polymer films were in the range of 100–300 mS cm^{-1} at elevated temperatures and high relative humidity. The SPTES copolymer membranes were also successfully integrated into MEAs and their overall performance was superior to that of the MEA fabricated from Nafion 117 as PEM. The SPTES polymers and copolymers have been demonstrated to be promising candidates for high temperature PEMs in fuel cell applications.

The post-synthesis sulfonation of polyphthalazinones (PPs), including poly(phthalazinone ether sulfone) (PPES), poly(phthalazinone ether ketone) (PPEK), and poly(phthalazinone ether sulfone ketone) (PPESK) and its proton conductivity properties of PEM are reviewed by Kim and Guiver. They have also synthesized the sulfonated poly(phthalazinone ether ketone) (SPPEK) co-polymers and sulfonated poly(phthalazinone ether sulfone) (SPPES) co-polymers containing pendant sodium sulfonate groups by direct copolymerization using bisphenol type monomers. The sulfonated poly(aryl ether ketones) (SPAEK-NA, where NA indicate for naphthalene structure of aryl) were synthesized by direct copolymerization of various sulfonated monomers containing larger rigid hexafluoroisopropylidene bisphenol (Bisphenol 6F) moiety, and these membranes show high proton conductivities, very near to that of commercial Nafion 117. Recently, sulfonated poly(aryl ether ether ketone ketone) (SPAEEKK) containing sulfonic acid groups far apart from polymer backbone was synthesized using sodium 6,7-dihydroxy-2-naphthalenesulfonate (DHNS), a commercially available and inexpensive naphthalene diol containing a sulfonic acid side group. The composite membranes containing inorganic materials, e.g., silica embedded sulfonated poly(phthalazinone ether sulfone ketone) (SPPESK) and SPPEK/silica hybrid membranes, were found suitable for a high methanol concentration in feed due to suppressed methanol crossover behavior.

A composite membrane with 5 phr (parts per hundred resin) silica nanoparticles in SPPEK was incorporated into MEA and its performance was tested. An open cell potential of 0.6 V and an optimum power density of 52.9 mW cm^{-2} were obtained at a current density of 264.6 mA^{-2}, which is better than the performance of Nafion 117. SPPEK membranes have better performance than Nafion 117 in terms of higher power density, ultimate current density, and higher optimal operating concentration of methanol in feed.

A new method for the preparation of PEMs is based on cross-linking and by thermally activated bridging of the polymer chains with polyatomic alcohols through condensation reaction with sulfonic acid functions. This was applied to sulfonated poly(ether ether ketone) (SPEEK) and some of the membranes exhibited conductivity higher than 2×10^{-2} S cm^{-1} at room temperature. The SPEEK may potentially find applications as PEM materials for fuel cells.

Membrane properties and fuel cell performances by: (1) using thick stacked Nafion membranes, (2) blending Nafion with poly(tetrafluoroethylene-co-hex-afluoropropylene (Teflon-FEP) or poly(tetrafluoroethylene-co-perfluorovinylpropyl ether) (Teflon-PFA), and (3) doping Nafion with polybenzimidazole (PBI) were addressed by Lin et al. for DMFC. (1) The methanol crossover flux decreased with membrane thickness, and MEAs with an intermediate number of Nafion-112 films with four or five layers worked best. The fuel cell power is lost due to excessive methanol crossover for thinner membranes, while for thicker membranes, there is a drop in power due to membrane resistance losses. (2) A series of membranes, 50–100 μm in thickness, were prepared from blends of Nafion and Teflon-FEP/-PFA by melt processing, by melt-mixing, and by extruding pellets of Nafion precursor in the sulfonyl fluoride form. When the membrane contained more than 50% FEP, fuel cell performance fell below that of Nafion 112 and 117, due to excessively high resistance losses, i.e., the membrane conductivity was too low. At 1 M methanol feed concentration (MFC), the 50 μm Nafion/FEP with 50 wt% FEP blended membrane worked as well as Nafion 117 (96 mW cm^{-2}) but this film outperformed at 5 M (with a maximum power density of 69 mW cm^{-2}, one and half times that of Nafion 117). At 10 M MFC, the 100 μm Nafion/FEP membrane worked best (59 mW cm^2), 6.5-times higher power density than Nafion 117) due to a significantly lower crossover. (3) At a given initial protonation degree of the Nafion powder, there was a decrease in conductivity and permeability with increasing PBI content. Membranes were less conductive and better methanol barriers for a given PBI content when the initial proton degree was high i.e., when there were more SO$_3$H groups in Nafion and complexation with PBI. The DMFC performance of PBI-doped membranes is superior to Nafion 117 at 1 and 5 M MFC. The MEA with a 3 wt% PBI membrane delivered the highest power density, which was greater than that with either of the two Nafion MEAs. As the membrane PBI content increased, its ohmic resistance increased and performance in a fuel cell decreased. The methanol crossover flux in all of the Nafion-PBI MEAs was lower than Nafion 117, even though the doped films were two to four times thinner than Nafion.

Due to the inappropriately high methanol permeability of commercial Nafion 117, most DMFC tests are performed with relatively dilute, typically 0.5 or 1 M, aqueous MFC. The Nafion-PBI membrane, 5 wt% PBI with Nafion polymer that was 40% in the protonated form, performed well at 1 and 5 M MFC. The PBI-doped Nafion films contain much less fluoro-polymer, two to four times less, as compared with commercial Nafion 117 membranes, yet PBI blended membrane performance is equal or superior to any Nafion material in a DMFC. A significant cost savings can be also realized when using the Nafion/Teflon blended films rather than Nafion 117 in a DMFC. For example, a 50 μm thick membrane with 50 wt% FEP blended Nafion membrane uses seven times less Nafion polymer, as compared with commercial Nafion 117 due to the reduction in membrane thickness and dilution of Nafion with FEP. For a 100 μm blended membrane, the higher power density at 10 M methanol and the decrease in Nafion polymer content realizes cost savings.

Zhu et al. broadly summarized polymer/inorganic composite membranes for application at elevated temperatures. Most polymer systems include Nafion, sulfonated poly(arylene ethers), polybenzimidazoles and others. The inorganic proton conductors involve silica, heteropolyacids, layered zirconium phosphates and liquid phosphoric acid. Most recently, Lin and Lin et al. developed a new type of Nafion/SiO$_2$ composite membranes with immobilized PTA via both in-situ and ex-situ sol-gel procedures. Initially, the silica is functionalized with aminopropyl triethoxylsilane (APTES) to form amine containing silica materials, and then, PTA is immobilized onto silica by ionic complexation with the amine groups in modified silica. The in-situ sol-gel reaction is carried out in the presence of Nafion, while the ex-situ sol-gel reaction was performed without Nafion. Although the in-situ strategy was good to immobilize PTA in the membrane, the obtained composite membranes were brittle at SiO$_2$ contents exceeding 10 wt%. It is speculated that competitive ionic complexion existed between Nafion and PTA to the amine groups on functionalized silica. The in-situ Nafion/APTES/PTA composite membrane showed much worse conductivity and performance than Nafion/PTA composite membrane at the same conditions, possibly because of poisoning of Pt catalysts of residual NH$_2$ contamination in the membranes.

Trogadas and Ramani summarized the modification of PEM membranes, including Nafion modified by zirconium phosphates, heteropolyacids, hydrogen sulfates, metal oxides, and silica. Membranes with sulfonated non-fluorinated backbones were also described. The base polymers polysulfone, poly(ether sulfone), poly(ether ether ketone), polybenzimidazole, and polyimide. Another interesting category is acid-base polymer blend membranes. This review also paid special attention to electrode designs based on catalyst particles bound by a hydrophobic polytetrafluoroethylene (PTFE) structure or hydrophilic Nafion, vacuum deposition, and electrodeposition method. Issues related to the MEA were presented. In their study on composite membranes, the effects of particle sizes, cation sizes, number of protons, etc., of HPA were correlated with the fuel cell performance. To promote stability of the PTA within the membrane matrix, the investigators have employed PTA supported on metal oxides such as silicon dioxide as additives to Nafion.

The effectiveness of carbon-filled polymer blends in improving simultaneously the electrical conductivity and the mechanical strength has been demonstrated by Shaw. Carbon-filled polymer blends with a triple-continuous structure, consisting of a binary (or ternary) polymer blend with carbon particles have great potential to provide injection moldable PEM fuel cell bipolar plates with superior electrical conductivity and sufficient mechanical properties. Four CNT-filled polymer blends have been investigated in the present study: (1) CNT-filled poly(ethylene terephthalate) (PET)/poly(vinylidene fluoride) (PVDF); (2) PET/polypropylene; (3) PET/Nylon 6,6; and (4) PET/high density polyethylene blends. The CNT-filled PET/PVDF blend exhibits improvement of 2,500% in electrical conductivity, 36% in tensile strength, and 325% in elongation over the CNT-filled PET with the same carbon loading. Such improvements have been related to the formation of a triple-continuous structure achieved through the forced segregation of CNT in the PET phase of the CNT-filled PET/PVDF blend. This CNT-filled PET phase offers an electrical short circuit for the composite, while the clean PVDF phase provides the strength and elongation for the composite. The distribution of CNTs in the polymer blend plays a very important role in both mechanical and electrical properties of carbon-filled polymer blends which are predominantly determined by the thermodynamic driving force when the polymer blend is in a fully molten state. The preferential location of CNTs in one of the continuous polymer phases in the polymer blend is highly desirable from the viewpoint of the mechanical and electrical properties.

Nasef reported the preparation of PTFE-graft-poly(styrene sulfonic acid), PTFE-g-PSSA, membranes with grafting degrees up to 36% using direct irradiation method with relatively low dose range (5–20 kGy), followed by sulfonation with chlorosulfonic acid. The membranes showed a good combination of physicochemical properties and ion conductivities up to 34 mS cm^{-1}. Interestingly, these membranes have surfaces predominated by hydrocarbon fraction originated from PSSA side chain despite achieving homogeneous grafting at degrees of grafting 24% or above, and performances were found to stand at few hundred hours in PEMFC operated at temperature of 50°C. The preparation of poly(tetrafluoroethylene-co-perfluorovinyl ether)-graft-PSSA (PFA-g-PSSA) membranes by grafting of styrene using direct irradiation followed by sulfonation was also reported. The highest degree of grafting (63%) was achieved upon using dichloromethane to dilute styrene (60 vol%) at a total dose of 30 kGy. The PFA-g-PSSA membrane surfaces were found to be predominated by hydrocarbon fraction originated from PSSA side chains despite achieving bulk grafting. However, the performances were found to be limited to few hundreds hours under fuel cell conditions at 50°C. The membrane stability was improved by cross-linking with divinylbenzene during styrene grafting and the obtained membranes maintained a combination of properties suitable for fuel cell applications. The preparation of radiation-grafted pore filled membranes was reported for DMFC by impregnating micro-porous structure of PVDF films with styrene followed by direct electron beam irradiation and subsequent sulfonation with chlorosulfonic acid/dichloromethane mixture. Membranes with degrees of grafting in the range of 8–45% were obtained with those having 40 and 45% polystyrene demonstrating excellent combinations of physico-chemical properties compared with Nafion 117. For instance, 45% grafted membrane achieved 61 mS cm^{-1} conductivity (compared with 53 mS cm^{-1} for Nafion 117) and five folds lower methanol permeability (0.7×10^{-6} cm^2 s^{-1}) than Nafion 117 (3.5×10^{-6} cm^2 s^{-1}), under the same conditions.

The blend concepts for fuel cells membranes, basically importance of blend microstructure and its integrity to PEM, were presented by Kerres. Differently cross-linked blend membranes were prepared from commercial arylene main-chain polymers from the classes of poly(ether ketone)s and poly(ether sulfones) modified with sulfonate groups, sulfinate cross-linking groups and basic nitrogen-groups. Following six types of membranes have been prepared: (1) van der Waals/dipole-dipole blends by mixing a polysulfonate with unmodified PSU (commercial Udel, bisphenol A polysulfone), which showed a heterogeneous morphology, leading to extreme swelling and even dissolution of the sulfonated component at elevated temperatures. (2) Hydrogen bridge blends by mixing a polysulfonate with a polyamide or a poly(ether imide), showing a partially heterogeneous morphology, also leading to extreme swelling/dissolution of the sulfonated blend component at elevated temperatures. (3) Acid-base blends by mixing a polysulfonate with a polymeric N-base (self-developed/commercial), showing excellent stability and good fuel cell performance up to 100°C for PEMFC and 130°C for DMFC. (4) Covalently cross-linked (blend) membranes by either mixing of a polysulfonate with a polysulfinate or by preparation of a poly(sulfinate sulfonate), followed by reaction of the sulfinate groups in solution with a dihalogeno compound under sulphur-alkylation, which showed effective suppression of swelling without H^+-conductivity loss. The membranes showed also good performance for PEMFC up to 100°C and for DMFC up to 130°C. (5) Covalent-ionically cross-linked blend membranes by mixing polysulfonates with polysulfinates and polybases or by mixing a polysulfonate with a polymer carrying both sulfinate and basic nitrogen-groups. The covalent-ionically cross-linked membranes were tested in DMFC up to 110°C and showed also a good performance. (6) Differently cross-linked organic-inorganic blend composite membranes via different procedures. The best results were obtained with blend membranes having a layered zirconium phosphate phase and these membranes were transparent, and show good proton conductivity and stability. Application of one of these composite membranes to a PEMFC yielded good performance up to 115°C.

The development of organic-inorganic materials for hybrid membranes is elaborately examined by Nunes. According to Nunes, inorganic particles with higher aspect ratios, defined as (length/width) of a filler, will lower permeability of the membrane. The addition of silane, a functionalized inorganic surface, improves the interaction between the particle and the host polymer. The modification of silica (SiO_2), or zirconium oxide (ZrO_2) particles with silanes, containing organic functionalized groups, opens covalently linking of the particles to the main chain. The treatment with amino silanes and further with carbodiimidazole, which then reacts with part of the sulfonic groups of the polymer matrix is effective for fuel cell membrane performance.

Poly(acrylic acid) (PAA), poly(acrylic acid-co-vinyl sulfonic acid) (PAAVS) or poly(acrylamide tert-butyl sulfonic acid) (PATBS) was used as the filling polymer, and PTFE, high-density cross-linked polyethylene (CLPE) or a polyimide (PI) was used as the substrate by Yamaguchi to prepare pore-filling membranes. PAAVS and PATBS have sulfonic acid groups of 0.7 and 4.5 mmol gm^{-1} of polymer, respectively.

Below a 60% filling ratio, the methanol permeability decreased with an increasing pore-filling ratio for the samples with both substrates. The filling polymer density was low and some leakage flux occurred through the membranes. Above an 80% filling ratio, the methanol permeation was almost independent of the filling ratio because filling electrolyte almost filled the pores. If a single polyelectrolyte is used as the filling polymer with different substrates and pore-filling ratios, the relationship between the proton conductivity and the inverse of the methanol permeability would be a single line. The order of performance would be: PAA < PAAVS < PATBS, and this order corresponds to the sulfonic acid content. Therefore, the better performance membranes can be obtained by combining a filling polymer having a high strong-acid-group content with a mechanically strong matrix to suppress the swelling. At the same time, it is possible to achieve a balance in the methanol permeability and proton conductivity that is suitable for realizing a DMFC by controlling the substrate strength and the filling ratio. The CLPE and PI substrates can suppress membrane swelling and the change in membrane area between the dry and swollen states is negligible for pore-filling membranes containing those substrates. The PATBS filling polymer having a high sulfonic acid content shows a high proton conductivity of 0.15 S/cm^{-1} at 25°C. The relationship between the proton conductivity and the methanol permeability of a single pore-filling polymer can be controlled by changing the substrate strength and pore-filling ratio. The DMFC performances of MEA with a Nafion 117 membrane, were extremely low at methanol fuel concentrations above 15 wt%. In contrast, a MEA using the pore-filling membrane gave high DMFC performance with 16 wt% methanol in aqueous solution and the results were not significantly different from the results obtained using an 8 wt% methanol fuel system. Even when 32 wt% (10 M) aqueous methanol fuel was employed, approximately 50 mW cm^{-2} was obtained at 50°C using the MEA with the pore-filling membrane. Although the pore-filling membrane used had a thickness of only 23 µm, the membrane showed high resistance to methanol permeation, compared with the 200 µm thick Nafion 117 membrane. The MEA was fabricated using pore-filling electrolyte membranes and its fuel cell performance in PEMFC operation was measured using pure hydrogen gas as a fuel. The membrane was prepared using a CLPE film as the porous substrate and PATBS as the filling electrolyte. The fuel cell performance of the MEA containing the CLPE-PATBS membrane was examined as a single cell at ambient pressure and 60°C. It was found that the MEA containing the CLPE-PATBS membrane achieved a relatively high fuel cell performance during H$_2$-O$_2$ PEMFC operation.

The pore-filling membranes can reduce methanol crossover in a wide range of methanol concentration due to the suppression effect of the substrate matrix. The MEA using pore-filling electrolyte membranes successfully generated electricity and showed excellent fuel cell performance with a high concentration methanol fuel. To reduce the methanol crossover using the new type of membrane a high concentration methanol of 32 wt% (10 M) fuel can be directly used. This results in high energy density DMFCs that can be compared with current battery devices. Furthermore, the pore-filling membrane concept can be applied to higher temperature PEMFCs using different substrates and filling polymer materials having thermal and electrochemical durability. In addition, using an inorganic substrate, such as silica, makes it possible to further enhance the thermal stability and a thin fragile ceramic substrate can be used for an integrated MEA system. As a filling polymer, nano-dispersed zirconia hydrogen phosphate with poly(sulfonated arylene ether) can be used to produce high proton conductivity with low humidity.

Zaidi et al. have compared various techniques for determining the methanol permeability through membranes. Assessment of these techniques has shown that electrochemical techniques were the most accurate method. The potentiometry has additional advantages of better reproducibility of results and convenience as compared with other techniques, for example, cyclic voltammetry, chronoamperometry. Zaidi et al. also elucidated modification of Nafion membrane via composite membrane for better performance of fuel cell. In order to reduce methanol permeability, sulfonated poly(ether ether ketone), SPEEK, membranes were blended with appropriate amount of inorganic materials such as boron orthophosphate (BPO_4), Mobil composite material (MCM41), ultrastable Y ($SiO_2 / Al_2O_3 > 5$) -zeolite (USY-zeolite), SiO_2, ZrO_2, etc., which act both as a barrier to methanol permeation as well as enhancer for the conductivity of the membranes. The inorganic network can remarkably reduce the methanol permeation through the SPEEK membranes even if heteropolyacids such as TPA, MPA, etc., are incorporated into the SPEEK polymer matrix, which usually increases water and methanol permeation. Incorporating different proportions of solid proton conductors: TPA and MPA loaded MCM41, and TPA and MPA loaded USY-zeolite into the SPEEK polymer matrix provides composite membranes with low methanol permeability. The reduction in the methanol permeability is better achieved with lower loadings than with higher loadings of the solid inorganic materials (TPA- or MPA-loaded MCM41/USY-zeolite). These composite membranes offer less expensive alternative, have low methanol permeability, high proton conductivity, and high thermal stability. The modification of Nafion membranes and particularly the development of membranes based on alternative polymers have achieved a certain degree of success, in which membranes have been developed with permeability values of 10–70 times lower than the pure Nafion membranes.

Chang et al. comprehensively examined the critical issues of core technologies for commercialization of DMFC including catalyst technology, membrane modification, MEA, stack configuration, designing of market research, etc. The technical roles of the PEM in DMFC, including modification of Nafion and perfluorinated, partially fluorinated, hydrocarbon membranes, and composite membranes with organic and inorganic materials, were critically examined. Technology roadmap of DMFC for membranes and mobile applications were also discussed.

17.4   Future Directions of the Research Agenda

The fuel cell membranes offer potentially non-thermal full energy efficiency conversion as an alternative energy device. As mentioned, the requirement for the successful substitutes of Nafion membrane is low methanol permeability and overall stability of fuel cell performance. Intense R&D is going on in regard with novel material development and formation of new membranes to fulfill the above goals. In particular, the following subjects are highlighted as indicating the directions in the future R&D activities.

Although substantial progress has been made using the concept of carbon-filled polymer blends having a triple-continuous structure, so far blends with relatively low carbon concentrations are studied. As a result, their electrical conductivities are still far below the desire expected values (larger than 100 S cm^{-1}) for the application of PEMFC bipolar plates. Therefore, it is worth investigating the two imperative topics: (1) whether such a triple-continuous structure can still be injection molded using polymer blends with high carbon concentrations (more than 30 vol% of carbon), and (2) whether the polymer blends with high carbon concentrations with triple-continuous structures still possess superior electrical conductivity and mechanical properties if injection molding is possible.

The radiation-grafted membranes using various fluoro-polymer films are moderately suitable for PEMs application, however, only few membranes reached the stability level required for long-term performance. The development of these membranes require enhancement of their long-term stability levels equal to (or greater than) 5,000 h, which is not viable with the current technology. It is recommended to study in details the following approaches: (1) to use PTFE and/or copolymer of fluorinated ethylene propylene (FEP) as starting materials and graft substituted or fluorinated styrene monomers, followed by sulfonation to establish the grafting parameters as well as its correlations between compositions and properties; (2) to develop novel monomers containing sulfonic acid groups, like styrene sodium sulfonate monomer, and then, graft them to the fluorinated backbone, which could make breakthrough in improving the PEMs stability. This would avoid aggressive sulfonation reaction that causes degradation in crystalline structure of the starting polymer films; and (3) to expand new generation of radiation grafted membranes from novel polymeric materials having sulfonic acid groups directly attached to the irradiated polymer backbone without using any monomers. During the fuel cell operation, delaminating eventually happens due to the lack of interfacial properties (i.e., poor adhesion between the membrane and electrode) of MEA. It is extremely necessary to optimize MEA making conditions for improving fuel cell performance with each membrane. In addition, the durability of radiation-grafted membranes under dynamic operation conditions should be further addressed focusing on degradation and failure modes of membrane in a similar mechanistic way.

Until now most of membrane material and membrane development was based on the dense homogeneous structure. Attempts should be made to examine the effect of asymmetric [70] or composite [71] structures on the performance of PEM. Considering that only a small amount of polymeric additive is required for surface coating of the membrane, the future direction of research and development efforts should be focused on the development of new methods of membrane surface coating and surface modification. In this context, the approach of surface modifying macromolecules (SMMs) might be interesting. When two macromolecules are blended, one of the macromolecular components tends to migrate toward the surface to reduce the surface energy. By controlling the amount of the migrating component, the surface can be modified by blending only a small quantity of macromolecules. Recently tailor-made hydrophilic surface modifying macromolecules (LSMMs) comprised of polyurethane (PU) segment-blocked copolymer with hydroxy end-groups were synthesized, and used as additives to prepare asymmetric filtration membranes [72]. The SMMs with fluorohydrocarbon end-groups and PU or poly(urethane urea) (PUU) with sulfonic acid containing segment (aromatic, for example, 4,4'-diamino-2,2'-biphenyldisulfonic acid disodium salt; 3,6-dihydroxynaphthalene 2,7-disulfonic acid disodium salt; etc., and/or aliphatic with various spacer lengths, for example, sodium 1,5-dihydroxypentane-3-sulfonate; sodium 1,8-diaminooctane-3-sulfonate; etc.) would act as novel additives for fuel cell composite membranes. The hydrophobic layer containing fluoro-hydrocarbon end-groups stay at the top layer which prevents the methanol permeation, however, protons easily pass through the sulfonate groups in the hydrophilic segment which makes a water pore channel.

Another novel approach is the use of electrospun nano-fiber membrane (ENM) for fuel cell. Polymeric material having sulfonic acid groups has been fabricated into ENM. The electro-spinning process is capable of producing fibers in the sub-micron to nano-scale range. The use of ENMs in PEMFC was proposed by Ramakrishna et al. [73] due to their high water content and reduced cost. Very recently, ENM has been successfully applied for water filtration [74], and a new attempt was made to electro-spin nanofibers the surface of which was modified by blending LSMM having poly(ethylene glycol) end-groups [75]. It would be an interesting research project to fabricate composite polymeric ENMs having sulfonic acid end-groups and/or PU block-segment that contain sulfonic acid groups. ENMs can also be used as substrates for pore-filled membranes. It is also suggested that ENM could be made using polymeric conductive materials, such as polypyrrole, polyaniline, polythiophene, poly(3-octylthiophene), etc. ENM could fit for the revolution of the energy generation in the future world.

17.5 Conclusion

Many new polymers have been synthesized and tested for their proton conductivity, methanol permeability, thermal as well as mechanical stability, electrode-membrane interface connectivity, etc., aiming at improvement in membrane performance for fuel cell applications. These efforts seem to continue with insightful vision and strong commitment in the future. However, only a handful of polymers are currently being used as the materials for commercial applications, and they are not necessarily the polymers of the best performance properties. This is mainly due to the cost factor that governs the present membrane market. The fuel cell performance of membranes is, on the other hand, known primarily ruled by the various factors, mainly, membrane fuel permeability, electrode-membrane adhesion (or compatibility), thermal and mechanical stabilities. The knowledge of the effects of these factors on the fuel cell performance will allow utilizing fully the potential of polymers for membrane materials, which will open up new promising avenues for further research and development of the areas.

Acknowledgments The authors thank the Taylor & Francis, Inc., T. A. Trabold, Minichannels in polymer electrolyte membrane fuel cells, *Heat Transfer Eng.* **26**(3), 3–12 (2005), and the Elsevier, K. Sundmacher, L. K. Rihko-Struckmann, and V. Galvita, *Catal. Today* **104**(2–4), 185–199 (2005); and the Elsevier, K. D. Kreuer, *J. Membr. Sci.* **185**(1), 29–39 (2001), for allowing us to use their figures.

References

1. W. R. Grove, On voltaic series and the combination of gases by platinum, *Phil. Mag. Ser. 3* **14**, 127–130 (1839).
2. S. Srinivasan, *Fuel Cells: Fundamentals to Applications* (Springer-Verlag, New York, 2005).
3. E. Spohr, Proton transport in polymer electrolyte fuel cell membranes, in: *Ionic Soft Matter: Modern Trends in Theory and Applications*, edited by D. Henderson, M. Holovko, and A. Trokhymchuk (Springer, Dordrecht, Netherlands, 2005), pp. 361–379.
4. J. Larminie, and A. Dicks, *Fuel Cell Systems Explained* (Wiley, Chichester, UK, 2003).
5. a) B. C. H. Steele, and A. Heinzel, Materials for fuel-cell technologies, *Nature* **414**(6861), 345–352 (2001); b) B. C. H. Steele, Material science and engineering: The enabling technology for the commercialization of fuel cell systems, *J. Mater. Sci.* **36**(5), 1053–1068 (2001).
6. L. Carrette, K. A. Friedrich, and U. Stimming, Fuel cells-fundamentals and applications, *Fuel Cells* **1**(1), 5–39 (2001).
7. M. Kunimatsu, T. Shudo, and Y. Nakajima, Study of performance improvement in a direct methanol fuel cell, *JSAE Rev.* **23**(1), 21–26 (2002).
8. G. Alberti, and M. Casciola, Composite membranes from medium-temperature PEM fuel cells, *Annu. Rev. Mater. Res.* **33**, 129–154 (2003).
9. J. Rozière, and D. J. Jones, Non-fluorinated polymer materials for proton exchange membrane fuel cells, *Annu. Rev. Mater. Res.* **33**, 503–555 (2003).
10. K. Scott, and A. K. Shukla, Polymer electrolyte membrane fuel cells: Principles and advances, *Rev. Environ. Sci. Bio/Tech.* **3**(3), 273–280 (2004).
11. M. A. Hickner, H. Ghassemi, Y. S. Kim, B. R. Einsla, and J. E. McGrath, Alternative polymer systems for proton exchange membranes (PEMs), *Chem. Rev.* **104**(10), 4587–4612 (2004).
12. R. Dillon, S. Srinivasan, A. S. Aricò, and V. Antonucci, International activities in DMFC R&D: Status of technologies and potential applications, *J. Power Sources* **127**(1–2), 112–126 (2004).
13. A. Hayashi, T. Kosugi, and H. Yoshida, Evaluation of polymer electrolyte fuel cell application technology R&Ds by GERT analysis, *Int. J. Hydrogen Energy* **30**(9), 931–941 (2005).
14. B. Smitha, S. Sridhar, and A. A. Khan, Solid polymer electrolyte membranes for fuel cell applications - A review, *J. Membr. Sci.* **259**(1–2), 10–26 (2005).
15. J. Meier-Haack, A. Taeger, C. Vogel, K. Schlenstedt, W. Lenk, and D. Lehmann, Membranes from sulfonated block copolymers for use in fuel cells, *Sep. Purif. Technol.* **41**(3), 207–220 (2005).
16. C. Iojoiu, M. Maréchal, F. Chabert, and J.-Y. Sanchez, Mastering sulfonation of aromatic polysulfones: Crucial for membranes for fuel cell application, *Fuel Cells* **5**(3), 344–354 (2005).
17. T. A. Trabold, Minichannels in polymer electrolyte membrane fuel cells, *Heat Transfer Eng.* **26**(3), 3–12 (2005).

18. K. Sundmacher, L. K. Rihko-Struckmann, and V. Galvita, Solid electrolyte membrane reactors: Status and trends, *Catal. Today* **104**(2–4), 185–199 (2005).

19. a) H. H. Gibbs, and R. N. Griffin, Fluorocarbon sulfonyl fluorides, E. I. du Pont de Nemours and Company, US Patent **3 041 317**, 26 June 1962; b) D. J. Connolly, and W. F. Gresham, Fluorocarbon vinyl ether polymers, E. I. du Pont de Nemours and Company, US Patent **3 282 875**, 1 Nov 1966; c) R. Beckerbauer, Unsaturated α-hydroperfluoroalkylsulfonyl fluoride, E. I. du Pont de Nemours and Company, US Patent **3 714 245**, 30 Jan 1973; d) P. N. Walmsley, Composite cation exchange membrane and use thereof in electrolysis of an alkali metal halide, E. I. du Pont de Nemours and Company, US Patent **3 909 375**, 30 Sep 1975; e) W. G. Grot, Electrolysis cell using cation exchange membranes of improved permselectivity, E. I. du Pont de Nemours and Company, US Patent **4 026 783**, 31 May 1977.

20. a) T. Kuwata, and S. Yoshikawa, Cation permselective membranes, Asahi Glass Co. Ltd., US Patent **3 086 947**, 23 Apr 1963; b) Y. Oda, M. Suhara, and E. Endo, Process for producing alkali metal hydroxide, Asahi Glass Co. Ltd., US Patent **4 065 366**, 27 Dec 1977.

21. a) N. Seko, Y. Yamakoshi, H. Miyauchi, M. Fukumoto, K. Kimoto, I. Watanabe, T. Hane, and S. Tsushima, Cation exchange membrane preparation and use thereof, Asahi Kasei Kogyo Kabushiki Kaisha, US Patent **4 151 053**, 24 Apr 1979; b) K. Kimoto, H. Miyauchi, J. Ohmura, M. Ebisawa, and T. Hane, Novel fluorinated copolymer with trihydro fluorosulfonyl fluoride pendent groups and preparation thereof, Asahi Kasei Kogyo Kabushiki Kaisha, US Patent **4 329 435**, 11 May 1982.

22. a) B. R. Ezzell, W. P. Carl, and W. A. Mod, Novel polymers having acid functionality, The Dow Chemical Co., US Patent **4 330 654**, 18 May 1982; b) B. R. Ezzell, W. P. Carl, and W. A. Mod, Sulfonic acid electrolytic cell membranes, The Dow Chemical Co., US Patent **4 417 969**, 29 Nov 1983.

23. a) B. Bahar, A. R. Hobson, J. A. Kolde, and D. Zuckerbrod, Ultra-thin integral composite membrane, W. L. Gore & Associates, Inc., US Patent **5 547 551**, 20 Aug 1996; b) B. Bahar, A. R. Hobson, and J. A. Kolde, Integral composite membrane, W. L. Gore & Associates, Inc., US Patent **5 599 614**, 4 Feb 1997; c) B. Bahar, A. R. Hobson, and J. A. Kolde, Electrode apparatus containing an integral composite membrane, W. L. Gore & Associates, Inc., US Patent **5 635 041**, 3 Jun 1997.

24. a) J. Wei, C. Stone, and A. E. Steck, Trifluorostyrene and substituted trifluorostyrene copolymeric compositions and ion-exchange membrane formed therefrom, Ballard Power Systems Inc., US Patent **5 422 411**, 6 Jun 1995; b) C. Stone, and A. E. Steck, Graft polymeric membranes and ion-exchange membranes formed therefrom, Ballard Power Systems Inc., US Patent **6 359 019**, 19 Mar 2002.

25. a) S. G. Ehrenberg, J. Serpico, G. E. Wnek, and J. N. Rider, Fuel cell incorporating novel ion-conducting membrane, Dais Corporation, US Patent **5 468 574**, 21 Nov 1995; b) S. G. Ehrenberg, J. M. Serpico, G. E. Wnek, and J. N. Rider, Fuel cell incorporating novel ion-conducting membrane, Dais Corporation, US Patent **5 679 482**, 21 Oct 1997.

26. a) K. A. Mauritz, and R. B. Moore, State of understanding of Nafion, *Chem. Rev.* **104**(10), 4535–4585 (2004); b) S. Banerjee, and D. E. Curtin, Nafion® perfluorinated membranes in fuel cells, *J. Fluorine Chem.* **125**(8), 1211–1216 (2004).

27. a) W. T. Grubb Jr., Fuel cell, General Electric Company, US Patent **2 913 511**, 17 Nov 1959; b) L. W. Niedrach, Fuel cell, General Electric Company, US Patent **3 134 697**, 26 May 1964; c) R. B. Hodgdon Jr., Cation exchange fuel cell, General Electric Company, US Patent **3 484 293**, 16 Dec 1969.

28. a) K. D. Kreuer, On the development of proton conducting polymer membranes for hydrogen and methanol fuel cells, *J. Membr. Sci* **185**(1), 29–39 (2001); b) M. Ise, Ph. D. Thesis, University Stuttgart, Stuttgart, Germany, 2000.

29. R. A. Zoppi, I. V. P. Yoshida, and S. P. Nunes, Hybrids of perfluorosulfonic acid ionomer and silicon oxide by sol-gel reaction from solution: Morphology and thermal analysis, *Polymer* **39**(6–7), 1309–1315 (1997).

30. K. A. Mauritz, Organic-inorganic hybrid materials: Perfluorinated ionomers as sol-gel polymeri-
    zation templates for inorganic alkoxides, *Mater. Sci. Eng. C* **6**(2–3), 121–133 (1998).
31. S. Malhotra, and R. Datta, Membrane-supported nonvolatile acidic electrolytes allow higher
    temperature operation of proton-exchange membrane fuel cells, *J. Electrochem. Soc.* **144**(2),
    L23–L26 (1997).
32. S. M. J. Zaidi, S. D. Mikhailenko, G. P. Robertson, M. D. Guiver, and S. Kaliaguine, Proton
    conducting composite membranes from polyether ether ketone and heteropolyacids for fuel
    cell applications, *J. Membr. Sci.* **173**(1), 17–34 (2000).
33. B. Tazi, and O. Savadogo, Parameters of PEM fuel-cells based on new membranes fabricated
    from Nafion®, silicotungstic acid and thiophene, *Electrochim. Acta* **45**(25–26), 4329–4339
    (2000).
34. P. Genova-Dimitrova, B. Baradie, D. Foscallo, C. Poinsignon, and J. Y. Sanchez, Ionomeric
    membranes for proton exchange membrane fuel cell (PEMFC): Sulfonated polysulfone asso-
    ciated with phosphatoantimonic acid, *J. Membr. Sci.* **185**(1), 59–71 (2001).
35. S. M. Haile, Materials for fuel cells, *Mater. Today* **6**(3), 24–29 (2003).
36. S. Song, and P. Tsiakaras, Recent progress in direct ethanol proton exchange membrane fuel
    cells (DE-PEMFCs), *Appl. Catal. B* **63**(3–4), 187–193 (2006).
37. V. S. Silva, A. Mendes, L. M. Maderia, and S. P. Nunes, Proton exchange membranes for
    direct methanol fuel cells: Properties critical study concerning methanol crossover and proton
    conductivities, *J. Membr. Sci.* **276**(1–2), 126–134 (2006).
38. C. Iojoiu, F. Chabert, M. Maréchal, N. El. Kissi, J. Guindet, and J.-Y. Sanchez, From polymer
    chemistry to membrane elaboration: A global approach of fuel cell polymeric electrolytes,
    J. Power Sources **153**(2), 198–209 (2006).
39. J. A. Asensio, and P. Gómez-Romero, Recent developments on proton conducting poly(2,5-
    benzimidazole) (ABPBI) membranes for high temperature polymer electrolyte membrane fuel
    cells, *Fuel Cells* **5**(3), 336–343 (2005).
40. J. Kerres, A. Ullrich, M. Hein, V. Gogel, K. A. Friedrich, and L. Jörissen, Cross-linked
    polyaryl blend membranes for polymer electrolyte fuel cells, *Fuel Cells* **4**(1–2), 105–112
    (2004).
41. D. S. Kim, H. B. Park, J. W. Rhim, and Y. M. Lee, Preparation and characterization of
    crosslinked PVA/SiO₂ hybrid membranes containing sulfonic acid groups for direct methanol
    fuel cell applications, *J. Membr. Sci.* **240**(1–2), 37–48 (2004).
42. G. K. Surya Prakash, M. C. Smart, Q. -J. Wang, A. Atti, V. Pleynet, B. Yang, K. McGrath,
    G. A. Olah, S. R. Narayanan, W. Chum, T. Valdez, and S. Surampudi, High efficiency direct
    methanol fuel cell based on poly(styrenesulfonic acid) (PSSA)-poly(vinylidene fluoride)
    (PVDF) composite membranes, *J. Fluorine Chem.* **125**(8), 1217–1230 (2004).
43. M. Schuster, T. Rager, A. Noda, K. D. Kreuer, and J. Maier, About the choice of the proto-
    genic group in PEM separator materials for intermediate temperature, low humidity operation:
    A critical composition of sulfonic acid, phosphonic acid and imidazole functionalized model
    compounds, *Fuel Cells* **5**(3), 355–365 (2005).
44. L. Xiao, H. Zhang, T. Jana, E. Scanlon, R. Chen, E.-W. Choe, L. S. Ramanathan, S. Yu, and B. C.
    Benicewicz, Synthesis and characterization of pyridine-based polybenzimidazoles for high tem-
    perature polymer electrolyte membrane fuel cell applications, *Fuel Cells* **5**(2), 287–295 (2005).
45. R. P. Raffaelle, B. J. Landi, J. D. Harris, S. G. Bailey, and A. F. Hepp, Carbon nanotubes for
    power applications, *Mater. Sci. Eng. B* **116**(3), 233–243 (2005).
46. N. Rajalakshmi, H. Ryu, M. M. Shaijumon, and S. Ramaprabhu, Performance of polymer
    electrolyte membrane fuel cells with carbon nanotubes as oxygen reduction catalyst support
    material, *J. Power Sources* **140**(2), 250–257 (2005).
47. W. Li, X. Wang, Z. Chen, M. Waje, and Y. Yan, Carbon nanotube film by filtration as cathode
    catalyst support for proton-exchange membrane fuel cell, *Langmuir* **21**(21), 9386–9389
    (2005).
48. Y. Gao, G. P. Robertson, M. D. Guiver, S. D. Mikhailenko, X. Li, and S. Kaliaguine, Low-
    swelling proton-conducting copoly(aryl ether nitrile)s containing naphthalene structure with
    sulfonic acid groups *meta* to the ether linkage, *Polymer* **47**(3), 808–816 (2006).

49. S. Li, Z. Zhou, M. Liu, W. Li, J. Ukai, K. Hase, and M. Nakanishi, Synthesis and properties of imidazole-grafted hybrid inorganic-organic polymer membranes, *Electrochim. Acta* **51**(7), 1351–1358 (2006).

50. A. Taniguchi, and K. Yasuda, Highly water-proof coating of gas flow channels by plasma polymerization for PEM fuel cells, *J. Power Sources* **141**(1), 8–12 (2005).

51. M. Shen, S. Roy, J. W. Kuhlmann, K. Scott, K. Lovell, and J. A. Horsfall, Grafted polymer electrolyte membrane for direct methanol fuel cell, *J. Membr. Sci.* **251**(1–2), 121–130 (2005).

52. J. Chen, M. Asano, T. Yamaki, and M. Yoshida, Improvement of chemical stability of polymer electrolyte fuel cell membranes by grafting of new substituted styrene monomers into ETFE films, *J. Mater. Sci.* **41**(4), 1289–1292 (2006).

53. Z.-G. Shao, H. Xu, M. Li, and I.-M. Hsing, Hybrid Nafion-inorganic oxides membrane doped with heteropolyacids for high temperature operation of proton exchange membrane fuel cell, *Solid State Ionics* **177**(7–8), 779–785 (2006).

54. M. L. Di Vona, D. Marani, C. D'Ottavi, M. Trombetta, E. Traversa, I. Beurroies, P. Knauth, and S. Licoccia, A simple new route to covalent organic/inorganic hybrid proton exchange polymeric membranes, *Chem. Mater.* **18**(1), 69–75 (2006).

55. S. Reichman, T. Duvdevani, A. Aharon, M. Philosoph, D. Golodnitsky, and E. Peled, A novel PTFE-based proton-conductive membrane, *J. Power Sources* **153**(2), 228–233 (2006).

56. S. Swier, V. Ramani, J. M. Fenton, H. R. Kunz, M. T. Shaw, and R. A. Weiss, Polymer blends based on sulfonated poly(ether ketone ketone) and poly(ether sulfone) as proton exchange membranes for fuel cells, *J. Membr. Sci.* **256**(1–2), 122–133 (2005).

57. a) S. M. J. Zaidi, Preparation and characterization of composite membranes using blends of SPEEK/PBI with boron phosphate, *Electrochim. Acta* **50**(24), 4771–4777 (2005); b) S. M. J. Zaidi, and M. I. Ahmad, Novel SPEEK/heteropolyacids loaded MCM-41 composite membranes for fuel cell applications, *J. Membr. Sci.* **279**(1–2), 548–557 (2006).

58. Y. Yang, and S. Holdcroft, Synthetic strategies for controlling the morphology of proton conducting polymer membranes, *Fuel Cells* **5**(2), 171–186 (2005).

59. M. A. Smit, A. L. Ocampo, M. A. Espinosa-Medina, and P. J. Sebastián, A modified Nafion membrane with in situ polymerized polypyrrole for the direct methanol fuel cell, *J. Power Sources* **124**(1), 59–64 (2003).

60. S. Moravcová, Z. Cílová, and K. Bouzek, Preparation of a novel composite material based on a Nafion® membrane and polypyrrole for potential application in a PEM fuel cell, *J. Appl. Electrochem.* **35**(10), 991–997 (2005).

61. F. Xu, C. Innocent, B. Bonnet, D. J. Jones, and J. Rozière, Chemical modification of perfluorosulfonated membranes with pyrrole for fuel cell application: Preparation, characterization and methanol transport, *Fuel Cells* **5**(3), 398–405 (2005).

62. J. Shah, J. W. Brown, E. M. Buckley-Dhoot, and A. J. Bandara, The use of a phenylpyrazine liquid crystalline material with a liquid crystalline solvent mediator as an ion-selective electrode, *J. Mater. Chem.* **10**(12), 2627–2628 (2000).

63. H. Wolf, and M. Willert-Porada, Electrically conductive LCP-carbon composite with low carbon content for bipolar plate application in polymer electrolyte membrane fuel cell, *J. Power Sources* **153**(1), 41–46 (2006).

64. C. Sanchez, B. Julián, P. Belleville, and M. Popall, Applications of hybrid organic-inorganic nanocomposites, *J. Mater. Chem.* **15**(35–36), 3559–3592 (2005).

65. N. H. Jalani, K. Dunn, and R. Datta, Synthesis and characterization of Nafion®–MO_2 (M = Zr, Si, Ti) nanocomposite membranes for higher temperature PEM fuel cells, *Electrochim. Acta* **51**(3), 553–560 (2005).

66. C. S. Karthikeyan, S. P. Nunes, L. A. S. A. Prado, M. L. Ponce, H. Silva, B. Ruffmann, and K. Schulte, Polymer nanocomposite membranes for DMFC application, *J. Membr. Sci.* **254**(1–2), 139–146 (2005).

67. J.-M. Thomassin, C. Pagnoulle, G. Caldarella, A. Germain, and R. Jérôme, Contribution of nanoclays to the barrier properties of a model proton exchange membrane for fuel cell application, *J. Membr. Sci.* **270**(1–2), 50–56 (2006).

68. I. Stamatin, A. Morozan, K. Scott, A. Dumitru, S. Vulpe, and F. Nastase, Hybrid membranes for fuel cells based on nanometer YSZ and polyacrylonitrile matrix, *J. Membr. Sci.* **277**(1–2), 1–6 (2006).

69. T. R. Farhat, and P. T. Hammond, Designing a new generation of proton-exchange membranes using layer-by-layer deposition of electrolytes, *Adv. Funct. Mater.* **15**(6), 945–954 (2005).

70. H. Pei, L. Hong, and J. Y. Lee, Embedded polymerization driven asymmetric PEM for direct methanol fuel cells, *J. Membr. Sci.* **270**(1–2), 169–178 (2006).

71. S. Ren, C. Li, X. Zhao, Z. Wu, S. Wang, G. Sun, Q. Xin, and X. Yang, Surface modification of sulfonated poly(ether ether ketone) membranes using Nafion solution for direct methanol fuel cells, *J. Membr. Sci.* **247**(1–2), 59–63 (2005).

72. a) D. Rana, T. Matsuura, R. M. Narbaitz, and C. Feng, Development and characterization of novel hydrophilic surface modifying macromolecule for polymeric membranes, *J. Membr. Sci.* **249**(1–2), 103–112 (2005); b) D. Rana, T. Matsuura, and R. M. Narbaitz, Novel hydrophilic surface modifying macromolecules for polymeric membranes: Polyurethane ends capped by hydroxy group, *J. Membr. Sci.* **282**(1–2), 205–216 (2006).

73. S. Ramakrishna, K. Fujihara, W.-E. Teo, T. Yong, Z. Ma, and R. Ramaseshan, Electrospun nanofibers: Solving global issues, *Mater. Today* **9**(3), 40–50 (2006).

74. a) R. Gopal, S. Kaur, Z. Ma, C. Chan, S. Ramakrishna, and T. Matsuura, Electrospun nanofibrous filtration membrane, *J. Membr. Sci.* **281**(1–2), 581–586 (2006); b) R. Gopal, S. Kaur, C. Y. Feng, C. Chan, S. Ramakrishna, S. Tabe, and T. Matsuura, Electrospun nanofibrous polysulfone membranes as pre-filters: Particulate removal, *J. Membr. Sci.* **289**(1–2), 210–219 (2007).

75. S. Kaur, D. Rana, G. Singh, W. J. Ng, S. Ramakrishna, and T. Matsuura, 2008 (unpublished results).

Index